INTERNATIONAL SERIES OF MONOGRAPHS IN
NATURAL PHILOSOPHY
GENERAL EDITOR: D. TER HAAR

VOLUME 80

PLASMA ELECTRODYNAMICS 2

PLASMA ELECTRODYNAMICS

VOLUME 2. NON-LINEAR THEORY AND FLUCTUATIONS

A. I. Akhiezer, I. A. Akhiezer,

R. V. Polovin, A. G. Sitenko and K. N. Stepanov

Translated by D. ter Haar

PERGAMON PRESS

OXFORD · NEW YORK · TORONTO
SYDNEY · PARIS · BRAUNSCHWEIG

U. K.	Pergamon Press Ltd., Headington Hill Hall, Oxford OX3 0BW, England
U. S. A.	Pergamon Press Inc., Maxwell House, Fairview Park, Elmsford, New York 10523, U.S.A.
CANADA	Pergamon of Canada Ltd., 207 Queen's Quay West, Toronto 1, Canada
AUSTRALIA	Pergamon Press (Aust.) Pty. Ltd., 19a Boundary Street, Rushcutters Bay, N.S.W. 2011, Australia
FRANCE	Pergamon Press SARL, 24 rue des Ecoles, 75240 Paris, Cedex 05, France
WEST GERMANY	Pergamon Press GMbH, 3300 Braunschweig, Postfach 2923, Burgplatz 1, West Germany

First edition 1975
Library of Congress Catalog Card No. 74-3323

Printed in Hungary
ISBN 0 08 018016 7

Contents

CONTENTS

Contents of Volume 1. Linear Theory

Preface

THE properties of plasmas as a specific state of matter are to an important extent determined by the fact that there are between the particles which constitute the plasma electromagnetic forces which act over macroscopic distances. Processes occurring in a plasma are therefore as a rule accompanied by the excitation of electromagnetic fields which play a fundamental role in the way these processes develop.

The electromagnetic interactions which extend over macroscopic distances show up first of all in the occurrence in the plasma of collective oscillations in which a large number of particles takes part simultaneously. The existence of these specific collective electromagnetic oscillations is just as much a characteristic of a plasma as a specific state of matter as, for instance, the crystalline ordering is for the solid state of matter.

This explains the place occupied in plasma physics by plasma electrodynamics, that is, the theory of electromagnetic fields in a plasma—and, in the first instance, the theory of electromagnetic oscillations of a plasma—and the theory of macroscopic electrical and magnetic properties of a plasma.

Such problems as the theory of magnetic traps, the problem of plasma heating by external fields or currents, and the theory of instabilities in a non-uniform plasma belong also to the field of plasma electrodynamics in its widest sense. We shall not consider these problems in the present book—not because they are not important; to the contrary, they are of great importance. We restrict ourselves to an exposition of the theory of the electromagnetic properties of a uniform plasma as this theory is the basis of the whole of plasma electrodynamics.

Although there are several monographs (see, for example, Alfvén, 1950; Artsimovich, 1963; Akhiezer, Akhiezer, Polovin, Sitenko, and Stepanov, 1967; Vedenov, 1965; Ginzburg, 1970; Cowling, 1957; Kulikovskiĭ and Lyubimov, 1962; Leontovich, 1965, 1966, 1967, 1968, 1970; Silin and Rukhadze, 1961; Spitzer, 1956; Stix, 1962; Tsytovich, 1970) devoted to the problems of plasma electrodynamics, we decided all the same to write yet another book on this topic having in mind to give the theory of both low- and high-frequency oscillations—without restricting ourselves to small amplitude oscillations only—from a unified point of view and to give the basic fundamental applications of this theory.

The book starts with an exposition of the general methods of describing a plasma. Chapter 1 is devoted to this problem; in this chapter we construct the BBGKY-hierarchy of kinetic equations, introduce the self-consistent field, and introduce the Vlasov kinetic equation and the Landau collision integral. We give an account of Boltzmann's H-theorem as applied to a plasma and study the problem of the relaxation of a plasma. Finally, in that chapter we elucidate the transition from a kinetic to a hydrodynamic description of a plasma and derive the equations of magneto-hydrodynamics.

The methods for describing a plasma which we have discussed allow us then to start a detailed study of both low- and high-frequency plasma oscillations.

We start with the theory of low-frequency oscillations in the case of frequent collisions when the concise hydrodynamic description of a plasma suffices. Chapters 2 and 3 are devoted to the low-frequency oscillations.

We give in Chapter 2 the linear theory of magneto-hydrodynamic waves. We define there phase velocities, damping, and polarization of different waves and study the conic refraction of magneto-hydrodynamic waves and the excitation of these waves, as well as the problem of the formation of lacunae when two-dimensional excitations propagate from a point source. Finally, we study the characteristics of magneto-hydrodynamic flow.

After that we turn to non-linear magneto-hydrodynamic waves, both simple waves and shock waves (Chapter 3). Here we study the distortion of the profile of a simple wave leading to the formation of discontinuities. We integrate the equations for simple waves and, in particular, we evaluate the Riemann invariants.

Then follows an exposition of the theory of shock waves. We prove the Zemplén theorem and study simple and shock waves in relativistic magneto-hydrodynamics. We put the problem of the evolutionarity and structure of shock waves. Finally, we solve the problem of the formation and splitting-up of an arbitrary discontinuity in magneto-hydrodynamics.

Having studied magneto-hydrodynamic waves for the case of frequent collisions we turn to a consideration of another limiting case—oscillations in a collisionless plasma. Chapters 4 and 5 deal with this problem. In the first of these chapters we give the theory of oscillations in an unmagnetized plasma, and in the second one the theory of oscillations in a collisionless plasma in an external magnetic field.

Chapter 4 starts with an exposition of the theory of oscillations in a collisionless plasma in the hydrodynamic approximation and then these oscillations are studied using a kinetic equation. The spectra of both the high- and the low-frequency oscillations (Langmuir waves and ion-sound waves) are studied in detail. We consider the collisionless (Landau) damping of the oscillations and we solve the problem of the anomalous skin effect.

In Chapter 5 we study in detail the spectra and damping of oscillations in a collisionless magneto-active plasma. At the start of the chapter we consider oscillations in a "cold" magneto-active plasma. Then we determine the dielectric tensor of a magneto-active plasma, using a kinetic equation, and we introduce a dispersion relation for electromagnetic waves, taking spatial dispersion, caused by the thermal motion of the electrons and ions in the plasma, into account. We find the frequencies and damping rates (Cherenkov and cyclotron damping) of practically all branches of the oscillations which can propagate in a magneto-active plasma with a Maxwell particle velocity distribution—the ordinary, fast and slow extra-ordinary, fast magneto-sound, and Alfvén waves, fast and slow ion-sound oscillations, electron-sound oscillations in a non-isothermal plasma, and different branches of electron and ion cyclotron waves.

Having studied the oscillation spectra in an equilibrium plasma, we turn to the study of oscillations in a non-equilibrium, uniform plasma (Chapter 6).

First of all we study the interaction of a beam of charged particles with the oscillations of an unmagnetized plasma and show that the plasma-beam system is unstable, that is, that the interaction between the beam particles and the plasma oscillations leads to an exponential growth in time of a small initial perturbation. We then find the growth rates for different kinds of oscillations, consider the problem of the stability of a plasma in an electric field, and study the excitation of non-potential (electromagnetic) waves in a plasma with anisotropic particle velocity distributions. We study the interaction of charged particle currents with slow waves in a magneto-active plasma (the particles in the currents are characterized either by isotropic or by anisotropic distribution functions). Finally, we consider the excitation of electromagnetic waves in a plasma by currents of relativistic particles.

Having studied the interaction of charged particle currents with the plasma, we elucidate the general criteria of the stability of different particle distributions in a plasma. We consider separately an unmagnetized plasma and a plasma in an external magnetic field. We solve the problem of the two-beam instability.

Concluding Chapter 6 we study the general problem of the nature of the instability, give a definition of absolute and convective instabilities, and establish criteria for those two kinds of instability. We also establish criteria for the amplification and blocking of waves and, finally, consider the global instability caused by the reflection of waves from the system boundaries.

The problem of the interaction between charged particle currents and the plasma is related to the problem of the behaviour of a partially ionized plasma in an external electric field. As the stationary states of such a plasma are characterized by a directed motion of the electrons relative to the ions there can arise in such a plasma an instability analogous to the beam instability of a collisionless plasma.

Having studied the interaction of charged particle currents with the plasma we consider the oscillations of a partially ionized plasma in an external electric field (both with and without an external magnetic field). This problem is treated in Chapter 7. We derive there the kinetic equation describing the electron component of a partially ionized plasma in external electric and magnetic fields and we determine the stationary electron distribution function in such a plasma (Druyvesteyn–Davydov distribution). We then study high-frequency (transverse electromagnetic and Langmuir), ion-sound, and magneto-sound oscillations and show that ion-sound and magneto-sound oscillations in an external electric field turn out to be growing oscillations. Finally, we study in Chapter 7 the peculiar oscillations of a partially ionized plasma—the ionization-recombination oscillations in which not only the charged particle density, but also their total number changes.

In Chapter 8 we turn again to the study of a completely ionized plasma. In this chapter we study the non-linear oscillations in such a plasma (in contrast to Chapters 4 to 6 in which we restricted our considerations to oscillations with a small amplitude). We discuss here non-linear high-frequency waves in a cold plasma, Langmuir waves in a non-relativistic plasma, and longitudinal, transverse, and coupled longitudinal-transverse waves in a relativistic plasma. We study non-linear waves in a plasma in which the average electron energy appreciably exceeds the average ion energy (ion-sound and magneto-sound waves of finite amplitude) and consider both simple (Riemann) waves and waves with a stationary profile (periodic, isolated, and quasi-shock waves with an oscillatory structure). We show that the nature of simple and stationary waves depends greatly on the electron velocity distribution. Finally, we study in Chapter 8 non-linear low-frequency waves in a cold magneto-active plasma.

Chapter 9 is devoted to a study of oscillations in the quasi-linear approximation in which the simplest non-linear effect is taken into account—the influence of oscillations on resonance particles. We first consider the interaction between resonance particles with longitudinal oscillations of an unmagnetized plasma and we give the derivation of the basic equation of a quasi-linear theory—the particle diffusion equation in velocity space. We then consider the quasi-linear relaxation process which leads to the formation of a plateau on the distribution function of resonance particles and study the quasi-linear wave transformation. In the same chapter we consider the quasi-linear theory of the interaction between resonance particles and the oscillations of a magneto-active plasma and study the problem of the quasi-linear relaxation of wave packets for the cases of cyclotron and Cherenkov resonance. Finally, we consider the influence of collisions on the quasi-linear relaxation process and on the damping of oscillations.

Having expounded the quasi-linear theory, which describes the effects of the first approximation in terms of the plasma wave energy, we turn to a study of the processes of higher order in the energy of the oscillations: the interaction between waves and waves and the non-linear interaction of waves and particles. This is the subject of Chapter 10 in which we obtain a kinetic equation for waves which takes into account three-wave processes and the non-linear interaction between waves and particles (sometimes called the non-linear Landau damping). We then study turbulent processes in which Langmuir waves take part: their interaction with ion sound and the decay instability and non-linear damping of Langmuir waves, and we study in detail ion-sound turbulence which occurs in a plasma with a directed motion of electrons relative to ions. Finally, we consider the interaction between Alfvén and magneto-sound waves in a magneto-active plasma.

The last three chapters of the book deal with the theory of fluctuations and of the wave and particle-scattering processes in a plasma caused by the fluctuations.

We give in Chapter 11 the theory of electromagnetic fluctuations in a plasma. We start with the derivation of the general fluctuation-dissipation relation which establishes a connection between the spectral distribution of the fluctuations and the energy dissipation in the medium; we use this relation to determine the fluctuations first in an equilibrium and then in a two-temperature plasma, both for an unmagnetized plasma and for a plasma in a magnetic field.

We then develop the theory of fluctuations in a non-equilibrium plasma and a kinetic theory of fluctuations; we find the fluctuations in the particle-distrubution function and consider the critical fluctuations near the instability limits of the plasma and study the fluctuations in the plasma-beam system. We elucidate how one can proceed to a hydrodynamic theory of fluctuations and, finally, we study fluctuations in a partially ionized plasma in an electric field.

Chapter 12 is devoted to the theory of scattering processes and the transformation of waves in a plasma. We study here the scattering of electromagnetic waves in an unmagnetized plasma and determine the spectral distribution of the scattered radiation. We consider critical opalescence connected with the scattering of waves in a plasma near the limits of instability, and we study the transformation of transverse and longitudinal waves in an unmagnetized plasma and also the spontaneous emission in an non-equilibrium plasma. We give the theory of incoherent reflection of electromagnetic waves from a plasma. We study scattering and transformation of waves in a magnetoactive plasma, in a partially ionized plasma in an external electric field, and in a turbulent plasma. Finally, we discuss echo effects in a plasma; these are connected with undamped oscillations of particle distribution functions in a plasma.

In Chapter 13 we study the scattering of charged particles in a plasma. We determine here the polarization energy losses when charged particles move in a plasma; we find the energy losses caused by the fluctuations of the field in the plasma, and we determine the coefficients of dynamic friction in diffusion. We also study the propagation of charged particles through a magneto-active plasma and the interaction of charged particles with a non-equilibrium plasma, as well as the scattering of particles by critical fluctuations and the interaction between particles and a turbulent plasma.

We are well aware that the problems considered by us do not cover the complete theory, even of a uniform plasma, and that we have not given equal weight to the different problems. However, this is apparently unavoidable when writing a relatively large book. A very apt quotation comes from one of the best books on elementary particle theory (Bernstein, 1968): "No doubt another physicist writing the same book would have emphasized different aspects of the subject or would have treated the same aspects differently. One of the few pleasures in writing such a book is that the author can present the subject as he would like to see it presented ... and if this encourages someone else to write a better book, then the present author will be among its most enthusiastic readers."

The authors express their gratitude for assitance and useful remarks to V. F. Aleksin, V. V. Angeleĭko, A. S. Bakaĭ, A. B. Mikhaĭlovskiĭ, S. S. Moiseev, V. A. Oraevskiĭ, J. R. Ross and V. P. Silin.

Preface to the English Edition

"PLASMA ELECTRODYNAMICS", which is here brought before the English-speaking public, is devoted to the theory of collective oscillations in a plasma—strictly speaking, of a uniform plasma.

We have endeavoured to collect here the most important aspects of the theory of plasma oscillations and decided therefore to present not only the theory of oscillations in a collisionless plasma, but also the theory of magneto-hydrodynamic waves. Of course, our considerations include both linear oscillations and large amplitude oscillations.

The book is an expanded and extended version of our booklet *Collective Oscillations in a Plasma*, the English edition of which appeared in 1967.

Although we restricted our considerations solely to a uniform plasma, nevertheless the material referring to the electromagnetic properties of such a plasma is so extensive that we considered it appropriate to split the English edition into two volumes. The first volume contains the theory of magneto-hydrodynamical waves and the theory of linear oscillations of a collisionless plasma. The second volume contains the theory of non-linear waves in a collisionless plasma, including the quasi-linear theory, the theory of plasma turbulence, and the theory of electromagnetic fluctuations in a plasma.

The publication of an English edition of our book would have been impossible without the active participation of D. ter Haar: not only was it his initiative which led to the publication in England, but he also undertook the translation of the book, which—as far as we can judge with our knowledge of the English language—is excellent. Both for this and for his many useful comments we want to thank Professor ter Haar most sincerely.

A. I. AKHIEZER
I. A. AKHIEZER
R. V. POLOVIN
A. G. SITENKO
K. N. STEPANOV

CHAPTER 8

Non-Linear Waves in a Collisionless Plasma

8.1. Non-linear High-frequency Waves in a Cold Plasma

8.1.1. NON-LINEAR NON-RELATIVISTIC LANGMUIR OSCILLATIONS

In the preceding chapters we have considered small oscillations in a collisionless plasma. We now turn to a study of finite-amplitude waves in such a plasma. We shall start with considering longitudinal electron oscillations in a plasma, neglecting thermal effects (Akhiezer and Lyubarskiĭ, 1951). In this case the state of the plasma is characterized by the electron density $n \equiv n(\boldsymbol{r}, t)$ rather than by the distribution function and instead of the kinetic equation we can use a hydrodynamic equation to determine the electron velocity $\boldsymbol{v} \equiv \boldsymbol{v}(\boldsymbol{r}, t)$:

$$\frac{d\boldsymbol{v}}{dt} \equiv \frac{\partial \boldsymbol{v}}{\partial t}+(\boldsymbol{v}\cdot\nabla)\,\boldsymbol{v} = -\frac{e}{m_e}\boldsymbol{E}, \tag{8.1.1.1}$$

where $\boldsymbol{E}$ is the electrical field, satisfying the Maxwell equations

$$\operatorname{div}\boldsymbol{E} = 4\pi e(n_0-n), \tag{8.1.1.2}$$

with $-e$ the electron charge, m_e the electron mass, and n_0 the ion density; we shall assume the ions to be fixed and we shall neglect the action of the magnetic field, assuming that $v \ll c$. Adding to these equations the equation of continuity,

$$\frac{\partial n}{\partial t}+\operatorname{div}(n\boldsymbol{v}) = 0, \tag{8.1.1.3}$$

we obtain a complete set of equations which determine the state of the plasma.

Let us consider one-dimensional motion along the z-axis. We can then write eqns. (8.1.1.1) to (8.1.1.3) in the form

$$\frac{dv}{dt} = -\frac{e}{m_e}E, \quad \frac{dE}{dt} \equiv \frac{\partial E}{\partial t}+v\frac{\partial E}{\partial z} = 4\pi e n_0 v,$$

where $v \equiv v_z$, $v_x = v_y = 0$. Differentiating the first equation with respect to t and substituting the result into the second equation we obtain the following equation for the non-linear velocity oscillations in Lagrangian coordinates (Polovin, 1957)

$$\frac{d^2v}{dt^2}+\omega_p^2 v = 0 \tag{8.1.1.4}$$

where ω_p is the Langmuir frequency,

$$\omega_p = \sqrt{\frac{4\pi e^2 n_0}{m_e}}.$$

This equation is the same as the equation for small amplitudes and leads to the important conclusion that the frequency of the non-linear Langmuir oscillations is independent of their amplitude (Akhiezer and Lyubarskiĭ, 1951).

One can easily generalize eqn. (8.1.1.4) to the case when there is a constant magnetic field applied to the plasma; in that case also the frequency of the oscillations is independent of the amplitude (Vedenov, 1958; Stepanov, 1963c).

We emphasize that the proof that the amplitude of the oscillations is independent of the amplitude is valid only in the case of a cold plasma and when the ion motion is neglected: taking the thermal motion of the electrons and the motion of the ions into account leads to an amplitude-dependence of the frequency of the oscillations (Bohm and Gross, 1949a; Sizonenko and Stepanov, 1965; Repalov and Khizhnyak, 1968; Wilhelmsson, 1961).†

Let us consider in more detail the non-linear Langmuir waves, which appear in a cold plasma, with a density and velocity which are functions of the following combination of the variables t and z:

$$\tau = t-z/V,$$

where V is a constant which is the velocity of propagation of the wave. Denoting differentiation with respect to the variable τ by a prime, we get from (8.1.1.1) to (8.1.1.3):

$$Vn'-(nv)' = 0, \quad Vv'-vv' = -\frac{e}{m_e}\varphi', \quad -\varphi'' = 4\pi e(n_0-n)V^2, \tag{8.1.1.5}$$

where φ is the scalar potential, $E = -\partial\varphi/\partial z$.

From the first of equations (8.1.1.5) it follows that

$$n = \frac{A}{V-v},$$

where A is an integration constant. As the electron density must equal the ion density, $n = n_0$, when there are no oscillations ($v = 0$), we have $A = n_0 V$.

Integrating then the second of eqns. (8.1.1.5), we get

$$V-v = \sqrt{\frac{2e\varphi}{m_e}}. \tag{8.1.1.6}$$

† We note that the statement by Amer (1958) that the frequency of the non-linear Langmuir oscillations in a cold plasma is amplitude-dependent is based upon an error: the author commutes the operators $\partial/\partial z$ and $d/dt \equiv (\partial/\partial t)+v(\partial/\partial z)$ (see Derfler, 1961), which is not allowed, as v is a function of z. The amplitude-dependence of the frequency found by Sturrock (1957) is also based upon an error: in this paper the integration constant is chosen in such a way that the average velocity $\bar{v}$ vanishes, while it is the average current $\bar{j}$ which vanishes.

Using these relations we can write the third of eqns. (8.1.1.5) in the form

$$\varphi'' + 4\pi e n_0 V^2 \left[1 - \frac{V}{\sqrt{\dfrac{2e\varphi}{m_e}}}\right] = 0.$$

We can easily integrate this equation, after multiplying it by φ' (Akhiezer and Lyubarskiĭ, 1951; see also Smerd, 1955)

$$\omega_p \tau = \arcsin \frac{1}{C}\left[\sqrt{\left(\frac{2e\varphi}{m_e V^2}\right)} - 1\right] - \sqrt{\left\{C^2 - \left[\sqrt{\left(\frac{2e\varphi}{m_e V^2}\right)} - 1\right]^2\right\}}, \qquad (8.1.1.7)$$

where C is an integration constant. Hence it follows that the frequency of the non-linear oscillations is the same as the Langmuir frequency ω_p and that the potential φ varies within the limits

$$\frac{m_e V^2}{2e}(1-C)^2 \leqslant \varphi \leqslant \frac{m_e V^2}{2e}(1+C)^2.$$

Equation (8.1.1.6) shows that the quantity C is the maximum of the ratio of the particle velocity to the wave propagation velocity,

$$-C \leqslant \frac{v}{V} \leqslant C.$$

8.1.2. EQUATIONS DESCRIBING NON-LINEAR WAVES IN A RELATIVISTIC PLASMA WHEN THERE ARE NO THERMAL EFFECTS

We now generalize the results obtained in the previous subsection to the case of a relativistic plasma. As before, we shall assume the plasma to be sufficiently cold; as far as the nature of the oscillations is concerned, we shall not assume them necessarily to be longitudinal.

The basic equations to determine the electron velocity $\boldsymbol{v}$, their density n, and the fields $\boldsymbol{E}$ and $\boldsymbol{B}$ now have the form

$$\left.\begin{aligned} &\frac{\partial \boldsymbol{p}}{\partial t} + (\boldsymbol{v}\cdot\nabla)\boldsymbol{p} = -e\boldsymbol{E} - \frac{e}{c}[\boldsymbol{v} \wedge \boldsymbol{B}], \\ &\operatorname{curl} \boldsymbol{E} = -\frac{1}{c}\frac{\partial \boldsymbol{B}}{\partial t}, \quad \operatorname{curl} \boldsymbol{B} = \frac{1}{c}\frac{\partial \boldsymbol{E}}{\partial t} - \frac{4\pi}{c} e n \boldsymbol{v}, \\ &\operatorname{div} \boldsymbol{E} = 4\pi e(n_0 - n), \quad \operatorname{div} \boldsymbol{B} = 0, \end{aligned}\right\} \qquad (8.1.2.1)$$

where n_0 is the equilibrium electron density which is equal to the density of the ions which we assume to be infinitely heavy and fixed in space, while $\boldsymbol{p}$ is the electron momentum

$$\boldsymbol{p} = \frac{m_e \boldsymbol{v}}{\sqrt{[1 - (v^2/c^2)]}}.$$

The problem which will occupy us lies in a general study of the wave motions of a relativistic cold plasma, that is, such motions of the electrons in which the variable quantities are functions of a single combination $(\boldsymbol{i}\cdot\boldsymbol{r})-Vt$, with $\boldsymbol{i}$ a constant unit vector and V a constant, rather than separately of $\boldsymbol{r}$ and t. The meaning of this kind of solutions lies in the fact that they are plane waves propagating in the direction $\boldsymbol{i}$ with velocity V.

Indicating differentiation with respect to the argument $(\boldsymbol{i}\cdot\boldsymbol{r})-Vt$ by a prime, we can write eqns. (8.1.2.1) in the form

$$[\boldsymbol{i}\wedge\boldsymbol{E}'] = \beta\boldsymbol{B}', \tag{8.1.2.2}$$

$$[\boldsymbol{i}\wedge\boldsymbol{B}'] = -\beta\boldsymbol{E}'-\frac{4\pi}{c}en\boldsymbol{v}, \tag{8.1.2.3}$$

$$(\boldsymbol{i}\cdot\boldsymbol{B}') = 0, \tag{8.1.2.4}$$

$$(\boldsymbol{i}\cdot\boldsymbol{E}') = -4\pi e(n-n_0), \tag{8.1.2.5}$$

$$[(\boldsymbol{i}\cdot\boldsymbol{v})-V]\boldsymbol{p}' = -e\boldsymbol{E}-\frac{e}{c}[\boldsymbol{v}\wedge\boldsymbol{B}], \tag{8.1.2.6}$$

where $\beta = V/c$. Integrating (8.1.2.2) we get

$$\boldsymbol{B} = \frac{1}{\beta}[\boldsymbol{i}\wedge\boldsymbol{E}]+\boldsymbol{B}_0, \tag{8.1.2.7}$$

where $\boldsymbol{B}_0$ is the external magnetic field strength, which acts upon the plasma. If there is no such field and there are natural oscillations in the plasma, we have

$$\boldsymbol{B} = \frac{1}{\beta}[\boldsymbol{i}\wedge\boldsymbol{E}].$$

In that case

$$(\boldsymbol{i}\cdot\boldsymbol{B}) = 0, \quad (\boldsymbol{E}\cdot\boldsymbol{B}) = 0. \tag{8.1.2.8}$$

In other words, if there is no external magnetic field $\boldsymbol{B}_0$, the variable magnetic field $\boldsymbol{B}$ is transverse and at right angles to the electrical field.

It follows from (8.1.2.3) and (8.1.2.8) that

$$n = \frac{n_0V}{V-(\boldsymbol{i}\cdot\boldsymbol{v})}. \tag{8.1.2.9}$$

Since $n_0 > 0$, we have $(\boldsymbol{i}\cdot\boldsymbol{v}) < V$, that is, the component of the electron velocity along the direction of the propagation of the wave is always less than the wave velocity.

Taking the vector product of (8.1.2.6) with $\boldsymbol{i}$ and using (8.1.2.7), we find

$$\boldsymbol{B} = \frac{c}{e}[\boldsymbol{i}\wedge\boldsymbol{p}']+\frac{V\boldsymbol{B}_0-\boldsymbol{v}(\boldsymbol{i}\cdot\boldsymbol{B}_0)}{V-(\boldsymbol{i}\cdot\boldsymbol{v})}. \tag{8.1.2.10}$$

After this taking the vector product of (8.1.2.3) and $\boldsymbol{i}$ we get

$$\boldsymbol{B}' = -\frac{4\pi}{c}\frac{en}{\beta^2-1}[\boldsymbol{i}\wedge\boldsymbol{v}]. \tag{8.1.2.11}$$

It follows from (8.1.2.10) and (8.1.2.11) that

$$[\boldsymbol{i}\wedge\boldsymbol{p}]''+\frac{4\pi e^2n}{(\beta^2-1)c^2}[\boldsymbol{i}\wedge\boldsymbol{v}] = -\frac{e}{c}\left[\frac{V\boldsymbol{B}_0-\boldsymbol{v}(\boldsymbol{i}\cdot\boldsymbol{B}_0)}{V-(\boldsymbol{i}\cdot\boldsymbol{v})}\right]'. \tag{8.1.2.12}$$

Taking the dot product of (8.1.2.6) and $\boldsymbol{i}$ and using (8.1.2.5) we find

$$\left\{[(\boldsymbol{i}\cdot\boldsymbol{v})-V](\boldsymbol{i}\cdot\boldsymbol{p}')+([\boldsymbol{i}\wedge\boldsymbol{v}]\cdot[\boldsymbol{i}\wedge\boldsymbol{p}])'+e\beta\frac{(\boldsymbol{i}\cdot[\boldsymbol{v}\wedge\boldsymbol{B}_0])}{V-(\boldsymbol{i}\cdot\boldsymbol{v})}\right\}' = 4\pi e^2 n_0 \frac{(\boldsymbol{i}\cdot\boldsymbol{v})}{V-(\boldsymbol{i}\cdot\boldsymbol{v})}. \quad (8.1.2.13)$$

Equations (8.1.2.12) and (8.1.2.13) determine the transverse and the longitudinal components of the electron velocity in the general case when the external magnetic field $\boldsymbol{B}_0$ is non-vanishing (Akhiezer and Polovin, 1956).

If we assume that the vector $\boldsymbol{i}$ is along the z-axis and introduce the dimensionless momentum $\boldsymbol{\rho} = \boldsymbol{p}/m_e c$ and the dimensionless velocity $\boldsymbol{u} = \boldsymbol{v}/c$, we can write eqns. (8.1.2.12) and (8.1.2.13) in the form

$$\left.\begin{aligned}
&\frac{d^2\varrho_x}{d\tau^2}+\frac{\omega_p^2\beta^2}{\beta^2-1}\frac{\beta}{\beta-u_z}u_x+\beta\frac{d}{d\tau}\frac{\beta\omega_{Bey}-u_y\omega_{Bez}}{\beta-u_z}=0,\\
&\frac{d^2\varrho_y}{d\tau^2}+\frac{\omega_p^2\beta^2}{\beta^2-1}\frac{\beta}{\beta-u_z}u_y-\beta\frac{d}{d\tau}\frac{\beta\omega_{Bex}-u_x\omega_{Bez}}{\beta-u_z}=0,\\
&\frac{d}{d\tau}\left\{(u_z-\beta)\frac{d\varrho_z}{d\tau}+u_x\frac{d\varrho_x}{d\tau}+u_y\frac{d\varrho_y}{d\tau}+\frac{\beta^2}{\beta-u_x}(u_x\omega_{Bey}-u_y\omega_{Bex})\right\}=\omega_p^2\frac{\beta^2u_z}{\beta-u_z},
\end{aligned}\right\} \quad (8.1.2.14)$$

where

$$\tau = t-\frac{(\boldsymbol{i}\cdot\boldsymbol{r})}{V}, \quad \omega_p^2 = \frac{4\pi e^2 n_0}{m_e}, \quad \omega_{Be} = \frac{eB_0}{m_e c}.$$

When there is no external field $\boldsymbol{B}_0$ eqns. (8.1.2.14) take the form

$$\left.\begin{aligned}
&\frac{d^2\varrho_x}{d\tau^2}+\frac{\omega_p^2\beta^2}{\beta^2-1}\frac{\beta u_x}{\beta-u_z}=0,\\
&\frac{d^2\varrho_y}{d\tau^2}+\frac{\omega_p^2\beta^2}{\beta^2-1}\frac{\beta u_y}{\beta-u_z}=0,\\
&\frac{d}{d\tau}\left\{(u_z-\beta)\frac{d\varrho_z}{d\tau}+u_x\frac{d\varrho_x}{d\tau}+u_y\frac{d\varrho_y}{d\tau}\right\}=\omega_p^2\frac{\beta^2u_z}{\beta-u_z},
\end{aligned}\right\} \quad (8.1.2.15)$$

or

$$\left.\begin{aligned}
&\frac{d^2\varrho_x}{d\tau^2}+\frac{\omega_p^2\beta^2}{\beta^2-1}\frac{\beta\varrho_x}{\beta\sqrt{(1+\varrho^2)}-\varrho_z}=0,\\
&\frac{d^2\varrho_y}{d\tau^2}+\frac{\omega_p^2\beta^2}{\beta^2-1}\frac{\beta\varrho_y}{\beta\sqrt{(1+\varrho^2)}-\varrho_z}=0,\\
&\frac{d^2}{d\tau^2}[\beta\varrho_z-\sqrt{(1+\varrho^2)}]+\frac{\omega_p^2\beta^2\varrho_z}{\beta\sqrt{(1+\varrho^2)}-\varrho_z}=0.
\end{aligned}\right\} \quad (8.1.2.16)$$

The first two eqns. (8.1.2.15) for the transverse components of the velocity clearly allow bounded solutions only in the case when $\beta > 1$, that is, $V > c$. As far as the third eqn. (8.1.2.15) for the longitudinal velocity component u_z is concerned, it allows bounded solutions for any value of β, provided $u_x = u_y = 0$. Such purely longitudinal motions are, however, unstable due to the coupling between longitudinal and transverse motions in the case when $\beta < 1$.

Introducing instead of the momentum components the new variables

$$\xi = \varrho_x \sqrt{(\beta^2-1)}, \quad \eta = \varrho_y \sqrt{(\beta^2-1)}, \quad \zeta = \beta\varrho_z - \sqrt{(1+\varrho^2)},$$

we can write eqns. (8.1.2.16) in the form

$$\left.\begin{aligned} \frac{d^2\xi}{d\tau^2} + \frac{\omega_{\mathrm{pl}}^2\beta^2}{\beta^2-1}\frac{\beta\xi}{\sqrt{[\beta^2-1+\xi^2+\eta^2+\zeta^2]}} &= 0, \\ \frac{d^2\eta}{d\tau^2} + \frac{\omega_{\mathrm{pl}}^2\beta^2}{\beta^2-1}\frac{\beta\eta}{\sqrt{[\beta^2-1+\xi^2+\eta^2+\zeta^2]}} &= 0, \\ \frac{d^2\zeta}{d\tau^2} + \frac{\omega_{\mathrm{pl}}^2\beta^2}{\beta^2-1}\frac{\beta\zeta}{\sqrt{[\beta^2-1+\xi^2+\eta^2+\zeta^2]}} + \frac{\omega_{\mathrm{pl}}^2\beta^2}{\beta^2-1} &= 0. \end{aligned}\right\} \tag{8.1.2.17}$$

These equations can be obtained from the Lagrangian

$$L = \frac{1}{2}\left[\left(\frac{d\xi}{d\tau}\right)^2 + \left(\frac{d\eta}{d\tau}\right)^2 + \left(\frac{d\zeta}{d\tau}\right)^2\right] - \frac{\omega_{\mathrm{pl}}^2\beta^2}{\beta^2-1}[\beta\sqrt{(\beta^2-1+\xi^2+\eta^2+\zeta^2)}+\zeta]. \tag{8.1.2.18}$$

The general problem of the relativistic wave motions of a plasma in the case when there is no external magnetic field is thus equivalent to the problem of the non-relativistic motion of a particle with a mass equal to unity in a field with a potential energy

$$U = \frac{\omega_{\mathrm{pl}}^2\beta^2}{\beta^2-1}[\beta\sqrt{(\beta^2-1+\xi^2+\eta^2+\zeta^2)}+\zeta]. \tag{8.1.2.18'}$$

From the form of the Lagrangian we obtain immediately the conservation laws for the energy W and the angular momentum M:

$$\frac{1}{2}\left[\left(\frac{d\xi}{d\tau}\right)^2 + \left(\frac{d\eta}{d\tau}\right)^2 + \left(\frac{d\zeta}{d\tau}\right)^2\right] + \frac{\omega_{\mathrm{pl}}^2\beta^2}{\beta^2-1}[\beta\sqrt{(\beta^2-1+\xi^2+\eta^2+\zeta^2)}+\zeta] = W,$$

$$\xi\frac{d\eta}{d\tau} - \eta\frac{d\xi}{d\tau} = M.$$

If we change from the variables ξ, η, and ζ to the fields $\boldsymbol{E}$ and $\boldsymbol{B}$ and the velocity $\boldsymbol{v}$, these conservation laws take the following form:

$$\frac{\beta}{8\pi}(E^2+B^2) + \frac{\beta n_0 m_e c^2}{\sqrt{[1-(v^2/c^2)]}} - \frac{(\boldsymbol{i}\cdot[\boldsymbol{E} \wedge \boldsymbol{B}])}{4\pi} = \text{constant}, \quad (\boldsymbol{p}\cdot\boldsymbol{B}) = \text{constant}. \tag{8.1.2.19}$$

We note that these equations are direct consequences of the basic equations. Indeed, taking the dot product of (8.1.2.3) and $\boldsymbol{B}'$ one checks easily that $(\boldsymbol{v}\cdot\boldsymbol{B}') = 0$, that is, $(\boldsymbol{p}\cdot\boldsymbol{B}')=0$. Also, taking the dot product of (8.1.2.6) with $\boldsymbol{B}$, we get $(\boldsymbol{p}'\cdot\boldsymbol{B}) = 0$. Adding these two results we obtain the second of eqns. (8.1.2.19). The first eqn. (8.1.2.19) can be obtained in a similar way.

8.1.3. LONGITUDINAL WAVES IN A RELATIVISTIC PLASMA

Let us now turn to a study of longitudinal non-linear waves in a relativistic plasma, neglecting thermal effects. Putting $u_x = u_y = 0$ in (8.1.2.14), we get

$$\frac{d}{d\tau}\left\{(u-\beta)\frac{d\varrho}{d\tau}\right\} = \frac{\omega_p^2\beta^2 u}{\beta-u},$$

where $u = u_z$ and $\varrho = \varrho_z$ are the dimensionless electron velocity and momentum. Expressing the momentum in terms of the velocity we can write this equation in the form

$$\frac{d^2}{d\tau^2}\frac{1-\beta u}{\sqrt{(1-u^2)}} = \frac{\omega_p^2\beta^2 u}{\beta-u}.$$

Multiplying it by $(d/d\tau)\{(1-\beta u)/(1-u^2)^{1/2}\}$ and integrating, we get

$$\frac{1}{2}\left[\frac{d}{d\tau}\frac{1-\beta u}{\sqrt{(1-u^2)}}\right]^2 = \beta^2\omega_p^2\left[C-\frac{1}{\sqrt{(1-u^2)}}\right], \tag{8.1.3.1}$$

where C is an integration constant. Putting $C = (1-u_m^2)^{-1/2}$ we see that u lies in the range $-u_m \le u \le u_m$. Integrating (8.1.3.1) we get (Akhiezer and Polovin, 1955)

$$\int \frac{(\beta-u)\,du}{(1-u^2)^{3/2}\sqrt{[(1-u_m^2)^{-1/2}-(1-u^2)^{-1/2}]}} = \sqrt{(2)}\beta\omega_p\tau. \tag{8.1.3.2}$$

This formula solves our problem in principle, expressing u as a function of $\tau = t-z/V$.

It is clear that u is a periodic function of τ. Its period, which we shall denote by T is determined by the equation

$$2\int_{-u_m}^{u_m} \frac{(\beta-u)\,du}{(1-u^2)^{3/2}\sqrt{[(1-u_m^2)^{-1/2}-(1-u^2)^{-1/2}]}} = \sqrt{(2)}\beta\omega_p T. \tag{8.1.3.3}$$

Introducing the frequency $\omega = 2\pi/T$ instead of the period, we get

$$\omega = \frac{\pi\omega_p}{I(u_m)\sqrt{2}}, \quad I(u_m) = \int_0^{u_m} \frac{du}{(1-u^2)^{5/4}\sqrt{\{\sqrt{[(1-u^2)/(1-u_m^2)]}-1\}}}. \tag{8.1.3.4}$$

One can obtain simple formulae in two limiting cases—small and large velocity amplitudes.† In the first case, when $u_m \ll 1$, the frequency is equal to

$$\omega = \omega_p(1-\tfrac{3}{16}u_m^2), \quad u_m \ll 1. \tag{8.1.3.5a}$$

† It is interesting to note that in the electron frame of reference the frequency of the non-linear oscillations is equal to (Francis, 1960):

$$\omega_p' = \sqrt{\frac{4\pi e^2 n_0}{m_e'}}, \quad m_e' = \frac{m_e}{\sqrt{[1-(v^2/c^2)]}}.$$

Equation (8.1.3.4) in the laboratory frame of reference can be obtained through a transformation of the frequency. Wang (1963) has studied the influence of the thermal spread in electron velocities on relativistic longitudinal non-linear waves.

In the second limiting case when $1-u_m \ll 1$ the frequency is equal to

$$\omega = \frac{\pi\omega_p[1-u_m^2]^{1/4}}{2\sqrt{2}}, \qquad 1-u_m \ll 1. \tag{8.1.3.5b}$$

As $u_m \to 0$ the frequency tends to ω_p, as should be the case. As $u_m \to 1$, the frequency tends to zero. This is connected with the fact that as $u_m \to 1$, the effective mass of the electron tends to infinity.

In the general case, for intermediate values of u_m, the integral determining τ in (8.1.3.2) and the period of the oscillations can be expressed in terms of elliptical functions (Akhiezer and Polovin, 1956; Cavaliere, 1962).

Once we know u we can easily determine the particle density and the electrical field which is in the direction of the wave propagation. Using (8.1.2.6) and (8.1.2.9) we find

$$n(\tau) = \frac{n_0\beta}{\beta-u}, \quad eE(\tau) = \pm m_e\omega_p c\sqrt{\left[2\left\{\frac{1}{\sqrt{(1-u_m^2)}}-\frac{1}{\sqrt{(1-u^2)}}\right\}\right]}. \tag{8.1.3.6}$$

The maximum value of the field is proportional to u_m for small velocities and determined by the equation

$$eE_m = m_e\omega_p c\sqrt{(2)}(1-u_m^2)^{-1/4}, \tag{8.1.3.7}$$

if $1-u_m \ll 1$.

8.1.4. TRANSVERSE WAVES IN A RELATIVISTIC PLASMA

For purely transverse oscillations $\varrho_z = 0$ and the third eqn. (8.1.2.16) gives

$$p^2 = \text{constant}.$$

We obtain the same result also when $\boldsymbol{B}_0$ is different from zero, provided $\boldsymbol{B}_0$ lies along the direction of the wave propagation.

Putting $\varrho_z = 0$ in the first two eqns. (8.1.2.16) we get

$$\varrho_x = \varrho\cos\omega\tau, \qquad \varrho_y = \varrho\sin\omega\tau,$$

where

$$\omega = \omega_p\frac{\beta}{\sqrt{(\beta^2-1)}}\frac{1}{(1+\varrho^2)^{1/4}}. \tag{8.1.4.1}$$

From this it follows that the wave velocity β can be written in the form

$$\beta = \frac{1}{\sqrt{\varepsilon}},$$

where

$$\varepsilon = 1-\frac{\omega_p'^2}{\omega^2}, \quad \omega_p' = \sqrt{\frac{4\pi e^2 n_0}{m_e'}}, \quad m_e' = \frac{m_e}{\sqrt{[1-(v^2/c^2)]}},$$

with v the electron velocity.

In the case of purely transverse waves the electrons move therefore in circles with an angular velocity ω and only waves with circular polarization are possible.† This is connected with the fact that if the amplitude of the oscillations is large it is impossible to have a superposition of two waves with different circular polarizations because of the non-linearity of the equations. For small amplitudes when the oscillations are linear such a superposition is possible and this leads to the occurrence of linearly polarized transverse waves.

Using eqns. (8.1.2.10) (for $\boldsymbol{B}_0 = 0$) and (8.1.4.1) one can easily show that the magnetic field $\boldsymbol{B}$ is parallel to the electron momentum and is determined by the equations

$$eB_x = \frac{m_e c\omega}{\beta}\varrho_x = \frac{m_e c\omega\varrho}{\beta}\cos\omega\tau,$$

$$eB_y = \frac{m_e c\omega}{\beta}\varrho_y = \frac{m_e c\omega\varrho}{\beta}\sin\omega\tau.$$

The electrical field is according to (8.1.2.7) equal to

$$eE_x = -m_e c\omega\varrho\sin\omega\tau, \quad eE_y = m_e c\omega\varrho\cos\omega\tau.$$

If the external magnetic field is non-vanishing and its direction is the same as the direction of the propagation of the wave, the equations of motion for the transverse oscillations take the form:

$$\frac{d^2\varrho_x}{d\tau^2} - \omega_{Be}\frac{du_y}{d\tau} + \frac{\beta^2}{\beta^2-1}\omega_p^2 u_x = 0,$$

$$\frac{d^2\varrho_y}{d\tau^2} + \omega_{Be}\frac{du_x}{d\tau} + \frac{\beta^2}{\beta^2-1}\omega_p^2 u_y = 0,$$

where, as before, $u_x^2 + u_y^2 = \text{constant}$. Noting that

$$\varrho = \frac{u}{\sqrt{(1-u^2)}},$$

and putting $u_x = U\cos\omega\tau$, $u_y = U\sin\omega\tau$, we find the following expression for the frequency ω:

$$\omega = \sqrt{\left[\frac{1}{4}\omega_{Be}'^2 + \frac{\beta^2}{\beta^2-1}\omega_p'^2\right]} \pm \frac{1}{2}\omega'_{Be}, \tag{8.1.4.2}$$

whence

$$\frac{1}{\beta^2} = 1 - \frac{\omega_p'^2}{\omega^2 \pm \omega\omega'_{Be}},$$

where

$$\omega_p'^2 = \omega_p^2(1-u^2)^{1/2}, \quad \omega'_{Be} = \omega_{Be}(1-u^2)^{1/2}.$$

Let us now turn to a study of oscillations which are nearly transverse, for which the

† This conclusion loses its meaning in practice when $\beta \gg 1$ when because of the smallness of the longitudinal velocity components oscillations which are practically linearly polarized and transverse become possible (see below in Subsection 8.1.5). The solution (8.1.4.1) can be generalized to the case when the electron moves along a spiral and when ions take part in the motion (Wang and Lojko, 1963).

electron orbits are nearly circular. To do this we transform the basic eqns. (8.1.2.16), introducing instead of ϱ_x and ϱ_y new variables $\varrho_\perp$ and φ which are connected with ϱ_x and ϱ_y by the relation

$$\varrho_x + i\varrho_y = \varrho_\perp e^{i\varphi}.$$

The first two eqns. (8.1.2.16) can be written in the form

$$\frac{d^2\varrho_\perp}{d\tau^2} - \varrho_\perp \left(\frac{d\varphi}{d\tau}\right)^2 + \frac{\beta^2\omega_p^2}{\beta^2-1}\,\frac{\beta\varrho_\perp}{\beta[1+\varrho_\perp^2+\varrho_z^2]^{1/2}-\varrho_z} = 0,$$

$$2\varrho_\perp \frac{d\varrho_\perp}{d\tau}\frac{d\varphi}{d\tau} + \varrho_\perp^2 \frac{d^2\varphi}{d\tau^2} = 0.$$

Integration of the second of these equations gives

$$\varrho_\perp^2 \frac{d\varphi}{d\tau} = M,$$

where M is a constant. The first equation then becomes

$$\frac{d^2\varrho_\perp}{d\tau^2} - \frac{M^2}{\varrho_\perp^3} + \frac{\beta^2\omega_p^2}{\beta^2-1}\,\frac{\beta\varrho_\perp}{\beta[1+\varrho_\perp^2+\varrho_z^2]^{1/2}-\varrho_z} = 0. \tag{8.1.4.3}$$

Putting here $\varrho_z = 0$ we arrive at transverse oscillations with a constant value of $\varrho_\perp$. Denoting this value by ϱ_0, we get from (8.1.4.3)

$$M^2 = \frac{\omega_\perp^2 \varrho_0^4}{\sqrt{(1+\varrho_0^2)}}, \quad \omega_\perp^2 = \frac{\beta^2\omega_p^2}{\beta^2-1}. \tag{8.1.4.4}$$

Let us now consider small oscillations of the quantity $\varrho_\perp$ around the value ϱ_0. Putting $\varrho_\perp = \varrho_0 + \delta$, assuming that δ and ϱ_z are small compared to ϱ_0, and using (8.1.4.4), we get from (8.1.4.3)

$$\frac{d^2\delta}{d\tau^2} + \frac{\omega_\perp^2(4+3\varrho_\perp^2)}{(1+\varrho_0^2)^{3/2}}\,\delta + \frac{\omega_\perp^2\varrho_0\varrho_z}{\beta(1+\varrho_0^2)} = 0.$$

Performing a similar expansion in the third of eqns. (8.1.2.16) we get a second equation to determine δ and ϱ_z:

$$\beta\frac{d^2\varrho_z}{d\tau^2} - \frac{\varrho_0}{\sqrt{(1+\varrho_0^2)}}\frac{d^2\delta}{d\tau^2} + \omega_p^2 \frac{\beta\varrho_z}{\sqrt{(1+\varrho_0^2)}} = 0.$$

Putting $\delta = De^{i\omega\tau}$, $\varrho_z = Re^{i\omega\tau}$, we get the following equation to determine the frequencies ω of the coupled transverse-longitudinal oscillations:

$$\omega^4 - \frac{\omega_p^2(4\beta^2\varrho_0^2+5\beta^2-1)}{(\beta^2-1)(1+\varrho_0^2)^{3/2}}\,\omega^2 + \frac{\omega_p^4\beta^2(4+3\varrho_0^2)}{(\beta^2-1)(1+\varrho_0^2)^2} = 0,$$

whence (Akhiezer and Polovin, 1956)

$$\omega_{1,2} = \frac{\omega_\perp}{\sqrt{(1+\varrho_0^2)}}\left\{\frac{4\beta^2\varrho_0^2+5\beta^2-1}{2\beta^2\sqrt{(1+\varrho_0^2)}} \pm \left[\left(\frac{4\beta^2\varrho_0^2+5\beta^2-1}{2\beta^2\sqrt{(1+\varrho_0^2)}}\right)^2 - \frac{(\beta^2-1)(4+3\varrho_0^2)}{\beta^2}\right]^{1/2}\right\}^{1/2}. \tag{8.1.4.5}$$

We note some limiting cases. If $\varrho_0 \ll 1$, we have

$$\omega_1 = 2\omega_\perp, \quad \omega_2 = \omega_p. \tag{8.1.4.6}$$

If $\varrho_0 \gg 1$,

$$\omega_{1,2} = \frac{\omega_\perp}{\sqrt{\varrho_0}} \sqrt{\left\{2 \pm \frac{\sqrt{(3+\beta^2)}}{\beta}\right\}}. \tag{8.1.4.7}$$

If $\beta - 1 \ll 1$, the frequencies of the coupled oscillations are for any value of ϱ_0 equal to

$$\omega_1 = 0, \quad \omega_2 = \frac{2\omega_\perp}{(1+\varrho_0^2)^{1/4}}. \tag{8.1.4.8}$$

If $\beta \gg 1$, we have for any value of ϱ_0

$$\omega_1 = \omega_p \frac{(4+3\varrho_0^2)^{1/2}}{(1+\varrho_0^2)^{3/4}}, \quad \omega_2 = \frac{\omega_p}{(1+\varrho_0^2)^{1/4}}. \tag{8.1.4.9}$$

8.1.5. COUPLED LONGITUDINAL-TRANSVERSE WAVES IN A RELATIVISTIC PLASMA

In the preceding subsections we have considered waves in a plasma which were longitudinal, transverse, or nearly transverse. A study of the general case, which can be called the case of longitudinal-transverse waves, reduces to integrating eqns. (8.1.2.17) and is a very complicated problem for which we can find a solution in a few limiting cases, namely, for large β and for β close to unity.

Let us first consider the case $\beta \gg 1$, and we shall assume that β^2 and $\xi^2+\eta^2+\zeta^2$ are of the same order of magnitude. (If $\xi^2+\eta^2+\zeta^2 \ll \beta^2$, we are back in the small amplitude case, as in that case $\xi/\beta \sim \varrho_x, \eta/\beta \sim \varrho_y, \zeta/\beta \sim \varrho_z$.) We can now in the potential energy U which characterizes the motion of the plasma and which is given by eqn. (8.1.2.18′) drop the term ζ. The problem is thus reduced to integrating the equations of motion of a particle in a central field with a Lagrangian which according to (8.1.2.18) has the form

$$L = \frac{1}{2}\left[\left(\frac{d\xi}{d\tau}\right)^2 + \left(\frac{d\eta}{d\tau}\right)^2 + \left(\frac{d\zeta}{d\tau}\right)^2\right] - \omega_p^2\beta\sqrt{[\beta^2+\xi^2+\eta^2+\zeta^2]}.$$

Introducing instead of ξ, η, ζ, τ, and L the new variables X, Y, Z, θ, and $\mathcal{L}$:

$$X = \xi/\theta, \quad Y = \eta/\theta, \; Z = \zeta/\theta, \quad \theta = \omega_p\tau, \quad \mathcal{L} = \omega_p^2\beta^2 L,$$

we get

$$\mathcal{L} = \frac{1}{2}\left[\left(\frac{dX}{d\theta}\right)^2 + \left(\frac{dY}{d\theta}\right)^2 + \left(\frac{dZ}{d\theta}\right)^2\right] - \sqrt{(1+R^2)},$$

where $R^2 = X^2+Y^2+Z^2$. As the motion in a central field is planar, it is convenient to rotate the system of coordinates in such a way that the X, Y-plane is at right angles to the angular momentum. The Lagrangian then takes the form

$$\mathcal{L} = \frac{1}{2}\left[\left(\frac{dR}{d\theta}\right)^2 + R^2\left(\frac{d\varphi}{d\tau}\right)^2\right] - \sqrt{(1+R^2)},$$

where φ is the polar angle. If we write down the energy and angular momentum conservation laws,

$$\frac{1}{2}\left(\frac{dR}{d\theta}\right)^2+\frac{M^2}{2R^2}+\sqrt{(1+R^2)} = W, \quad R^2\frac{d\varphi}{d\theta} = M,$$

we get from integrating the equations (Akhiezer and Polovin, 1956)

$$\omega_p\tau = \int \frac{dR}{[2W-(M^2/R^2)-2\sqrt{(1+R^2)}]^{1/2}}. \tag{8.1.5.1}$$

The quantity R which occurs here is connected with the dimensionless momentum ϱ through the equation

$$R^2 = \varrho^2-\frac{2\varrho_z}{\beta}\sqrt{(1+\beta^2)}+\frac{1}{\beta^2}.$$

It follows from eqn. (8.1.5.1) that the quantity R oscillates between two values R_0 and R_1 which are connected with W and M through the equations

$$M^2 = \frac{2R_0^2R_1^2}{\sqrt{(1+R_0^2)}+\sqrt{(1+R_1^2)}}, \quad W = \frac{R_0^2+R_1^2+1+[(1+R_0^2)(1+R_1^2)]^{1/2}}{\sqrt{(1+R_0^2)}+\sqrt{(1+R_1^2)}}.$$

The frequency of the oscillations is equal to

$$\omega = \frac{\pi\omega_p}{I(R_0, R_1)\sqrt{2}}, \quad I(R_0, R_1) = \int_{R_0}^{R_1} \frac{dR}{[W-(M^2/2R^2)-\sqrt{(1+R^2)}]^{1/2}}. \tag{8.1.5.2}$$

If $R_0 = 0$, this integral is the same as the integral $I(u_m)$ which determines the frequency of the longitudinal oscillations, if we put $R_1 = u_m(1-u_m)^{-1/2}$. However, the case considered by us does not reduce to the case of purely longitudinal oscillations, which we studied earlier, when $R_0 = 0$ as now we may have oscillations which are practically linearly polaized in any direction. In particular, these may be oscillations which are nearly transverse linearly polarized oscillations for which ϱ_x is non-vanishing, $\varrho_y = 0$, and ϱ_z non-vanishing, but considerably less than ϱ_x, $\varrho_z \sim \varrho_x/\beta$. The possibility that such oscillations exist does not contradict the earlier given statement that purely transverse oscillations correspond to circular rather than linear polarization.

If $R_0 = R_1$, R is constant and the vector R describes a circle with an angular frequency equal to

$$\omega = \omega_p(1+R^2)^{-1/4}.$$

This equation agrees with eqn. (8.1.4.1) for the frequency of transverse oscillations, if in the latter we put $\beta \gg 1$. The oscillations considered now are in the case $R_0 = R_1$ nearly oscillations with circular polarization, but for them the plane of the oscillations is not necessarily at right angles to the direction of the wave propagation.

We now turn to a study of the case when β lies close to unity, $\beta-1 \ll 1$. We can in this case write the basic eqns. (8.1.2.16) in the form

$$\left.\begin{aligned} \frac{d^2\varrho_x}{d\theta^2}+\frac{\varrho_x}{\sqrt{(1+\varrho^2)}-\varrho_z} &= 0, \\ \frac{d^2\varrho_y}{d\theta^2}+\frac{\varrho_y}{\sqrt{(1+\varrho^2)}-\varrho_z} &= 0, \\ \frac{d^2}{d\theta^2}\left[\varrho_z-\sqrt{(1+\varrho^2)}\right]+\frac{(\beta^2-1)\,\varrho_z}{\sqrt{(1+\varrho^2)}-\varrho_z} &= 0, \end{aligned}\right\} \tag{8.1.5.3}$$

where $\theta = \omega_p\tau(\beta^2-1)^{-1/2}$. Neglecting the last term in the third equation we get

$$\sqrt{(1+\varrho^2)}-\varrho_z = C^2, \tag{8.1.5.4}$$

where C is a constant. The first two eqns. (8.1.5.3) then take the form

$$\frac{d^2\varrho_x}{d\theta^2}+\frac{\varrho_x}{C^2} = 0, \quad \frac{d^2\varrho_y}{d\theta^2}+\frac{\varrho_y}{C^2} = 0,$$

whence

$$\varrho_x = R_x \cos\frac{\theta}{C}, \quad \varrho_y = R_y \sin\frac{\theta}{C}. \tag{8.1.5.5}$$

Substituting these expressions into (8.1.5.4) we find

$$\varrho_z = \frac{1}{4C^2}\left[R_x^2+R_y^2-2(C^4-1)-(R_x^2-R_y^2)\cos\frac{2\theta}{C}\right]. \tag{8.1.5.6}$$

There is a relation between C, R_x, and R_y which follows from the fact that the average value of nv_z vanishes. (This last condition follows from the vanishing of the average values of $\boldsymbol{E}'$ and $\boldsymbol{B}'$; see eqn. (8.1.2.3).) Noting that $u_z = \varrho_z(1+\varrho^2)^{-1/2}$ and using (8.1.2.9) and (8.1.5.4) we get for $\beta-1 \ll 1$

$$nu_z \approx \frac{n_0 u_z}{1-u_z} = \frac{n_0\varrho_z}{\sqrt{(1+\varrho^2)}-\varrho_z} = \frac{n_0}{C^2}\,\varrho_z.$$

As the average value of nu_z vanishes, it follows from this that also the average value of the quantity ϱ_z vanishes. We can therefore put in (8.1.5.6)

$$R_x^2+R_y^2 = 2(C^4-1), \quad C^2 = \sqrt{[1+\tfrac{1}{2}(R_x^2+R_y^2)]}.$$

Finally ϱ_x, ϱ_y, and ϱ_z become (Akhiezer and Polovin, 1956)

$$\varrho_x = R_x\cos\omega\tau, \quad \varrho_y = R_y\sin\omega\tau, \quad \varrho_z = \frac{(R_x^2-R_y^2)\cos 2\omega\tau}{4\sqrt{[1+\frac{1}{2}(R_x^2+R_y^2)]}}, \tag{8.1.5.7}$$

where

$$\omega = \omega_p(\beta^2-1)^{-1/2}\,[1+\tfrac{1}{2}(R_x^2+R_y^2)]^{-1/4}. \tag{8.1.5.8}$$

These results are in agreement with eqns. (8.1.4.1) and (8.1.4.8) which describe waves which are nearly transverse waves with circular polarization. Indeed, when $R_x \approx R_y$ the frequency of the oscillations of the quantity R_z is the same as the value given by eqn. (8.1.4.8).

Let us finally consider the case of high energies when we have the inequality

$$\xi^2+\eta^2+\zeta^2 \gg \beta^2-1,$$

and where the value of β remains arbitrary ($\beta > 1$). The Lagrangian describing the motion of the plasma can in this case be written in the form

$$L = \frac{1}{2}\left[\left(\frac{d\xi}{d\tau}\right)^2+\left(\frac{d\eta}{d\tau}\right)^2+\left(\frac{d\zeta}{d\tau}\right)^2\right]-\frac{\omega_{\mathrm{p}}^2\beta^2}{\beta^2-1}\,[\beta\sqrt{(\xi^2+\eta^2+\zeta^2)}+\zeta].$$

Under the transformation

$$\xi = \mu\xi', \quad \eta = \mu\eta', \quad \zeta = \mu\zeta', \quad \tau = \sqrt{(\mu)}\tau',$$

where μ is an arbitrary constant, the Lagrangian is multiplied by μ. Hence it follows that if the motion

$$\xi = \xi(\tau), \quad \eta = \eta(\tau), \quad \zeta = \zeta(\tau)$$

is possible, one can also have the analogous motion

$$\xi' = \xi(\tau'), \quad \eta' = \eta(\tau'), \quad \zeta' = \zeta(\tau').$$

It follows in particular from this that the way the frequency of the oscillations depends on the quantity p_0 which characterizes the electron momentum is well defined; in fact, the frequency must be inversely proportional to the square root of p_0:

$$\omega = \frac{\text{constant}}{\sqrt{p_0}}. \tag{8.1.5.9}$$

This formula agrees with the earlier obtained expressions for the frequency in the high-energy region (eqns. (8.1.3.5b), (8.1.4.1), (8.1.4.7), and (8.1.5.8)).

8.2. Non-linear Waves in an Unmagnetized Two-temperature Plasma

8.2.1 EQUATIONS DESCRIBING A NON-LINEAR WAVE IN A QUASI-EQUILIBRIUM PLASMA

We now turn to a study of finite-amplitude waves in a quasi-equilibrium plasma, in which the electrons and ions are characterized by Maxwellian velocity distributions with different temperatures, the electron temperature being appreciably higher than the ion temperature. As we showed in Chapter 4, low-frequency oscillations, which in the long-wavelength region have a linear dispersion law (ion sound), can propagate in such a plasma. The phase velocity of an ion-sound wave is large compared to the average ion thermal velocity; to describe the motion of the ions in such a wave we can thus use the hydrodynamic equations with a self-consistent field

$$\frac{\partial \boldsymbol{u}}{\partial t}+(\boldsymbol{u}\cdot\nabla)\,\boldsymbol{u}+\frac{Ze}{m_i}\nabla\varphi = 0, \tag{8.2.1.1}$$

$$\frac{\partial n_i}{\partial t}+\operatorname{div}(n_i\boldsymbol{u}) = 0, \tag{8.2.1.2}$$

where n_i and $\boldsymbol{u}$ are the density and the hydrodynamical velocity of the ions, Ze and m_i their charge and mass, and φ the electrostatic potential which is connected with the electron and ion densities n_e and n_i through the Poisson equation

$$\nabla^2\varphi-4\pi e(n_e-Zn_i) = 0. \tag{8.2.1.3}$$

We must describe the electron component of the plasma by a kinetic equation for the electron distribution function $F(\boldsymbol{r}, \boldsymbol{v}, t)$. As the phase velocity of the ion sound is small compared with the electron thermal velocity we can in this kinetic equation neglect the term $\partial F/\partial t$ compared with $(\boldsymbol{v}\cdot\nabla)F$. If, moreover, the collision frequency ν is sufficiently small, so that we have the inequality $kv_e \gg \nu$ (although the inequality $\omega > \nu$ may not hold; k and ω are the wavenumber and frequency of the perturbation), we can neglect the collision integral and use the equation

$$(\boldsymbol{v}\cdot\nabla)\,F+\frac{e}{m_e}\left(\nabla\varphi\cdot\frac{\partial F}{\partial \boldsymbol{v}}\right) = 0. \tag{8.2.1.4}$$

For the case of a quasi-equilibrium plasma with a Maxwellian electron velocity distribution with a temperature T_e one can immediately write down the solution of eqn. (8.2.1.4):

$$F(\boldsymbol{r}, \boldsymbol{v}, t) = \exp\left[\frac{e\varphi(\boldsymbol{r}, t)}{T_e}\right]F_0(\boldsymbol{v}), \tag{8.2.1.5}$$

where $F_0(\boldsymbol{v})$ is the distribution function at a point where the potential $\varphi = 0$. Integrating this expression we obtain the electron density

$$n_e(\boldsymbol{r}, t) = n_{e0}\exp\left[\frac{e\varphi(\boldsymbol{r}, t)}{T_e}\right], \tag{8.2.1.6}$$

where n_{e0} is the electron density at the point where $\varphi = 0$.

The distribution (8.2.1.6) is, clearly, simply a Boltzmann distribution for electrons in an electrostatic field with a potential $\varphi(\boldsymbol{r}, t)$ which varies in space and time. The fact that the electron density and the electron distribution function depend on the position and the time only through the potential φ is connected with the fact that the phase velocity of the oscillations considered is small when compared with the electron thermal velocity; because of this it is possible to establish everywhere in space and at all times a local Boltzmann distribution for the electrons. We emphasize that it is not necessary for this that the condition $\omega\tau \ll 1$ is satisfied, where $\tau \sim \nu^{-1}$.

Equations (8.2.1.1), (8.2.1.2), (8.2.1.3), and (8.2.1.6) form a complete set of non-linear equations which describe the ion-sound oscillations in a quasi-equilibrium plasma. Apart from the non-linear relation (8.2.1.6) which connects the electron density n_e and the electrostatic potential φ the hydrodynamical eqn. (8.2.1.1) also contains non-linearity—in the term $(\boldsymbol{u}\cdot\nabla)\,\boldsymbol{u}$. Of course, if we restrict ourselves—as we did in Volume 1—to a study of weak perturbations of the plasma, we can easily by linearizing these equations obtain the formula $\omega = (1+r_D^2k^2)^{-1/2}\,kv_s$ which connects the frequency and wavenumber of the small-amplitude ion-sound oscillations.

In the case of long-wavelength perturbations ($kr_D \ll 1$, where r_D is the electron Debye radius), one can slightly simplify the set of equations which describe the ion sound. In that case we can neglect the spatial charge separation and assume that

$$n_e = Zn_i. \tag{8.2.1.7}$$

Using that relation and introducing the notation

$$p = n_eT_e, \tag{8.2.1.8}$$

we can write eqn. (8.2.1.1) in the form

$$m_in_i\left[\frac{\partial \boldsymbol{u}}{\partial t}+(\boldsymbol{u}\cdot\nabla)\,\boldsymbol{u}\right]+\nabla p = 0. \tag{8.2.1.9}$$

Equations (8.2.1.2), (8.2.1.8), and (8.2.1.9) are formally the same as the equations of ordinary hydrodynamics, if we put in them the adiabatic index $\gamma = 1$, that is, assume the temperature, rather than the entropy per unit mass, to be constant.

The long-wavelength oscillations of arbitrary amplitude in a quasi-equilibrium plasma with $T_e \gg T_i$ are thus described by the equations of isothermal hydrodynamics. We emphasize, however, the fact that this analogy is to a large extent formal. Equations (8.2.1.2), (8.2.1.8), and (8.2.1.9) describe oscillations with a frequency $\omega \gg \nu$, where ν is the collision frequency, and are the consequences of a kinetic equation without a collision integral, while the analogous equations of ordinary hydrodynamics refer to the case $\omega \ll \nu$—and can be obtained from the kinetic equation only if the collisions between the particles are taken into account.

We note that—as we shall show in Section 8.3—equations similar to (8.2.1.1) to (8.2.1.4) describe ion-sound oscillations also in the case of a non-equilibrium plasma in which the electron velocity distribution is non-Maxwellian. However, in the case of a non-equilibrium plasma these equations can not be reduced to equations of the kind (8.2.1.8) or (8.2.1.9), with some adiabatic index.

8.2.2. SIMPLE (RIEMANN) WAVES

In the preceding subsection we obtained the non-linear equations which describe the motion of a quasi-equilibrium plasma consisting of hot electrons and cold ions. Turning now to a study of different kinds of non-linear oscillations in such a plasma we shall, first of all, consider one-dimensional simple (Riemann) waves (Vedenov, Velikhov, and Sagdeev, 1961a). Apart from giving us the possibility to follow the evolution of an initial perturbation, a study of simple waves is of great interest especially because only the region of simple waves can (when there are no discontinuities) delimit an unperturbed plasma (see in that connection Chapter 3).

Simple waves can be excited only in the case of long-wavelength perturbations of the plasma ($kr_D \ll 1$) when the wavevector dependence of the phase velocity does not play an essential role; we can thus use for a study of these waves together with eqns. (8.2.1.1), (8.2.1.2), and (8.2.1.6) also eqn. (8.2.1.7). Choosing the x-axis in the direction of the wave propagation we can write these equations in the form

$$\frac{du_x}{dt}+\frac{Ze}{m_i}\frac{\partial\varphi}{\partial x}=0, \qquad \frac{d\boldsymbol{u}_\perp}{dt}=0, \tag{8.2.2.1}$$

$$\frac{dn_i}{dt}+n_i\frac{\partial u_x}{\partial x}=0, \tag{8.2.2.2}$$

$$n_e = Zn_i = n_{e0}\exp(e\varphi/T_e), \tag{8.2.2.3}$$

where $d/dt = \partial/\partial t+u_x\partial/\partial x$ while $\boldsymbol{u}_\perp$ is the component of the vector $\boldsymbol{u}$ which is at right angles to the direction of the wave propagation.

It is well known that in the case of simple waves one can write all quantities which characterize the plasma as functions of one of these quantities—say, n_i—which, in turn, is a function of x and t. The set of eqns. (8.2.2.1) to (8.2.2.3) then becomes a set of ordinary differential equations for the functions $u(n_i)$ and $\varphi(n_i)$, while the phase velocity V is determined from the condition that this set can be solved.

Introducing the notation $v_s = \sqrt{(ZT_e/m_i)}$ we get after a few transformations

$$\frac{du_x}{dn_i}=\varepsilon\frac{v_s}{n_i}, \qquad \frac{d\boldsymbol{u}_\perp}{dn_i}=0, \qquad \frac{d\varphi}{dn_i}=\frac{T_e}{en_i}, \qquad n_e = Zn_i, \tag{8.2.2.4}$$

$$V = u_x+\varepsilon v_s, \tag{8.2.2.5}$$

where $\varepsilon = +1$ ($\varepsilon = -1$), if the wave propagates in the direction of the positive (negative) x-axis.

The set of eqns. (8.2.2.4) enables us to study the direction of changes in the quantities which characterize the plasma and to follow the evolution of a finite-amplitude perturbation.

We note first of all that the quantities v_s and $\boldsymbol{u}_\perp$ are independent of n_i and are thus integrals of motion. The electron temperature and the electron distribution function, normalized to one particle, F/n_e, also remain unchanged during the propagation of the wave. Furthermore, in a compression wave the densities of both kinds of particles and the electrostatic potential increase; in a rarefaction wave these quantities decrease. In order to determine how the shape of the ion-sound wave changes when it propagates we must evaluate the derivative dV/dn_i. Using (8.2.2.5) and (8.2.2.4) and putting $\varepsilon = +1$ to fix the ideas we have

$$\frac{dV}{dn_i}=\frac{v_s}{n_i}. \tag{8.2.2.6}$$

We see that—as in ordinary hydrodynamics—the derivative dV/dn_i is positive; in other words, elements with a larger density move with a larger velocity. The wave profile therefore becomes steeper and steeper along compression sections and flatter and flatter along rarefaction sections.

It is well known that in ordinary hydrodynamics such a change in the profile leads to the occurrence of shock waves (see, for instance, Landau and Lifshitz, 1959). The Euler equation

—together with the continuity equation and the equation of state of the fluid—then enables us to follow the evolution of the perturbation until the gradients of the hydrodynamic quantities become so large that it becomes necessary to take dissipative processes into account. Ultimately a stationary shock wave is established thanks to these processes (see Chap- 3).

We have a somewhat different picture in a collisionless plasma because of the smallness of the dissipative processes. When the width of the leading front of a compression wave becomes equal to the electron Debye radius, eqn. (8.2.2.3) becomes inapplicable and it is necessary to take the dispersion of the ion sound into account. When the gradients increase further there may appear multiple-current flow (Vedenov, Velikhov, and Sagdeev, 1961a) or quasi-shock waves (Moiseev and Sagdeev, 1963; Sagdeev, 1966) along compression sections.

Let us dwell briefly on self-similar one-dimensional motions of a two-temperature plasma,† that is, motions such that all quantities characterizing the plasma depend on the coordinates and the time solely in the combination x/t. It is well known (see Chapter 3 in this connection) that the nature of the self-similar waves, like that of simple waves, is determined by the derivative dV/dn_i. According to (8.2.2.6) we have $dV/dn_i > 0$; self-similar waves in a quasi-equilibrium plasma thus are always rarefaction waves.

The problem of motions in a two-temperature plasma which arise when its volume is changed uniformly (see Landau and Lifshitz (1959) for the analogous hydrodynamical problem of a moving piston) is connected with the problem of self-similar waves.

We shall assume that the plasma fills the half-space $x > u_0 t$ which is bounded by a uniformly moving plane—such a boundary may, in particular, be a region of a very strong magnetic field. It is well known that, if there are no discontinuities, the motions which appear in a uniformly compressed or rarefied medium can only be self-similar waves. In the case considered self-similar waves which are rarefaction waves occur when the plasma expands ($u_0 < 0$).

Using (8.2.2.4) and (8.2.2.5) we can connect the changes in all quantities X which characterize the plasma in a self-similar wave with the "piston" velocity u_0. Introducing the notation

$$\Delta X = X_{x = u_0 t} - X_{x \to \infty},$$

and assuming for the sake of simplicity that $u_0 \ll v_s$, we find

$$\frac{\Delta n_i}{n_i} = \frac{\Delta n_e}{n_e} = \frac{u_0}{v_s}, \quad \Delta \boldsymbol{u}_\perp = 0, \quad \Delta\varphi = \frac{u_0 T_e}{e v_s}. \tag{8.2.2.7}$$

In concluding this subsection we note that when deriving the initial equations we did not take into account Landau damping for ion-sound waves (mathematically this is connected with neglecting the term $\partial F/\partial t$ in the kinetic equation for the electrons). Bearing in mind that the damping of small-amplitude ion-sound waves is proportional to the small parameter $\sqrt{(m_e/m_i)}$ (see Chapter 4), one sees easily that in order that non-linear effects—rather

† Gurevich, Pariĭskaya, and Pitaevskiĭ (1966) have studied self-similar waves in a collisionless plasma with an electron Boltzmann distribution for the case of hot ions.

than the damping of the sound—play the main role in the evolution of the perturbation, the amplitude of the perturbation Δn_e may not be too small,

$$\frac{\Delta n_e}{n_e} \gg \sqrt{\frac{m_e}{m_i}}.$$

8.2.3. PERIODIC AND SOLITARY WAVES

One-dimensional simple (Riemann) and self-similar waves, which were considered in the preceding subsection, are the simplest kinds of non-stationary motions of a quasi-equilibrium plasma. We saw that both kinds of waves occur in the case of long-wavelength perturbations of the plasma ($kr_D \ll 1$). Generally speaking, any long-wavelength perturbation must, because of the amplitude dependence of the propagation velocity of the perturbation, propagate in the plasma in the form of non-stationary waves.

The situation is different for short-wavelength perturbations. One must take the dispersion of ion sound into account on those sections of the wave where the characteristic length of the inhomogeneity is comparable to the electron Debye length; as a result the propagation velocity of the perturbation becomes dependent on the wavelength. The amplitude dependence and the wavelength dependence of the velocity of propagation of the perturbation may then compensate one another. When the dispersion of sound in a plasma is taken into account stationary waves may thus become a possibility; we shall now turn to a study of those waves.

In a one-dimensional stationary wave all quantities depend clearly on the coordinates and the time in the combination $x' \to x - Vt$, where V is a constant. Changing to a coordinate system which moves with the wave—that is, which moves with a velocity V with respect to the laboratory system—and using (8.2.1.1) and (8.2.1.2) we get

$$un_i = u_0 n_{i0}, \quad \tfrac{1}{2}m_i u^2 + Ze\varphi = \tfrac{1}{2}m_i u_0^2, \tag{8.2.3.1}$$

where u_0 and n_{i0} are the values of the velocity and of the ion density in a point where the potential φ vanishes (here and henceforth we shall drop the index x of the velocity component u_x). Equations (8.2.3.1) express clearly the fact that the number of ions and the total energy of an ion are integrals of motion. Together with eqn. (8.2.1.6) and the Poisson equation,

$$\frac{\partial^2 \varphi}{\partial x^2} = -4\pi e(Zn_i - n_e), \tag{8.2.3.2}$$

eqns. (8.2.3.1) form a complete set of equations which determine the quantities u, n_i, n_e, and φ in a one-dimensional stationary wave.

Solving these equations and introducing the electrical field $E = -\partial\varphi/\partial x$, we get

$$E^2 = -8\pi e \int_0^{\varphi} d\varphi \{Zn_i(\varphi) - n_e(\varphi)\}, \tag{8.2.3.3}$$

where the function $n_e(\varphi)$ is given by formula (8.2.1.6) while

$$n_i(\varphi) = n_{i0}\left[1 - \frac{2Ze\varphi}{m_i u_0^2}\right]^{-1/2}. \tag{8.2.3.4}$$

We have reckoned the potential φ from the point where $E = 0$; n_{e0} is the value of the electron density at that point.

Performing the integration in (8.2.3.3) we get

$$E^2 = 8\pi n_{i0} T_e \left[\frac{n_{e0}}{n_{i0}} \left\{ \exp\left(\frac{e\varphi}{T_e}\right) - 1 \right\} + \frac{m_i u_0^2}{T_e} \left\{ \sqrt{\left(1 - \frac{2Ze\varphi}{m_i u_0^2}\right)} - 1 \right\} \right]. \quad (8.2.3.5)$$

Knowing $E^2(\varphi)$ we easily find the equation which determines the potential φ as function of the coordinates and the time:

$$x - Vt = \pm \int \frac{d\varphi}{\sqrt{(E^2(\varphi))}}. \quad (8.2.3.6)$$

In the general case when $n_{e0} \neq Zn_{i0}$ eqns. (8.2.1.6), (8.2.3.4), (8.2.3.5), and (8.2.3.6) determine the distribution of the quantities which characterize the plasma in a one-dimensional periodic wave. (The potential distribution in such a wave is sketched in Fig. 8.2.1).

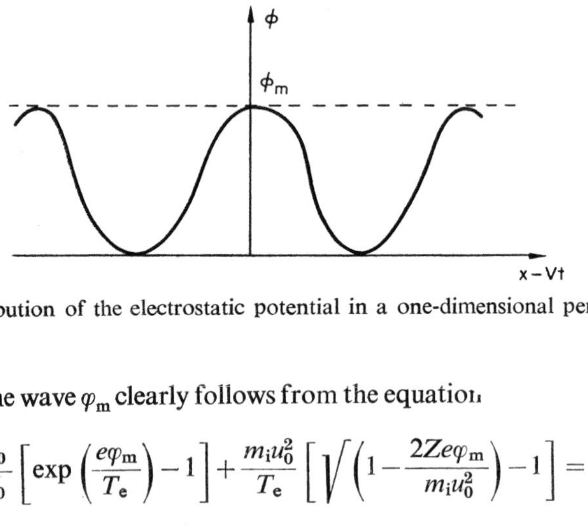

FIG. 8.2.1. The distribution of the electrostatic potential in a one-dimensional periodic stationary wave

The amplitude of the wave φ_m clearly follows from the equation

$$\frac{n_{e0}}{n_{i0}} \left[\exp\left(\frac{e\varphi_m}{T_e}\right) - 1 \right] + \frac{m_i u_0^2}{T_e} \left[\sqrt{\left(1 - \frac{2Ze\varphi_m}{m_i u_0^2}\right)} - 1 \right] = 0. \quad (8.2.3.7)$$

For the wavelength we get

$$\lambda = \frac{1}{\sqrt{(2\pi n_{i0} T_e)}} \int_0^{\varphi_m} \left\{ \frac{n_{e0}}{n_{i0}} \left[\exp\left(\frac{e\varphi}{T_e}\right) - 1 \right] + \frac{m_i u_0^2}{T_e} \left[\sqrt{\left(1 - \frac{2Ze\varphi}{m_i u_0^2}\right)} - 1 \right] \right\}^{-1/2} d\varphi. \quad (8.2.3.8)$$

In the special case when in some point the charge density and the electrical field vanish simultaneously ($n_{e0} = Zn_{i0}$), eqns. (8.2.1.6), (8.2.3.4), (8.2.3.5), and (8.2.3.6) describe a so-called *solitary wave*, which is a potential burst which moves uniformly across the plasma (Vedenov, Velikhov, and Sagdeev, 1961a; see also Sagdeev, 1966). A solitary wave is a degenerate case of a periodic wave: the latter changes into a solitary wave, as $\lambda \to \infty$. (We note that a similar situation occurs in ordinary hydrodynamics in the theory of wave propagation in a channel of finite depth where there is also the possibility of periodic and solitary waves (Korteweg and de Vries, 1895).)

Let us consider the structure of a solitary wave in somewhat more detail. (The potential distribution in a solitary wave is sketched in Fig. 8.2.2.) Bearing in mind that far from the crest of the wave—in the unperturbed region—the plasma is at rest (in the laboratory frame of reference), we have $u_0 = -V$. Using (8.2.3.6) we can easily determine the potential φ far from the crest of the wave ($|x-Vt| \to \infty$):

$$\varphi \propto \exp\left[\frac{-|x-Vt|}{r_D}\sqrt{\left(1-\frac{v_s^2}{V^2}\right)}\right], \tag{8.2.3.9}$$

where $v_s = \sqrt{(ZT_e/m_i)}$ is the ion sound velocity and $r_D = \sqrt{(4\pi e^2 n_{e0}/T_e)}$ the electron Debye radius. We see that, in the laboratory frame of reference, a solitary wave moves with super-

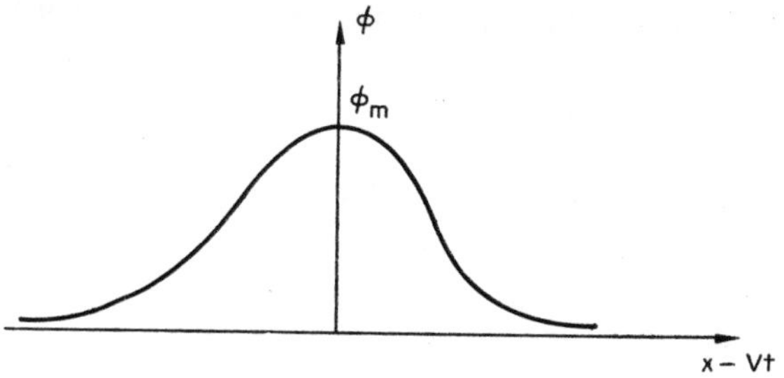

FIG. 8.2.2. The distribution of the electrostatic potential in a solitary compression wave.

sonic velocity ($V > v_s$) while the perturbation far from the crest of the wave decreases exponentially.

Putting $n_{e0} = Zn_{i0}$, $u_0 = -V$ in (8.2.3.7) we get the equation which determines the amplitude of the potential in the solitary wave:

$$1+\frac{v_s^2}{V^2}-\sqrt{\left(1-\frac{2Ze\varphi_m}{m_i V^2}\right)}-\frac{v_s^2}{V^2}\exp\left(\frac{e\varphi_m}{T_e}\right) = 0. \tag{8.2.3.10}$$

One can easily show that this equation can only be solved if $\varphi_m > 0$. Noting that the amplitude of the electron and ion densities n_{em} and n_{im} are, according to (8.2.1.6) and (8.2.3.4), connected with φ_m through the relations

$$n_{em} = n_{e0}\exp\left(\frac{e\varphi_m}{T_e}\right), \qquad n_{im} = \frac{n_{i0}}{\sqrt{\left(1-\frac{2Ze\varphi_m}{m_i V^2}\right)}}, \tag{8.2.3.11}$$

we see that $n_{em} > n_{e0}$ and $n_{im} > n_{i0}$. A solitary wave in a quasi-equilibrium plasma with a Maxwellian electron velocity distribution is therefore always a compression wave.

There is, according to (8.2.3.10), a unique relation between the wave amplitude and its propagation velocity (Vedenov, Velikhov, and Sagdeev, 1961a)

$$V^2 = \frac{1}{2}v_s^2\left[\exp\left(\frac{e\varphi_m}{T_e}\right)-1\right]^2\left\{\exp\left(\frac{e\varphi_m}{T_e}\right)-1-\frac{e\varphi_m}{T_e}\right\}^{-1}. \tag{8.2.3.12}$$

If the amplitude of the solitary wave is small ($e\varphi_m \ll T_e$) its propagation speed tends, according to (8.2.3.12) to the ion sound velocity v_s. The potential distribution is in that case determined by the formula

$$\varphi = \cosh^{-2}\left\{\frac{x-Vt}{2r_D}\left(1-\frac{v_s^2}{V^2}\right)^{1/2}\right\}\varphi_m, \quad \varphi_m = \frac{3T_e}{2e}\left(1-\frac{v_s^2}{V^2}\right). \quad (8.2.3.13)$$

We note that eqn. (8.2.3.10) has a solution only provided φ_m is not too large. This means that a solitary wave cannot have an arbitrarily large amplitude. The maximum possible value of the amplitude of a solitary wave can be determined from the relation $\frac{1}{2}m_i V^2 = Ze\varphi_m$, which is the condition that the ions at the crest of the wave completely lose their kinetic energy—for larger values of φ_m the ions would not be able to cross the wave potential barrier which is formed. Solving this equation together with eqn. (8.2.3.12) we get

$$e\varphi_m \approx 1.3T_e, \quad V \approx 1.6v_s.$$

In concluding this subsection we note that the solitary waves (sometimes called *solitons*) play apparently a special role in the propagation of perturbations in a collisionless plasma. Computer experiments (Zabusky and Kruskal, 1965; Berezin and Karpman, 1967), which used non-linear equations of the Korteweg–de Vries type, have shown that a perturbation of an arbitrary shape is, generally speaking, unstable and decays after some time into a succession of solitons which move one after another.

8.2.4. QUASI-SHOCK WAVES

In the case of solitary waves the state of the plasma behind the wave is the same as that in front of the leading wave front: both as $x-Vt \to -\infty$ and as $x-Vt \to \infty$ the plasma is at rest in the laboratory frame of reference while both the electrical field and the charge density in it are equal to zero. There is another kind of stationary wave possible in a collisionless plasma in which the state of the plasma behind the wave is not the same as that in front of the wave—the so-called *quasi-shock waves* (Sagdeev, 1966). These waves are similar to shock waves in that respect. However, in contrast to ordinary shock waves whose structure is in an essential way determined by dissipative effects, quasi-shock waves occur also when there is no dissipation and the width of the front of these waves can be considerably smaller than the mean free path of the particles in the plasma—in particular, in a plasma without a magnetic field the width of the front of a quasi-shock wave is of the same order of magnitude as the Debye radius.

A quasi-shock wave occurs when for some reason or other the symmetry of the spatial distribution of the physical quantities in the plasma is broken. In the case of a quasi-equilibrium plasma with a Maxwellian electron velocity distribution (Moiseev and Sagdeev, 1963) the reason may be the reflection of ions from the potential barrier formed by the wave. The leading front of the perturbation has in that case the same shape as the leading front of a solitary wave, but the perturbation does not decrease exponentially behind the crest of the wave, but oscillates. The profile of such a quasi-shock wave is sketched in Fig. 8.2.3.

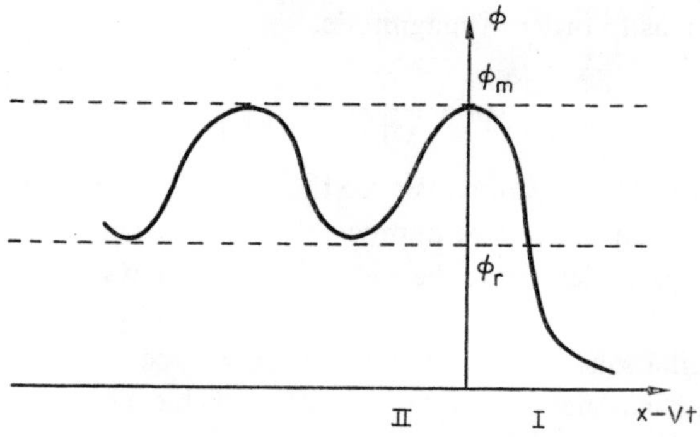

FIG. 8.2.3. The distribution of the electrostatic potential in a quasi-shock compression wave.

Assuming for the sake of simplicity that the density of the reflected ions Δn_i is small ($\Delta n_i \ll n_i$), we can clearly determine the potential distribution in the leading front of the wave (region I), substituting $n_{e0} = Zn_{i0}$ and $u_0 = -V$ into eqns. (8.2.3.5) and (8.2.3.6). The result is

$$x-Vt = \frac{1}{\sqrt{(8\pi n_{e0} T_e)}} \int\limits_0^{\varphi} d\varphi \left\{ \exp\left(\frac{e\varphi}{T_e}\right) - 1 + \frac{V^2}{v_s^2}\left[\sqrt{\left(1-\frac{2Ze\varphi}{m_i V^2}\right)} - 1\right]\right\}^{-1/2}. \quad (8.2.4.1)$$

The maximum of the potential φ_m and the maxima of the electron and ion densities n_{em} and n_{im} are determined as before from the eqns. (8.2.3.10) and (8.2.3.11). We see that, as in the case of a solitary wave, $n_{im} > n_{i0}$; we can therefore call the quasi-shock wave in a plasma with a Maxwellian electron velocity distribution a *quasi-shock compression wave*.

In region II we have

$$x-Vt = \frac{1}{\sqrt{(8\pi n_{em} T_e)}} \int\limits_0^{\varphi-\varphi_m} d\varphi \left\{ \exp\left(\frac{e\varphi}{T_e}\right) - 1 + Z\frac{n_{im}-\Delta n_i}{n_{em}} \frac{u_m^2}{v_s^2}\left[\sqrt{\left(1-\frac{2Ze\varphi}{m_i u_m^2}\right)} - 1\right]\right\}^{-1/2}$$

(8.2.4.2)

where $u_m^2 = V^2 - 2Ze\varphi_m/m_i$. We find for the minimum value of the potential φ_r in the region of the oscillations and for the wavelength of the oscillations λ

$$\exp\left[\frac{e(\varphi_r-\varphi_m)}{T_e}\right] - 1 + Z\frac{n_{im}-\Delta n_i}{n_{em}}\frac{u_m^2}{v_s^2}\left\{\sqrt{\left[1-\frac{2Ze(\varphi_r-\varphi_m)}{m_i u_m^2}\right]} - 1\right\} = 0, \quad (8.2.4.3)$$

$$\lambda = \frac{1}{\sqrt{(2\pi n_{em} T_e)}} \int\limits_{\varphi_r-\varphi_m}^{0} d\varphi \left\{ \exp\left(\frac{e\varphi}{T_e}\right) - 1 + Z\frac{n_{im}-\Delta n_i}{n_{em}}\frac{u_m^2}{v_s^2}\left[\sqrt{\left(1-\frac{2Ze\varphi}{m_i u_m^2}\right)} - 1\right]\right\}^{-1/2}.$$

(8.2.4.4)

The quasi-shock wave thus connects two different states of the plasma—the unperturbed plasma in front of the wave and the plasma with periodic oscillations of E, n_e, n_i, φ, and so on, behind the wave front. If we now understand by the width of the wave front x_0 the dimension of the region in space which separates these two states of the plasma, we find, according to (8.2.4.1), as to order of magnitude,

$$x_0 \sim \frac{r_D}{\sqrt{[1-(v_s^2/V^2)]}},$$

where r_D is the electron Debye radius. We see that the width of the front of a quasi-shock wave in a collisionless plasma can be appreciably smaller than the mean free path of the particles which we know to determine the width of the shock wave in ordinary gas-dynamics (see Chapter 3).

Of course, the oscillations in a quasi-shock wave will be slowly damped as a result of dissipative effects. We shall not consider in detail the problem of the damping of the oscillations.

8.3. Non-linear Waves in an Unmagnetized Non-equilibrium Plasma

8.3.1. EQUATIONS DESCRIBING A NON-LINEAR WAVE IN A NON-EQUILIBRIUM PLASMA

Weakly damped ion-sound oscillations can propagate not only in a two-temperature plasma but also in a plasma with an arbitrary (not necessarily Maxwellian) velocity distribution provided the average energy of the random motion of the electrons is considerably higher than the average energy of the random motion of the ions. As before, the small-amplitude ion-sound oscillations are then characterized by the dispersion law

$$\omega = \frac{kv_s}{\sqrt{(1+k^2r_D^2)}}, \tag{8.3.1.1}$$

where

$$v_s = r_D\sqrt{\frac{4\pi e^2Z^2n_{i0}}{m_i}},$$

while r_D is the electron Debye radius which is connected with the electron distribution function $F(\boldsymbol{v})$ through the equation

$$r_D = \left|\frac{4\pi e^2}{m_e}\int\left(\boldsymbol{k}\cdot\frac{\partial F}{\partial \boldsymbol{v}}\right)\frac{d^3v}{(\boldsymbol{k}\cdot\boldsymbol{v})}\right|^{-1/2}; \tag{8.3.1.2}$$

n_{i0} and $n_{e0} = Zn_{i0}$ are as before the unperturbed values of the ion and electron densities.

The nature of the dispersion of the small-amplitude ion-sound oscillations in a plasma with a non-Maxwellian velocity distribution does therefore not differ at all from the nature of the dispersion of the analogous oscillations in a plasma with a Maxwellian distribution; we need only understand by the electron temperature the quantity

$$T_e = m_e n_{e0}\left|\int\left(\boldsymbol{k}\cdot\frac{\partial F}{\partial \boldsymbol{v}}\right)\frac{d^3v}{(\boldsymbol{k}\cdot\boldsymbol{v})}\right|^{-1}. \tag{8.3.1.3}$$

The situation is different in the case of non-linear ion-sound oscillations: we shall show in a moment that the nature of those turns out to depend in an essential way on the form of the electron distrubution function.

For a study of the non-linear ion-sound oscillations we can as before use eqns. (8.2.1.1) to (8.2.1.4) where we understand by $F(\boldsymbol{v})$ the non-Maxwellian electron velocity distribution. In the case of one-dimensional oscillations—and we shall restrict our considerations to that case—one can easily find a solution of the kinetic eqn. (8.2.1.4). Using the notation $F(\boldsymbol{v}) = F(v_x^2; \boldsymbol{v}_\perp)$ we get

$$F(v_x^2; \boldsymbol{v}_\perp; x, t) = F\left(v_x^2 - \frac{2e}{m_e}\varphi(x, t); \boldsymbol{v}_\perp\right); \tag{8.3.1.4}$$

the x-axis corresponds to the direction of the wave propagation.

Integrating (8.3.1.4) over the electron velocities we get the electron "equation of state",

$$n_e = n_e(\varphi), \tag{8.3.1.5}$$

which for a non-equilibrium plasma replaces eqn. (8.2.1.6). Together with the Poisson eqn. (8.2.3.2) and eqns. (8.2.2.1) and (8.2.2.2) the "equation of state" (8.3.1.5) forms a complete set of equations which describe non-linear one-dimensional motions of a plasma with an arbitrary velocity distribution.

8.3.2. SIMPLE WAVES

Turning now to a study of simple (Riemannian) waves in a non-equilibrium plasma (Akhiezer, 1965b) we first of all note that, as in the case of a plasma with a Maxwellian electron distribution, simple waves can only be excited in the case of long-wavelength perturbations of the plasma ($kr_D \ll 1$). In that case we can neglect space charge in the plasma and assume that $n_e = Zn_i$. We can thus for a study of simple waves use eqns. (8.2.2.1), (8.2.2.2.), and the "equation of state"

$$n_e = Zn_i = n_e(\varphi). \tag{8.3.2.1}$$

Bearing in mind that in the case of simple waves all quantities characterizing the plasma can be written as functions of one of them (for instance, n_i), we get a set cf ordinary differential equations for the functions $u(n_i)$ and $\varphi(n_i)$ and the phase velocity V is determined from the condition that this set can be solved. Using the notation

$$v_s^2 = \frac{eZ^2 n_i}{m_i D^{(1)}}, \quad D^{(j)} = \frac{\partial^j n_e(\varphi)}{\partial \varphi^j}, \quad j = 0, 1, 2, \ldots, \tag{8.3.2.2}$$

we get

$$V = u_x + \varepsilon v_s, \quad \frac{du_x}{dn_i} = \varepsilon \frac{v_s}{n_i}, \quad \frac{d\boldsymbol{u}_\perp}{dn_i} = 0, \quad \frac{d\varphi}{dn_i} = \frac{m_i v_s^2}{eZn_i}, \quad n_e = Zn_i, \tag{8.3.2.3}$$

where $\varepsilon = \pm 1$, depending on whether the wave moves in the direction of the positive or negative x-axis.

The set of eqns. (8.3.2.3) together with eqn. (8.3.1.4) enables us to study the way in which the quantities characterizing the plasma—including the electron distribution function—change and to follow the evolution of a finite-amplitude perturbation.

First of all we note that the densities of both kinds of particles as well as the electrostatic potential increase along compression sections while they decrease along rarefaction sections.

Let us follow the change of the electron distribution function in a simple wave. Noting that $d\varphi/dn_i > 0$ and using (8.3.1.4) we show easily that for those velocities for which $v_x \partial F/\partial v_x < 0$ the number of electrons with velocities in the range $\boldsymbol{v}, \boldsymbol{v}+d\boldsymbol{v}$ increase in a compression wave and decrease in a rarefaction wave. On the other hand, for those values of $\boldsymbol{v}$ for which $v_x \partial F/\partial v_x > 0$ the number of electrons with velocities in the range $\boldsymbol{v}, \boldsymbol{v}+d\boldsymbol{v}$ increases in a rarefaction wave and decreases in a compression wave. In particular, if the initial electron velocity distribution has apart from the maximum at $v_x = 0$ also a narrow spike which occupies a small velocity region, the spike will shift into the region of larger (smaller) values of $|v_x|$ during the motion of a compression (rarefaction) wave.

To determine how the shape of an ion-sound wave changes we need to evaluate the derivative dV/dn_i. Using (8.3.2.3) and putting $\varepsilon = +1$ to fix the ideas, we get

$$\frac{dV}{dn_i} = \frac{v_s}{2n_i}\left[3 - \frac{n_e D^{(2)}}{(D^{(1)})^2}\right]. \tag{8.3.2.4}$$

Depending on the shape of the electron distribution function the quantity dV/dn_i can be positive, negative, zero, or changing sign, that is, positive for some and negative for other values of the potential φ. (We remember that both in ordinary and in magneto-hydrodynamics the quantity dV/dn_i is always positive.)

If $dV/dn_i > 0$ (normal case), elements with a larger density move—as in ordinary hydrodynamics—with a larger velocity; as a result there can appear discontinuities along compression sections.† Self-similar waves are in this case rarefaction waves. Such a possibility occurs in particular when the electron velocity distribution is Maxwellian (see preceding section) and for a step-function distribution

$$F(\boldsymbol{v}) = \begin{cases} 0 & (v > v_0(n_i)) \\ \text{constant} & (v < v_0(n_i)). \end{cases}$$

If $dV/dn_i = 0$ for all values of the potential all elements will move with the same velocity during the wave motion so that the wave profile is not distorted and no discontinuities appear. Using (8.3.2.3) one can verify that the ion-sound velocity and the quantity $\sqrt{\varphi}$ are in that case inversely proportional to the density

$$v_s n_i = \text{constant}, \quad \varphi n_i^2 = \text{constant}.$$

The case $dV/dn_i = 0$ is, in particular, realized for a distribution of the form $F(\boldsymbol{v}) \propto \{v^2 + v_0^2(n_i)\}^{-2}$.

If $dV/dn_i < 0$ (anomalous case) elements with larger densities move with a lower velocit Discontinuities therefore appear along rarefaction sections: self-similar waves are compression waves. Such a situation can be realized, in particular, if the distribution function in the main velocity range has the form

$$F(\boldsymbol{v}) \propto \{v^2 + v_0^2(n_i)\}^{-\alpha},$$

† We apply the term discontinuity to narrow regions in which the gradients of the quantities which characterize the plasma are so large that the original relation $n_e = Zn_i$ becomes inapplicable. If $kr_D \gtrsim 1$ it is necessary to take the ion-sound dispersion into account. When the gradients increase further in these regions multiple-current flow or shock waves may occur.

where $\frac{3}{2} < \alpha < 2$. We note that in that case the sound velocity v_s increases in a rarefaction wave and decreases in a compression wave.

Let us finally consider the case when the quantity dV/dn_i can be both positive and negative, depending on the value of the potential φ. To fix the ideas we shall assume that $dV/dn_i > 0$ when $\varphi < \varphi_1$ and $dV/dn_i < 0$ when $\varphi > \varphi_1$ where φ_1 is some critical value of the potential. When a compression wave moves through such a plasma the "peak" of the wave, that is those parts for which $n_i > n_{i1}$, where n_{i1} is given by the equations $\varphi(n_{i1}) = \varphi_1$, will lag behind the "base" of the wave, that is those parts for which $n_i < n_{i1}$; if such a wave leads to a discontinuity the density in that point can therefore not exceed n_{i1}. When a rarefaction wave propagates there is also the possibility that a discontinuity may occur; in that case the density in that point cannot be less than n_{i1}.

If, on the other hand, $dV/dn_i < 0$ when $\varphi < \varphi_2$ and $dV/dn_i > 0$ when $\varphi > \varphi_2$, one can easily check that the density in a discontinuity cannot exceed n_{i2} if the discontinuity develops from a rarefaction wave and cannot be less than n_{i2} if it develops from a compression wave (the critical value of the density n_{i2} follows from the equation $\varphi(n_{i2}) = \varphi_2$).

Both possibilities are, in particular, realized if the electron velocity distribution is a superposition of two, a "hot" and a "cold", Maxwell distributions:

$$F(\boldsymbol{v}) = \nu_1(n_i) \exp\left(\frac{-m_e v^2}{2T_1}\right) + \nu_2(n_i) \exp\left(\frac{-m_e v^2}{2T_2}\right), \quad T_1 \gg T_2.$$

Let us now consider briefly the problem of the motion in a non-equilibrium plasma which occurs when it is uniformly compressed or expanding—the analogy to the hydrodynamical piston problem. It is well known that the motion which occurs in a uniformly compressed (or expanding) plasma can only be self-similar waves—when there are no discontinuities. If $dV/dn_i > 0$, a self-similar wave—which in this case is a rarefaction wave—occurs when the plasma is expanding ($u_0 < 0$, where u_0 is the "piston" velocity). If $dV/dn_i < 0$, the self-similar wave, which in this case is a compression wave occurs when the plasma is compressed ($u_0 > 0$).

Using (8.3.1.4) and (8.3.2.3) we can connect the change ΔX in the quantities which characterize the plasma X in a self-similar wave with the "piston" velocity u_0. In particular, we have when $u_0 \ll v_s$

$$\frac{\Delta n_e}{n_e} = \frac{\Delta n_i}{n_i} = \frac{u_0}{v_s}, \quad \Delta \boldsymbol{u}_\perp = 0, \quad \Delta\varphi = \frac{m_i v_s u_0}{eZ},$$

$$\Delta v_s = \frac{1}{2} u_0 \left[1 - \frac{n_e D^{(2)}}{(D^{(1)})^2}\right], \quad \Delta F(\boldsymbol{v}) = -\frac{m_i v_s u_0}{m_e Z} \frac{\partial F(\boldsymbol{v})}{v_x \partial v_x}. \qquad (8.3.2.5)$$

8.3.3. STATIONARY WAVES

When long-wavelength ion-sound oscillations with a linear dispersion law propagate through the plasma the wave profile will, in general, change with time—an exception is the case of a plasma with an electron distribution function of the form $\{v^2 + v_0^2\}^{-2}$ for which a sound wave can propagate without distortion of its shape.

Along sections of the wave with a characteristic inhomogeneity length of the order of the electron Debye radius the dispersion of ion sound becomes appreciable and as a result waves with a stationary profile become possible: solitary, periodic, and quasi-shock waves. We shall now turn to the study of the propagation of such waves in a non-equilibrium plasma (Akhiezer and Borovik, 1967).

In the case of a non-equilibrium plasma the velocity distribution, the ion density, and the electrostatic potential in a one-dimensional stationary wave are—in the frame of reference which moves with the wave—clearly determined by the same eqns. (8.2.3.1) and (8.2.3.2) as in the case of a plasma with a Maxwellian distribution. Adding to these equations the "equation of state" (8.3.1.5), $n_e = n_e(\varphi)$, we get the complete set of equations which describe such a wave.

One can easily show by solving these equations that the distribution of the physical quantities in a stationary wave is determined in the case of a non-equilibrium plasma by the same eqns. (8.2.3.3), (8.2.3.4), and (8.2.3.6) as in the case of a plasma with a Maxwellian electron velocity distribution, provided we understand by $n_e(\varphi)$ the function which is obtained by integrating the non-Maxwellian distribution (8.3.1.4) over the velocities.

In the general case when $n_{e0} \neq Zn_{i0}$, these relations determine the distribution of the quantities which characterize the plasma in a one-dimensional periodic wave. The wave amplitude φ_m is clearly determined from the equation

$$\int_0^{\varphi_m} d\varphi\, [n_e(\varphi) - Zn_i(\varphi)] = 0, \tag{8.3.3.1}$$

and the wavelength is equal to

$$\lambda = \frac{1}{\sqrt{(2\pi e)}} \int_0^{\varphi_m} d\varphi \left\{ \int_0^{\varphi} d\varphi' [n_e(\varphi') - Zn_i(\varphi')] \right\}^{-1/2}. \tag{8.3.3.2}$$

In the special case when the charge density and the electrical field vanish in some point ($n_{e0} = Zn_{i0}$) eqns. (8.2.3.3), (8.2.3.4), and (8.2.3.6) describe a solitary wave, the structure of which we shall now consider in some detail.

We note first of all that far from the crest of the wave ($|x - Vt| \to \infty$) the potential distribution in a solitary wave is in the case of a non-equilibrium plasma determined, according to these relations, by the same formula (8.2.3.9) as in the case of a plasma with a Maxwellian distribution; we need merely in the case of a non-equilibrium plasma use for the electron Debye radius r_D eqn. (8.3.1.2) or

$$r_D = [4\pi e D^{(1)}]^{-1/2}, \tag{8.3.3.3}$$

which is equivalent to it.

We see that in the case of a non-equilibrium plasma a solitary wave moves—in the laboratory frame of reference—with a supersonic velocity ($V > v_{s0}$, where v_{s0} is the ion sound velocity when $n_i = n_{i0}$) while the perturbation far from the crest of the wave falls off exponentially.

Putting $n_{e0} = Zn_{i0}$, $u_0 = -V$ into (8.3.3.1) and (8.2.3.4) we get the equation which determines the amplitude φ_m of the potential in the solitary wave:

$$\int_0^{\varphi_m} d\varphi \left[n_e(\varphi) - n_{e0} \left\{ 1 - \frac{2Ze\varphi}{m_i V^2} \right\}^{-1/2} \right] = 0. \tag{8.3.3.4}$$

In the case of a Maxwellian electron velocity distribution this equation, which in this case takes the form (8.2.3.10), has a solution $\varphi_m > 0$ as was pointed out in the preceding section. Clearly, the ion and electron densities then increase together with the potential (solitary compression wave).

However, it is possible to have such an electron velocity distribution that eqn. (8.3.3.4) has a solution which lies in the region of negative values of the potential, $\varphi_m < 0$, rather than in the region of positive values. The ion and electron densities in the solitary wave then decrease (solitary rarefaction wave). This case is, in particular, realized for distributions which in the main velocity region have the form $(v^2+v_0^2)^{-\alpha}$ with $\frac{3}{2} < \alpha < 2$.

We shall now show that a solitary compression wave is formed in the case of distribution functions which lead to an increase of the gradients along compression regions (normal case), while a solitary rarefaction wave is formed in the case of distribution functions leading to an increase in the gradients along rarefaction regions (anomalous case).

To do this, we note that the integrand in eqn. (8.3.3.4) vanishes when $\varphi = 0$ and when $\varphi = \varphi_1$, where φ_1 lies within the range $0, \varphi_m$. The quantity

$$A = \frac{d}{d\varphi}\left\{\frac{1}{n_e^2(\varphi)} - \frac{1}{n_{e0}^2}\left[1 - \frac{2Ze\varphi}{m_i V^2}\right]\right\}$$

therefore vanishes for some value $\varphi = \varphi_2$ which lies within the interval $0, \varphi_1$. Bearing in mind that because of the condition $V > v_{s0}$ the quantity A is negative for $\varphi = 0$, we have

$$(\operatorname{sgn} \varphi_m) \frac{d^2}{d\varphi^2} [n_e(\varphi)]^{-2} > 0. \qquad (8.3.3.5)$$

On the other hand, the profile of a simple wave is distorted in the normal direction, if expression (8.3.2.4) is positive, and in the anomalous direction, if that expression is negative, and according to (8.3.3.5) this expression has the same sign as φ_m†.

If for some reason the symmetry of the spatial distribution of the physical quantities in the plasma is broken, there appears a wave with a leading front of the same shape as the leading front of a solitary wave, but behind the crest of the wave the perturbation does not fall off exponentially, but oscillates (quasi-shock wave).

In the case of a solitary compression wave a cause for the violation of the symmetry of the spatial distribution of physical quantities and for the appearance of a quasi-shock compression wave may be the reflection of ions from the potential barrier formed by the wave. In the case of a solitary rarefaction wave one may have similarly reflection of electrons from the potential barrier formed by the wave; as a result we get a quasi-shock rarefaction wave.

The profile of a quasi-shock rarefaction wave is sketched in Fig. 8.3.1. The distribution of physical quantities in the leading front of the wave (region I) is, as before, described by eqns. (8.2.3.3), (8.2.3.4), and (8.2.3.6)—with $n_{e0} = Zn_{i0}$ and $u_0 = -V$—while the potential

† The function $d^2[n_e(\varphi)]^{-2}/d\varphi^2$ is assumed to be monotonic; in the case of non-monotonic functions the gradients in a simple wave can increase both on compression and on rarefaction sections, depending on the magnitude of the perturbation, as was pointed out in the preceding section.

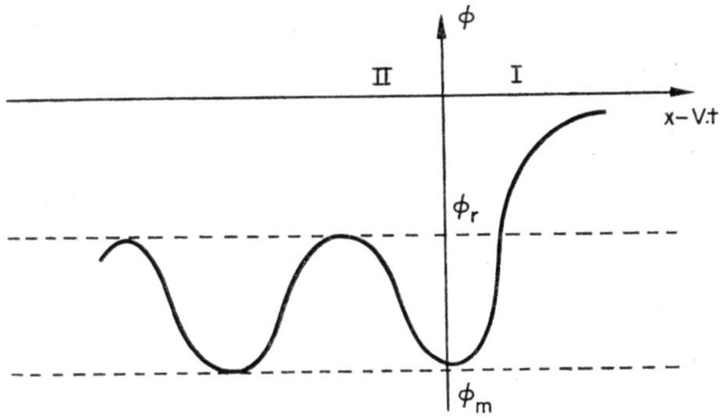

FIG. 8.3.1. The distribution of the electrostatic potential in a quasi-shock rarefaction wave.

minimum φ_m is determined from eqn. (8.3.3.4). In region II eqn. (8.2.3.3) must be replaced by

$$E^2 = 8\pi e \int_{\varphi_m}^{\varphi} d\varphi[n_e^+(\varphi)-Zn_i(\varphi)], \tag{8.3.3.6}$$

where $n_e-n_e^+$ is the density of the reflected electrons. Hence we get for the maximum value φ_r of the potential in the region of the oscillations and for the wavelength λ of the oscillations the equations

$$\int_{\varphi_m}^{\varphi_r} d\varphi\,[n_e^+(\varphi)-Zn_i(\varphi)] = 0, \tag{8.3.3.7}$$

$$\lambda = \frac{1}{\sqrt{(2\pi e)}} \int_{\varphi_m}^{\varphi_r} d\varphi \left\{ \int_0^{\varphi} d\varphi'[n_e^+(\varphi')-Zn_i(\varphi')] \right\}^{-1/2}. \tag{8.3.3.8}$$

A quasi-shock compression wave—for which clearly $\varphi_m > \varphi_r > 0$—has a similar structure. To determine the quantities characterizing that wave, we must make the following substitution in eqns. (8.3.3.6) to (8.3.3.8):

$$n_e^+-n_e \to Z(n_i-n_i^+),$$

where $n_i-n_i^+$ is the density of the reflected ions. Of course, the oscillations in both kinds of quasi-shock waves will be slowly damped due to dissipative effects.

We noted earlier that a solitary and therefore also a quasi-shock wave moves with a supersonic velocity, $V > v_{s0}$, with respect to the unperturbed plasma. However, with respect to the plasma behind the crest of the wave a quasi-shock wave moves with a velocity less than the sound velocity, $u < v_s$. Indeed, the expression on the right-hand side of eqn. (8.3.3.6) is positive and vanishes when $\varphi = \varphi_m$ and $\varphi = \varphi_r$; this expression must therefore have a negative second derivative. On the other hand, that derivative is equal to $8\pi e D^{(1)}\{1-(v_s^2/u^2)\}$.

8.3.4. MULTIPLE-CURRENT FLOW IN A NON-EQUILIBRIUM PLASMA

Let us now consider non-linear multiple-current flow in a non-equilibrium plasma in which the average electron energy appreciably exceeds the average ion energy (Akhiezer and Borovik, 1967). We noted earlier that such a flow may arise when finite-amplitude waves propagate in a plasma which initially only contained a single ion current. Moreover, multiple-current flow may arise when solitary or quasi-shock waves propagate in the plasma. In particular, in the case of a quasi-shock compression wave the ions reflected from the potential barrier formed by the wave may form three currents in front of the wave front. In the case of a solitary (or quasi-shock) rarefaction wave multiple-current flow is possible inside the perturbed region and is produced by the ions which are trapped by the wave.

A peculiar feature of a plasma with several ion currents is the possibility that ion-sound oscillations may become unstable in it. We shall show below that two cases may then be realized: either the wave moves away from the spatial boundary of the instability region, or there appears a discontinuity at the boundary of that region.

We shall use for a description of multiple-current flow in a plasma the hydrodynamical equations with a self-consistent field for each kind of ions. Restricting ourselves to one-dimensional motion we have

$$\frac{\partial u_j}{\partial t}+u_j\frac{\partial u_j}{\partial x}+\frac{eZ_j}{M_j}\frac{\partial \varphi}{\partial x}=0, \qquad (8.3.4.1)$$

$$\frac{\partial n_j}{\partial t}+\frac{\partial}{\partial x}(n_j u_j)=0, \qquad (8.3.4.2)$$

where n_j and u_j are the density and hydrodynamic velocity of the jth kind of ion ($j = 1, 2, \ldots, N$), and M_j and eZ_j are the ion mass and charge; we chose the x-axis in the direction of the wave propagation, and here and henceforth we shall drop the index x of the velocity component u_x. Together with the Poisson equation

$$\frac{\partial^2\varphi}{\partial x^2}-4\pi e\left[n_e-\sum_{j=1}^{N} Z_j n_j\right]=0, \qquad (8.3.4.3)$$

and the electron "equation of state" (8.3.1.5), eqns. (8.3.4.1) and (8.3.4.2) form a complete set of equations describing multiple-current flow of the plasma.

In the case of large-scale motions when the distance over which the quantities characterizing the plasma change appreciably is large compared to the electron Debye radius, we can neglect the first term in eqn. (8.3.4.3). Equations (8.3.4.1) to (8.3.4.3) then allow solutions in the form of simple waves corresponding to such motions of the plasma for which the perturbations of all quantities propagate with the same velocity.

It is well known that in the case of simple waves all quantities which characterize the plasma can be written as functions of one of them which, in turn, is a function of x and t. Choosing the potential φ as this quantity we get a set of ordinary differential equations for the functions $n_j = n_j(\varphi)$, $u_j = u_j(\varphi)$:

$$\frac{dn_j}{d\varphi}=\frac{eZ_j}{M_j}\frac{n_j}{(V-u_j)^2}; \quad \frac{du_j}{d\varphi}=\frac{eZ_j}{M_j}\frac{1}{V-u_j}; \quad n_e=\sum_{j=1}^{N} Z_j n_j, \qquad (8.3.4.4)$$

where $V \equiv V(\varphi)$ is the phase velocity of the wave. It is determined from the condition that the set of equations can be solved, that is, from the equation

$$\Phi(V)=1, \qquad (8.3.4.5)$$

where

$$\Phi(V) = \sum_{j=1}^{N} \frac{v_j^2}{(V-u_j)^2}, \quad v_j^2 = \frac{eZ_j^2 n_j}{M_j D^{(1)}}; \tag{8.3.4.6}$$

here we use again the notation $D^{(i)} = \partial^i n_e / \partial \varphi^i$.

The dispersion eqn. (8.3.4.5) is an algebraic equation of degree $2N$, where N is the number of currents; it has $2N$ roots of which two—the smallest and the largest—are always real while the remaining $2(N-1)$ roots can be either real or imaginary. We show the function $\Phi(V)$ in Fig. 8.3.2.

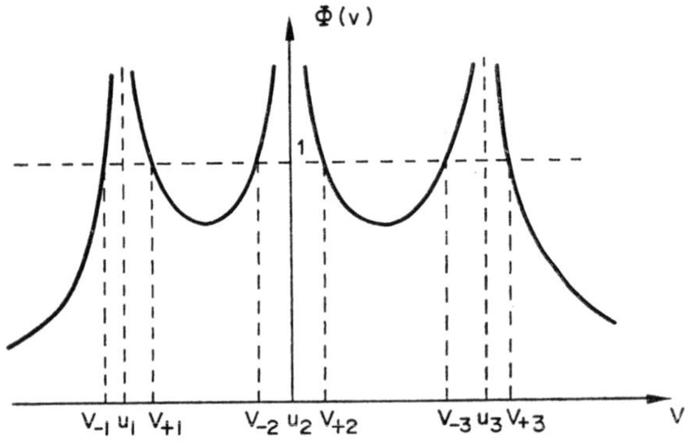

FIG. 8.3.2. The function $\Phi(V)$ for the case of three ion currents. The equation $\Phi(V) = 1$ determines the phase velocities of six oscillation branches, V_{-1}, V_{+1}, V_{-2}, V_{+2}, V_{-3}, and V_{+3}.

If all $2N$ roots of the dispersion eqn. (8.3.4.5) are real the ion-sound oscillations will be stable. There can then propagate in the plasma $2N$ kinds of simple waves which are characterized by different phase velocities $V_{\pm j} (j = 1, 2, \ldots, N)$. (We denote by $V_{\pm j}$ the roots of the dispersion eqn. (8.3.4.5) which lie closest to u_j, and we take $V_{-j} < u_j < V_{+j}$.)

In each of the simple waves the potential φ, the electron density n_e, and the densities n_j of the ions corresponding to the each of N currents increase along compression sections; along rarefaction sections these quantities decrease.

We note that it follows from eqns. (8.3.4.4) that if the phase velocity of one of the waves lies close to u_j the ions of the jth current interact particularly strongly with that wave.

Let us determine how the shape of an ion-sound wave changes when it propagates through the plasma. Using (8.3.4.4) to (8.3.4.6) we find

$$\frac{dV}{d\varphi} = \frac{D^{(1)}}{n_e} \left[-\frac{\partial \Phi}{\partial V} \right]^{-1} \left\{ 3G - \frac{n_e D^{(2)}}{[D^{(1)}]^2} \right\}, \tag{8.3.4.7}$$

where

$$G = n_e \sum_{j=1}^{N} \frac{v_j^4}{Z_j n_j (V-u_j)^4}. \tag{8.3.4.8}$$

The denominator in (8.3.4.7) is always positive for waves with the phase velocity V_{+j} and negative for waves with the phase velocity V_{-j}; however, the enumerator can have both signs, depending on the nature of the electron distribution function, and hence on the form of the "equation of state" $n_e = n_e(\varphi)$.

One sees easily that $G \geq 1$—the case $G = 1$ is realized if there is only one ion current in the plasma, $N = 1$. Therefore, in the case of electron distributions for which $n_e D^{(2)}\{D^{(1)}\}^{-2} < 3$, the enumerator in equation (8.3.4.7) is always positive. For an analysis of how the profile of a wave moving with the velocity V_{+j} or V_{-j} changes in that case it is convenient to change to a frame of reference in which the ions of the jth kind are at rest. In this frame elements with a larger density move in both above-mentioned waves with a velocity which has a larger absolute magnitude. Discontinuities—or additional ion currents—can therefore appear, as in ordinary hydrodynamics, along compression sections; self-similar waves are rarefaction waves.

We shall call the case $3G - n_e D^{(2)}\{D^{(1)}\}^{-2} > 0$ the normal case. The normal case is realized, in particular, for a Maxwellian electron velocity distribution (Taniuti, Yajima, and Outi, 1966). We showed in Subsection 8.3.2 that there exists also a class of distribution functions for which $n_e D^{(2)}\{D^{(1)}\}^{-2} > 3$ so that the enumerator in (8.3.4.7) can be negative. Discontinuities can then occur along rarefaction sections rather than along compression sections, as in the normal case; self-similar waves are compression waves (anomalous case; Akhiezer and Borovik, 1967). This case can be realized, in particular, for distributions which in the main velocity region have the form $(v^2 + v_0^2)^{-\alpha}$, where $\frac{3}{2} < \alpha < 2$.

We emphasize that in the normal (anomalous) case discontinuities occur along those sections of a wave with phase velocity $V_{\pm j}$ which are compression (rarefaction) sections from the point of view of an observer who moves with the jth ion current. Of course, the concept of compression or rarefaction sections is not invariant under a transformation to a moving frame of reference. For instance, compression sections appear as rarefaction sections (and vice versa) to an observer moving with a velocity larger than V_{+j}, in particular, moving with the $j+1$st ion current.

If there are complex roots among the roots of the dispersion eqn. (8.3.4.5) ion-sound oscillations in the plasma are unstable. One sees easily that when one approaches the boundary of the stability region the phase velocities of any two kinds of waves approach one another: $V_{j-1} \to V_{-j} \to V_c$ where the boundary of the stability region and the critical phase velocity are determined from the equations

$$\Phi(V_c) = 1, \quad \Phi'(V_c) = 0; \tag{8.3.4.9}$$

the prime on the function Φ indicates differentiation with respect to V.

It follows, in particular, that in a plasma containing two ion currents the stability condition has the form

$$(u_1 - u_2)^2 \geqslant u_c^2 \tag{8.3.4.10}$$

where

$$u_c = [v_1^{2/3} + v_2^{2/3}]^{3/2}, \tag{8.3.4.10'}$$

while near the boundary of the stability region

$$V_{1,-2} = V_c \mp \frac{1}{\sqrt{3}} (u_c v_1 v_2)^{1/3} \left[\left(\frac{u_1 - u_2}{u_c} \right)^2 - 1 \right]^{1/2}, \quad V_c = u_c^{-2/3}[u_1 v_2^{2/3} + u_2 v_1^{2/3}]. \tag{8.3.4.11}$$

In order to study the evolution of a wave near the boundary of the stability region we note that the quantity

$$W = \pm \Phi'(V_{\pm j}),$$

which is positive in the stability region and vanishes on its boundary, is—like all quantities in a simple wave—a function of $x-Vt$. There can therefore not occur an instability in a simple wave if the plasma were initially stable in the whole of space; the spatial boundary of the instability region $x = x_c(t)$, determined by the equation

$$W(x_c(t), t) = 0,$$

is carried along the characteristics.

Bearing in mind that

$$\frac{dW}{d\varphi} = \pm\Phi''(V_{\pm j})\frac{dV}{d\varphi}, \quad \Phi'' > 0,$$

one sees easily that the quantity $dW/d\varphi$ is positive (negative) in the normal (anomalous) case. If, therefore, initially in the normal (anomalous) case the spatial boundary of the stability region terminates along a compression (rarefaction) section, a discontinuity will develop later on on the boundary. If, however, the boundary of the stability region terminates on a rarefaction (compression) region, the distance between the crest of the simple wave and the point $x_c(t)$ will increase with time.

We note that on the boundary of the stability region the quantity $dV/d\varphi$, which characterizes the rate of change of the profile of a simple wave—and, in particular, determines the moment when a discontinuity will appear—becomes infinite according to (8.3.4.7) and (8.3.4.9).

Let us now consider the case of a plasma through which a low-density ion current passes ($n_1 \gg n_2$). Four kinds of simple waves can propagate in such a plasma. Two of them have phase velocities close to the ion-sound velocity in a plasma with one kind of ions, $V_{\pm 1} = u_1 \pm v_1$. In agreement with (8.3.2.4) we have for such waves

$$\frac{dV_{\pm 1}}{d\varphi} = \pm\frac{D^{(1)}v_1}{2n_e}\left\{3-\frac{n_e D^{(2)}}{[D^{(1)}]^2}\right\}. \tag{8.3.4.12}$$

Depending on the form of the electron distribution function discontinuities can thus occur either along compression sections (normal case) or along rarefaction sections (anomalous case).

There are two other kinds of waves which have phase velocities close to u_2:

$$V_{\pm 2} = u_2 \pm v_2\left[1-\frac{v_1^2}{(u_2-u_1)^2}\right]^{-1/2}. \tag{8.3.4.13}$$

Equations (8.3.4.4) then become

$$\left.\begin{aligned}\frac{dn_1}{d\varphi} &= \frac{eZ_1 n_1}{M_1(u_2-u_1)^2}, & \frac{dn_2}{d\varphi} &= \frac{eZ_2 n_2}{M_2 v_2^2}\left[1-\frac{v_1^2}{(u_2-u_1)^2}\right],\\ \frac{du_1}{d\varphi} &= \frac{eZ_1}{M_1(u_2-u_1)}, & \frac{du_2}{d\varphi} &= \pm\frac{eZ_2}{M_2 v_2}\left[1-\frac{v_1^2}{(u_2-u_1)^2}\right].\end{aligned}\right\} \tag{8.3.4.14}$$

From (8.3.4.7) we get for the derivatives $dV_{\pm 2}/d\varphi$

$$\frac{dV_{\pm 2}}{d\varphi} = \pm\frac{3D^{(1)}v_2}{2Z_2 n_2}\left[1-\frac{v_1^2}{(u_2-u_1)^2}\right]^{1/2}. \tag{8.3.4.15}$$

The quantity $dV/d\varphi$ which characterizes the rate at which the shape of the wave changes is large in waves of this kind (proportional to $n_2^{-1/2}$) while, independent of the form of the electron distribution function, the wave profile can change only in the usual direction (normal case).

We note that an ion current with a small density interacts very intensively with such waves. In particular, $dn_2/dn_1 \gg n_2/n_1$.

8.4. Non-linear Waves in a Magneto-active Plasma with Hot Electrons

8.4.1 EQUATIONS DESCRIBING A NON-LINEAR WAVE IN A PLASMA IN A MAGNETIC FIELD

In the preceding section we considered non-linear waves in an unmagnetized non-equilibrium plasma with hot electrons. We shall now study non-linear waves in a plasma with hot electrons in an external constant and uniform magnetic field (Akhiezer and Borovik, 1968). We shall consider both equilibrium and non-equilibrium plasmas.

As before, for the description of the ion motion in a magneto-active plasma with hot electrons we can start from the hydrodynamical equations with a self-consistent field and for the description of the electron motion from the kinetic equation for the electron distribution function $\boldsymbol{F}(\boldsymbol{v})$:

$$\left.\begin{aligned}\frac{\partial \boldsymbol{u}}{\partial t}+(\boldsymbol{u}\cdot\nabla)\,\boldsymbol{u}-\frac{Ze}{m_i c}[\boldsymbol{u}\wedge\boldsymbol{B}_0]+\frac{Ze}{m_i}\nabla\varphi=0,\\ \frac{\partial n_i}{\partial t}+\operatorname{div}(n_i\boldsymbol{u})=0,\end{aligned}\right\} \tag{8.4.1.1}$$

$$(\boldsymbol{v}\cdot\nabla)\,F-\frac{e}{m_e c}\left([\boldsymbol{v}\wedge\boldsymbol{B}_0]\cdot\frac{\partial F}{\partial \boldsymbol{v}}\right)+\frac{e}{m_e}\left(\nabla\varphi\cdot\frac{\partial F}{\partial \boldsymbol{v}}\right)=0, \tag{8.4.1.2}$$

where $\boldsymbol{B}_0$ is the external magnetic field and Ze the ion charge. Adding the Poisson eqn. (8.2.1.3) to eqns. (8.4.1.1) and (8.4.1.2) we obtain a complete set of equations which describe the non-linear motions of a non-equilibrium plasma in an external magnetic field.

In what follows we shall be interested in the case of strong magnetic fields, when

$$\frac{v_A}{c} \gg Z\frac{m_e}{m_i}kr_D,$$

where $v_A = B_0/\sqrt{(4\pi n_i m_i)}$ is the Alfvén velocity, r_D the electron Debye radius, and k^{-1} a length over which the quantities characterizing the plasma change appreciably. In this case—often called the case of strongly magnetized electrons—it is convenient to change from the kinetic eqn. (8.4.1.2) to the equation obtained by averaging over the fast electron rotation:

$$\left[\frac{\partial}{\partial z}+\frac{e}{m_e}\frac{\partial\varphi}{\partial z}\frac{\partial}{v_z\,\partial v_z}\right]f=0, \tag{8.4.1.3}$$

where f is the distribution function integrated over the electron velocity components at right angles to the magnetic field,

$$f(v_z^2) = \int F(v)\, dv_x\, dv_y, \tag{8.4.1.4}$$

and where the z-axis is taken in the direction of the external magnetic field.

Solving equation (8.4.1.3) we find

$$f(v_z^2; \boldsymbol{r}, t) = f\left(v_z^2 - \frac{2e}{m_e}\varphi(\boldsymbol{r}, t)\right). \tag{8.4.1.5}$$

If we then integrate this equation over the electron velocity we get the electron "equation of state":

$$n_e = n_e(\varphi), \tag{8.4.1.6}$$

which connects the electron density with the electrostatic potential.

The set of eqns. (8.4.1.1), (8.2.1.3), and (8.4.1.6) describes the non-linear motions in a plasma with an arbitrary electron velocity distribution, provided the average electron energy is considerably larger than the average ion energy, while the form of the "equation of state" (8.4.1.6) depends on the actual form of the electron distribution function.

Let us first of all consider the case of a not too strong magnetic field, when $v_A/c \ll kr_D$ (but, as before, $v_A/c \gg kr_D Zm_e/m_i$). In this case—usually called the case of weakly magnetized ions—we can neglect the term involving the magnetic field in the first of eqns. (8.4.1.1). If we are interested in one-dimensional motions of the plasma and introduce a new variable $\xi = V^{-1}(\boldsymbol{r}\cdot\boldsymbol{V})$, we can write eqns. (8.4.1.1) and (8.2.1.3) in the form

$$\left(\frac{\partial}{\partial t} + u\frac{\partial}{\partial \xi}\right)u + \frac{Ze}{m_i}\frac{\partial \varphi}{\partial \xi} = 0, \quad \frac{\partial n_i}{\partial t} + \frac{\partial}{\partial \xi}(n_i u) = 0, \tag{8.4.1.7}$$

$$\frac{\partial^2 \varphi}{\partial \xi^2} - 4\pi e(n_e - Zn_i) = 0, \tag{8.4.1.8}$$

where $u = V^{-1}(\boldsymbol{u}\cdot\boldsymbol{V})$ and $\boldsymbol{V}$ is the wave velocity.

The set of eqns. (8.4.1.6) to (8.4.1.8) is completely analogous to the set of equations describing the non-linear motion of a non-equilibrium plasma when there is no magnetic field. The only essential difference between these sets of equations lies in the form of the "equations of state": in magnetic fields, satisfying the condition $v_A/c \gg kr_D Zm_e/m_i$, the quantity $n_e(\varphi)$ is a functional of the one-dimensional electron distribution function f rather than of the total distribution function F, as in the case when $\boldsymbol{B}_0 = 0$.

We studied the non-linear motion of a non-equilibrium plasma when there is no magnetic field in detail in Section 8.3. The just-mentioned analogy enables us to transfer directly all conclusions of Section 8.3 to the case of a plasma in a not too strong magnetic field, $v_A/c \ll kr_D$. In particular, the nature of the non-linear waves in such a plasma will be different in the normal case, when the quantity

$$Q = 3 - \frac{nD^{(2)}}{[D^{(1)}]^2} \tag{8.4.1.9}$$

is positive, from the anomalous case when the quantity Q is negative.

In the normal case discontinuities (or multiple-current flow) will occur in a simple wave along compression sections; self-similar waves are rarefaction waves. Solitary waves are in this case compression waves. As a result of the reflection of ions from the potential barrier formed by such a wave a quasi-shock compression wave may occur. The normal case is, in particular, realized for a Maxwellian electron velocity distribution.

In the anomalous case discontinuities (or multiple-current flow) occur in a simple wave along rarefaction sections; self-similar waves are compression waves. Solitary waves are in this case rarefaction waves. As a result of the reflection of electrons from the potential barrier formed by such a wave a quasi-shock rarefaction wave may occur. The anomalous case is, in particular, realized if the electron distribution in the main velocity range has the form

$$f \propto (v_z^2+v_0^2)^{-\alpha},$$

where $\frac{1}{2} < \alpha < 1$.

Apart from the normal and the anomalous cases one can also have the intermediate case $Q = 0$—as when there is no magnetic field. In that case the wave profile is not distorted while the wave propagates and no discontinuities occur. The case $Q = 0$ is, in particular, realized if the electron distribution function in the main velocity range has the form

$$f \propto (v_z^2+v_0^2)^{-1}.$$

8.4.2. SIMPLE MAGNETO-SOUND WAVES

Let us now consider non-linear waves for the case of a strong magnetic field, $v_A/c \gg kr_D$. In that case—usually called the case of strongly magnetized ions—it is convenient to change from the hydrodynamical eqns. (8.4.1.1) to the equations which are averaged over the fast rotation of the ions:

$$\left(\frac{\partial}{\partial t}+u_z\frac{\partial}{\partial z}\right)u_z+\frac{Ze}{m_i}\frac{\partial\varphi}{\partial z} = 0,$$
$$\frac{\partial n_i}{\partial t}+\frac{\partial}{\partial z}(n_i u_z) = 0, \qquad [\boldsymbol{B}_0 \wedge \nabla](\varepsilon_t+Ze\varphi) = 0, \tag{8.4.2.1}$$

where ε_t is the average value of the kinetic energy of the ion cyclotron rotation,

$$\varepsilon_t = \tfrac{1}{2}m_i(\overline{u^2}-u_z^2).$$

These equations together with the Poisson eqn. (8.4.1.8) and the electron "equation of state" (8.4.1.6) form a complete set of equations describing the non-linear motions in a plasma with hot electrons in a strong magnetic field (non-linear magneto-sound waves).

In the case of large-scale motions ($kr_D \ll 1$) we can neglect the first term in (8.4.1.8). Equations (8.4.1.6), (8.4.1.8), and (8.4.2.1) then have solutions in the form of simple waves. As we have already said several times, in the case of simple waves all quantities characterizing the plasma can be written as functions of one of them, for instance, the ion density n_i which, in turn, is a function of ξ and t. Equations (8.4.2.1) then turn into a set of ordinary differential equations for the functions $\varphi(n_i)$, $u_z(n_i)$, and $\varepsilon_t(n_i)$:

$$\frac{du_z}{dn_i} = \frac{v_s}{n_i}, \quad \frac{d\varepsilon_t}{dn_i} = -\frac{m_i v_s^2}{n_i}, \quad \frac{d\varphi}{dn_i} = \frac{m_i v_s^2}{eZn_i}, \tag{8.4.2.2}$$

while the phase velocity $V = V(n_i)$ is determined from the condition that these equations have a solution,

$$V = (u_z + v_s) \cos \theta, \tag{8.4.2.3}$$

where v_s is the ion-sound velocity, given by formula (8.3.2.2) and θ the angle between the direction of wave propagation and that of the magnetic field (to fix the ideas we assume that $\theta < \pi/2$).

We can use eqns. (8.4.2.2) and (8.4.2.3) to study the direction in which the quantities which characterize the plasma change and to follow the evolution of a finite-amplitude perturbation. We note, first of all, that along compression sections the potential φ increases, as well as the electron and ion densities; along rarefaction sections these quantities decrease. As far as ε_t is concerned, this quantity increases along rarefaction and decreases along compression sections.

To determine how the shape of a magneto-sound wave changes when it moves in the plasma, we evaluate the derivative dV/dn_i. Using (8.4.2.2) and (8.4.2.3) we get

$$\frac{dV}{dn_i} = \frac{v_s}{2n_i} Q \cos \theta, \tag{8.4.2.4}$$

where the quantity Q is defined by eqn. (8.4.1.9). Depending on the sign of the quantity Q, that is, depending on the form of the electron distribution function, the derivative dV/dn_i can be positive (normal case), negative (anomalous case), or zero. (We remind ourselves that both in ordinary and in magneto-hydrodynamics the derivative dV/dn_i is always positive.)

If $dV/dn_i > 0$, elements with a larger density move, as in magneto-hydrodynamics, with a higher velocity; discontinuities (or multiple-current flow) therefore occur along compression sections. Self-similar waves are then rarefaction waves.

If $dV/dn_i = 0$ all elements move with the same velocity during the propagation of the wave; the wave profile is thus not distorted and no discontinuities occur.

Finally, if $dV/dn_i < 0$, elements with larger density move with a smaller velocity. Discontinuities (or multiple-current flow) occur along rarefaction sections and self-similar waves are compression waves.

We draw attention to the fact that the sign of the derivative dV/dn_i is, according to (8.4.2.4), the same as the sign of the quantity Q which determines the direction in which the profile of a simple wave is distorted in the case of a not very strong magnetic field. The direction in which the profile of a simple wave is distorted is thus determined solely by the form of the electron distribution function and is independent of the strength of the magnetic field.

8.4.3. STATIONARY MAGNETO-SOUND WAVES

We have seen that when long-wavelength magneto-sound waves with a linear dispersion law propagate in a plasma the wave profile changes with time—except when the electron distribution function has the form $f \propto (v_z^2 + v_0^2)^{-1}$; in that case the magneto-sound wave propagates without distortion of its shape.

Along those sections of the wave where the characteristic size of the inhomogeneity is comparable with the electron Debye radius, $kr_D \approx 1$, we must take the sound dispersion, described by the first term in eqn. (8.4.1.8), into account. It is well known that when the sound dispersion is taken into account waves with a stationary profile, solitary, periodic, or quasi-shock waves, become a possibility.

Turning to a study of stationary magneto-sound waves we note, first of all, that one can easily obtain a solution of eqns. (8.4.2.1) in closed form in the case of such waves. Using (8.4.1.8) we get a complete set of equations describing a stationary wave:

$$\left.\begin{aligned} &n_i(V-u_z\cos\theta) = n_{i0}V, \quad \frac{1}{2}m_i\left(u_z-\frac{V}{\cos\theta}\right)^2+Ze\varphi = \frac{m_iV^2}{2\cos^2\theta}, \\ &\varepsilon_t+Ze\varphi = \varepsilon_0, \quad \frac{\partial^2\varphi}{\partial\xi^2} = 4\pi e[n_e(\varphi)-Zn_i], \end{aligned}\right\} \tag{8.4.3.1}$$

where V is the wave velocity, θ the angle between the directions of the wave propagation and the magnetic field, n_{i0} and ε_0 the values of the ion density and of the average energy of the ion cyclotron rotation in the point where $u_z = 0$ (we have taken the zero of the electrostatic potential to be at the point where $u_z = 0$).

Solving these equations, we get

$$\left(\frac{\partial\varphi}{\partial\xi}\right)^2 = 8\pi e\int_0^{\varphi} d\varphi[n_e(\varphi)-Zn_i(\varphi)], \tag{8.4.3.2}$$

where

$$n_i(\varphi) = n_{i0}\left[1-\frac{2Ze\varphi\cos^2\theta}{m_iV^2}\right]^{-1/2}. \tag{8.4.3.3}$$

By integrating (8.4.3.2) we get an implicit equation determining φ as a function of the coordinates and the time:

$$\xi-Vt = \frac{\pm 1}{\sqrt{(8\pi e)}}\int d\varphi\left\{\int_0^{\varphi} d\varphi'[n_e(\varphi')-Zn_i(\varphi')]\right\}^{-1/2}. \tag{8.4.3.4}$$

In the general case when $n_{e0} \neq Zn_{i0}$ this equation determines the distribution of the potential in a periodic magneto-sound wave. The amplitude φ_m of such a wave and its wavelength λ are determined by eqns. (8.3.3.1) and (8.3.3.2) in which we must substitute expression (8.4.3.3) for the function $n_i(\varphi)$.

In the case when the total charge vanishes ($n_{e0} = Zn_{i0}$) at the same points as the velocity u_z, $\lambda \to \infty$, and the periodic wave degenerates into a solitary wave. One sees easily that all equations which describe a solitary magneto-sound wave can be obtained from the corresponding equations describing a solitary ion-sound wave when there is no magnetic field by making in those equations the substitution

$$V \to \frac{V}{|\cos\theta|}.$$

We can therefore at once use the results of the preceding section to reach various conclusions about the structure of a solitary magneto-sound wave.

Firstly, the perturbation in a solitary magneto-sound wave falls off exponentially far from the crest of the wave—as $|\xi - Vt| \to \infty$,

$$\varphi \propto \exp\left\{-\frac{|\xi - Vt|}{r_D}\left[1-\frac{v_{s0}^2\cos^2\theta}{V^2}\right]^{1/2}\right\}, \qquad (8.4.3.5)$$

where v_{s0} is the value of the ion-sound velocity as $|\xi - Vt| \to \infty$ and r_D is the electron Debye radius defined by eqn. (8.3.3.3). From this one can, in particular, conclude easily that a solitary wave moves relative to the unperturbed plasma with a velocity which exceeds the velocity of small-amplitude magneto-sound waves, $V > v_s |\cos\theta|$.

Furthermore, the amplitude φ_m of a solitary magneto-sound wave is clearly determined by the equation

$$\int_0^{\varphi_m} d\varphi\left\{n_e(\varphi) - Zn_{i0}\left[1-\frac{2Ze\varphi\cos^2\theta}{m_i V^2}\right]^{-1/2}\right\} = 0. \qquad (8.4.3.6)$$

In the case of a Maxwellian electron distribution this equation has a solution in the range of positive φ-values, $\varphi_m > 0$. Clearly, both the potential and the ion and electron densities increase in such a wave (solitary compression wave).

One can also have such electron velocity distributions that eqn. (8.4.3.6) has a solution with a negative rather than a positive value of φ_m. The ion and electron densities in a solitary wave then decrease (solitary rarefaction wave). This case is, in particular, realized for distributions which in the main velocity range have the form

$$f \propto (v_z^2 + v_0^2)^{-\alpha},$$

with $\frac{1}{2} < \alpha < 1$.

One can show that in a plasma in a magnetic field a solitary compression wave occurs for electron distributions leading to increasing gradients along compression sections (normal case) and a solitary rarefaction wave occurs in the case of electron distributions leading to increasing gradients along rarefaction sections (anomalous case). The proof of this statement is completely analogous to the proof given in Section 8.3 for the case of an unmagnetized plasma; we shall therefore not give the proof here.

If for some reason or other the symmetry of the spatial distribution of the physical quantities in the plasma is broken, a wave appears with a leading front of the same shape as the leading front of a solitary wave, but with the perturbation behind the crest of the wave oscillating rather than falling off exponentially (quasi-shock wave). As in the case when there is no magnetic field, in the case of a solitary compression (rarefaction) wave the cause of the breaking of the symmetry of the spatial distribution of the physical quantities which leads to the appearance of quasi-shock compression (rarefaction) waves can be the reflection of ions (electrons) from the potential barrier formed by the wave. All equations characterizing a quasi-shock magneto-sound wave can be obtained from the corresponding relations in Section 8.3 provided we substitute there the function $n_i(\varphi)$ defined by eqn. (8.4.3.3); we shall therefore not give here the corresponding expressions.

To conclude this subsection we note that a solitary wave—and thus also a quasi-shock wave—moves with respect to the unperturbed plasma with a velocity which is larger than the velocity of a small-amplitude magneto-sound wave, $V > v_{s0} |\cos\theta|$. However, the quasi-shock wave moves with respect to the plasma behind the wave front with a velocity smaller than the magneto-sound velocity.

8.5. Non-linear Low-frequency Waves in a Cold Plasma in a Magnetic Field

8.5.1. EQUATIONS DESCRIBING A NON-LINEAR WAVE IN A COLD MAGNETO-ACTIVE PLASMA

In the preceding three sections we have studied waves in a plasma with hot electrons and we have shown that in the case of large-scale perturbations the wave profile is distorted when the wave propagates and that this leads to the appearance of discontinuities (or multiple-current flow). In the short-wavelength region, however, perturbations in a plasma with hot electrons can propagate in the form of waves with a stationary profile—periodic, solitary, or quasi-shock waves.

The possibility of the propagation of stationary waves was in that case connected in an essential way with the dispersion of the ion-sound (or magneto-sound) waves. The dispersion length of such oscillations—that is, the characteristic size of the inhomogeneity for which dispersion effects start to play an important role—is determined by the electron Debye radius; the width of the wave front (of a solitary or quasi-shock wave) in a plasma with hot electrons is of the same order of magnitude as the Debye radius.

We shall now consider non-linear low-frequency waves in a cold plasma in a magnetic field. (We considered high-frequency waves in a cold plasma in Section 8.1.) It is well known that in such a plasma we can have waves which have a linear dispersion law in the long-wavelength region and a phase velocity equal to the Alfvén velocity. We shall see that when such a wave propagates its profile is distorted and this may lead to the occurrence of discontinuities (or multiple-current flow). However, in the short-wavelength region perturbations in the form of waves with a stationary profile can propagate. As before, the possibility of the propagation of stationary waves is in an essential manner connected with dispersion effects. The dispersion length of oscillations in a cold plasma in a magnetic field is then determined by the equation

$$l = \frac{1}{\sqrt{(4\pi e^2 n_e)}} \operatorname{Max}\left\{\frac{B}{\sqrt{(4\pi n_e)}},\ \sqrt{(m_e c^2)}\right\},$$

where n_e is the electron density and B the magnetic field. The width of the wave front (of a solitary or quasi-shock wave) must be of the same order of magnitude as l.

We shall restrict ourselves to the study of non-linear waves propagating in a cold plasma ($T \ll B^2/8\pi n_e$, where T is the plasma temperature) at right angles to the magnetic field (Adlam and Allen, 1958; Sagdeev, 1958b; Gardner, Goertzel, Grad, Morawetz, Rose, and Rubin, 1958; Davies, Lüst, and Schlüter, 1958). We can then use the hydrodynamical equations for the electron and ion densities, n_e and n_i, and their velocities, $\boldsymbol{v}$ and $\boldsymbol{u}$:

$$\frac{\partial \boldsymbol{v}}{\partial t}+v_x\frac{\partial \boldsymbol{v}}{\partial x}+\frac{e}{m_e c}[\boldsymbol{v}\wedge\boldsymbol{B}]+\frac{e}{m_e}\boldsymbol{E}=0, \tag{8.5.1.1}$$

$$\frac{\partial n_e}{\partial t}+\frac{\partial}{\partial x}(n_e v_x)=0, \tag{8.5.1.2}$$

$$\frac{\partial \boldsymbol{u}}{\partial t}+u_x\frac{\partial \boldsymbol{u}}{\partial x}-\frac{Ze}{m_i c}[\boldsymbol{u}\wedge\boldsymbol{B}]-\frac{Ze}{m_i}\boldsymbol{E}=0, \tag{8.5.1.3}$$

$$\frac{\partial n_i}{\partial t}+\frac{\partial}{\partial x}(n_i u_x)=0, \tag{8.5.1.4}$$

where the x-axis is chosen to be in the direction of the wave propagation.

We shall first consider the case of a strong magnetic field when $B^2/4\pi n_e \gg m_e c^2$ (the case of a weak magnetic field, $B^2/4\pi n_e \ll m_e c^2$, will be considered in Subsection 8.5.3). In that case we get by solving eqn. (8.5.1.1)

$$\boldsymbol{v}=c\frac{[\boldsymbol{E}\wedge\boldsymbol{B}]}{B^2}. \tag{8.5.1.5}$$

One checks easily that when the condition

$$m_e c^2 \ll \frac{B^2}{4\pi n_i} \ll m_i c^2$$

is statisfied, we can neglect the Lorentz force in eqn. (8.5.1.3). Introducing the electrostatic potential φ which satisfies the Poisson equation,

$$\frac{\partial^2\varphi}{\partial x^2}-4\pi e(n_e-Zn_i)=0, \tag{8.5.1.6}$$

we find

$$\frac{\partial u}{\partial t}+u\frac{\partial u}{\partial x}+\frac{Ze}{m_i}\frac{\partial\varphi}{\partial x}=0; \tag{8.5.1.7}$$

here and henceforth we drop the index x from the velocity component u_x.

Equations (8.5.1.4), (8.5.1.6), and (8.5.1.7) are exactly the same as the analogous equations of Sections 8.2 and 8.3 describing the motion in a plasma when there is no magnetic field.

We shall now prove that eqns. (8.5.1.2) and (8.5.1.5) together with the Maxwell equations,

$$\frac{\partial E_y}{\partial x}=-\frac{1}{c}\frac{\partial B}{\partial t}, \tag{8.5.1.8}$$

$$\frac{\partial B}{\partial x}=\frac{4\pi e}{c}v_y n_e-\frac{1}{c}\frac{\partial E_y}{\partial t}, \tag{8.5.1.9}$$

can be reduced to an equivalent electron "equation of state", $n_e = n_e(\varphi)$; the z-axis has been chosen in the direction of the magnetic field.

Substituting (8.5.1.5) into (8.5.1.2) and comparing the resultant equation with eqn.

(8.5.1.8) we check easily that the electron density changes proportional to the magnetic field,

$$\frac{n_e}{B} = \text{constant}. \tag{8.5.1.10}$$

Using then (8.5.1.9) and (8.5.1.5) and noting that in the first of these equations we can neglect the term $(1/c)(\partial E_y/\partial t)$, we get

$$B - 4\pi e\varphi\frac{n_e}{B} = \text{constant}. \tag{8.5.1.11}$$

Comparison of (8.5.1.10) and (8.5.1.11) then leads to the electron "equation of state":

$$n_e(\varphi) = n_{e0}\left[1 + 4\pi e\varphi\frac{n_{e0}}{B_0^2}\right], \tag{8.5.1.12}$$

where n_{e0} and B_0 are the values of the electron density and the magnetic field at the point where the potential φ vanishes.

We have thus reduced the problem of the non-linear motion in a cold plasma in a strong magnetic field to the form which is completely analogous to the problem of the non-linear motion of a plasma with hot electrons and no magnetic field. (According to (8.5.1.12) the role of the average thermal electron energy is now played by the quantity $T_{\text{eff}} \sim B^2/4\pi n_e$.)

8.5.2. NON-LINEAR WAVES FOR THE CASE OF A STRONG MAGNETIC FIELD

Turning now to the study of non-linear oscillations of a cold plasma in a strong magnetic field ($B^2/4\pi n_e \gg m_e c^2$) we note first of all that according to (8.5.1.12) and (8.3.3.3) the dispersion length for such oscillations is equal to

$$l = \frac{B}{4\pi e n_e}. \tag{8.5.2.1}$$

We draw attention to the fact that, according to (8.5.1.10), the dispersion length l is an integral of motion.

In the case of large-scale perturbations ($kl \ll 1$), when we can neglect the first term in eqn. (8.5.1.6), eqns. (8.5.1.4), (8.5.1.6), (8.5.1.7), and (8.5.1.12) allow solutions in the form of simple (Riemannian) waves. Introducing the Alfvén velocity

$$v_A = \frac{B}{\sqrt{(4\pi n_i m_i)}}, \tag{8.5.2.2}$$

which in the case considered plays the same role as the ion-sound velocity v_s in the case of a plasma with hot electrons (compare (8.3.2.2)) we get equations which are satisfied by the functions $u(n_i)$, $\varphi(n_i)$, $n_e(n_i)$, $B(n_i)$, and $E_e(n_i)$ in the case of a simple wave

$$\frac{du}{dn_i} = \frac{v_A}{n_i}, \quad \frac{d\varphi}{dn_i} = \frac{m_i v_A^2}{eZn_i}, \quad n_e = Zn_i, \quad \frac{dB}{dn_i} = \frac{B}{n_i}, \quad \frac{dE_y}{dn_i} = \frac{u+v_A}{c}\frac{B}{n_i}; \tag{8.5.2.3}$$

to fix the ideas we assume that the wave propagates in the direction of the positive x-axis. The phase velocity of the wave is connected with the Alfvén velocity through the equation

$$V = u + v_A. \tag{8.5.2.4}$$

We showed in Section 8.3 that the direction in which the front of a simple wave is distorted is determined by the sign of the quantity $Q = 3 - n_e D^{(2)}\{D^{(1)}\}^{-2}$. Using the "equation of state" (8.5.1.12) we have $Q = 3$. The profile of a simple wave propagating through a cold plasma is therefore always distorted in the normal direction, which means that the gradients of the quantities characterizing the plasma increase along compression sections and this leads to the appearance of discontinuities (or multiple-current flow) along such sections. Self-similar waves in such a plasma are rarefaction waves.

Let us now consider the propagation of stationary waves in a cold plasma in a magnetic field. We can obtain all the equation which describe such waves from the corresponding equations of Subsection 8.3.3 by making there the substitutions

$$v_s \to v_A, \quad r_D \to l,$$

and by using for $n_e(\varphi)$ the function determined by eqn. (8.5.1.12). Of course, solitary waves are in this case compression waves. In such waves not only the electron and ion densities increase, but also the magnetic field and the electrostatic potential.

Using (8.3.3.4), (8.5.1.10), and (8.5.1.11) we can establish a relation between the velocity V of the solitary wave and its amplitude; in fact, if B_m is the amplitude of the magnetic field, we have

$$V = v_{A0}\frac{B_m + B_0}{2B_0}, \tag{8.5.2.5}$$

where B_0 and v_{A0} are the values of the magnetic field and the Alfvén velocity in the unperturbed region (as $|x - Vt| \to \infty$).

In concluding this subsection we note that the amplitude of a solitary wave—and hence also its velocity—cannot be arbitrarily large. Indeed, the expression under the square root in eqn. (8.4.3.3) for the ion density must be non-negative. Substituting (8.5.2.5) into this equation, using (8.5.1.10) and (8.5.1.11), and bearing in mind that we have noted earlier that $B_m \geq B_0$, we find

$$B_0 \leq B_m \leq 3B_0, \quad v_{A0} \leq V \leq 2v_{A0}. \tag{8.5.2.6}$$

8.5.3. NON-LINEAR WAVES IN THE CASE OF A WEAK MAGNETIC FIELD

Let us now consider non-linear waves in a cold plasma in a weak magnetic field ($T \ll B^2/4\pi n_e \ll m_e c^2$). We can again use in this case the hydrodynamical eqns. (8.5.1.1) to (8.5.1.4) and the Maxwell equations. Noting that we can neglect the appearance of space charge in the plasma when $B^2/4\pi n_e \ll m_e c^2$ and neglecting corrections proportional to the small parameter Zm_e/m_i we can write these equations in the form

$$n_e = Zn_i,\ v_x = u_x, \tag{8.5.3.1}$$

$$\frac{dn_i}{dt}+n_i\frac{\partial u_x}{\partial x} = 0, \tag{8.5.3.2}$$

$$\frac{du_x}{dt} = -\frac{e}{m_e}\left(E_x+\frac{v_y}{c}B\right), \tag{8.5.3.3}$$

$$E_x+\frac{v_y}{c}B = \frac{Zm_e}{m_i c}v_y B, \tag{8.5.3.4}$$

$$\frac{dv_y}{dt}+\frac{e}{m_e}\left(E_y-\frac{u_x}{c}B\right) = 0, \tag{8.5.3.5}$$

$$\frac{\partial E_y}{\partial x} = -\frac{1}{c}\frac{\partial B}{\partial t}, \tag{8.5.3.6}$$

$$\frac{\partial B}{\partial x} = \frac{4\pi Ze}{c}n_e v_y, \tag{8.5.3.7}$$

where $d/dt = \partial/\partial t+u_x\partial/\partial x$ and where we have as before chosen the x-axis in the direction of the wave propagation and the z-axis in the direction of the magnetic field.

Substituting eqns. (8.5.3.4) and (8.5.3.7) into eqn. (8.5.3.3) we get

$$m_i n_i\frac{du_x}{dt} = -\frac{\partial}{\partial x}\frac{B^2}{8\pi}. \tag{8.5.3.8}$$

The equations of motion (8.5.3.5) and (8.5.3.8) (the second of these is clearly the Euler equation in which the magnetic pressure $B^2/8\pi$ plays the role of the pressure) form together with the continuity eqn. (8.5.3.2) and the Maxwell eqns. (8.5.3.6) and (8.5.3.7) a complete set of equations describing non-linear motions in a plasma in a weak magnetic field.

Let us first of all consider the propagation of long-wavelength perturbations, $k^2 \ll 4\pi e^2 n_e/m_e c^2$, through the plasma. In that case we have, neglecting the first term on the left-hand side of eqn. (8.5.3.5),

$$E_y = \frac{u_x}{c}B. \tag{8.5.3.9}$$

The set of eqns. (8.5.3.2), (8.5.3.6), (8.5.3.8), and (8.5.3.9) allow solutions in the form of simple waves corresponding to such plasma oscillations for which the quantities characterizing the plasma can be written as functions of one of them—for which we choose the ion density n_i. Assuming to fix the ideas that the simple wave propagates in the direction of the positive x-axis, we have

$$\frac{du_x}{dn_i} = \frac{v_A}{n_i},\quad \frac{dB}{dn_i} = \frac{B}{n_i},\quad \frac{dE_y}{dn_i} = \frac{u_x+v_A}{c}\frac{B}{n_i}, \tag{8.5.3.10}$$

and the phase velocity V of the wave is connected with the Alfvén velocity v_A through eqn. (8.5.2.4).

In order to follow the change of the profile of a simple wave when it propagates we evaluate the derivative dV/dn_i:

$$\frac{dV}{dn_i} = \frac{3v_A}{2n_i}. \tag{8.5.3.11}$$

As the derivative dV/dn_i is positive, gradients of the quantities which characterize the plasma in a simple wave increase along compression sections and this leads to the appearance of discontinuities (or multiple-current flow) along these sections. Self-similar waves in a cold plasma are rarefaction waves.

Let us now consider stationary waves in a cold plasma in a weak magnetic field. Eliminating from eqns. (8.5.3.2), (8.5.3.5) to (8.5.3.8) all quantities bar the magnetic field B, we get

$$l^2\left[\frac{B^2-B_0^2}{8\pi m_i n_{i0}V^2}-1\right]\frac{\partial}{\partial x}\left\{\left[\frac{B^2-B_0^2}{8\pi m_i n_{i0}V^2}-1\right]\frac{\partial B}{\partial x}\right\}+\left[\frac{B^2-B_0^2}{8\pi m_i n_{i0}V^2}-1\right]B+B_0=0, \quad (8.5.3.12)$$

where n_{i0} and B_0 are the values of the ion density and the magnetic field at the point where $u_x = 0$, V is the wave velocity, and

$$l=\sqrt{\frac{m_e c^2}{4\pi Z e^2 n_{i0}}}. \quad (8.5.3.13)$$

In the general case, when the derivative $\partial B/\partial x$ is non-vanishing when $B = B_0$, eqn. (8.5.3.12) determines the distribution of the magnetic field in a periodic wave. In the special case when $B-B_0$ and $\partial B/\partial x$ vanish in the same point, the wavelength tends to infinity and the periodic wave degenerates into a solitary wave.

Integrating eqn. (8.5.3.12) we get an equation implicitly determining the magnetic field distribution in a solitary wave as a function of the variable $x-Vt$:

$$x-Vt=\pm\int\frac{dB}{B-B_0}\left[\frac{B^2-B_0^2}{8\pi m_i n_{i0}V^2}-1\right]\left\{1-\frac{(B+B_0)^2}{16\pi m_i n_{i0}V^2}\right\}^{-1/2}. \quad (8.5.3.14)$$

We draw attention to the fact that the width of a solitary wave in a plasma in a weak magnetic field is, according to (8.5.3.13) and (8.5.3.14), of the order of $(m_e c^2/4\pi e^2 n_e)^{1/2}$, whereas the width of a solitary wave in the case of a strong magnetic field is, according to (8.5.2.1), of the order of $B/4\pi e n_e$.

Using (8.5.3.14) one checks easily that the amplitude B_m of a solitary wave and its velocity V are related to one another in the case of a weak magnetic field through the same relation (8.5.2.5) as in the case of a strong magnetic field. The quantities B_m and V can then not be arbitrarily large and, as before, satisfy the inequalities (8.5.2.6).

We have thus considered non-linear waves propagating in a cold plasma at right angles to the magnetic field. One can in a similar way consider non-linear waves propagating in a cold plasma at some other angle to the magnetic field (Sagdeev, 1961, 1966; Karpman, 1964b; Kazantsev, 1963; Berezin and Karpman, 1964). Without discussing the structure of such waves in detail we draw attention to one of its important features. We saw in Chapter 5 that the phase velocity of waves propagating in a cold plasma at an angle to the magnetic field which was different from $\pi/2$ increased with increasing wavenumber rather than decreased as in the case of waves propagating in a cold plasma at right angles to the magnetic field or in the case of a plasma with hot electrons. Such a behaviour of the phase velocity in its dependence on the wavenumber leads to the fact that solitary waves propagating in a cold plasma at an angle which is not close to $\pi/2$ turn out to be solitary rarefaction waves, rather than solitary compression waves.

CHAPTER 9

Theory of Plasma Oscillations in the Quasi-linear Approximation

9.1. Quasi-linear Theory of the Oscillations of an Unmagnetized Plasma

9.1.1. THE QUASI-LINEAR APPROXIMATION

When studying small (linear) oscillations in a plasma we split the distribution function into two terms—a non-oscillating part (the initial distribution function) and a small correction to it which oscillates with the frequency of the plasma oscillations—and we then assumed that the non-oscillating part was not connected at all with the oscillations occurring in the plasma. In actual fact, however, plasma oscillations affect the non-oscillating part of the distribution function and this effect increases with increasing amplitude of the oscillations.

When the amplitude of the oscillations increases one also violates the basic property of linear oscillations—the independence of the propagation of oscillations with different wave-vectors and frequencies (superposition principle) since processes involving the interactions between different oscillations begin to play an ever larger part.

We now turn to a study of these so-called non-linear processes. The simplest of the non-linear processes, which we shall first of all study, is the reaction of the plasma oscillations on the non-oscillating part of the distribution function. We shall here assume the amplitude of the oscillations to be so small that we can for the time being neglect wave-wave interactions, that is, that we can neglect the violation of the superposition principle.

The theory of plasma oscillations in which we take into account the effect of the oscillations on the non-oscillating part of the distribution function but still assume that the principle of the superposition of oscillations is valid is called the quasi-linear theory and this kind of approach to a study of plasma oscillations is called the *quasi-linear approximation.*

As the amplitude of the oscillations is assumed to be so small that in fact we can in the quasi-linear approximation consider the influence of the plasma oscillations only on the distribution of the resonance particles, that is, the particles for which the velocity component w along the direction of propagation lies close to the phase velocity ω/k,

$$w \approx \frac{\omega(\boldsymbol{k})}{k},$$

where $\omega(\boldsymbol{k})$ is the frequency of the plasma oscillations with wave vector $\boldsymbol{k}$. We know that these particles are important both for the damping and for the building up of oscillations as they interact strongly with the plasma oscillations. Non-resonance particles practically do not exchange energy with the waves; we may thus assume that their distribution is not affected by the plasma oscillations.

To obtain the basic equations of the quasi-linear theory we turn to the kinetic equation for the distribution function $F = F(\boldsymbol{r}, \boldsymbol{v}, t)$ with a self-consistent field $\boldsymbol{E} = \boldsymbol{E}(\boldsymbol{r}, t)$,

$$\frac{\partial F}{\partial t} + (\boldsymbol{v} \cdot \nabla) F + \frac{e}{m}\left(\boldsymbol{E} \cdot \frac{\partial F}{\partial \boldsymbol{v}}\right) = 0, \tag{9.1.1.1}$$

where for the sake of simplicity we have assumed that there is one kind of particles with charge e and mass m and that the oscillations are one-dimensional,

$$\boldsymbol{E} = -\nabla \varphi, \quad \nabla^2 \varphi = -4\pi e\left(\int F d^3\boldsymbol{v} - n_0\right). \tag{9.1.1.2}$$

As we assume that the superposition principle is valid for the field we can write the electrical field $\boldsymbol{E}$ as a wavepacket,

$$\boldsymbol{E} = \sum_{\boldsymbol{k}} \boldsymbol{E}_{\boldsymbol{k}}(t)\, e^{i(\boldsymbol{k}\cdot\boldsymbol{r}) - i\omega(\boldsymbol{k})\, t}, \tag{9.1.1.3}$$

where the frequency of the oscillations $\omega(\boldsymbol{k})$ is determined by the dispersion relation of the linear theory, while $\boldsymbol{E}_{\boldsymbol{k}}(t)$ is the complex amplitude of the oscillation which varies slowly with time. This change can be caused either by Landau damping or by a growth connected with the non-monotonic nature of the change in the initial distribution function, but in all cases we require that the damping rate or the growth rate of the oscillations $\gamma(\boldsymbol{k})$ is sufficiently small,

$$|\gamma(\boldsymbol{k})| \ll \omega(\boldsymbol{k}), \quad |\gamma(\boldsymbol{k})| \ll k v_T,$$

where v_T is the mean thermal particle velocity.

Apart from assuming that $|\gamma|$ is small, we make two more assumptions about the nature of the wavepacket (9.1.1.3). First of all, we shall assume that the phases $\delta(\boldsymbol{k})$ of the different complex amplitudes $\boldsymbol{E}_{\boldsymbol{k}}(t)$ are independent and random. This means that each amplitude when averaged over this phase vanishes,

$$\langle \boldsymbol{E}_{\boldsymbol{k}}(t) \rangle = 0,$$

where the pointed brackets $\langle \ldots \rangle$ indicate averaging over $\delta(\boldsymbol{k})$. Moreover, it follows from this assumption that

$$\langle (\boldsymbol{E}_{\boldsymbol{k}}(t) \cdot \boldsymbol{E}^*_{\boldsymbol{k}'}(t)) \rangle = |\boldsymbol{E}_{\boldsymbol{k}}(t)|^2\, \delta_{\boldsymbol{k},\boldsymbol{k}'}.$$

Similar formulae are also valid for the amplitudes of the potential $\varphi_{\boldsymbol{k}}(t)$:

$$\langle \varphi_{\boldsymbol{k}}(t) \rangle = 0, \quad \langle \varphi_{\boldsymbol{k}}(t) \varphi^*_{\boldsymbol{k}'}(t') \rangle = \varphi_{\boldsymbol{k}}(t')\, \varphi^*_{\boldsymbol{k}'}(t')\, \delta_{\boldsymbol{k},\boldsymbol{k}'}.$$

Secondly, we shall assume that the wavepacket (9.1.1.3) is sufficiently narrow (in $\boldsymbol{k}$-space) so that the number of resonance particles, interacting with the wavepacket, is considerably smaller than the total number of particles

$$\int_{\Delta w} dw \int d^2\boldsymbol{v}_\perp F(\boldsymbol{r}, \boldsymbol{v}, t) \ll n_0,$$

where $\Delta w = \Delta(\omega(\boldsymbol{k})/k)$ is the range of phase velocities corresponding to the range of $\boldsymbol{k}$-

values, $\Delta \boldsymbol{k}$, in the wavepacket (9.1.1.3). This assumption is made in order that we may assume that the frequency $\omega(\boldsymbol{k})$ is formally determined by the same dispersion relation as in the linear theory.

We now substitute the field (9.1.1.3) into the kinetic eqn. (9.1.1.1) and average it over the phases. To do this we split off from the distribution function the term

$$f_0 \equiv \langle F \rangle, \tag{9.1.1.4}$$

which is the value of the distribution function when averaged over the phases—it is called the *background* distribution function; we thus put

$$F = f_0 + f. \tag{9.1.1.5}$$

As $\langle f \rangle = 0$, we get the following equation for f_0:

$$\frac{\partial f_0}{\partial t} + (\boldsymbol{v} \cdot \nabla) f_0 + \frac{e}{m} \left\langle \left(\boldsymbol{E} \cdot \frac{\partial f}{\partial \boldsymbol{v}} \right) \right\rangle = 0, \tag{9.1.1.6}$$

while f, in turn, satisfies the equation

$$\frac{\partial f}{\partial t} + (\boldsymbol{v} \cdot \nabla) f + \frac{e}{m} \left(\boldsymbol{E} \cdot \frac{\partial f_0}{\partial \boldsymbol{v}} \right) + \frac{e}{m} \left(\boldsymbol{E} \cdot \frac{\partial f}{\partial \boldsymbol{v}} \right) - \frac{e}{m} \left\langle \left(\boldsymbol{E} \cdot \frac{\partial f}{\partial \boldsymbol{v}} \right) \right\rangle = 0.$$

We can in this equation neglect the last two terms which are responsible for non-linear interactions between the waves which are not taken into account in the quasi-linear approximation

$$\frac{\partial f}{\partial t} + (\boldsymbol{v} \cdot \nabla) f + \frac{e}{m} \left(\boldsymbol{E} \cdot \frac{\partial f_0}{\partial \boldsymbol{v}} \right) = 0. \tag{9.1.1.7}$$

We shall assume the plasma to be uniform so that the background distribution function will be independent of the coordinates and its equation becomes

$$\frac{\partial f_0}{\partial t} = -\frac{e}{m} \left\langle \left(\boldsymbol{E} \cdot \frac{\partial f}{\partial \boldsymbol{v}} \right) \right\rangle. \tag{9.1.1.8}$$

Our problem now consists in using (9.1.1.7) to express f in terms of f_0,

$$f = f\{f_0\},$$

and substituting this expression into (9.1.1.8) to obtain an equation containing only f_0.

One sees easily that in this way we get a second-order differential equation,

$$\frac{\partial f_0}{\partial t} = \sum_{i,j} \frac{\partial}{\partial v_i} D_{ij} \frac{\partial f_0}{\partial v_j}, \tag{9.1.1.9}$$

where $D_{ij} = D_{ij}(\boldsymbol{v}, t)$ is a function of the velocity and the time; eqn. (9.1.1.9) has the form of a diffusion equation—in velocity space. This is the reason why the D_{ij} are called the *diffusion coefficients in velocity space*.

Our problem thus consists in finding the diffusion coefficients under the above-mentioned assumptions about the properties of the wavepacket (9.1.1.3). To do this we expand the function $f(\boldsymbol{r}, \boldsymbol{v}, t)$ in a Fourier series (or integral):

$$f(\boldsymbol{r}, \boldsymbol{v}, t) = \sum_k f_k(\boldsymbol{v}, t)\, e^{i(\boldsymbol{k}\cdot\boldsymbol{r})-i\omega(k)\,t},$$

where the $f_k(\boldsymbol{v}, t)$ are slowly varying functions of the time,

$$\frac{\partial f_k}{\partial t}+i[(\boldsymbol{k}\cdot\boldsymbol{v})-\omega(\boldsymbol{k})]f_k = i\frac{e}{m}\varphi_k(t)\left(\boldsymbol{k}\cdot\frac{\partial f_0}{\partial \boldsymbol{v}}\right).$$

Integrating this equation and assuming that the initial perturbation of the distribution function $f(\boldsymbol{r}, \boldsymbol{v}, t)$ vanishes, we get

$$f_k(\boldsymbol{v}, t) = \frac{ie}{m} e^{i[\omega(k)-(\boldsymbol{k}\cdot\boldsymbol{v})]\,t} \int\limits_0^t dt'\, e^{i[(\boldsymbol{k}\cdot\boldsymbol{v})-\omega(\boldsymbol{k})]\,t'}\varphi_k(t')\left(\boldsymbol{k}\cdot\frac{\partial f_0(\boldsymbol{v}, t')}{\partial \boldsymbol{v}}\right). \qquad (9.1.1.10)$$

After substituting this expression into the Poisson equation, we get

$$\varphi_k(t) = \frac{4\pi i e^2}{mk}\int\limits_0^t dt'\, I(t, t')\,\varphi_k(t') \qquad (9.1.1.11)$$

where

$$I(t, t') = \int\limits_{-\infty}^{+\infty} dw\, e^{i[kw-\omega(\boldsymbol{k})](t'-t)}\frac{\partial f_0(w, t')}{\partial w}, \quad f_0(w, t) = \int d^2\boldsymbol{v}_\perp f_0(\boldsymbol{v}, t).$$

One sees easily that for sufficiently large values of t ($\omega t \gg 1$) the main contribution to the integral in (9.1.1.11) comes from t'-values close to t:

$$t \leqslant t' \leqslant t+\Delta t, \qquad \Delta t \ll t.$$

Indeed, the contribution of non-resonance particles to the integral $I(t, t')$ for t'-values not close to t ($|\,\omega-kw\,|\,(t-t') \gg 1$) will be small because of the fast oscillations of the integrand. As far as the contribution to $I(t, t')$ from the resonance particles for t' not close to t is concerned, it will be proportional to the range Δw occupied by the resonance particles in w-space and it can also be neglected because the number of resonance particles is small. We need therefore consider only t'-values close to t for the integration over t'.

Choosing the range Δt over which t' moves such that

$$\omega(\boldsymbol{k})\,\Delta t \gg 1, \quad kv_T\,\Delta t \gg t, \quad |\gamma(\boldsymbol{k})|\,\Delta t \ll 1, \quad \Delta t \ll \tau(w), \qquad (9.1.1.12)$$

where

$$\gamma(\boldsymbol{k}) \equiv \frac{\partial}{\partial t}\ln\varphi_k(t'), \quad \frac{1}{\tau(w)} \equiv \frac{\partial}{\partial t}\ln\frac{\partial f_0(w, t)}{\partial w}, \qquad (9.1.1.13)$$

we get

$$\int\limits_{t-\Delta t}^{t} dt'\,\varphi_k(t')\frac{\partial f_0(w, t')}{\partial w}e^{i[kw-\omega(\boldsymbol{k})](t'-t)}$$

$$\approx \varphi_k(t)\frac{\partial f_0(w, t)}{\partial w}\int\limits_{t-\Delta t}^{t} dt'\,e^{i[kw-\omega(\boldsymbol{k})-i\gamma(\boldsymbol{k})-\{i/\tau(w)\}](t'-t)}.$$

Using the above inequalities we have

$$\int_{t-\Delta t}^{t} dt'\, e^{i[kw-\omega(k)-i\gamma(k)-i/\tau(w)](t'-t)}$$

$$= \frac{-i}{kw-\omega(\boldsymbol{k})-i\gamma(\boldsymbol{k})-i/\tau(w)}\left[1-e^{-i[kw-\omega(k)-i\gamma(k)-i/\tau(w)]\,\Delta t}\right]$$

$$\approx \mathcal{P}\frac{-i}{kw-\omega(\boldsymbol{k})-i\gamma(\boldsymbol{k})-i/\tau(w)}+\pi\delta\{\omega(\boldsymbol{k})-kw\}\bigg], \tag{9.1.1.14}$$

where $\mathcal{P}$ is the principal-value symbol. Equation (9.1.1.11) thus becomes

$$1-\sum_\alpha \frac{4\pi e_\alpha^2}{m_\alpha k^2}\mathcal{P}\int \frac{1}{kw-\omega(\boldsymbol{k})-i\gamma(\boldsymbol{k})-i/\tau(w)}\frac{\partial f_{\alpha 0}(w,t)}{\partial w}\,dw$$

$$-\sum_\alpha i\pi\frac{4\pi e_\alpha^2}{m_\alpha k^2}\frac{\partial f_{\alpha 0}}{\partial w}\bigg|_{w=\omega(k)/k} = 0. \tag{9.1.1.15}$$

The main contribution to the integral over w in this equation comes from the non-resonance particles for which the quantity $\partial f_{\alpha 0}/\partial w$ is practically independent of time so that we can neglect in eqn. (9.1.1.15) the quantity $1/\tau(w)$. The equation which we then obtain is the same as the dispersion equation for the longitudinal, weakly damped, oscillations of an unmagnetized plasma.

Under the assumptions made ($|\gamma| \ll \omega$, $|\gamma| \ll kv_T$) we can thus use the dispersion equation from the linear theory, replacing in it the initial distribution function by the background distribution function.

Let us now transform eqn. (9.1.1.8). Substituting in the right-hand side of this eqn. $\boldsymbol{E}(\boldsymbol{r}, t)$ and $f(\boldsymbol{r}, \boldsymbol{v}, t)$ in their Fourier representation and using eqn. (9.1.1.10) we can write the equation in the form

$$\frac{\partial f_0}{\partial t} = \frac{e^2}{m^2}\left(\frac{\partial}{\partial \boldsymbol{v}}\cdot\left\langle \sum_{\boldsymbol{k},\boldsymbol{k}'} \boldsymbol{k}\varphi_{\boldsymbol{k}}^*(t)\int_0^t dt'\, e^{i[(\boldsymbol{k}'\cdot\boldsymbol{v})-\omega(\boldsymbol{k}')](t'-t)}\varphi_{\boldsymbol{k}'}(t')\left(\boldsymbol{k}'\cdot\frac{\partial f_0(\boldsymbol{v},t')}{\partial \boldsymbol{v}}\right)\right\rangle\right).$$

Averaging over the phases and changing from a sum to an integral we get

$$\frac{\partial f_0}{\partial t} = \sum_i \frac{\partial}{\partial v_i} J_i, \tag{9.1.1.16}$$

where

$$J_i(\boldsymbol{v}, t) = \sum_j \frac{e^2}{m^2}\int d^2\boldsymbol{k}_\perp \int_{-\infty}^{+\infty} dk_{||} k_i k_j \varphi_{\boldsymbol{k}}^*(t)\int_0^t dt'\, e^{i[k_{||}v-\omega(\boldsymbol{k})](t'-t)}\varphi_{\boldsymbol{k}}(t')\frac{\partial f_0(\boldsymbol{v},t')}{\partial v_j}, \tag{9.1.1.17}$$

where $k_{||}$ and $\boldsymbol{k}_\perp$ are the wavevector components parallel and at right angles to the particle velocity $\boldsymbol{v}$.

The main contribution to the integral over t' in (9.1.1.17) comes only from t'-values close to t. Indeed, if t' is not close to t, the main contribution in the integration over $k_{||}$ comes from $k_{||}$-values close to k_0 where the wavevector $k_{||} = k_0$ is determined from the equation $k_{||}v = \omega(\boldsymbol{k})$. Expanding the expression in the exponent in (9.1.1.17) in a series near $k_{||} = k_0$, we find

$$\int_{-\infty}^{+\infty} dk_{||}k_ik_j\varphi_{\boldsymbol{k}}^*(t)\,\varphi_{\boldsymbol{k}}(t')\,e^{i[k_{||}v-\omega(\boldsymbol{k})](t'-t)}$$

$$\approx \int_{-\infty}^{+\infty} d\Delta k_{||}k_ik_j\varphi_{\boldsymbol{k}}^*(t)\,\varphi_{\boldsymbol{k}}(t')\exp\left[i\left(v-\left.\frac{\partial\omega(\boldsymbol{k})}{\partial k_{||}}\right|_{k_{||}=k_0}\right)\Delta k_{||}(t'-t)\right].$$

As t' is not close to t this integral is negligibly small because of the fast oscillations of the integrand $(|\,v-(\partial\omega(\boldsymbol{k})/\partial k_{||})\,|_{k_{||}=k_0}\,\Delta k_{||}(t'-t)\,| \gg 1)$.

We still must evaluate the integrals over t' in (9.1.1.17) when $t' \approx t$. Choosing in that case the integration range Δt such that conditions (9.1.1.12) are satisfied we can, as in (9.1.1.14), put

$$\int_0^t dt'\,\varphi_{\boldsymbol{k}}(t')\frac{\partial f_0(\boldsymbol{v}, t')}{\partial v_j}\,e^{i[k_{||}v-\omega(\boldsymbol{k})](t'-t)}$$

$$\approx \varphi_{\boldsymbol{k}}(t)\frac{\partial f_0(\boldsymbol{v}, t)}{\partial v_j}\left\{\mathcal{P}\frac{-i}{(\boldsymbol{k}\cdot\boldsymbol{v})-\omega(\boldsymbol{k})-i\gamma(\boldsymbol{k})-i/\tau(\boldsymbol{v})}+\pi\delta\{\omega(\boldsymbol{k})-(\boldsymbol{k}\cdot\boldsymbol{v})\}\right\},$$

where

$$\frac{1}{\tau(\boldsymbol{v})} \equiv \frac{\partial}{\partial t}\ln\frac{\partial f_0(\boldsymbol{v}, t)}{\partial v_j}.$$

Substituting this expression into the right-hand side of eqn. (9.1.1.17) and using the fact that the principal value integral vanishes as $\gamma(\boldsymbol{k}) \to 0$ and $\tau(\boldsymbol{v}) \to \infty$ because the integrand is an odd function, we find that eqn. (9.1.1.16) has the same form as the diffusion eqn. (9.1.1.9) where the diffusion coefficients D_{ij} are determined by the equations

$$D_{ij} = \pi\frac{e^2}{m^2}\int d^3\boldsymbol{k}k_ik_j\,|\varphi_{\boldsymbol{k}}(t)|^2\,\delta\{(\boldsymbol{k}\cdot\boldsymbol{v})-\omega(\boldsymbol{k})\}, \tag{9.1.1.18}$$

or

$$D_{ij}(\boldsymbol{v}, t) = \pi\frac{e^2}{m^2}\int d^2\boldsymbol{k}_\perp \left.\frac{k_ik_j\,|\varphi_{\boldsymbol{k}}(t)|^2}{|\,v-\partial\omega(k)/\partial k_{||}|}\right|_{k_{||}v=\omega(k)}.$$

Equation (9.1.1.9) determines the change in the averaged distribution function of the resonance particles under the influence of high-frequency oscillations.† The fact that it refers only to resonance particles is connected with the presence of the δ-function in eqn. (9.1.1.18).

We have already noted that eqn. (9.1.1.9) has the form of a diffusion equation in velocity space; the diffusion occurring under the influence of random oscillations. One must add to

† It was established by Romanov and Fillipov (1961), Vedenov, Velikhov, and Sagdeev (1961 a, b; 1962), and Drummond and Pines (1962).

eqn. (9.1.1.9) an equation determining the time-dependence of the field amplitudes:

$$\frac{d\,|E_k(t)|^2}{dt} = 2\gamma(\boldsymbol{k})\,|E_k(t)|^2, \tag{9.1.1.19}$$

where the growth (damping) rate of the oscillations is, according to (9.1.1.15), determined by the equation

$$\gamma(\boldsymbol{k}) = -\sum_\alpha \frac{\pi e_\alpha^2}{m_\alpha k}\frac{\partial f_{\alpha 0}(w)}{\partial w}\bigg|_{w=\omega(k)/k}\left[\sum_\beta \frac{e_\beta^2}{m_\beta}\frac{\partial}{\partial\omega(\boldsymbol{k})}\mathcal{P}\int\frac{\partial f_{\beta 0}}{\partial w}\frac{dw}{kw-\omega(\boldsymbol{k})}\right]^{-1}. \tag{9.1.1.20}$$

The set of eqns. (9.1.1.9), and (9.1.1.18) to (9.1.1.20) for the functions $f_{\alpha 0}(\boldsymbol{v}, t)$ and $|E_k(t)|^2$ determines the change in the background distribution function and also the damping (or growth) of the oscillations in the quasi-linear approximation.

9.1.2. QUASI-LINEAR RELAXATION

We shall show that as a result of the diffusion of the particles due to interaction with the waves the system of resonance particles and plasma oscillations reaches a stationary state as $t \to \infty$. This process is called *quasi-linear relaxation.* To show this we multiply both sides of eqn. (9.1.1.9) by $f_{\alpha 0}$ and integrate over the whole of velocity space:

$$\frac{\partial}{\partial t}\frac{1}{2}\int f_{\alpha 0}^2\,d^3v = -\pi\frac{e_\alpha^2}{m_\alpha^2}\sum_k\int d^3v\,|\varphi_k(t)|^2\left(\boldsymbol{k}\cdot\frac{\partial f_{\alpha 0}}{\partial \boldsymbol{v}}\right)^2\delta\{\omega(\boldsymbol{k})-(\boldsymbol{k}\cdot\boldsymbol{v})\}. \tag{9.1.2.1}$$

The right-hand side of this equation is negative, that is, the quantity

$$\sigma(t) \equiv \int d^3v\, f_{\alpha 0}^2$$

decreases and tends to a constant value as $t \to \infty$. Therefore

$$\frac{d\sigma}{dt}\bigg|_{t\to\infty} \to 0,$$

whence it follows, in turn, that as $t \to \infty$ the right-hand side of eqn. (9.1.2.1) also vanishes. As the integrands in (9.1.2.1) are non-negative, we must have the following equality as $t \to \infty$:

$$|\varphi_k|^2\left(\boldsymbol{k}\cdot\frac{\partial f_{\alpha 0}}{\partial \boldsymbol{v}}\right)^2\bigg|_{\omega(k)=(k\cdot v)} = 0.$$

Therefore, as $t \to \infty$ we must have either

$$|\varphi_k|^2 = 0, \tag{9.1.2.2}$$

or

$$\left(\boldsymbol{k}\cdot\frac{\partial f_{\alpha 0}}{\partial \boldsymbol{v}}\right)\bigg|_{\omega(k)=(k\cdot v)} = 0. \tag{9.1.2.3}$$

In other words, the oscillations are either damped, or a plateau is formed on the distribution function:

$$f_{\alpha 0}(\boldsymbol{v}_\perp, w, t)\,|_{t\to\infty} = f_{\alpha 0}(\boldsymbol{v}_\perp), \tag{9.1.2.4}$$

that is, the distribution function does not change along the direction of the wave propagation.†

We note that if the volume in velocity space occupied by the resonance particles is rather large, it is impossible to form a plateau in that volume, as it would require too much energy. In that case the oscillations are either damped or the oscillation spectrum becomes one-dimensional while along the direction of the oscillation a plateau is formed on the distribution function as $t \to \infty$.

In conclusion we note that the quasi-linear relaxation process is accompanied by an increase in the entropy of the resonance particles:

$$S = -\int d^3v \ln \frac{f_{\alpha 0}}{A}, \quad A = \left(\frac{m_\alpha}{2\pi\hbar}\right)^3. \tag{9.1.2.5}$$

Indeed, if we substitute expression (9.1.1.9) for $\partial f_{\alpha 0}/\partial t$ into the equation

$$\frac{dS}{dt} = -d^3v \left[1 + \ln \frac{f_{\alpha 0}}{A}\right] \frac{\partial f_{\alpha 0}}{\partial t},$$

and integrate by parts, we get

$$\frac{dS}{dt} = \pi \frac{e_\alpha^2}{m_\alpha^2} \int d^3v \int d^3k \, |\varphi_k|^2 \frac{1}{f_{\alpha 0}} \left(k \cdot \frac{\partial f_{\alpha 0}}{\partial v}\right)^2 \delta\{\omega(k) - (k \cdot v)\}.$$

The integrand in this expression is positive, and hence $dS/dt > 0$.

We note that the increase in entropy during the quasi-linear relaxation of the distribution function does not contradict what we said in Subsection 1.4.3 about the self-consistent field in itself not leading to an increase in entropy. Indeed, the change to the quasi-linear approximation is connected with an additional averaging over the phases of the random oscillations. Moreover, we have here defined the entropy in terms of the background (averaged) distribution function, while the total distribution function entered into the definition (1.4.3.1).

9.1.3. RELAXATION OF ONE-DIMENSIONAL WAVEPACKETS

We can follow in detail the quasi-linear relaxation process in the case of one-dimensional oscillations and find the final state when a plateau is formed on the distribution function by the oscillation energy (Vedenov, Velikhov, and Sagdeev, 1961a, 1962; Drummond and Pines, 1962). For the sake of simplicity we restrict ourselves to considering only Langmuir oscillations. In that case the basic equations describing the quasi-linear relaxation of the distribution function and the field amplitudes are

$$\frac{\partial f_{e0}}{\partial t} = \frac{\partial}{\partial w}\left(\mathcal{E} B \frac{\partial f_{e0}}{\partial w}\right), \tag{9.1.3.1}$$

$$\frac{\partial \mathcal{E}}{\partial t} = A\mathcal{E} \frac{\partial f_{e0}}{\partial w}, \tag{9.1.3.2}$$

† Vedenov, Velikhov, and Sagdeev (1961 a, b; 1962) and Drummond and Pines (1962) indicated the possibility of the formation of a plateau on the distribution function for the case of one-dimensional oscillations; the proof given here is due to Sizonenko and Stepanov (1966; Bernstein and Engelman, 1966).

where

$$A = \pi\omega_{pe}w^2/n_0, \quad B = \frac{4\pi^2 e^2}{n_0 m_e^2 w},$$

$\mathscr{E}(k, t)$ is the spectral density of the energy of the oscillations,

$$\mathscr{E}(k, t) = \frac{|E_k(t)|^2}{4\pi},$$

while the wavenumber k and the particle velocity w are connected through the relation $\omega_{pe} = kw$.

Let the spectral density of the oscillation energy initially be given:

$$\mathscr{E}(k, t) = \begin{cases} \mathscr{E}_0(k) & \text{when} \quad k_1 < k < k_2, \\ 0 & \text{when} \quad k < k_1 \text{ or } k > k_2, \end{cases}$$

as well as the background electron distribution function

$$f_{e0}(w, t)\,|_{t=0} = f_0(w), \quad w_1 < w < w_2,$$

where $w_1 = \omega_{pe}/k_2$ and $w_2 = \omega_{pe}/k_1$. If $\partial f_0/\partial w < 0$, the oscillations will be damped and the magnitude of the derivative of the distribution function $|\partial f_{e0}/\partial w|$ will decrease until a plateau is formed and the exchange of energy between the waves and the particles ceases (see Fig. 9.1.1). In the points $w = w_{1,2}$ the distribution function is discontinuous. We note,

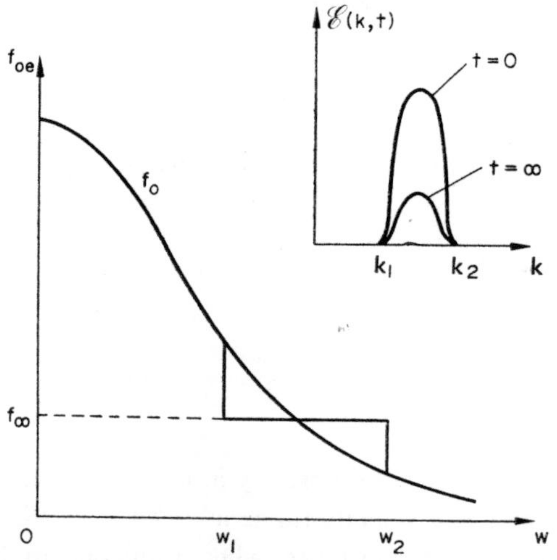

FIG. 9.1.1. Formation of a "plateau" on the distribution function of the resonance particles through the absorption of energy from a wavepacket.

however, that the shape of the distribution function in this figure corresponds to a total neglect of binary collisions. Taking these collisions into account—even though they are very rare—leads to a "smearing out" of the distribution function so that it changes continuously near the points $w = w_{1,2}$. (We refer to Subsection 9.1.4 for details.)

Let us find the oscillation energy in the final state with a plateau. Substituting expression (9.1.3.2) for $\partial f_{e0}/\partial w$ into (9.1.3.1) we find

$$\frac{\partial}{\partial t}\left[f_{e0}-\frac{\partial}{\partial w}\frac{B}{A}\mathcal{E}\right]=0,$$

that is,

$$f_{e0}-\frac{\partial}{\partial w}\frac{B}{A}\mathcal{E}=f_0(w)-\frac{\partial}{\partial w}\frac{B}{A}\mathcal{E}_0=\text{constant}. \tag{9.1.3.3}$$

Hence we find for the spectral density of the energy of the oscillations

$$\mathcal{E}(k,t)=\mathcal{E}_0(k)+\frac{A}{B}\int_{w_1}^{\omega_{pe}/k}dw(f_{e0}-f_0). \tag{9.1.3.4}$$

In particular, we have as $t\to\infty$

$$\mathcal{E}_\infty(k)=\mathcal{E}_0(k)+\frac{A}{B}\int_{w_1}^{\omega_{pe}/k}dw(f_\infty-f_0), \tag{9.1.3.5}$$

where $\mathcal{E}_\infty(k)=\mathcal{E}(k,t)\,|_{t\to\infty}$.

The height f_∞ of the plateau can be found from the condition that the number of resonance particles be conserved,

$$\int_{w_1}^{w_2}dwf_{e0}(w,t)=\int_{w_1}^{w_2}dw\,f_0(w),$$

whence we find as $t\to\infty$

$$f_\infty=\frac{1}{w_2-w_1}\int_{w_1}^{w_2}dwf_0(w). \tag{9.1.3.6}$$

In the relaxation process the resonance particles clearly gain energy:

$$\Delta\varepsilon=\int_{w_1}^{w_2}dw\frac{m_e}{2}(f_\infty-f_0)\,w^2.$$

The relaxation proceeds somewhat differently in the case when the initial distribution is unstable, that is, when there is a velocity range $w^*<w<w_0$ where $\partial f_0/\partial w>0$. Such an electron distribution is realized, for instance, when a low density beam passes through the plasma (see Fig. 9.1.2). In this case, oscillations with phase velocities in the interval $w^*<\omega_{pe}/k<w_0$ will grow in time, while oscillations with phase velocities outside that range will be damped. However, due to the deformation of the distribution function the instability region of the oscillations will widen with time (see the dotted curve in Fig. 9.1.2) and finally a plateau will be formed.

The energy of the oscillations and the height of the plateau are again given by eqns. (9.1.3.5) and (9.1.3.6), but the position of the points w_1 and w_2 are no longer determined by

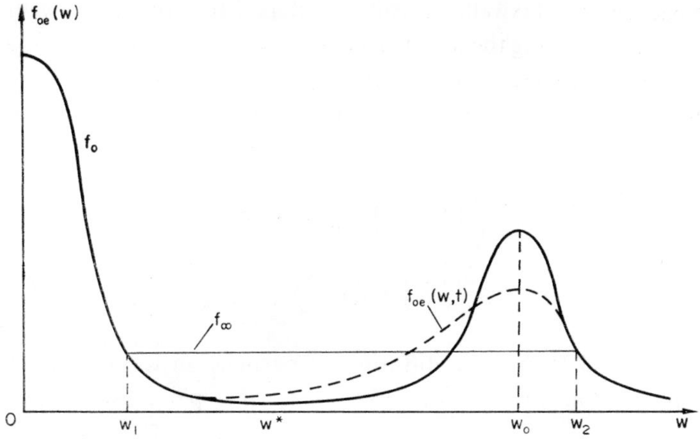

FIG. 9.1.2. Formation of a plateau on the distribution function of the resonance electrons when oscillations are excited by an electron beam.

the position of the boundaries of the original perturbation but by the condition that the function $f_e(w, t)$ be continuous:

$$f_\infty = f_0(w_1), \quad f_\infty = f_0(w_2).$$

9.1.4. EFFECT OF THE COULOMB COLLISIONS ON THE QUASI-LINEAR RELAXATION AND LANDAU DAMPING OF THE LANGMUIR OSCILLATIONS

The state of the plasma with a plateau on the distribution function is, of course, not an equilibrium state, but a quasi-stationary one, as binary (Coulomb) collisions between particles will tend to produce a Maxwellian distribution. Even though very rare, Coulomb collisions will appreciably affect the collisionless absorption of waves when the diffusion of resonance particles in velocity space caused by Coulomb collisions is comparable with the diffusion caused by particle-wave interactions. Let us consider this problem in more detail.

It is convenient to write the Landau collision integral (1.3.2.7), which we must now add to the right-hand side of eqn. (9.1.1.9), in the following form (Trubnikov, 1965)

$$\left[\frac{\partial f_{\alpha 0}}{\partial t}\right]_c = -\frac{2\pi e_\alpha^2 L}{m_\alpha}\sum_\beta e_\beta^2\left\{-\frac{8\pi}{m_\beta}f_{\alpha 0}f_{\beta 0}+\frac{1}{m_\alpha}\sum_{i,j}\frac{\partial^2 f_{\alpha 0}}{\partial v_i \partial v_j}\frac{\partial^2 \Psi_\beta}{\partial v_i \partial v_j}\right.$$
$$\left.+\left[\frac{1}{m_\alpha}-\frac{1}{m_\beta}\right]\sum_{i,j}\frac{\partial f_{\alpha 0}}{\partial v_i}\frac{\partial^3 \Psi_\beta}{\partial v_i \partial v_j^2}\right\}, \qquad (9.1.4.1)$$

where

$$\Psi_\beta(\boldsymbol{v}) = \int d^3v' f_{\beta 0}(\boldsymbol{v}')\,|\,\boldsymbol{v}-\boldsymbol{v}'\,|,$$

while L is the Coulomb logarithm.

For narrow one-dimensional wavepackets when the number of resonance particles is small we can neglect the collisions of resonance particles with each other. One can in that case simplify the collision integral (9.1.4.1) bearing in mind that the distribution function for the non-resonance particles is Maxwellian, while it differs little from a Maxwellian one for the resonance particles. Neglecting the quantities $\partial(f_{\alpha 0}-f_{\mathrm{M}})/\partial v_{\perp} \sim (f_{\alpha 0}-f_{\mathrm{M}})/v_{\alpha}$ as compared to $\partial(f_{\alpha 0}-f_{\mathrm{M}})/\partial w \sim (f_{\alpha 0}-f_{\mathrm{M}})/\Delta w$, we can write the equation for the electron background distribution function in the case of one-dimensional Langmuir oscillations, taking collisions into account, in the form

$$\frac{\partial f_{\mathrm{e}0}}{\partial t} = \frac{\partial}{\partial w}\left(D\frac{\partial f_{\mathrm{e}0}}{\partial w}\right) + \frac{\partial}{\partial w}\left[D_{\mathrm{c}}\frac{\partial(f_{\mathrm{e}0}-f_{\mathrm{M}})}{\partial w}\right], \tag{9.1.4.2}$$

$$D_{\mathrm{c}} = \nu_{\mathrm{c}}\frac{2v_{\mathrm{e}}^3}{w^3}(v_{\perp}^2+v_{\mathrm{e}}^2), \qquad \nu_{\mathrm{c}} = \frac{2\pi e^4 L n_0}{m_{\mathrm{e}}^2 v_{\mathrm{e}}^3}. \tag{9.1.4.3}$$

It follows from (9.1.4.3) that the diffusion of particles in the resonance region, when binary Coulomb collisions are taken into account, takes place over a characteristic time (Vedenov, 1962, 1967)

$$\tau = \frac{1}{\nu_{\mathrm{c}}}\left(\frac{\Delta w}{v_{\mathrm{e}}}\right)^2 \ll \tau_{\mathrm{c}} = \frac{1}{\nu_{\mathrm{c}}},$$

which is appreciably less than the average relaxation time τ_{c}.

The magnitude of the derivative $\partial f_{\mathrm{e}0}/\partial w$ and, hence, the rate at which energy is absorbed from the oscillations are now determined by two competing processes—plateau formation due to diffusion and Maxwellization due to collisions.

In the quasi-stationary state the diffusion due to interaction with the waves is balanced by the diffusion due to collisions—such a situation may, for instance, be established in the case of forced oscillations when there is an external source of oscillations or in the case of free oscillations with a sufficiently large initial noise level. In that case we get, neglecting in eqn. (9.1.4.2) the term $\partial f_{\mathrm{e}0}/\partial t$

$$\frac{\partial f_{\mathrm{e}0}}{\partial w} = \frac{\partial f_{\mathrm{M}}}{\partial w}\frac{1}{1+D/D_{\mathrm{c}}}. \tag{9.1.4.4}$$

We note that the quantity D_{c} occurring here depends on $v_{\perp}$.

The magnitude of the damping rate, in agreement with (9.1.1.20) and taking eqn. (9.1.4.4) into account, is determined by the equation (Rowlands, Sizonenko, and Stepanov, 1966)

$$\gamma = \gamma_0 \Phi(y), \tag{9.1.4.5}$$

where γ_0 is the damping rate of the Langmuir oscillations in a plasma with a Maxwellian electron velocity distribution, while

$$\Phi(y) = 1 + y e^{y+1/2}\,\mathrm{Ei}\,(-y-\tfrac{1}{2}), \quad \mathrm{Ei}\,(-x) = -\int_x^{\infty} dt\,\frac{e^{-t}}{t}, \quad y = \frac{D}{D_{\mathrm{c}0}}, \quad D_{\mathrm{c}0} = \nu_{\mathrm{c}}\frac{4v_{\mathrm{e}}^5}{w^3}. \tag{9.1.4.6}$$

The decrease in the damping rate caused by the plateau-formation effect and characterized by the monotonically decreasing function $\Phi(y)$ thus depends on the ratio $D/D_{\mathrm{c}0}$ of the diffusion coefficient of resonance electrons caused by the interaction with waves to the characteristic diffusion coefficient caused by collisions, $D_{\mathrm{c}0}$.

For a small-amplitude wavepacket ($y \ll 1$) the damping rate of the Langmuir oscillations is the same as γ_0. In the case of a large-amplitude wavepacket ($y \gg 1$) the quantity $\partial f_{\mathrm{e}0}/\partial w$

decreases roughly y times as much as $\partial f_M/\partial w$, while the damping rate decreases in the same ratio:

$$\gamma = \frac{3}{2y}\gamma_0.$$

9.2. Quasi-linear Theory of the Oscillations of a Magneto-active Plasma

9.2.1. BASIC EQUATIONS

In the preceding section we developed a quasi-linear theory for the oscillations of an unmagnetized plasma. We shall now consider the oscillations of a magneto-active plasma in the quasi-linear approximation. To do this we need study the reaction of the plasma oscillations on the resonance particles in a magnetic field. We saw in Chapter 5 that the nature of the interaction of resonance particles in a plasma in a magnetic field is appreciably different from the case of an unmagnetized plasma, as cyclotron resonances become important as well as the Cherenkov resonance.

As in the preceding section we shall consider the interaction of the resonance particles with a small-amplitude wavepacket with random phases which slowly grows (or is damped) with time. We shall again assume the number of resonance particles to be small. We shall, however, not restrict ourselves, as we did in the preceding section, to longitudinal oscillations alone, but we shall consider the general case of electromagnetic waves, which are not necessarily longitudinal.

We put

$$F_\alpha(\boldsymbol{r}, \boldsymbol{v}, t) = f_{\alpha 0}(\boldsymbol{v}, t) + f_\alpha(\boldsymbol{r}, \boldsymbol{v}, t), \tag{9.2.1.1}$$

where $f_{\alpha 0} = \langle F_\alpha \rangle$ is the average value of the distribution function which we shall assume to be independent of the spatial coordinates, while f_α is the oscillating part of the distribution function ($\langle f_\alpha \rangle = 0$). We expand the function f_α and the electrical and magnetic field strengths in Fourier series in the spatial coordinates:

$$\left.\begin{aligned} f_\alpha(\boldsymbol{r}, \boldsymbol{v}, t) &= \sum_{\boldsymbol{k},j} f_{\boldsymbol{k}}(\boldsymbol{v}, t)\, e^{i(\boldsymbol{k}\cdot\boldsymbol{r}) - i\omega^{(j)}(\boldsymbol{k})t}, \\ \boldsymbol{E}(\boldsymbol{r}, t) &= \sum_{\boldsymbol{k},j} \boldsymbol{E}_{\boldsymbol{k}}^{(j)}(t)\, e^{i(\boldsymbol{k}\cdot\boldsymbol{r}) - i\omega^{(j)}(\boldsymbol{k})t}, \quad \boldsymbol{B}(\boldsymbol{r}, t) = \sum_{\boldsymbol{k},j} \boldsymbol{B}_{\boldsymbol{k}}^{(j)}(t)\, e^{i(\boldsymbol{k}\cdot\boldsymbol{r}) - i\omega^{(j)}(\boldsymbol{k})t}, \end{aligned}\right\} \tag{9.2.1.2}$$

where the summations are over all values of the wavevector $\boldsymbol{k}$ and over all branches j of the oscillations, while $\omega^{(j)}(\boldsymbol{k})$ is the eigenfrequency of the jth branch of oscillations with wavevector $\boldsymbol{k}$.

Let us now obtain the equation for the background distribution function $f_{\alpha 0}(\boldsymbol{v}, t)$ taking into account the reaction of the oscillations on the resonance particles and also the equations which determine the change with time of the amplitudes $\boldsymbol{E}_{\boldsymbol{k}}^{(j)}(t)$ and $\boldsymbol{B}_{\boldsymbol{k}}^{(j)}(t)$. Substituting expression (9.2.1.1) into the kinetic equation

$$\frac{\partial F_\alpha}{\partial t} + (\boldsymbol{v}\cdot\nabla) F_\alpha - \omega_{B\alpha}\frac{\partial F_\alpha}{\partial \phi} + \frac{e_\alpha}{m_\alpha}\left(\left\{\boldsymbol{E} + \frac{1}{c}[\boldsymbol{v}\wedge\boldsymbol{B}]\right\}\cdot\frac{\partial F_\alpha}{\partial \boldsymbol{v}}\right) = 0,$$

where ϕ is the azimuthal angle in velocity space, and averaging over the phases we get

$$\frac{\partial f_{\alpha 0}}{\partial t} - \omega_{B\alpha} \frac{\partial f_{\alpha 0}}{\partial \phi} = -\frac{e_\alpha}{m_\alpha} \left\langle \left(\left\{ \boldsymbol{E} + \frac{1}{c} [\boldsymbol{v} \wedge \boldsymbol{B}] \right\} \cdot \frac{\partial f_\alpha}{\partial \boldsymbol{v}} \right) \right\rangle . \tag{9.2.1.3}$$

The oscillating part f_α of the distribution function is determined from the equation

$$\frac{\partial f_\alpha}{\partial t} + (\boldsymbol{v} \cdot \nabla) f_\alpha + \frac{e_\alpha}{m_\alpha} \left(\left\{ \boldsymbol{E} + \frac{1}{c} [\boldsymbol{v} \wedge \boldsymbol{B}] \right\} \cdot \frac{\partial f_{\alpha 0}}{\partial \boldsymbol{v}} \right) - \omega_{B\alpha} \frac{\partial f_\alpha}{\partial \phi} = 0, \tag{9.2.1.4}$$

in which we have dropped the non-linear terms

$$\frac{e_\alpha}{m_\alpha} \left(\left\{ \boldsymbol{E} + \frac{1}{c} [\boldsymbol{v} \wedge \boldsymbol{B}] \right\} \cdot \frac{\partial f_\alpha}{\partial \boldsymbol{v}} \right) - \frac{e_\alpha}{m_\alpha} \left\langle \left(\left\{ \boldsymbol{E} + \frac{1}{c} [\boldsymbol{v} \wedge \boldsymbol{B}] \right\} \cdot \frac{\partial f_\alpha}{\partial \boldsymbol{v}} \right) \right\rangle ,$$

responsible for non-linear wave-wave interactions.

Bearing in mind that oscillations with different values of $\boldsymbol{k}$ and j are statistically independent, we get from (9.2.1.3)

$$\frac{\partial f_{\alpha 0}}{\partial t} - \omega_{B\alpha} \frac{\partial f_{\alpha 0}}{\partial \phi} = Q(\boldsymbol{v}, t), \tag{9.2.1.5}$$

where

$$Q(\boldsymbol{v}, t) = -\frac{e_\alpha}{m_\alpha} \left(\frac{\partial}{\partial \boldsymbol{v}} \cdot \sum_{k,j} \left\{ \boldsymbol{E}_k^{(j)*} + \frac{1}{c} [\boldsymbol{v} \wedge \boldsymbol{B}_k^{(j)*}] \right\} f_k^{(j)} \right) .$$

The Fourier component $f_k^{(j)}(\boldsymbol{v}, t)$ of the oscillating part of the distribution function is here, in accordance with (9.2.1.4) determined by the equation

$$\frac{\partial f_k^{(j)}}{\partial t} + i[(\boldsymbol{k} \cdot \boldsymbol{v}) - \omega^{(j)}(\boldsymbol{k})] f_k^{(j)} - \omega_{B\alpha} \frac{\partial f_k^{(j)}}{\partial \psi} = R(\boldsymbol{k}, \boldsymbol{v}), \tag{9.2.1.6}$$

where

$$R(\boldsymbol{k}, \boldsymbol{v}) = -\frac{e_\alpha}{m_\alpha} \left(\left\{ \boldsymbol{E}_k^{(j)*} + \frac{1}{c} [\boldsymbol{v} \wedge \boldsymbol{B}_k^{(j)*}] \right\} \cdot \frac{\partial f_{\alpha 0}}{\partial \boldsymbol{v}} \right) ,$$

$\psi = \phi - \varphi$, with φ the azimuthal angle in wavevector space with the polar axis parallel to $\boldsymbol{B}_0$.

We shall write the background distribution function in the form of a Fourier series in the angular variable ϕ:

$$f_{\alpha 0}(\boldsymbol{v}, t) = \sum_{l=-\infty}^{+\infty} f_{\alpha 0}^{(l)}(v_{\|}, v_{\perp}, t)\, e^{il\phi}, \tag{9.2.1.7}$$

where $v_\perp$ and $v_\|$ are the components of the particle velocity $\boldsymbol{v}$ at right angles and parallel to the magnetic field $\boldsymbol{B}_0$.

Introducing a coordinate system with unit basis vectors $\boldsymbol{e}_1$, $\boldsymbol{e}_2$, and $\boldsymbol{e}_3$, where $\boldsymbol{e}_3$ is parallel to $\boldsymbol{B}_0$, $\boldsymbol{e}_1$ parallel to the wavevector component $\boldsymbol{k}_\perp = [\boldsymbol{B}_0 \wedge [\boldsymbol{k} \wedge \boldsymbol{B}_0]]/B_0^2$, which is at right angles to $\boldsymbol{B}_0$, and $\boldsymbol{e}_2 = [\boldsymbol{e}_3 \wedge \boldsymbol{e}_1]$, we can write eqn. (9.2.1.6) in the following form:

$$\frac{\partial f_k^{(j)}}{\partial t} + i[k_\| v_\| - \omega^{(j)}(\boldsymbol{k}) + k_\perp v_\perp \cos\psi] f_k^{(j)} - \omega_{B\alpha} \frac{\partial f_k^{(j)}}{\partial \psi} = R. \tag{9.2.1.8}$$

The function R is here defined by the formula

$$R = \sum_{l=-\infty}^{+\infty} R^{(l)}(k_{||}, k_{\perp}, v_{||}, v_{\perp}, t)\, e^{il\psi}. \tag{9.2.1.9}$$

Let us now find the solution of this equation. To do this, we write the function $f_k^{(j)}$ in the form

$$f_k^{(j)} = u_k(v)\, e^{i\lambda \sin\psi}, \quad \lambda \equiv \frac{k_{\perp} v_{\perp}}{\omega_{B\alpha}}.$$

We then get for the function $u_k(v)$ the equation

$$\frac{\partial u_k}{\partial t} + i[k_{||}v_{||} - \omega^{(j)}(\boldsymbol{k})]\, u_k - \omega_{B\alpha}\frac{\partial u_k}{\partial \psi} = P,$$

where

$$P = R\, e^{-i\lambda \sin\psi}.$$

Putting

$$u_k(v) = \sum_{l=-\infty}^{+\infty} u_k^{(l)}(v_{||}, v_{\perp}, t)\, e^{il\psi},$$

we get hence an equation for the function $u_k^{(l)}$

$$\frac{\partial u_k^{(l)}}{\partial t} + i[k_{||}v_{||} - l\omega_{B\alpha} - \omega^{(j)}(\boldsymbol{k})]\, u_k^{(l)} = P^{(l)},$$

where

$$\begin{aligned}
P^{(l)} &= \frac{1}{2\pi}\int_0^{2\pi} d\psi\, e^{-i\lambda\sin\psi - il\psi} R \\
&= -\frac{e_\alpha}{m_\alpha}\sum_{l'=-\infty}^{+\infty} e^{il'\varphi}(-1)^{l-l'}\left\{\left(E_{k3}^{(j)}\frac{\partial f_{\alpha0}^{(l')}}{\partial v_{||}} - il' f_{\alpha0}^{(l')}\frac{1}{c}B_{k3}^{(j)}\right)J_{l-l'}(\lambda)\right. \\
&\quad -\left[\left(E_{k1}^{(j)} - \frac{v_{||}}{c}B_{k2}^{(j)}\right)\frac{\partial f_{\alpha0}^{(l')}}{\partial v_{\perp}} + \frac{v_{\perp}}{c}B_{k2}^{(j)}\frac{\partial f_{\alpha0}^{(l')}}{\partial v_{||}} + \left(E_{k2}^{(j)} + \frac{v_{||}}{c}B_{k1}^{(j)}\right)\frac{il'}{v_{\perp}}f_{\alpha0}^{(l')}\right]\frac{(l-l')\,J_{l-l'}(\lambda)}{\lambda} \\
&\quad \left. +i\left[\left(E_{k2}^{(j)} + \frac{v_{||}}{c}B_{k1}^{(j)}\right)\frac{\partial f_{\alpha0}^{(l')}}{\partial v_{\perp}} - \frac{v_{\perp}}{c}B_{k1}^{(j)}\frac{\partial f_{\alpha0}^{(l')}}{\partial v_{||}} - \left(E_{k1}^{(j)} - \frac{v_{||}}{c}B_{k2}^{(j)}\right)\frac{il'}{v_{\perp}}f_{\alpha0}^{(l')}\right]J'_{l-l'}(\lambda)\right\}.
\end{aligned} \tag{9.2.1.10}$$

Integrating this equation we find

$$\begin{aligned}
u_k^{(l)}(t) &= u_k^{(l)}(0)\exp\left[-i\{k_{||}v_{||} - l\omega_{B\alpha} - \omega^{(j)}(\boldsymbol{k})\}\, t\right] \\
&\quad + \int_0^t dt'\, R^{(l)}(t')\exp\left[i\{k_{||}v_{||} - l\omega_{B\alpha} - \omega^{(j)}(k)\}(t'-t)\right],
\end{aligned}$$

where $u_k^{(l)}(0)$ is the initial value of $u_k^{(l)}(t)$.

The Fourier components of the oscillating part of the distribution function thus have the following form:

$$f_k^{(j)}(\boldsymbol{v}, t) = f_{k1}^{(j)}(\boldsymbol{v}, t) + f_{k2}^{(j)}(\boldsymbol{v}, t), \tag{9.2.1.11}$$

where

$$\left.\begin{aligned} f_{k1}^{(j)} &= \sum_{l=-\infty}^{+\infty} \exp\{i(\lambda \sin\psi + l\psi) - i[k_{\|}v_{\|} - l\omega_{B\alpha} - \omega^{(j)}(\boldsymbol{k})]t\}\, u_k^{(l)}(0) \\ f_{k2}^{(j)} &= \sum_{n=-\infty}^{+\infty} e^{i\lambda\sin\psi + il\psi} \int_0^{t'} dt' P^{(l)}(k_{\|}, k_{\perp}, v_{\|}, v_{\perp}, t') \exp\{i[k_{\|}v_{\|} - l\omega_{B\alpha} - \omega^{(j)}(\boldsymbol{k})](t'-t)\}. \end{aligned}\right\} \tag{9.2.1.12}$$

Let us now turn to eqn. (9.2.1.5) for the background distribution function. Using eqn. (9.2.1.7) we find

$$\frac{\partial f_{\alpha 0}^{(l)}}{\partial t} - il\omega_{B\alpha} f_{\alpha 0}^{(l)} = Q^{(l)}(v_{\|}, v_{\perp}, t), \tag{9.2.1.13}$$

where

$$Q^{(l)} = \frac{1}{2\pi}\int_0^{2\pi} d\phi\, Q(\boldsymbol{v}, t)\, e^{-il\phi}.$$

Hence it follows that

$$f_{\alpha 0}^{(0)}(v_{\|}, v_{\perp}, t) = f_{\alpha 0}^{(0)}(v_{\|}, v_{\perp}, 0) + \int_0^t dt'\, Q^{(0)}(v_{\|}, v_{\perp}, t'),$$

$$f_{\alpha 0}^{(l)}(v_{\|}, v_{\perp}, t) = \left[f_{\alpha 0}^{(l)}(v_{\|}, v_{\perp}, 0) - i\frac{Q^{(l)}(v_{\|}, v_{\perp}, 0)}{l\omega_{B\alpha}}\right]\exp(il\omega_{B\alpha}t) + i\frac{Q^{(l)}(v_{\|}, v_{\perp}, t)}{l\omega_{B\alpha}}.$$

If the initial perturbation is such that

$$f_{\alpha 0}^{(l)}(v_{\|}, v_{\perp}, t) \approx i\frac{Q^{(l)}(v_{\|}, v_{\perp}, 0)}{l\omega_{B\alpha}}, \tag{9.2.1.14}$$

the functions $f_{\alpha 0}^{(l)}(v_{\|}, v_{\perp}, t)\ (l \neq 0)$ and hence also the distribution function $f_{\alpha 0}(v, t)$ will not contain terms $\propto \exp(il\omega_{B\alpha}t)$ which oscillate fast with time so that we can neglect the quantities $f_{\alpha 0}^{(l)} \approx iQ^{(l)}/l\omega_{B\alpha}$ when $|\omega_{B\alpha}|\, t \gg 1$ as compared to $f_{\alpha 0}^{(0)} \approx Q^{(0)}t$. In the case considered we may thus, when condition (9.2.1.14) is satisfied, assume the background distribution function to be independent of the azimuthal angle ϕ, that is,

$$f_{\alpha 0}(\boldsymbol{v}, t) \approx f_{\alpha 0}(v_{\|}, v_{\perp}, t) = f_{\alpha 0}^{(0)}(v_{\|}, v_{\perp}, t).$$

Thanks to this property of the background distribution function we can simplify the quantities $P^{(l)}$ occurring in (9.2.1.12)

$$\begin{aligned} P^{(l)} = -\frac{e_\alpha}{m_\alpha}(-1)^l \Bigg\{ & E_{k3}^{(j)} \frac{\partial f_{\alpha 0}}{\partial v_{\|}} J_l(\lambda) - \left[\left(E_{k1}^{(j)} - \frac{v_{\|}}{c} B_{k2}^{(j)}\right)\frac{\partial f_{\alpha 0}}{\partial v_{\perp}} + \frac{v_{\perp}}{c} B_{k2}^{(j)} \frac{\partial f_{\alpha 0}}{\partial v_{\|}}\right]\frac{l}{\lambda} J_l(\lambda) \\ & + i\left[\left(E_{k2}^{(j)} + \frac{v_{\|}}{c} B_{k1}^{(j)}\right)\frac{\partial f_{\alpha 0}}{\partial v_{\perp}} - \frac{v_{\perp}}{c} B_{k1}^{(j)} \frac{\partial f_{\alpha 0}}{\partial v_{\|}}\right] J_l'(\lambda)\Bigg\}. \end{aligned} \tag{9.2.1.15}$$

The Fourier components of the magnetic field, $B_k^{(j)}$, can by means of the equation

$$\text{curl}\, \boldsymbol{E} = -\frac{1}{c}\frac{\partial \boldsymbol{B}}{\partial t}$$

be expressed in terms of the Fourier components of the electrical field:

$$B_k^{(j)} = \sum_{l,m} \frac{c\varepsilon_{ilm} k_l E_{km}^{(j)}}{\omega^{(j)}(\boldsymbol{k}) + i\gamma_m^{(j)}(\boldsymbol{k})}, \tag{9.2.1.16}$$

where ε_{ilm} is the antisymmetric third-rank unit tensor and the $\gamma_m^{(j)}(\boldsymbol{k}, t)$ are the momentaneous growth rates of the mth component of the electrical field amplitude,

$$\gamma_m^{(j)} = \gamma_m^{(j)\prime} + i\gamma_m^{(j)\prime\prime} \equiv \frac{\partial}{\partial t} \ln E_{km}^{(j)}(t). \tag{9.2.1.17}$$

Let us now find the expression for the current density in the quasi-linear approximation. Writing the current density as a Fourier series in the spatial variables

$$\boldsymbol{j} = \sum_{k,j} \boldsymbol{j}_k^{(j)}(t)\, e^{i(\boldsymbol{k}\cdot\boldsymbol{r}) - i\omega^{(j)}(\boldsymbol{k})t},$$

and using eqn. (9.2.1.11) we can split the Fourier components of the current density into two terms

$$\boldsymbol{j}_k^{(j)} = \boldsymbol{j}_{k1}^{(j)} + \boldsymbol{j}_{k2}^{(j)},$$

where

$$\boldsymbol{j}_{k1} = \sum_\alpha \sum_{l=-\infty}^{+\infty} e_\alpha \int d^3\boldsymbol{v} e^{i\lambda \sin\psi + il\psi} \boldsymbol{v} u_k^{(l)}(v_{||}, v_\perp, 0) \exp\{-i[k_{||}v_{||} - l\omega_{B\alpha} - \omega^{(j)}(\boldsymbol{k})]\, t\},$$

$$\boldsymbol{j}_{k2} = \sum_\alpha \sum_{l=-\infty}^{+\infty} e_\alpha \int d^3\boldsymbol{v} e^{i\lambda \sin\psi + il\psi} \boldsymbol{v} \int_0^t dt' P^{(l)}(v_{||}, v_\perp, t') \exp\{i[k_{||}v_{||} - l\omega_{B\alpha} - \omega^{(j)}(\boldsymbol{k})]\, (t'-t)\}, \tag{9.2.1.18}$$

while the function $P^{(l)}(v_{||}, v_\perp, t)$ is determined by eqn. (9.2.1.15).

The term $\boldsymbol{j}_{k1}$ is caused by the initial perturbation of the oscillating part of the distribution function. This term plays the role of an external current and does not affect the dispersion equation. For the sake of simplicity we shall therefore assume in what follows that $\boldsymbol{j}_{k1} = 0$.

Let us now turn to an evaluation of the quantity $\boldsymbol{j}_{k2}$ which is given by eqn. (9.2.1.18). The integrals over $v_{||}$ which occur in that expression have the form

$$\int_{-\infty}^{+\infty} dv_{||} S(v_{||}, t') \exp\{i[k_{||}v_{||} - l\omega_{B\alpha} - \omega^{(j)}(\boldsymbol{k})]\, (t'-t)\}, \tag{9.2.1.19}$$

where

$$S(v_{||}, t') = E_{kl}^{(j)}(t') \frac{\partial f_{\alpha 0}(v_\perp, v_{||}, t')}{\partial v_{||,\perp}}.$$

The main contribution to this integral comes from $v_{||}$-values close to $v_{||\,\text{res}} = \{\omega^{(j)}(\boldsymbol{k}) + l\omega_{B\alpha}\}/k_{||}$, provided t' is not too close to t ($|\, k_{||}v_{||} - l\omega_{B\alpha} - \omega^{(j)}(\boldsymbol{k})\,|\, (t-t') \gg 1$) and the whole integral will be proportional in that case to the interval $\Delta v_{||}$, that is, proportional to the number of resonance particles; we can then neglect the integral.

If t' lies close to t, the region of $v_{||}$-values where $|\, k_{||}v_{||} - l\omega_{B\alpha} - \omega^{(j)}(\boldsymbol{k})\,|\, (t-t') \gg 1$ will not contribute to the integral (9.2.1.19) because of the fast oscillations of the integrand.

There remains the evaluation of the integrals occurring in (9.2.1.18) when $t' \approx t$ in the region $v_{||} \approx v_{||\,\text{res}}$, where $|\,k_{||}v_{||} - l\omega_{B\alpha} - \omega^{(j)}(\boldsymbol{k})\,|\,(t-t') \lesssim 1$. Choosing the range Δt of the integration over t' $(t-\Delta t < t' < t)$ such that

$$|\,\gamma_l^{(j)}(\boldsymbol{k})|\,\Delta t \ll 1, \quad \Delta t \ll \tau_{||,\perp}(\boldsymbol{v}),$$

and

$$\omega^{(j)}(\boldsymbol{k})\,\Delta t \gg 1, \quad k_{||}v_\alpha\,\Delta t \gg 1,$$

we get

$$\int\limits_{t-\Delta t}^{t} dt' E_{kl}^{(j)}(t') \frac{\partial f_{\alpha 0}(t')}{\partial v_{||,\perp}} \exp\{i[k_{||}v_{||} - l\omega_{B\alpha} - \omega^{(j)}(\boldsymbol{k})]\,(t'-t)\}$$

$$\approx -iE_{kl}^{(j)}(t)\frac{\partial f_{\alpha 0}(t)}{\partial v_{||,\perp}}\left\{\mathcal{P}\frac{1}{k_{||}v_{||} - l\omega_{B\alpha} - \omega^{(j)}(\boldsymbol{k}) - i\gamma_l^{(j)}(\boldsymbol{k}) - i/\tau_{||,\perp}(\boldsymbol{v})} + i\pi\delta\{\omega^{(j)}(\boldsymbol{k}) + l\omega_{B\alpha} - k_{||}v_{||}\}\right\}, \tag{9.2.1.20}$$

where the $\tau_{||,\perp}$ are the relaxation times of the background distribution function,

$$\frac{1}{\tau_{||,\perp}(\boldsymbol{v}, t)} \equiv \frac{\partial}{\partial t}\ln\frac{\partial f_{\alpha 0}(v_{||}, v_\perp, t)}{\partial v_{||,\perp}},$$

and $\mathcal{P}$ indicates that in the integration over $v_{||}$ one must take the principal value integral.

Using eqn. (9.2.1.20) we can write the Fourier components $\boldsymbol{j}_k = \boldsymbol{j}_{k2}$ of the current density in the form

$$j_{kl}^{(j)} = \sum \sigma_{lm}(t)\, E_{km}^{(j)}(t) \tag{9.2.1.21}$$

where

$$\sigma_{lm} = \sum_\alpha \sigma_{lm}^{(\alpha)},$$

while $\sigma_{lm}^{(\alpha)}$ is the contribution of particles of the αth kind of the plasma conductivity tensor In the system of coordinates in which the 3-axis is parallel to $\boldsymbol{B}_0$ and the 1-axis lies in the plane through the vectors $\boldsymbol{k}$ and $\boldsymbol{B}_0$ the quantities $\sigma_{lm}^{(\alpha)}$ have the form

$$\sigma_{ij}^{(\alpha)} = \frac{1}{2}\,i\omega_{p\alpha}^2 \sum_{l=-\infty}^{+\infty} \int v_\perp\, dv_\perp\, dv_{||}\xi_{ij}^{(l)} R_j f_{\alpha 0} + \frac{i\omega_{p\alpha}^2}{4\pi[\omega^{(j)} + i\gamma_3^{(j)}(\boldsymbol{k})]}\left[1 + \int d^3v\,\frac{v_{||}^2}{v_\perp}\frac{\partial f_{\alpha 0}}{\partial v_\perp}\right] b_i b_j,$$

where $b_i = B_{0i}/B_i$,

$$\xi_{ij}^{(l)} = \begin{bmatrix} v_\perp \dfrac{l^2}{\lambda^2} J_l^2 & iv_\perp J_l^2 & v_{||}\dfrac{l}{\lambda} J_l^2 \\ -iv_\perp J_l^2 & v_\perp J_l'^2 & -iv_{||}J_l J_l' \\ v_{||}\dfrac{l}{\lambda} J_l^2 & iv_{||}J_l J_l' & \dfrac{v_{||}^2}{v_\perp} J_l' \end{bmatrix}, \tag{9.2.1.22}$$

$J_l = J_l(\lambda)$ and

$$\hat{R}_m f_{\alpha 0} = \frac{1}{k_{||}v_{||} + l\omega_{B\alpha} - \omega^{(j)}(\boldsymbol{k}) - i\gamma_m^{(j)}(\boldsymbol{k}) - i/\tau_\perp(\boldsymbol{v})}\left[1 - \frac{k_{||}v_{||} - i\delta_{3m}/\tau_\perp(\boldsymbol{v})}{\omega^{(j)}(\boldsymbol{k}) + i\gamma_m^{(j)}(\boldsymbol{k})}\right]\frac{\partial f_{\alpha 0}}{\partial v_{||}}$$

$$+ \frac{(v_\perp/v_{||})[k_{||}v_{||} - i\delta_{3m}/\tau_{||}(\boldsymbol{v})]}{[k_{||}v_{||} + l\omega_{B\alpha} - \omega^{(j)}(\boldsymbol{k}) - i\gamma^{(j)}(\boldsymbol{k}) - i/\tau_{||}(\boldsymbol{v})]\,[\omega^{(j)}(\boldsymbol{k}) + i\gamma_m^{(j)}(\boldsymbol{k})]}\frac{\partial f_{\alpha 0}}{\partial v_{||}}. \tag{9.2.1.23}$$

Using eqns. (9.2.1.21) we get from the Maxwell equations

$$\sum_m \Lambda_{lm} E^{(j)}_{km} = 0 \tag{9.2.1.24}$$

where

$$\Lambda_{lm} = \frac{c^2(k_l k_m - k^2\,\delta_{lm})}{[\omega^{(j)}(\boldsymbol{k})+i\gamma_l^{(j)}(\boldsymbol{k})]\,[\omega^{(j)}(\boldsymbol{k})+i\gamma_m^{(j)}(\boldsymbol{k})]} + \varepsilon_{lm}, \tag{9.2.1.25}$$

$$\varepsilon_{lm} = \delta_{lm} + \frac{4\pi i}{\omega^{(j)}(\boldsymbol{k})+i\gamma_l^{(j)}(\boldsymbol{k})}\sigma_{lm}. \tag{9.2.1.26}$$

The quantities ε_{lm} and σ_{lm} defined by eqns. (9.2.1.26) and (9.2.1.22) are the dielectric permittivity and the conductivity tensor of a magneto-active plasma in the quasi-linear approximation. These quantities differ appreciably from the corresponding quantities of the linear theory. The dfference consists in that in the expressions for σ_{lm} and ε_{lm} in the quasi-linear theory the time-dependent background distribution function $f_{\alpha 0}(\boldsymbol{v}, t)$ occurs rather than the initial distribution function, and also that in expressions (9.2.1.22) and (9.2.1.26) for σ_{lm} and ε_{lm} the quantities $\omega^{(j)}(\boldsymbol{k})+i\gamma^{(j)}_{l,m}(\boldsymbol{k})$ and $\omega^{(j)}(\boldsymbol{k})+i\gamma^{(j)}_{l,m}(\boldsymbol{k})+i/\tau_{||,\perp}$, which contain the momentaneous growth rates $\gamma_1^{(j)}$, $\gamma_2^{(j)}$, $\gamma_3^{(j)}$ and the relaxation times $\tau_{||}(\boldsymbol{v})$ and $\tau_\perp(\boldsymbol{v})$, occur, while in the linear theory there has occurred only the complex frequency $\omega = \omega^{(j)}(\boldsymbol{k})+i\gamma^{(j)}(\boldsymbol{k})$ which contains one growth rate, $\gamma^{(j)}(\boldsymbol{k})$, which is independent of time.

The set of eqns. (9.2.1.24) determines the time-dependence of the electrical field amplitude $E^{(j)}_{kl}(t)$. If the anti-Hermitean terms in the tensor ε_{lm} are small compared to the Hermitean parts, the ratio of the electrical field components $E^{(j)}_{kl}(t)/E^{(j)}_{km}(t)$ will, according to (9.2.1.24), be determined solely by the Hermitean parts of the tensor ε_{ij} which are independent of the detailed behaviour of the background distribution function $f_{\alpha 0}(\boldsymbol{v}, t)$ in the region occupied by the resonance particles and, hence, the ratio $E^{(j)}_{kl}(t)/E^{(j)}_{km}(t)$ will be independent of the time. This means that the quantities $E^{(j)}_{kl}(t)$ change with time in the same way, that is, the momentaneous growth rates $\gamma_l^{(j)}(\boldsymbol{k})$ are in this case all equal. It is also clear that the momentaneous growth rates $\gamma_l^{(j)}(\boldsymbol{k})$ are equal to one another also in the case of irrotational oscillations when $\boldsymbol{E}^{(j)}_k = -i\boldsymbol{k}\varphi^{(j)}_{\boldsymbol{k}}$. In these cases the equation

$$\Lambda = \operatorname{Det} \Lambda_{lm} = 0,$$

in which we can neglect the quantities $1/\tau_{||,\perp}$ because the number of resonance particles is small, is the same as the dispersion equation of the linear theory.

In the general case, however, the different components of the electrical field strength amplitude have different momentaneous growth rates $\gamma_k^{(j)}(\boldsymbol{k})$ since the anti-Hermitean terms of the tensor ε_{ij}, which in an essential way depend on the values of the quantities $\partial f_{\alpha 0}/\partial v_{||}$ and $\partial f_{\alpha 0}/\partial v_\perp$, which can change with time because of the particle diffusion due to interaction with the waves will enter in the ratio $E^{(j)}_{kl}(t)/E^{(j)}_{km}$.

Let us now transform eqn. (9.2.1.5) for the background distribution function. We can neglect the contribution from the function $f^{(j)}_{k1}(\boldsymbol{v})$ to the right-hand side of this equation as it is proportional to the number of resonance particles. Using eqn. (9.2.1.12) for $f^{(j)}_{k2}(\boldsymbol{v})$

we then get from eqn. (9.2.1.5)

$$\frac{\partial f_{\alpha 0}}{\partial t} - \omega_{B\alpha}\frac{\partial f_{\alpha 0}}{\partial \phi} = -\frac{e_\alpha}{m_\alpha}\left(\frac{\partial}{\partial \boldsymbol{v}}\cdot\sum_{k,j}\left\{\boldsymbol{E}_k^{(j)*}+\frac{1}{c}[\boldsymbol{v}\wedge \boldsymbol{B}_k^{(j)*}]\right\}\right)$$

$$\times\sum_{l=-\infty}^{+\infty} e^{i\lambda\sin\psi+il\psi}\int_0^t dt' P^{(l)}(t')\exp\{i[k_{||}v_{||}-l\omega_{B\alpha}-\omega^{(j)}(\boldsymbol{k})](t'-t)\}. \quad (9.2.1.27)$$

On the right-hand side of this equation we have an integral over $k_{||}$ of the form

$$I \equiv \int dk_{||} f(k_{||}, t')\exp\{i[k_{||}v_{||}-l\omega_{B\alpha}-\omega^{(j)}(\boldsymbol{k})](t'-t)\}, \quad (9.2.1.28)$$

where $f(k_{||}, t)$ is proportional to $E_{kl}^{(j)}(t)\,E_{km}^{(j)}(t)\;\partial f_{\alpha 0}(\boldsymbol{v}, t)/\partial v_{||,\perp}$. If t' does not lie close to t, the integrand in (9.2.1.28) oscillates rapidly and does not contribute to the integral unless $k_{||}v_{||}-l\omega_{B\alpha}-\omega^{(j)}(\boldsymbol{k})$ lies close to zero. The main contribution to this integral comes from the region of resonance values $k_{||}\approx k_0$, where k_0 is determined from the condition

$$\omega^{(j)}(\boldsymbol{k})-l\omega_{B\alpha}-k_{||}v_{||}=0, \quad k_{||}=k_0.$$

For narrow wavepackets ($\Delta k_{||}/k_0 \ll 1$) we can write the integral (9.2.1.28) in the form

$$I \approx \int dk_{||} f(k_{||}, t')\exp\left\{i\left[v_{||}-\frac{\partial\omega^{(j)}(\boldsymbol{k})}{\partial k_{||}}\right]_{k_{||}=k_0}(k_{||}-k_0)(t'-t)\right\}.$$

If the condition

$$\left|\left[v_{||}-\frac{\partial\omega^{(j)}(\boldsymbol{k})}{\partial k_{||}}\right]_{k_{||}=k_0}(k_{||}-k_0)(t'-t)\right| \gg 1 \quad (9.2.1.29)$$

holds, we can take the function $f(k_{||}, t')$ from under the integral sign and put $k_{||}=k_0$. The remaining integral is nearly $\delta(t'-t)$ and therefore does not contribute to the integral over t' in (9.2.1.27) (t' does not lie close to t). As long as we also consider times $\Delta t = t-t'$ small compared to $1/\gamma^{(j)}(\boldsymbol{k})$, it is necessary that the condition

$$\frac{\Delta k_{||}}{k_0} \gg \left.\frac{|\gamma_m^{(j)}(\boldsymbol{k})|}{k_{||}\,|v_{||}-\partial\omega^{(j)}(\boldsymbol{k})/\partial k_{||}|}\right|_{k_{||}=k_0} \quad (9.2.1.30)$$

is satisfied, in order that condition (9.2.1.29) can hold. When these conditions are satisfied the main contribution to the integral over t' in (9.2.1.27) comes therefore from values $t'\approx t$. Bearing this in mind, applying to the evaluation of the integrals over $k_{||}$ and t' in (9.2.1.27) the same considerations as for the evaluation of the current density $\boldsymbol{j}_k = \boldsymbol{j}_{k2}$, and neglecting terms of order γ/ω and $1/\omega\tau_{||,\perp}$ we get finally for the function

$$f_{\alpha 0}(v_{||}, v_{\perp}, t) = \frac{1}{2\pi}\int_0^{2\pi} d\phi\, f_{\alpha 0}(\boldsymbol{v}, t) \approx f_{\alpha 0}(\boldsymbol{v}, t)$$

the equation (which has been derived by Yakimenko (1963), Andronov and Trakhtengerts

(1964), Shapiro and Shevchenko (1962), and Sizonenko and Stepanov (1967b, 1968))

$$\frac{\partial f_{\alpha 0}}{\partial t} = \pi \frac{e_\alpha^2}{m_\alpha^2} \sum_j \sum_{l=-\infty}^{+\infty} \int d^3k \frac{1}{v_\perp} \hat{R} \left\{ v_\perp \left| E_{k1}^{(j)} \frac{l}{|\lambda|} J_l(|\lambda|) + i\eta_\alpha E_{k2}^{(j)} J_l'(|\lambda|) \right.\right.$$
$$\left.\left. + E_{k3}^{(j)} \frac{v_\parallel}{v_\perp} J_l(|\lambda|) \right|^2 (\hat{R} f_{\alpha 0}) \right\} \delta\{\omega^{(j)}(\boldsymbol{k}) - l\,|\omega_{B\alpha}| - k_\parallel v_\parallel\}, \qquad (9.2.1.31)$$

where

$$\eta_\alpha = \frac{e_\alpha}{|e_\alpha|}, \quad \hat{R} = \left[1 - \frac{k_\parallel v_\parallel}{\omega^{(j)}(\boldsymbol{k})}\right] \frac{\partial}{\partial v_\perp} + \frac{k_\parallel v_\perp}{\omega^{(j)}(\boldsymbol{k})} \frac{\partial}{\partial v_\parallel}. \qquad (9.2.1.32)$$

The set of eqns. (9.2.1.31) and (9.2.1.24) for the functions $f_{\alpha 0}(v_\parallel, v_\perp, t)$ and $E_{km}^{(j)}$ ($m = 1, 2, 3$) determine the damping (or growth) of the electromagnetic field and the relaxation of the averaged distribution function in the quasi-linear approximation in a plasma in an external magnetic field.

To conclude this subsection we note that condition (9.2.1.30), which was assumed to hold when we derived eqn. (9.2.1.31), is not satisfied for oscillations with a linear (or nearly linear) dispersion law, $\omega^{(j)} \approx k_\parallel v$. Examples of such waves are the Alfvén wave ($\omega = k_\parallel v_A$), the slow magneto-sound wave in a low pressure plasma ($\omega = k_\parallel v_s$, $\varkappa_\alpha = 8\pi n_0 T_\alpha / B_0^2 \ll 1$), and the fast magneto-sound wave propagating almost parallel to the magnetic field ($\omega \approx k v_A$, $k_\parallel \approx k$) in a low density plasma ($\varkappa_\alpha \ll 1$). The equation for the background distribution function for these oscillations is non-local in the time (Sizonenko and Stepanov, 1967b, 1968), that is, the rate of change of the function $f_{\alpha 0}(v_\parallel, v_\perp, t)$ at time t is determined by the values of the electrical field amplitude and of the derivative $\partial f_{\alpha 0}/\partial v_\parallel$ at all earlier times. We can easily obtain this equation from (9.2.1.27), retaining on the right-hand side only the term with $l = 0$:

$$\frac{\partial f_{\alpha 0}}{\partial t} = \frac{e_\alpha^2}{m_\alpha^2} \frac{\partial}{\partial v_\parallel} \int d^3k \left[E_{k3}^{(j)} J_0(|\lambda|) + i\eta_\alpha E_{k2}^{(j)} J_0'(|\lambda|) \frac{k_\parallel v_\perp}{\omega^{(j)}(\boldsymbol{k})} \right]^*$$
$$\times \int_0^t dt' \left[E_{k3}^{(j)}(t')\, J_0(|\lambda|) + i\eta_\alpha E_{k2}^{(j)}(t')\, J_0'(|\lambda|) \frac{k_\parallel v_\perp}{\omega^{(j)}(\boldsymbol{k})} \right] \frac{\partial f_{\alpha 0}(v, t')}{\partial v_\parallel} \exp\{i[k_\parallel v_\parallel - \omega^{(j)}(\boldsymbol{k})](t' - t)\}.$$

9.2.2. QUASI-LINEAR RELAXATION IN A MAGNETO-ACTIVE PLASMA

The reaction of the oscillations on the resonance particles leads to a diffusion of the particles in velocity space parallel and at right angles to the magnetic field. Let us now consider the quasi-linear relaxation in a magneto-active plasma (Andronov, and Trakhtengerts, 1964; Sizonenko and Stepanov, 1966; Rowlands, Sizonenko, and Stepanov, 1966). We shall restrict our discussion to oscillations for which the anti-Hermitean parts in the tensor ε_{ij} are small compared with the Hermitean parts while the momentaneous growth rates of the different components of the electrical field are the same.

One can easily check that during the quasi-linear relaxation in a magneto-active plasma, as in the case of unmagnetized oscillations, the entropy of the resonance particles increases.

Indeed, using the definition (9.1.2.5) of the entropy and eqn. (9.2.1.31) we find

$$\frac{dS}{dt} = -\int d^3\boldsymbol{v}\left[1+\ln\frac{f_{\alpha 0}}{A}\right]\frac{\partial f_{\alpha 0}}{\partial t} = \pi\frac{e_\alpha^2}{m_\alpha^2}\sum_{l=-\infty}^{+\infty}\sum_j\int d^3\boldsymbol{k}\int d^3\boldsymbol{v}\frac{1}{f_{\alpha 0}}(\hat{R}f_{\alpha 0})^2$$

$$\times\left|E_{k1}^{(j)}\frac{l}{|\lambda|}J_l(|\lambda|)+i\eta_\alpha E_{k2}^{(j)}J_l'(|\lambda|)+E_{k3}^{(j)}\frac{v_{||}}{v_\perp}J_l(|\lambda|)\right|^2\delta\{\omega^{(j)}-l|\omega_{B\alpha}|-k_{||}v_{||}\}\geqslant 0. \tag{9.2.2.1}$$

We shall now show that the system of oscillations and resonance particles which we are considering relaxes to a stationary state. To do that we multiply eqn. (9.2.1.31) by $f_{\alpha 0}$ and integrate over the velocities. Integrating by parts we get

$$\frac{\partial}{\partial t}\int d^3v\,f_{\alpha 0}^2 = -\frac{2\pi e_\alpha^2}{m_\alpha^2}\sum_j\sum_{l=-\infty}^{+\infty}\int d^3v\int d^3k(\hat{R}f_{\alpha 0})^2$$

$$\times\left|E_{k1}^{(j)}\frac{l}{|\lambda|}J_l(|\lambda|)+i\eta_\alpha E_{k2}^{(j)}J_l'(|\lambda|)+E_{k3}^{(j)}\frac{v_{||}}{v_\perp}J_l(|\lambda|)\right|^2\delta\{\omega^{(j)}-l|\omega_{B\alpha}|-k_{||}v_{||}\}. \tag{9.2.2.2}$$

As the right-hand side of this equation is negative, the positive quantity

$$\sigma(t) = \int d^3v\,f_{\alpha 0}^2$$

decreases and as $t\to\infty$ tends to a constant limit. The left-hand side of the equation therefore tends to zero as $t\to\infty$. It then follows from (9.2.2.2) that as $t\to\infty$ either

$$\left|E_{k1}^{(j)}\frac{l}{|\lambda|}J_l(|\lambda|)+i\eta_\alpha E_{k2}^{(j)}J_l'(|\lambda|)+E_{k3}^{(j)}\frac{v_{||}}{v_\perp}J_l(|\lambda|)\right|^2 = 0, \tag{9.2.2.3}$$

or

$$(\hat{R}f_{\alpha 0})_{\omega^{(j)}(\boldsymbol{k})\,=\,l|\omega_{B\alpha}|+k_{||}v_{||}} = 0. \tag{9.2.2.4}$$

Bearing in mind that the amplitudes $E_{km}^{(j)}$ are interconnected through the linear relation (9.2.1.24) we can conclude that if condition (9.2.2.3) is satisfied, $|E_{km}^{(j)}| = 0$ as $t\to\infty$, that is, the oscillations are damped.

If, however, in the final state the energy of the oscillations is non-vanishing, condition (9.2.2.4) must be satisfied:

$$\left[l|\omega_{B\alpha}|\frac{\partial f_{\alpha 0}}{\partial v_\perp}+k_{||}v_\perp\frac{\partial f_{\alpha 0}}{\partial v_{||}}\right]_{\omega^{(j)}(\boldsymbol{k})\,=\,l|\omega_{B\alpha}|+k_{||}v_{||}} = 0. \tag{9.2.2.5}$$

In that case one speaks of the establishment of a "generalized" plateau on the distribution function $f_{\alpha 0}$.

As the anti-Hermitean terms in the tensor ε_{ij} which determine the growth rate $\gamma^{(j)}(\boldsymbol{k})$ are integrals over the velocities of expressions proportional to $(\hat{R}f_{\alpha 0})\,\delta(\omega^{(j)}-l|\omega_{B\alpha}|-k_{||}v_{||})$ the growth rate $\gamma^{(j)}(\boldsymbol{k})$ will tend to zero as condition (9.2.2.4) is satisfied, that is, in that case the exchange of energy between the resonance particles and the waves vanishes.

The quantity $k_{||}$ in eqn. (9.2.2.5) is, generally speaking, arbitrary so that this equation can be satisfied only when the two conditions

$$\frac{\partial f_{\alpha 0}}{\partial v_{||}} = 0,\quad \frac{\partial f_{\alpha 0}}{\partial v_\perp} = 0$$

are satisfied simultaneously. In the general case of three-dimensional wavepackets there will thus be established a three-dimensional plateau in the whole of the velocity region Ω_v occupied in the $v_{||}$, $v_\perp$-plane by the resonance particles.

The size of this region as far as $v_{||}$ is concerned is determined by the condition $\omega^{(j)}(\boldsymbol{k}) = k_{||}v_{||}+l\,|\,\omega_{B\alpha}\,|$ for all possible values of $\boldsymbol{k}$, l, j for which $\boldsymbol{E}_{\boldsymbol{k}}^{(j)} \neq 0$. For given l and j this condition determines a region of resonance particles $\Omega_v^{(l,j)}$. If the regions $\Omega_v^{(l,j)}$ do not intersect, in each of them its own "plateau" will be established.

If the initial distribution function $f_{\alpha 0}(v_{||}, v_\perp, 0)$ is such that oscillations with frequency $\omega^{(j)}(\boldsymbol{k})$ are unstable, either the initial level of the oscillations of the jth branch is large, but the other branches of the oscillations have small growth rates, or their initial amplitude is small and one can consider solely the development of oscillations of frequency $\omega^{j)}(\boldsymbol{k})$. As a result a generalized plateau may be formed on the distribution function $f_{\alpha 0}$ in the region $\Omega_v^{l,j)}$. It can happen that a distribution function with a plateau in the region $\Omega_v^{(l,j)}$ is unstable with regard to oscillations with frequency $\omega^{(j')}(\boldsymbol{k}')$. In that case a second stage quasi-linear relaxation starts during which the oscillations of the j'th branch grow and in which there is a further deformation of the distribution function. In that case one talks about a *quasi-linear wave-transformation* (this possibility was indicated by Andronov and Trakhtengerts (1964)). As the result of such a two-stage relaxation in the final state either the oscillations will be damped or a generalized plateau is formed in the whole region Ω_v for which $\boldsymbol{E}_{\boldsymbol{k}}^{(j)}$ and $\boldsymbol{E}_{\boldsymbol{k}}^{(j')}$ are non-vanishing.

One can also speak of a quasi-linear wave-transformation in the case when in the first stage the oscillations of a branch are damped (grow) leading to the establishment of a plateau in a region $\Omega_v^{(l,j)}$ of resonance particles for which $\omega^{(j)}(\boldsymbol{k}) = k_{||}v_{||}+l\,|\,\omega_{B\alpha}\,|$ while in the second stage there is a slower growth of the same branch of oscillations through resonance particles for which $\omega^{(j)}(\boldsymbol{k}') = k'_{||}v_{||}+l'\,|\,\omega_{B\alpha}\,|$ with $l' = l$ and $\boldsymbol{k}' = \boldsymbol{k}$ while the regions $\Omega_v^{(l,j)}$ and $\Omega_v^{(l'j')}$ intersect.

Let us, as an example, consider the problem of the passage of a low-density electron beam through a "cold" plasma in a strong magnetic field ($|\,\omega_{Be}\,| \gg \omega_{pe}$) (Shapiro and Shevchenko, 1968). Such a beam can excite low-frequency longitudinal oscillations with frequencies

$$\omega = \omega_\infty^{(1)}(\theta) \equiv \omega_+(k_{||}, \boldsymbol{k}_\perp) \approx |\,\omega_{Be}\,| \left[1+\frac{1}{2}\sin^2\theta\,\frac{\omega_{pe}^2}{\omega_{Be}^2}\right],$$

$$\omega = \omega_\infty^{(2)}(\theta) \equiv \omega_-(k_{||}, \boldsymbol{k}_\perp) \approx \omega_{pe}\cos\theta = \omega_{pe}\frac{k_{||}}{k}.$$

The growth-rate of the oscillations with frequency ω_+ is lower by a factor $|\,\omega_{Be}\,|/\omega_{pe}$ than the growth rate of the oscillations with frequency ω_- (see eqn. (6.2.2.4)) and we can neglect the excitation of that branch of oscillations.

The growth rate of the oscillations with frequency ω_- is a maximum for the case of Cherenkov excitation,

$$\omega_-(k_{||}, \boldsymbol{k}_\perp) = k_{||}v_{||},$$

while those oscillations which propagate in the direction of the magnetic field grow fastest.

In the first stage of the passage of a low-density beam through a plasma we shall thus see the excitation of almost one-dimensional longitudinal oscillations ($k_{||} \gg k_\perp$) with frequency

$\omega_- \approx \omega_{pe}$. The diffusion of resonance particles due to these oscillations leads to the formation of a plateau on the longitudinal velocity part of the electron distribution function, $\partial f_{e0}/\partial v_{\|} = 0$ for $v_{\|} = \omega_-/k_{\|}$, and to a limit to the growth of the Cherenkov instability. At the end of the first stage the electron distribution function will have the shape shown by the dashed line in Fig. 9.2.2 at the end of this subsection and in the plateau region will depend only on $v_\perp$, $f_{e0} = f_\infty(v_\perp)$.

In the next stage we have the slower cyclotron excitation of oscillations with frequency $\omega = \omega_-$ under anomalous Doppler effect conditions,†

$$\omega_-(k'_{\|}, k'_\perp) = k'_{\|}v_{\|} + l\,|\,\omega_{Be}\,|, \qquad l = -1, -2, \ldots,$$

and, as $\omega_- \ll |\,\omega_{Be}\,|$, the resonance particles, $v_{\|} = l\omega_{Be}/k'_{\|}$, are in the plateau region if $k'_{\|} \gg k_{\|}$, where $k_{\|}$ is the longitudinal wavevector of the oscillations with frequency ω_- which grow through Cherenkov resonance.

Indeed, the basic equations of the quasi-linear theory take the following form for longitudinal oscillations of a plasma in a magnetic field:

$$\frac{\partial f_{e0}}{\partial t} = \pi \frac{e^2}{m_e^2} \int k_\perp \, dk_\perp \sum_{l=-\infty}^{+\infty} \left[\frac{\partial}{\partial v_{\|}} k_{\|} + \frac{l\,|\,\omega_{Be}\,|}{v_\perp} \frac{\partial}{\partial v_\perp}\right] \left[k_{\|} \frac{\partial f_{e0}}{\partial v_{\|}} + \frac{l\,|\,\omega_{Be}\,|}{v_\perp} \frac{\partial f_{e0}}{\partial v_\perp}\right] \frac{|\,\varphi_k\,|^2 J_l^2(\lambda)}{|\,v_{\|} - v_g\,|} \tag{9.2.2.6}$$

where

$$\frac{\partial\,|\,\varphi_k\,|^2}{\partial t} = 2\gamma\,|\,\varphi_k\,|^2, \quad \gamma = \sum_{l=-\infty}^{+\infty} \gamma_l(k_{\|}, k_\perp),$$

$$\gamma_l = \frac{\pi}{2} \frac{\omega_-^3}{k^2 k_{\|} n_0} \int d^2 v_\perp \, J_l^2(\lambda) \left[k_{\|} \frac{\partial f_{e0}}{\partial v_{\|}} + \frac{l\,|\,\omega_{Be}\,|}{v_\perp} \frac{\partial f_{e0}}{\partial v_\perp}\right]_{\omega_- = k_{\|}v_{\|} + l\,|\,\omega_{Be}\,|}, \tag{9.2.2.7}$$

and $v_g = \partial\omega_-/\partial k_{\|}$ is the group velocity. In obtaining eqn. (9.2.2.6) we assumed that the quantity $|\,\varphi_k\,|^2$ is independent of the azimuthal angle φ which is clearly satisfied if the initial value of $|\,\varphi_k\,|^2$ were independent of the angle φ.

Changing from $v_{\|}$, $v_\perp$ to new variables $v_{\|}$, w,

$$w(l, k_\perp) = v_\perp^2 + v_{\|}^2 - 2 \int \frac{\omega_-}{k_{\|}} \, dv_{\|},$$

we can write eqns. (9.2.2.6) and (9.2.2.7) in the form

$$\frac{\partial f_{e0}}{\partial t} = \pi \frac{e^2}{m_e^2} \int k_\perp \, dk_\perp \sum_{l=-\infty}^{+\infty} \frac{\partial}{\partial v_{\|}} \left[|\,\varphi_k\,|^2 \frac{k_{\|}^2 J_l^2(\lambda)}{|\,v_{\|} - v_g\,|} \frac{\partial f_{e0}}{\partial v_{\|}}\right], \tag{9.2.2.8}$$

$$\gamma_l(k_{\|}, k_\perp) = \frac{\pi}{2} \frac{\omega_-^3 \cos^2\theta}{k\,k_{\|} n_0} \int d^2 v_\perp J_l^2(\lambda) \frac{\partial f_{e0}}{\partial v_{\|}}. \tag{9.2.2.9}$$

The derivatives with respect to $v_{\|}$ are taken here for $w(l, k_\perp) =$ constant. It follows from eqn. (9.2.2.8) that the diffusion of the resonance particles proceeds along the lines $w(l, k_\perp)$ $=$ constant.

† Kadomtsev and Pogutse (1968) indicated the possibility of the excitation of Langmuir and ion-sound oscillations of a plasma in a strong magnetic field under anomalous Doppler effect conditions by a group of "run-away" electrons for which the condition $\partial f_{e0}(v_{\|}, v_\perp)/\partial v_{\|} > 0$ is not satisfied.

Bearing in mind that

$$v_\perp^2 = w(l, k'_\perp)+2\int \frac{\omega_-(k'_{||}, k'_\perp)}{k'_{||}}\, dv_{||}-v_{||}^2,$$

where $v_{||} = \omega_-(k_{||}, k_\perp)/k_{||}$, we find that in the state with a plateau on the distribution function

$$\left[\frac{\partial f_{e0}}{\partial v_{||}}\right]_{w(l, k'_\perp) = \text{constant}} = \frac{\partial}{\partial v_{||}} f_\infty \left[w(l, k'_\perp)+2\int \frac{\omega_-(k'_{||}, k'_\perp)}{k'_{||}}\, dv_{||}-v_{||}^2\right]$$

$$= 2\left[\frac{\omega_-(k'_{||}, k'_\perp)}{k'_{||}}-v_{||}\right]\frac{df_\infty}{dv_\perp^2}.$$

Using this formula we find the growth rate of the oscillations with frequency $\omega_-(k'_{||}, k'_\perp)$ for cyclotron excitation when there is a plateau on the distribution function:

$$\gamma_l(k'_{||}, k'_\perp) = \pi \frac{\omega_-(k'_{||}, k'_\perp)}{(k'_{||})^3 n_0} [v_{\rm ph}(k'_{||}, k'_\perp)-v_{\rm ph}(k_{||}, k_\perp)] \int d^2v_\perp J_l^2(\lambda)\frac{df_\infty}{dv_\perp^2},$$

where the phase velocities are determined by the relations

$$v_{\rm ph}(k'_{||}, k'_\perp) = \frac{\omega_-(k'_{||}, k'_\perp)}{k'_{||}}, \quad v_{\rm ph}(k_{||}, k_\perp) = \frac{\omega_-(k_{||}, k_\perp)}{k_{||}} = v_{||}.$$

As $v_{\rm ph}(k'_{||}, k'_\perp) = v_{||}+l\,|\,\omega_{Be}\,|/k'_{||}$, the growth rate γ_l will be positive for distribution functions $f_\infty(v_\perp^2)$ which decrease with increasing $v_\perp$, if $l < 0$, that is, the formation of a plateau due to Cherenkov excitation leads, indeed, to cyclotron excitation of oscillations under anomalous Doppler effect conditions.

When oscillations with frequency ω_- are cyclotron excitated there occurs particle diffusion along the lines $w(l, k'_\perp) =$ constant, that is, both along and at right angles to the magnetic field (see Fig. 9.2.1). On these lines

$$\frac{dv_\perp}{dv_{||}} = -\frac{v_{||}-[\omega_-(k'_{||}, k'_\perp)/k'_{||}]}{v_\perp} < 0,$$

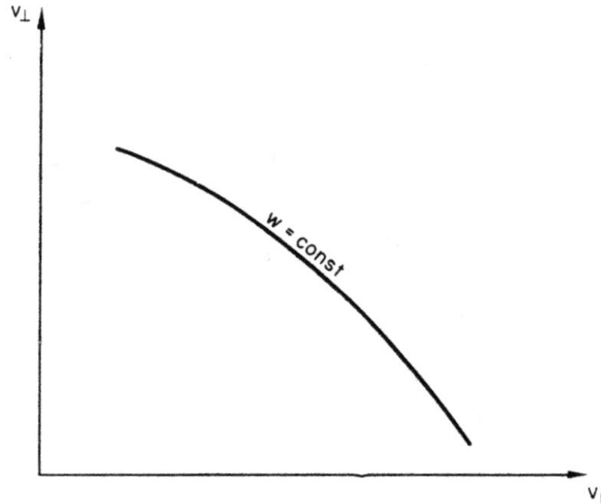

FIG. 9.2.1. Lines along which resonance particles diffuse when there is cyclotron excitation of longitudinal plasma oscillations by an electron beam under anomalous Doppler effect conditions.

and the diffusion is accompanied by an increase in the energy of the transverse motion of the particles in the beam and a decrease in the energy of their longitudinal motion, that is, by an incline in the plateau ($\partial f_{e0}/\partial v_{||} < 0$; see Fig. 9.2.2).

On the other hand, when $\partial f_{e0}/\partial v_{||} < 0$, oscillations which were excited in the first stage will be slowly damped. As a result these oscillations die down and the energy of the transverse motion of the particles in the beam increases considerably when we take into account the decrease in their translational motion.

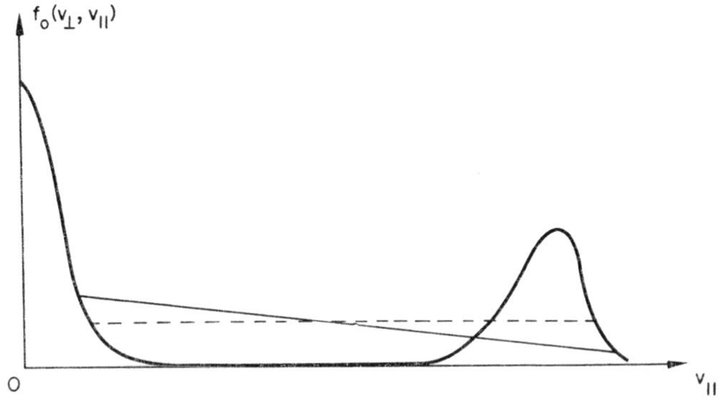

FIG. 9.2.2. Change of the electron distribution function during two-stage quasi-linear relaxation when an electron beam passes through a plasma in a strong magnetic field.

9.2.3. RELAXATION OF ONE-DIMENSIONAL WAVEPACKETS IN A MAGNETO-ACTIVE PLASMA

We shall consider in more detail the quasi-linear relaxation of one-dimensional wavepackets propagating at an angle θ to the magnetic field $\boldsymbol{B}_0$ (Rowlands, Sizonenko, and Stepanov, 1966). We shall assume that the spectral density of the wavepacket considered is non-zero in a narrow range of wavevectors, $k_1 < k < k_2$ with $k_2 - k_1 \ll k$.

The resonance regions $\Omega_v^{(l,j)}$ correspond to particles velocities along the magnetic field in the range

$$v_{\min}(l, j) < v_{||} < v_{\max}(l, j),$$

where

$$v_{\max(\min)} = \text{Max (Min)} \frac{\omega^{(j)}(\boldsymbol{k}) - l\,|\,\omega_{B\alpha}\,|}{k \cos \theta}.$$

It is clear that two regions $\Omega_v^{(l,j)}$ and $\Omega_v^{(l',j')}$ with $l, j \neq l', j'$ will not intersect when $v_{\max}(l', j') < v_{\min}(l, j)$ or $v_{\max}(l, j) < v_{\min}(l', j')$. In the remainder of this subsection we shall consider only the case when there are no intersections between the regions $\Omega_v^{(l,j)}$ so that in the right-hand side of eqn. (9.2.1.31) we can retain only one term in the sums over l and j.

First of all we shall study the Cherenkov interaction of resonance particles with one-dimensional wavepackets. Equation (9.2.1.31) then becomes

$$\frac{\partial f_{\alpha 0}}{\partial t} = \frac{\partial}{\partial v_{||}} D(v_{||}, v_\perp) \frac{\partial f_{\alpha 0}}{\partial v_{||}}, \tag{9.2.3.1}$$

where the diffusion coefficient D is equal to

$$D(v_{||}, v_{\perp}) = \pi \frac{e_\alpha^2}{m_\alpha^2}\left[\frac{|E_3^{(j)}J_0(\lambda)+i\eta_\alpha E_{k2}^{(j)}J_0'(|\lambda|)(v_\perp/v_{||})|^2}{|v_{||}\cos\theta - v_g|}\right]_{\omega^{(j)}(k,\theta)=k_{||}v_{||}}, \quad v_g = \frac{\partial\omega^{(j)}(k,\theta)}{\partial k}. \tag{9.2.3.2}$$

It follows from eqn. (9.2.3.1) that the particle diffusion due to interaction with waves under Cherenkov resonance conditions occurs only in the direction of the magnetic field. However, as the diffusion coefficient depends on $v_\perp$, the function $f_{\alpha 0}(v_{||}, v_\perp, t)$ will change at a different rate for different values of $v_\perp$ because of particle diffusion along $\boldsymbol{B}_0$.

If initially, $t = 0$, the function $f_{\alpha 0}$ were a product,

$$f_{\alpha 0}(v_{||}, v_\perp, 0) = f_M(v_\perp) f_{\alpha 0}(v_{||}, 0),$$

eqn. (9.2.3.1) will have in the case of longitudinal oscillations, $\boldsymbol{E}_k^{(j)} = -i\boldsymbol{k}\varphi_k^{(j)}$, a solution of the form

$$f_{\alpha 0}(v_{||}, v_\perp, t) = f_M(v_\perp) f_{\alpha 0}(v_{||}, J_0^2(\lambda)\, t). \tag{9.2.3.3}$$

It is clear that in that case the function $f_{\alpha 0}(v_{||}, v_\perp, t)$ will for long-wavelength oscillations ($|\lambda| = k_\perp v_\perp/|\omega_{B\alpha}| \ll 1$, $J_0(\lambda) \approx 1$) split into a product of distribution functions with respect to $v_{||}$ and $v_\perp$ for all t.

Let us now consider the interaction of resonance particles with one-dimensional wave-packets under cyclotron resonance conditions when $\omega^{(j)}(\boldsymbol{k}) = k_{||}v_\perp + l|\omega_{B\alpha}|$. Introducing instead of $v_{||}$ and $v_\perp$ new variables,

$$w_{1,2} = \frac{v_{||}^2}{\omega^{(j)}(\boldsymbol{k}) - l|\omega_{B\alpha}|} \pm \frac{v_\perp^2}{l|\omega_{B\alpha}|}, \tag{9.2.3.4}$$

we can write eqn. (9.2.1.31) in the form of a one-dimensional diffusion equation

$$\frac{\partial f_{\alpha 0}}{\partial t} = \frac{\partial}{\partial w_1} D(w_1, w_2)\frac{\partial f_{\alpha 0}}{\partial w_2}, \tag{9.2.3.5}$$

where

$$D = \frac{16\pi e_\alpha^2}{m_\alpha^2}\frac{|E_{k3}^{(j)}v_{||}J_l(|\lambda|) + i\eta_\alpha E_{k2}^{(j)}v_\perp J_l'(|\lambda|) + E_{k1}^{(j)}(l)|\omega_{B\alpha}|/k_\perp)J_l(|\lambda|)|^2}{[\omega^{(j)}(\boldsymbol{k})]^2|v_g - v_{||}\cos\theta|^2}. \tag{9.2.3.6}$$

Under cyclotron resonance conditions the diffusion in velocity space proceeds thus along the lines $w_2(v_{||}, v_\perp) = \text{constant}$.

The diffusion coefficient D given by (9.2.3.6) can vanish in several points

$$w_1 = q_m(w_2), \quad m = 0, 1, \ldots,$$

where $q_{m+1} > q_m$. The number of particles in the range $q_m < w_1 < q_{m+1}$ is conserved,

$$\frac{\partial}{\partial t}\int_{q_m}^{q_{m+1}} f_{\alpha 0}(w_1, w_2, t)\, dw_1 = D\frac{\partial f_{\alpha 0}}{\partial w_1}\Bigg|_{w_1 = q_m}^{w_1 = q_{m+1}} = 0,$$

and hence

$$\int_{q_m}^{q_{m+1}} dw_1 f_{\alpha 0} = \int_{q_m}^{q_{m+1}} dw_1 f_{\mathrm{M}},$$

where f_M is the initial (Maxwell) distribution function.

If the energy of the oscillations in the final state is non-zero, it follows from eqn. (9.2.3.5) that $\partial f_{\alpha 0}/\partial w_1 = 0$, that is, a plateau is formed along the lines $w_2 =$ constant:

$$f_{\alpha 0} = f_\infty(w_2).$$

If the initial velocity distribution function is Maxwellian, we have

$$f_\infty = \frac{4 \exp [-\{\omega^{(j)} - l\,|\,\omega_{B\alpha}\,|\}\, w_2/4v_\alpha^2]}{(2\pi)^{3/2}(q_{m+1}-q_m)\, v_\alpha \omega^{(j)}} \{\exp[-q_m\omega^{(j)}/4v_\alpha^2] - \exp[-q_{m+1}\omega^{(j)}/4v_\alpha^2]\}. \quad (9.2.3.7)$$

Let us, as an example, consider electromagnetic waves with frequency $\omega^{(j)}(\boldsymbol{k}) = l\,|\,\omega_{B\alpha}\,|$, propagating in a "cold" plasma when $kv_\alpha \ll |\,\omega_{B\alpha}\,|$. Using eqn. (9.2.1.24) to express the components $E_{k1}^{(j)}$ and $E_{k3}^{(j)}$ in terms of $E_{k2}^{(j)}$ we find

$$D = 16\pi \frac{e_\alpha^2}{m_\alpha^2} \frac{v_\perp^2\,|\,E_{k2}^{(j)}\,|^2\,|\,\eta_\alpha J_l'(|\,\lambda\,|) + (l/|\,\lambda\,|)\, J_l(|\,\lambda\,|)\,\xi\,|^2}{\omega^{(j)2}\,|\,v_g - v_{||}\cos\theta\,|}, \quad (9.2.3.8)$$

where

$$\xi = \frac{\varepsilon_1 - n^2}{\varepsilon_2}, \quad n = \frac{kc}{\omega^{(j)}(\boldsymbol{k})}, \quad \varepsilon_1 = 1 - \sum_\alpha \frac{\omega_{p\alpha}^2}{\omega^{(j)2} - \omega_{B\alpha}^2}, \quad \varepsilon_2 = -\sum_\alpha \frac{\omega_{p\alpha}^2 \omega_{B\alpha}}{\omega^{(j)}[\omega^{(j)2} - \omega_{B\alpha}^2]}.$$

For the narrow wavepacket considered for which the velocity of the resonance particles along the magnetic field changes within the range $v_1 < v_{||} < v_2$, the variable w_1 changes for given w_2 within the limits determined from the inequalities

$$v_1^2 \leqslant \tfrac{1}{2}[\omega^{(j)}(\boldsymbol{k}) - l\,|\,\omega_{B\alpha}\,|](w_1 + w_2) \leqslant v_2^2. \quad (9.2.3.9)$$

The diffusion coefficient (9.2.3.8) vanishes in the point $v_\perp = 0$, that is, $w_1 = w_2$, and also in the points $v_\perp = v_\nu \gtrsim |\,\omega_{B\alpha}\,|/k_\perp \gg v_\alpha$, determined by the equations $\eta_\alpha J_l'(|\,\lambda\,|) + lJ_l(|\,\lambda\,|)\,\xi = 0$. However, if $|\,\frac{1}{2}\pi - \theta\,| \gg kv_\alpha/|\,\omega_{B\alpha}\,|$, the points $v_\perp = v_\nu$ do not fall in the range (9.2.3.9) for which $D \neq 0$—and we restrict ourselves to considering that case.

The point $w_1 = w_2$ lies in the range (9.2.3.9), if

$$\left.\begin{aligned} &\frac{(v_{||} - v_1)\, v_{||}}{v_\perp^2} > \frac{\omega^{(j)}(\boldsymbol{k}) - l\,|\,\omega_{B\alpha}\,|}{2l\,|\,\omega_{B\alpha}\,|} \quad \text{when} \quad \omega^{(j)} - l\,|\,\omega_{B\alpha}\,| > 0, \\ \text{or} \quad &\frac{(v_2 - v_{||})\, v_{||}}{v_\perp^2} > \frac{l\,|\,\omega_{B\alpha}\,| - \omega^{(j)}(\boldsymbol{k})}{2l\,|\,\omega_{B\alpha}\,|} \quad \text{when} \quad \omega^{(j)} - l\,|\,\omega_{B\alpha}\,| < 0. \end{aligned}\right\} \quad (9.2.3.10)$$

If these inequalities are not satisfied, we get in the final state when $\omega^{(j)} - l\,|\,\omega_{B\alpha}\,| > 0$

$$f_\infty = \frac{1}{(2\pi)^{3/2}\, v_\alpha^3} \exp\left\{-\frac{v_\perp + v_1^2}{2v_\alpha^2} + \frac{v_1(v_{||} - v_1)\, l\,|\,\omega_{B\alpha}\,|}{v_\alpha^2[\omega^{(j)} - l\,|\,\omega_{B\alpha}\,|]}\right\} \frac{1 - e^{-x}}{x},$$

and when $\omega^{(j)} - l\,|\,\omega_{B\alpha}\,| < 0$ (9.2.3.11)

$$f_\infty = \frac{1}{(2\pi)^{3/2}\, v_\alpha^3} \exp\left\{-\frac{v_\perp^2 + v_2^2}{2v_\alpha^2} + \frac{v_2(v_2 - v_{||})\, l\,|\,\omega_{B\alpha}\,|}{v_\alpha^2[l\,|\,\omega_{B\alpha}\,| - \omega^{(j)}]}\right\} \frac{1 - e^{-x}}{x},$$

where

$$x = \frac{\omega^{(j)}(\boldsymbol{k})(v_2 - v_1)}{k_{||} v_\alpha^2}.$$

In this case diffusion therefore leads to a redistribution of the particles only with respect to $v_{||}$ in a narrow range of velocities.

If, however, the point $w_1 = w_2$ lies in the interval (9.2.3.9), we get in the equilibrium state when $\omega^{(j)}(\boldsymbol{k}) - l \mid \omega_{B\alpha} \mid > 0$

$$f_\infty = \frac{2 \exp\left\{-\frac{v_{||}^2}{2v_\alpha^2} + \frac{v_\perp^2[\omega^{(j)}(\boldsymbol{k}) - l \mid \omega_{B\alpha} \mid]}{2l \mid \omega_{B\alpha} \mid v_\alpha^2}\right\}\left[1 - \exp\left\{-\frac{\omega^{(j)}}{2v_\alpha^2}\left[\frac{v_2^2 - v_{||}^2}{\omega^{(j)} - l \mid \omega_{B\alpha} \mid} + \frac{v_\perp^2}{l \mid \omega_{B\alpha} \mid}\right]\right\}\right]}{(2\pi)^{3/2} v_\alpha \omega^{(j)}(\boldsymbol{k})[\{(v_2^2 - v_{||}^2)/[\omega^{(j)} - l \mid \omega_{B\alpha} \mid]\} + \{v_\perp^2 / l \mid \omega_{B\alpha} \mid\}]},$$

and when $\omega^{(j)}(\boldsymbol{k}) - l \mid \omega_{B\alpha} \mid < 0$ (9.2.3.12)

$$f_\infty = \frac{2 \exp\left\{-\frac{v_{||}^2}{2v_\alpha^2} + \frac{v_\perp^2[\omega^{(j)}(\boldsymbol{k}) - l \mid \omega_{B\alpha} \mid]}{2l \mid \omega_{B\alpha} \mid v_\alpha^2}\right\}\left[1 - \exp\left\{-\frac{\omega^{(j)}}{2v_\alpha^2}\left[\frac{v_1^2 - v_{||}^2}{\omega^{(j)} - l \mid \omega_{B\alpha} \mid} + \frac{v_\perp^2}{l \mid \omega_{B\alpha} \mid}\right]\right\}\right]}{(2\pi)^{3/2} v_\alpha \omega^{(j)}(\boldsymbol{k})[\{(v_1^2 - v_{||}^2)/[\omega^{(j)} - l \mid \omega_{B\alpha} \mid]\} + \{v_\perp^2 / l \mid \omega_{B\alpha} \mid\}]}.$$

Let us consider the distributions obtained, $f_{\alpha 0} = f_\infty(v_{||}, v_\perp)$, in the final state with a "plateau" along the direction of $w_1(v_{||}, v_\perp)$, given by (9.2.3.11) and (9.2.3.12) in the particular cases of "narrow" $(\mid v_2 - v_1 \mid \ll v_\alpha(k_{||} v_\alpha / \mid \omega_{B\alpha} \mid))$ and "broad" $(\mid v_2 - v_1 \mid \gg v_\alpha(k_{||} v_\alpha / \mid \omega_{B\alpha} \mid))$ wavepackets.

The inequality (9.2.3.10) can not be satisfied for a "narrow" wavepacket when $v_\perp \approx v_\alpha$. It is clear from eqns. (9.2.3.11) that the diffusion of particles due to the interaction with the waves in this case leads to the formation of a "plateau" with respect to $v_{||}$ in a narrow range and does not change the distribution with respect to $v_\perp$ for not too small $v_\perp$-values.

The inequality (9.2.3.10) is for "broad" wavepackets satisfied for all $v_{||}$ and $v_\perp$, except $v_{||}$-values very close to v_1 and v_2. The diffusion of particles due to the interaction with waves leads in this case to a great change in the distribution functions both with respect to $v_{||}$ and with respect to $v_\perp$. For that reason it is just the case of "broad" wavepackets which is of most interest for the heating of a plasma by means of high-frequency electromagnetic fields.

Let us consider the relaxation process for "broad" wavepackets in somewhat more detail. In the initial stage we can on the right-hand side of eqn. (9.2.1.31) neglect the terms $k_{||}\partial/\partial v_{||} \sim k_{||}/\Delta v_{||}$ as compared to the terms $(l \mid \omega_{B\alpha} \mid / v_\perp)(\partial/\partial v_\perp) \sim l \mid \omega_{B\alpha} \mid / v_\alpha^2$ $(\Delta v_{||} = v_2 - v_1)$. Equation (9.2.1.31) then becomes

$$\frac{\partial f_{\alpha 0}}{\partial t} = \frac{1}{v_\perp} \frac{\partial}{\partial v_\perp}\left(D_\perp \frac{\partial f_{\alpha 0}}{\partial v_\perp}\right), \tag{9.2.3.13}$$

where

$$D_\perp = \pi \frac{e_\alpha^2}{m_\alpha^2} \frac{v_\perp \left| E_{k1}^{(j)} \frac{l}{\mid \lambda \mid} J_l(\mid \lambda \mid) + i\eta_\alpha E_{k2}^{(j)} J_l'(\mid \lambda \mid) + E_{k3}^{(j)} \frac{v_{||}}{v_\perp} J_l(\mid \lambda \mid)\right|^2}{\mid v_g - v_{||} \cos \theta \mid}. \tag{9.2.3.14}$$

Under the action of the electromagnetic field of "broad" wavepackets diffusion of particles with respect to $v_\perp$ occurs, therefore, caused by cyclotron acceleration of the particles.

For long wavelengths and resonance particles with a not too large transverse velocity ($|\lambda| \ll 1$), eqn. (9.2.3.13) can be simplified:

$$\frac{\partial f_{\alpha 0}}{\partial \tau} = \frac{\partial}{\partial x}\left(x^{l}\frac{\partial f_{\alpha 0}}{\partial x}\right), \tag{9.2.3.15}$$

where

$$x = \frac{v_\perp^2}{2v_\alpha^2}, \quad \tau = \int_0^t dt\, D, \quad D = \pi\frac{e_\alpha^2}{m_\alpha^2}\left(\frac{k_\perp v_\alpha}{\omega_{B\alpha}}\right)^{2|l|-2}\frac{|E_{k1}^{(j)}+i\eta_\alpha E_{k2}^{(j)}+(k_\perp v_{\|}/l\,|\omega_{B\alpha}|)E_{k3}^{(j)}|^2}{2^{|l|}[(|l|-1)!]^2\,|v_g - v_{\|}\cos\theta|\,v_\alpha^2}.$$

Let us write down the solution of eqn. (9.2.3.15) for $l = 1$, 2, and 4:

$$f_{\alpha 0} = \frac{1}{(2\pi)^{3/2}v_\alpha^3(1+\tau)}\exp\left[-\frac{v_{\|}^2}{2v_\alpha^2}-\frac{x}{1+\tau}\right], \quad l = 1; \tag{9.2.3.16}$$

$$\frac{\partial f_{\alpha 0}}{\partial x} = -\frac{\sqrt{\pi}\exp[-v_{\|}^2/2v_\alpha^2]}{(2\pi)^{3/2}v_\alpha^3 x\sqrt{\tau}}\int_0^\infty dx'\,e^{-x'}\exp\left\{-\frac{4}{\tau}\left[\tau-\ln\frac{x}{x'}\right]^2\ln\frac{x}{x'}\right\}, \quad l = 2; \tag{9.2.3.17}$$

$$\frac{\partial f_{\alpha 0}}{\partial x} = -\frac{\sqrt{\pi}\exp[-v_{\|}^2/2v_\alpha^2]}{(2\pi)^{3/2}v_\alpha^3 x^3\sqrt{\tau}}\int_0^\infty dx'\,e^{-x'}\left\{\exp\left[-\left(\frac{1}{x'}-\frac{1}{x}\right)^2\frac{1}{4\tau}\right]\right.$$
$$\left.+\exp\left[-\left(\frac{1}{x'}+\frac{1}{x}\right)^2\frac{1}{4\tau}\right]\right\}, \quad l = 4. \tag{9.2.3.18}$$

It follows from (9.2.3.16) that the distribution with respect to $v_\perp$ remains Maxwellian when time goes on, while the transverse temperature of the particles increases,

$$T_\perp(t) = T_\alpha + T_\alpha\int_0^t D\,dt.$$

The damping of the oscillations is in this case linear, that is,

$$|E_k^{(j)}(t)|^2 = |E_k^{(j)}(0)|^2\exp(-2\gamma_L t), \tag{9.2.3.19}$$

where γ_L is the damping rate determined by the formulae of the linear theory for a Maxwellian velocity distribution.

In the case of two-fold resonance, $\omega^{(j)}(\boldsymbol{k}) \approx 2|\omega_{B\alpha}|$, the damping rate of long-wavelength "broad" wavepackets is given by the relation

$$\gamma^{(j)}(\boldsymbol{k}, t) = -\gamma_0\,e^{8\tau}. \tag{9.2.3.20}$$

We can easily show, using this relation, that the damping of the wave amplitude will proceed as follows:

$$|E_k^{(j)}(t)|^2 = |E_k^{(j)}(0)|^2\frac{1+\zeta}{\exp[2\gamma_L(1+\zeta)\,t]+\zeta}, \tag{9.2.3.21}$$

where

$$\zeta = \frac{\pi\, e_\alpha^2\, k_\perp^2\, |E_{k2}^{(j)}(0)|^2\, |1+\xi|^2}{4m_\alpha^2 \gamma_L\, |v_g - v_{||} \cos\theta|\, \omega_{B\alpha}^2}.$$

It follows from (9.2.3.20) and (9.2.3.21) that the damping of broad wavepackets with frequency $\omega^{(j)}(k) \approx 2\,|\omega_{B\alpha}|$ proceeds in the quasi-linear theory faster than in the linear theory.

9.2.4. EFFECT OF COLLISIONS ON QUASI-LINEAR RELAXATION AND CHERENKOV AND CYCLOTRON DAMPING OF OSCILLATIONS

We showed in Subsection 9.1.4 that the Coulomb collisions of resonance particles with the other particles in the plasma can appreciably affect the change of the distribution function with time and the magnitude of the derivative $\partial f_{\alpha 0}/\partial w$ which determines the speed of the collisionless damping of the oscillations.

We shall now study the effect of the collisions of resonance particles on their diffusion in velocity space and on the collisionless Cherenkov and cyclotron absorption of electromagnetic waves in a magneto-active plasma (Rowlands, Sizonenko, and Stepanov, 1966). In the case of narrow wavepackets when the number of resonance particles is small (small range, $\Delta v_{||} = v_2 - v_1$, of velocities occupied by resonance particles: $\Delta v_{||} \ll v_{||}$ for $v_1 < v_{||} < v_2$) we can neglect the collision of resonance particles with one another. For resonance electrons with a velocity $v_{||}$ appreciably larger than the thermal ion velocity v_i we can write the collision integral (9.1.4.1) in the form

$$\left[\frac{\partial f_{e0}}{\partial t}\right]_c = \frac{\partial}{\partial v_{||}}\left[D_c \frac{\partial (f_{e0} - f_{eM})}{\partial v_{||}}\right]. \tag{9.2.4.1}$$

where f_{eM} is the Maxwellian electron distribution function while the diffusion coefficient D_c is equal to

$$D_c = \frac{v_e^3}{\tau_c}\left(\frac{v_\perp^2}{v^3} + \frac{\partial^2 \Psi_{eM}}{\partial v_{||}^2}\right), \tag{9.2.4.2}$$

$$\frac{1}{\tau_c} = \frac{2\pi\, e^4 n_0 L}{m_e^2 v_e^3}, \tag{9.2.4.3}$$

$$\Psi_{eM} = \int d^3v'\, f_{eM}(v')\,|v - v'| = \sqrt{\frac{2}{\pi}}\, v_e \exp\left(-\frac{v^2}{2v_e^2}\right) + \frac{v_e^2 + v^2}{v}\,\Phi\left(\frac{v}{\sqrt{2}\, v_e}\right),$$

$$\Phi(x) = \frac{2}{\sqrt{\pi}} \int_0^x dt \exp(-t^2). \tag{9.2.4.4}$$

The first term in (9.2.4.2) takes the collisions of the electrons with the ions into account and the second one those with the electrons themselves. If $v_{||} \gg v_e$, expression (9.2.4.2) takes the simple form

$$D_e = \frac{2v_e^3}{\tau_c v_{||}^3}(v_\perp^2 + v_e^2). \tag{9.2.4.5}$$

For resonance ions we may take into account merely their collisions with the other (non-resonance) ions and neglect the collisions with the electrons. In that case we get from (9.1.4.1)

$$\left[\frac{\partial f_{i0}}{\partial t}\right]_c = \frac{\partial}{\partial v_{||}}\left[D_c \frac{\partial(f_{i0}-f_{iM})}{\partial v_{||}}\right], \tag{9.2.4.6}$$

where

$$D_c = \frac{v_i^3}{\tau_c}\frac{\partial^2 \Psi_{iM}}{\partial v_{||}^2}, \quad \frac{1}{\tau_c} = \frac{2\pi e^4 n_0 L}{m_i^2 v_i^3}, \tag{9.2.4.7}$$

and Ψ_{iM} is determined by eqn. (9.2.4.4) in which we must replace v_e by v_i. When $v_{||} \gg v_i$ eqn. (9.2.4.7) becomes

$$D_c = \frac{1}{\tau_c}\frac{v_i^3}{v_{||}^3}(v_\perp^2+2v_i^2). \tag{9.2.4.8}$$

As in Subsection 9.1.4 we assumed, when deriving eqns. (9.2.4.1) to (9.2.4.8) that the electron and ion distribution functions differ little from Maxwellian distribution functions in the resonance region, while they are Maxwellian outside the resonance region, and we neglected the quantities $\partial(f_{\alpha 0}-f_{\alpha M})/\partial v_\perp \sim (f_{\alpha 0}-f_{\alpha M})/v_\alpha$ in comparison with $\partial(f_{\alpha 0}-f_{\alpha M})/\partial v_{||} \sim (f_{\alpha 0}-f_{\alpha M})/\Delta v_{||}$.

Let us first of all consider the effect of the collisions of resonance particles on the Cherenkov damping of narrow wavepackets propagating at an angle θ to the direction of the external magnetic field in a low-pressure plasma ($4\pi n_0 T_\alpha \ll B_0^2$). We shall only consider long-wavelength oscillations with a wavelength much larger than the Larmor radius of the particles. Examples of such waves are different branches of longitudinal waves, the Alfvén and the fast magneto-sound waves, considered in Chapter 5. As the phase velocity of these waves is much larger than the ion thermal velocity we can neglect the exponentially small Cherenkov damping due to the ions and consider only the Cherenkov damping due to the electrons.

Retaining in eqn. (9.2.1.31) for electrons only a single term with $l = 0$ and adding to the right-hand side of this equation the collision integral in the form (9.2.4.2) we write the equation for f_{e0} in the form

$$\frac{\partial f_{e0}}{\partial t} = \frac{\partial}{\partial v_{||}}\left[D\frac{\partial f_{e0}}{\partial v_{||}} + D_c\frac{\partial(f_{e0}-f_{eM})}{\partial v_{||}}\right], \tag{9.2.4.9}$$

where D_c is given by eqn. (9.2.4.3) and

$$D = \pi\frac{e^2}{m_e^2}\int d^3\boldsymbol{k}\left|E_{k3}^{(j)} + i\frac{v_\perp}{2v_{||}}E_{k2}^{(j)}\right|^2 \delta\{\omega^{(j)}(\boldsymbol{k}) - k_{||}v_{||}\}. \tag{9.2.4.10}$$

In a quasi-equilibrium state $\partial f_{e0}/\partial t \approx 0$ and when the diffusion of electrons through interaction with the waves is counterbalanced by collisions we get from eqn. (9.2.4.9)

$$\frac{\partial f_{e0}}{\partial v_{||}} = \frac{\partial f_{eM}}{\partial v_{||}}\frac{1}{1+(D/D_c)}. \tag{9.2.4.11}$$

The anti-Hermitean term in the dispersion equation for longitudinal oscillations,

$$\sum_{i,j} \varepsilon_{ij} k_i k_j = 0,$$

and, hence, the damping rate, is proportional to the integral

$$\int_0^\infty dv_\perp v_\perp \left.\frac{\partial f_{e0}}{\partial v_{||}}\right|_{v_{||}=\omega^{(j)}(\boldsymbol{k})/k_{||}}.$$

Using eqn. (9.2.4.11) we get the following expression for $\gamma^{(j)}(\boldsymbol{k})$:

$$\gamma^{(j)}(\boldsymbol{k}) = -\gamma_{\mathrm{L}}\Phi, \tag{9.2.4.12}$$

where γ^{L} is the damping rate of longitudinal oscillations for a plasma with a Maxwellian electron velocity distribution while

$$\Phi = \int_0^\infty \frac{e^{-x}dx}{1+\eta(x)}, \quad x = \frac{v_\perp^2}{2v_{\mathrm{e}}^2}, \quad \eta = \frac{D(v_{||},\, v_\perp)}{D_{\mathrm{c}}(v_{||},\, v_\perp)}. \tag{9.2.4.13}$$

If the phase velocity of the oscillations considered is appreciably larger than the electron thermal velocity ($v_{||} \gg v_{\mathrm{e}}$), $D/D_{\mathrm{c}} = y/(x+\frac{1}{2})$, where

$$y = \frac{D}{D_0}, \quad D_0 = \frac{4v_{\mathrm{e}}^5}{\tau_{\mathrm{c}} v_{||}^3}, \quad D = \pi\frac{e^3}{m_{\mathrm{e}}^2}\int d^3\boldsymbol{k}\,|E_{\boldsymbol{k}3}^{(j)}|^2\,\delta\{\omega^{(j)}-k_{||}v_{||}\}.$$

In that case

$$\Phi = 1 + ye^{1/2+y}\,\mathrm{Ei}(-\tfrac{1}{2}-y), \quad \mathrm{Ei}(-x) = -\int_x^\infty dt\,\frac{e^{-t}}{t}.$$

For large-amplitude oscillations ($y \gg 1$) we have $\Phi \approx 3/2y$ and in that case the damping rate is proportional to the collision frequency.

Equation (9.2.4.12) determines the Cherenkov damping rate of longitudinal plasma oscillations in the quasi-linear approximation, taking electron collisions into account. One can similarly easily obtain the Cherenkov damping rate of Alfvén and fast magneto-sound waves in a low-pressure plasma in the quasi-linear approximation, taking electron collisions into account in a quasi-stationary state when the diffusion of the electrons due to interaction with the waves is counterbalanced by collisions. To do this we must use eqn. (9.2.4.11) and the dispersion eqn. (5.4.1.4). We just write down the final result:

$$\gamma^{(j)}(\boldsymbol{k}) = -\frac{\sqrt{\pi}\,\omega_{\mathrm{pi}}^2 k_\perp^2 v_{\mathrm{s}}^2 Q}{2\omega^{(j)}\omega_{B\mathrm{i}}^2 \varepsilon_1 P}, \tag{9.2.4.14}$$

where

$$\left.\begin{aligned} &v_{\mathrm{s}} = \sqrt{\frac{T_{\mathrm{e}}}{m_{\mathrm{i}}}}, \quad \varepsilon_1 = \frac{\omega_{\mathrm{pi}}^2}{\omega_{B\mathrm{i}}^2 - \omega^{(j)2}}, \quad P = \frac{(1+\cos^2\theta)\,k^2c^2\omega_{B\mathrm{i}}^2}{\omega^{(j)2}(\omega_{B\mathrm{i}}^2-\omega^{(j)2})} - \varepsilon_1\left[2 - \frac{\omega^{(j)2}}{\omega_{B\mathrm{i}}^2}\right], \\ &Q = 2\left[\frac{k_{||}^2c^2}{\omega^{(j)2}} - \varepsilon_1\right]\Phi_1 + \frac{1}{\Psi^2+\pi\Phi_2^2}\left\{\frac{k_{||}^2c^2}{\omega^{(j)2}-\omega_{B\mathrm{i}}^2}\left[2\Phi_3 - \Phi_2 - \frac{k^2c^2\omega_{B\mathrm{i}}^2}{\omega^{(j)2}\omega_{\mathrm{pi}}^2}\Phi_2\right]\right. \\ &\qquad \left. + \left(\varepsilon_1 - \frac{k_{||}^2c^2}{\omega^{(j)2}}\right)[2\Psi^2\Phi_3 - \Psi^2\Phi_2 + \pi\Phi_2\Phi_3^2]\right\}, \end{aligned}\right\} \tag{9.2.4.15}$$

$$\Phi_1 = -\frac{\pi^{3/2}}{4v_e^2}\int_0^\infty dv_\perp v_\perp^5 \frac{\partial f_{e0}}{\partial v_{||}}, \quad \Phi_2 = -2\pi^{3/2}v_e^2\int_0^\infty dv_\perp v_\perp \frac{\partial f_{e0}}{\partial v_{||}},$$

$$\Phi_3 = -\pi^{3/2}\int_0^\infty dv_\perp v_\perp^3 \frac{\partial f_{e0}}{\partial v_{||}}, \quad \Psi = 1-2z_e \exp(-z_e^2)\int_0^{z_e} \exp(t^2)\, dt, \quad z_e = \frac{\omega^{(j)}(\boldsymbol{k})}{\sqrt{2}\, k_{||}v_e}.$$

If the phase velocity of the Alfvén and fast magneto-sound waves is much larger than the electron thermal velocity, eqn. (9.2.4.15) can be simplified:

$$Q = \frac{2\varepsilon_1\omega^{(j)2}z_e^3}{v_e^2\omega_{Bi}^2}\left[\Phi+\frac{k^2c^2\omega_{Bi}^2}{\omega^{(j)2}\omega_{pi}^2}\Phi-2\tilde{\Phi}\right]\exp(-z_e^2), \tag{9.2.4.16}$$

where $\Phi = \Phi(y)$ is given by formula (9.2.4.13) and

$$\tilde{\Phi}(y) = 1-y-y(y+\tfrac{1}{2})\, e^{y+1/2}\mathrm{Ei}(-y-\tfrac{1}{2}),$$

$y = D/D_0$, D is given by formula (9.2.4.10) in which the term $\propto E_{k2}^{(j)}$ has been dropped.

For weak fields, $D/D_0 \ll 1$, expression (9.2.4.14) goes over into equation (5.4.1.5) for the damping rate of the linear theory for a plasma with a Maxwellian electron velocity distribution. For strong fields, $D/D_0 \gg 1$, the damping rate (9.2.4.14) is smaller than the damping rate of the linear theory by a factor D/D_0.

Let us now study the effect of Coulomb collisions on the cyclotron damping of one-dimensional wavepackets. We shall restrict our considerations to narrow wavepackets, $\Delta v_{||} \ll v_\alpha(k_{||}v_\alpha/|\omega_{B\alpha}|) \ll v_\alpha$. In that case the diffusion of particles due to interaction with the waves is practically along the magnetic field. Neglecting particle diffusion in directions at right angles to the magnetic field we write eqn. (9.2.1.31), taking collisions into account, in the form (9.2.4.9), where

$$D = \frac{e_\alpha^2}{m_\alpha^2}\left(\frac{k_\perp v_\alpha}{\omega_{B\alpha}}\right)^{2l-4}\frac{v_\alpha^4|1+\xi|^2|E_{k2}^{(j)}|^2k_{||}^2}{2^{l+1/2}[(l-1)!]^2\omega^{(j)2}|v_g-v_{||}\cos\theta|}\left(\frac{v_\perp}{\sqrt{2}\, v_\alpha}\right)^{2l} \tag{9.2.4.17}$$

In the final state when the diffusion of particles due to interaction with the waves is counterbalanced by collisions, the magnitude of the derivative of the distribution function $\partial f_{\alpha0}/\partial v_{||}$ will be determined by eqn. (9.2.4.11). Bearing in mind that the anti-Hermitean terms in the tensor ε_{ij} and hence the damping rate are proportional to the integral

$$\int dv_\perp v_\perp^{2l+1}\frac{\partial f_{\alpha0}(v_{||}, v_\perp)}{\partial v_{||}},$$

and using eqn. (9.2.4.11), we get

$$\gamma^{(j)}(\boldsymbol{k}) = -\gamma_0 F, \tag{9.2.4.18}$$

where

$$F = \frac{1}{l!}\int_0^\infty dx\, x^l \frac{e^{-x}}{1+\eta}, \quad \eta = \frac{D}{D_c}, \quad x = \frac{v_\perp^2}{2v_\alpha^2}.$$

In the case of electron cyclotron resonance, $\omega^{(j)}(\boldsymbol{k}) \approx l|\omega_{Be}|$, the quantity η simplifies when $v_{||} \gg v_e$:

$$\eta = \eta_0 \frac{x^l}{1+2x}, \quad \eta_0 = \left.\frac{D(v_\perp)\,\tau_c v_{||}^3}{2v_e^5}\right|_{v_\perp = \sqrt{2}\,v_e}. \tag{9.2.4.19}$$

Using this expression for η we get when $\omega \approx |\omega_{Be}|$

$$F = x\,[3-2x+x(1-2x)\,e^x\,\mathrm{Ei}(-x)], \qquad v_{||} \gg v_e, \tag{9.2.4.20}$$

where $x = (1+\eta_0)^{-1}$. For weak fields ($\eta_0 \ll 1$) $F = 1$ and $\gamma = -\gamma_0$; for strong fields ($\eta_0 \gg 1$) the damping rate is diminished, $F \approx 3/\eta_0 \ll 1$.

If $\omega \approx 2|\omega_{Be}|$, we find easily, using (9.2.4.19), that

$$F = \frac{3}{2\eta_0}\left\{1-\frac{4}{3\eta_0}-\frac{2(\eta_0-2)}{\eta_0\sqrt{(1-\eta_0)}}\,[(x_1+p)\exp(-x_1)\,\mathrm{Ei}\,(x_1)-(x_2+p)\exp(-x_2)\,\mathrm{Ei}\,(x_2)]\right\}, \tag{9.2.4.21}$$

where

$$p = \frac{\eta_0-4}{4(\eta_0-2)}, \qquad x_{1,2} = \frac{-1\pm\sqrt{(1-\eta_0)}}{\eta_0}.$$

If $\eta_0 \ll 1$, it follows from this that $F \approx 1$, while for $\eta_0 \gg 1$, we have $F \approx 3/2\eta_0 \ll 1$. In the strong field region ($\eta_0 \gg 1$) the function F asymptotically approaches the value

$$F = \frac{3}{l!}\frac{1}{\eta_0}, \qquad \eta_0 \gg 1, \quad v_{||} \gg v_e. \tag{9.2.4.22}$$

For ion cyclotron resonance, $\omega^{(j)}(\boldsymbol{k}) \approx l\omega_{Bi}$ we have when $v_{||} \gg v_i$

$$\eta = \eta_0 \frac{x^l}{1+x}, \quad \eta_0 = \left.\frac{D(v_\perp)\,\tau_c v_{||}^3}{2v_i^5}\right|_{v_{||} = \sqrt{2}\,v_i} \tag{9.2.4.23}$$

Using this equation, we get for $\omega^{(j)}(\boldsymbol{k}) \approx \omega_{Bi}$ and $\omega^{(j)}(\boldsymbol{k}) \approx 2\omega_{Bi}$

$$F = x[2-x+e^x\,\mathrm{Ei}\,(-x)\,x(1-x)], \quad x = \frac{1}{1+\eta_0}, \quad l = 1; \tag{9.2.4.24}$$

$$F = \frac{1}{\eta_0}\left\{1-\frac{1}{2\eta_0}+\frac{2\eta_0-1}{\eta_0\sqrt{(1-4\eta_0)}}\,[(x_1+p)\exp(-x_1)\,\mathrm{Ei}\,(x_1)-(x_2+p)\exp(-x_2)\,\mathrm{Ei}\,(x_2)]\right\}, \quad l = 2; \tag{9.2.4.25}$$

where

$$x_{1,2} = \frac{-1\pm\sqrt{(1-4\eta_0)}}{2\eta_0}, \qquad p = \frac{\eta_0-1}{2\eta_0-1}.$$

In the strong field case ($\eta_0 \gg 1$) we get when $\omega^{(j)}(\boldsymbol{k}) \approx l\omega_{Bi}$ and $v_{||} \gg v_i$

$$F \approx 2/(l!\,\eta_0).$$

For narrow wavepackets the diffusion of particles due to interaction with the waves under cyclotron resonance conditions therefore always leads to a decrease in the damping rate and in strong fields the cyclotron damping is determined by the collisions,

$$\gamma^{(j)}(\boldsymbol{k}) \sim -\gamma_0/\eta_0,$$

that is, the damping rate is proportional to the collision frequency.

CHAPTER 10

Non-linear Wave-Particle Interactions

10.1. Kinetic Equation for the Waves

10.1.1. NON-LINEAR EQUATION FOR THE WAVE AMPLITUDE

The quasi-linear approximation considered in the preceding chapter takes, from all different non-linear processes, into account only processes of the lowest order in the plasma oscillation energy and especially the reaction of the oscillations on the averaged distribution function and also the effect of a slow change in the particle distribution function on the rate of growth of the oscillations.

When the amplitude of the oscillations increases it becomes necessary to take into account higher-order non-linear processes—the non-linear interaction of waves and the scattering of waves by the plasma particles. It is convenient to describe these processes by means of the so-called kinetic equation for the waves; we shall now turn to its derivation.†

For the sake of simplicity we shall restrict ourselves to the case of longitudinal plasma oscillations when there are no external fields; we shall start from the kinetic equations for the distribution functions F_α for particles of the αth kind (α = e, i) and the equations of electrostatics:

$$\frac{\partial F_\alpha}{\partial t}+(\boldsymbol{v}\cdot\nabla)F_\alpha-\frac{e_\alpha}{m_\alpha}\left(\nabla\varphi\cdot\frac{\partial F_\alpha}{\partial \boldsymbol{v}}\right)=0, \tag{10.1.1.1}$$

$$\nabla^2\varphi=-4\pi\sum_\alpha e_\alpha\int F_\alpha\, d^3v. \tag{10.1.1.2}$$

Here F_α is the exact distribution function, not the one averaged over fluctuations or over a small time interval,

$$F_\alpha(\boldsymbol{v},\boldsymbol{r},t)=\frac{1}{V}\sum_j \delta[\boldsymbol{r}-\boldsymbol{r}_j(t)]\,\delta[\boldsymbol{v}-\boldsymbol{v}_j(t)], \tag{10.1.1.3}$$

where $\boldsymbol{r}_j$ and $\boldsymbol{v}_j$ are the coordinates and velocity of the jth particle, while the summation is over all particles of the αth kind; V is the volume of the plasma and e_α and m_α are the charge and mass of a particle of the αth kind.

† Kadomtsev and Petviashvili (1963), Galeev and Karpman (1963; Karpman, 1964a), and Silin (1964b) developed a method to study non-linear processes in a plasma by means of a kinetic equation for waves. There is a large literature devoted to non-linear wave-particle interactions and this is reflected in a number of review articles and monographs (Kadomtsev, 1965; Galeev, Karpman, and Sagdeev, 1965; Tsytovich, 1970).

For this definition of the distribution function eqn. (10.1.1.1) is clearly exact and completely equivalent to the equations of motion

$$\dot{\boldsymbol{r}}_j(t) = \boldsymbol{v}_j(t), \quad \dot{\boldsymbol{v}}_j(t) = -\frac{e_\alpha}{m_\alpha} \nabla \varphi \bigg|_{\boldsymbol{r} = \boldsymbol{r}_j}. \tag{10.1.1.4}$$

Averaging eqn. (10.1.1.1) over initial conditions and bearing in mind that, if there is no external electric field (and this is the only case we shall consider here),

$$\langle \varphi \rangle = 0,$$

we get equations for the averaged distribution functions $F_\alpha^{(0)}$:

$$\frac{\partial F_\alpha^{(0)}}{\partial t} + (\boldsymbol{v} \cdot \nabla) F_\alpha^{(0)} = \frac{e_\alpha}{m_\alpha} \left(\frac{\partial}{\partial \boldsymbol{v}} \cdot \langle f_\alpha \nabla \varphi \rangle \right), \tag{10.1.1.5}$$

where $f_\alpha = F_\alpha - F_\alpha^{(0)}$ is the oscillating perturbation of the distribution function and the $\langle \ldots \rangle$ brackets indicate averaging.

Substracting eqn. (10.1.1.5) from eqn. (10.1.1.1) and changing to Fourier components,

$$f(\boldsymbol{k}, \omega) = \frac{1}{(2\pi)^4} \int e^{-i(\boldsymbol{k} \cdot \boldsymbol{r}) + i\omega t} f(\boldsymbol{r}, t)\, d^3\boldsymbol{r}\, dt,$$

we get the following equation for the function $f_\alpha(\boldsymbol{k}, \omega) \equiv f_\alpha$:

$$-i[\omega - (\boldsymbol{k} \cdot \boldsymbol{v})] f_\alpha(\boldsymbol{k}, \omega) - i \frac{e_\alpha}{m_\alpha} \varphi(\boldsymbol{k}, \omega) \left(\boldsymbol{k} \cdot \frac{\partial F_\alpha^{(0)}}{\partial \boldsymbol{v}} \right)$$

$$= i \frac{e_\alpha}{m_\alpha} \left(\frac{\partial}{\partial \boldsymbol{v}} \cdot \int \boldsymbol{k}' \{ \varphi(\boldsymbol{k}', \omega') f_\alpha(\boldsymbol{k} - \boldsymbol{k}', \omega - \omega') - \langle \varphi(\boldsymbol{k}', \omega') f_\alpha(\boldsymbol{k} - \boldsymbol{k}', \omega - \omega') \rangle \} \right) d^3\boldsymbol{k}'\, d\omega', \tag{10.1.1.6}$$

where $\varphi(\boldsymbol{k}, \omega)$ is the Fourier component of the electrostatic potential which is connected with the functions f_α through the formula

$$\varphi(\boldsymbol{k}, \omega) = \frac{4\pi}{k^2} \sum_\alpha e_\alpha \int f_\alpha(\boldsymbol{k}, \omega)\, d^3\boldsymbol{v}. \tag{10.1.1.7}$$

Now introducing the operator

$$g_\alpha(\boldsymbol{k}, \omega) = -\frac{e_\alpha}{m_\alpha} [\omega - (\boldsymbol{k} \cdot \boldsymbol{v}) + io]^{-1} \frac{\partial}{\partial \boldsymbol{v}}, \tag{10.1.1.8}$$

where o indicates an infinitesimal positive quantity, introduced to indicate how to go around the pole when $\omega = (\boldsymbol{k} \cdot \boldsymbol{v})$, we can write this equation in a more compact form:

$$f_\alpha(\boldsymbol{k}, \omega) - \varphi(\boldsymbol{k}, \omega)(\boldsymbol{k} \cdot \boldsymbol{g}_\alpha(\boldsymbol{k}, \omega)) F_\alpha^{(0)} = \int (\boldsymbol{k}' \cdot \boldsymbol{g}_\alpha(\boldsymbol{k}, \omega)) \{ \varphi(\boldsymbol{k}', \omega') f_\alpha(\boldsymbol{k} - \boldsymbol{k}', \omega - \omega')$$
$$- \langle \varphi(\boldsymbol{k}', \omega') f_\alpha(\boldsymbol{k} - \boldsymbol{k}', \omega - \omega') \rangle \} d^3\boldsymbol{k}'\, d\omega'. \tag{10.1.1.9}$$

Equation (10.1.1.9) is convenient for a study of non-linear processes in a plasma in the case of so-called weak-turbulence states, that is, such states of the plasma that the energy of the plasma oscillations is small compared to the particle energies. In that case we can use a series expansion in powers of the amplitude of the oscillations.

Usually weak-turbulence states arise in a plasma when oscillations grow with a growth rate—given by the linear theory—small compared to the frequency, $\gamma \ll \omega$. We note, however, that the inequality $\gamma \ll \omega$ is by no means a sufficient condition for the occurrence of weak turbulence. If, as occurs for so-called explosive instabilities (Rosenbluth, Coppi, and Sudan, 1968), non-linear effects lead to an increase in the growth rate with increasing oscillation amplitude, strong turbulence will develop in the plasma, even when the values of the growth rate given by the linear theory are small.

We can in the case of weak turbulence find the functions f_α in any (finite) approximation in the amplitude of the oscillations, by iterating eqn. (10.1.1.9). To take the interaction between the waves and the non-linear wave-particle interactions in the plasma into account —in the first non-vanishing approximation—it is sufficient in iterating eqn. (10.1.1.9) to restrict ourselves to terms which are cubic in the amplitude of the oscillations. Introducing for the sake of convenience the notation

$$g_\alpha \equiv g_\alpha(\boldsymbol{k}, \omega), \quad \varphi^n \equiv \varphi(\boldsymbol{k}_n, \omega_n), \quad g_\alpha^n \equiv g_\alpha(\boldsymbol{k}_n, \omega_n), \qquad n = 1, 2, 3, \ldots,$$

we get

$$\begin{aligned} f_\alpha(\boldsymbol{k}, \omega) = {} & (\boldsymbol{k}\cdot\boldsymbol{g}_\alpha) F_\alpha^{(0)} \varphi(\boldsymbol{k}, \omega) + \int (\boldsymbol{k}_1\cdot\boldsymbol{g}_\alpha)(\boldsymbol{k}_2\cdot\boldsymbol{g}_\alpha^2) F_\alpha^{(0)}\{\varphi^1\varphi^2 - \langle\varphi^1\varphi^2\rangle\}\, \delta(\boldsymbol{k}-\boldsymbol{k}_1-\boldsymbol{k}_2) \\ & \times \delta(\omega-\omega_1-\omega_2)\, d^3\boldsymbol{k}_1\, d^3\boldsymbol{k}_2\, d\omega_1\, d\omega_2 + \int (\boldsymbol{k}_1\cdot\boldsymbol{g}_\alpha)(\boldsymbol{k}_2\cdot\boldsymbol{g}_\alpha(\boldsymbol{k}-\boldsymbol{k}_1, \omega-\omega_1)(\boldsymbol{k}_3\cdot\boldsymbol{g}_\alpha^3)) \\ & \times F_\alpha^{(0)}\{\varphi^1\varphi^2\varphi^3 - \varphi^1\langle\varphi^2\varphi^3\rangle - \langle\varphi^1\varphi^2\varphi^3\rangle\}\, \delta(\boldsymbol{k}-\boldsymbol{k}_1-\boldsymbol{k}_2-\boldsymbol{k}_3)\, \delta(\omega-\omega_1-\omega_2-\omega_3) \\ & \times d^3\boldsymbol{k}_1\, d^3\boldsymbol{k}_2\, d^3\boldsymbol{k}_3\, d\omega_1\, d\omega_2\, d\omega_3. \end{aligned} \tag{10.1.1.10}$$

Substituting (10.1.1.10) into (10.1.1.7) and bearing in mind that the dielectric permittivity of the plasma is connected with the averaged particle distribution function through the relation

$$\varepsilon(\boldsymbol{k}, \omega) = 1 - \frac{4\pi}{k^2} \sum_\alpha e_\alpha \int (\boldsymbol{k}\cdot\boldsymbol{g}_\alpha(\boldsymbol{k}, \omega))\, F_\alpha^{(0)}\, d^3\boldsymbol{v},$$

we get the following equation for the electrostatic potential:

$$\begin{aligned} \varepsilon(\boldsymbol{k}, \omega)\varphi(\boldsymbol{k}, \omega) = {} & \int V(\boldsymbol{k}, \omega; \boldsymbol{k}_1, \omega_1)\{\varphi^1\varphi^2 - \langle\varphi^1\varphi^2\rangle\}\, \delta(\boldsymbol{k}-\boldsymbol{k}_1-\boldsymbol{k}_2)\, \delta(\omega-\omega_1-\omega_2)\, d^3\boldsymbol{k}_1\, d^3\boldsymbol{k}_2 \\ & \times d\omega_1\, d\omega_2 + \int V(\boldsymbol{k}, \omega; \boldsymbol{k}_1, \omega_1; \boldsymbol{k}_2, \omega_2)\{\varphi^1\varphi^2\varphi^3 - \varphi^1\langle\varphi^2\varphi^3\rangle - \langle\varphi^1\varphi^2\varphi^3\rangle\} \\ & \times \delta(\boldsymbol{k}-\boldsymbol{k}_1-\boldsymbol{k}_2-\boldsymbol{k}_3)\, \delta(\omega-\omega_1-\omega_2-\omega_3)\, d^3\boldsymbol{k}_1\, d^3\boldsymbol{k}_2\, d^3\boldsymbol{k}_3\, d\omega_1\, d\omega_2\, d\omega_3, \end{aligned} \tag{10.1.1.11}$$

where

$$V(\boldsymbol{k}, \omega; \boldsymbol{k}_1, \omega_1) = \frac{4\pi}{k^2} \sum_\alpha e_\alpha \int (\boldsymbol{k}_1\cdot\boldsymbol{g}_\alpha)(\boldsymbol{k}_2\cdot\boldsymbol{g}_\alpha^2)\, F_\alpha^{(0)}\, \delta(\boldsymbol{k}-\boldsymbol{k}_1-\boldsymbol{k}_2)\, \delta(\omega-\omega_1-\omega_2)\, d^3\boldsymbol{v}\, d^3\boldsymbol{k}_2\, d\omega_2,$$

$$\begin{aligned} V(\boldsymbol{k}, \omega; \boldsymbol{k}_1\omega_1; \boldsymbol{k}_2, \omega_2) = {} & \frac{4\pi}{k^2} \sum_\alpha e_\alpha \int (\boldsymbol{k}_1\cdot\boldsymbol{g}_\alpha)(\boldsymbol{k}_2\cdot\boldsymbol{g}(\boldsymbol{k}-\boldsymbol{k}_1, \omega-\omega_1))(\boldsymbol{k}_3\cdot\boldsymbol{g}_\alpha^3)\, F_\alpha^{(0)} \\ & \times \delta(\boldsymbol{k}-\boldsymbol{k}_1-\boldsymbol{k}_2-\boldsymbol{k}_3)\, \delta(\omega-\omega_1-\omega_2-\omega_3)\, d^3\boldsymbol{v}\, d^3\boldsymbol{k}_3\, d\omega_3. \end{aligned} \tag{10.1.1.12}$$

10.1.2. EQUATION FOR THE CORRELATION FUNCTION

The non-linear eqn. (10.1.1.11) for the electrostatic potential is a consequence of the dynamic eqns. (10.1.1.4)—or, what amounts to the same, of the kinetic eqn. (10.1.1.1)—and the equations of electrostatics. We draw attention to the fact that in deriving this equation we implicitly used the assumption of the random nature of the plasma oscillations. Indeed, when solving the kinetic equations we chose particular solutions of the kind $f_\alpha = (\boldsymbol{k}\varphi \cdot \boldsymbol{g}_\alpha) F_\alpha^{(0)}$ rather than general solutions, and as a result eqns. (10.1.1.9) did not contain the initial perturbation of the particle distribution functions. In other words, in deriving eqn. (10.1.1.11) we tacitly implied that the system is able to "forget" the explicit form of the initial perturbation so that the function φ is determined by the properties of the system itself rather than by the initial perturbance. We shall not dwell on an elucidation of the criteria for when such a "loss of memory" of the initial conditions (stochastization of the oscillations) occurs.†

As we are interested in the spectral distribution of random oscillations we can change from eqn. (10.1.1.11) to an equation which is more convenient for the description of such oscillations—the so-called kinetic equation for the waves. To do this we multiply eqn. (10.1.1.11) by $\varphi^*(\boldsymbol{k}', \omega')$ and average over the random phases:

$$\begin{aligned}\varepsilon(\boldsymbol{k}, \omega)\langle\varphi(\boldsymbol{k}, \omega)\,\varphi^*(\boldsymbol{k}', \omega')\rangle &= \int V(\boldsymbol{k}, \omega; \boldsymbol{k}_1, \omega_1)\langle\varphi^*(\boldsymbol{k}', \omega')\,\varphi(\boldsymbol{k}_1, \omega_1)\,\varphi(\boldsymbol{k}-\boldsymbol{k}_1, \omega-\omega_1)\rangle \\ &\times d^3\boldsymbol{k}_1\,d\omega_1 + \int V(\boldsymbol{k}, \omega; \boldsymbol{k}_1\omega_1; \boldsymbol{k}_2, \omega_2)\{\langle\varphi^*(\boldsymbol{k}', \omega')\,\varphi^1\varphi^2\varphi^3\rangle - \langle\varphi^*(\boldsymbol{k}', \omega')\,\varphi^1\rangle\langle\varphi^2\varphi^3\rangle\} \\ &\times \delta(\boldsymbol{k}-\boldsymbol{k}_1-\boldsymbol{k}_2-\boldsymbol{k}_3)\,\delta(\omega-\omega_1-\omega_2-\omega_3)\,d^3\boldsymbol{k}_1\,d^3\boldsymbol{k}_2\,d^3\boldsymbol{k}_3\,d\omega_1\,d\omega_2\,d\omega_3. \qquad (10.1.2.1)\end{aligned}$$

This equation connects the pair correlation function $\langle\varphi^1\varphi^2\rangle$ with the triple correlation function $\langle\varphi^1\varphi^2\varphi^3\rangle$ and the quadruple correlation function $\langle\varphi^1\varphi^2\varphi^3\varphi^4\rangle$. The triple correlation function can then be reduced to the quadruple one if we use (10.1.1.11) and substitute in it the quantity φ up to second-order terms. As far as the quadruple correlation function is concerned, for random oscillations with completely uncorrelated phases it can be written as a sum of products of binary correlation functions,

$$\langle\varphi^1\varphi^2\varphi^3\varphi^4\rangle = \langle\varphi^1\varphi^2\rangle\langle\varphi^3\varphi^4\rangle + \langle\varphi^1\varphi^3\rangle\langle\varphi^2\varphi^4\rangle + \langle\varphi^1\varphi^4\rangle\langle\varphi^2\varphi^3\rangle. \qquad (10.1.2.2)$$

As a result eqn. (10.1.2.1) becomes a non-linear equation for a single function—the pair correlation function $\langle\varphi^1\varphi^2\rangle$.

Noting that the function $\langle\varphi^1\varphi^2\rangle$ is proportional to $\delta(\boldsymbol{k}_2+\boldsymbol{k}_1)\,\delta(\omega_2+\omega_1)$, we can introduce the correlator $I(\boldsymbol{k}, \omega)$ defined by the relation

$$\langle\varphi(\boldsymbol{k}, \omega)\,\varphi^*(\boldsymbol{k}', \omega')\rangle = \delta(\boldsymbol{k}-\boldsymbol{k}')\,\delta(\omega-\omega')\,I(\boldsymbol{k}, \omega). \qquad (10.1.2.3)$$

Using then (10.1.1.11) and (10.1.2.2) we can write eqn. (10.1.2.1) in the form

$$\begin{aligned}\varepsilon(\boldsymbol{k}, \omega)\,I(\boldsymbol{k}, \omega) &= I(\boldsymbol{k}, \omega)\int U(\boldsymbol{k}, \omega; \boldsymbol{k}_1, \omega_1)\,I(\boldsymbol{k}_1, \omega_1)\,d^3\boldsymbol{k}_1\,d\omega_1 + I(\boldsymbol{k}, \omega) \\ &\times \int \frac{v(\boldsymbol{k}, \omega; \boldsymbol{k}_1, \omega_1)\,v(\boldsymbol{k}_1, \omega_1; \boldsymbol{k}, \omega)}{\varepsilon(\boldsymbol{k}-\boldsymbol{k}_1, \omega-\omega_1+io)}\,I(\boldsymbol{k}_1, \omega_1)\,d^3\boldsymbol{k}_1\,d\omega_1 + \frac{1}{2\varepsilon^*(\boldsymbol{k}, \omega)} \\ &\times \int |\,v(\boldsymbol{k}, \omega; \boldsymbol{k}_1, \omega_1)\,|^2\,I(\boldsymbol{k}_1, \omega_1)\,I(\boldsymbol{k}-\boldsymbol{k}_1, \omega-\omega_1)\,d^3\boldsymbol{k}_1\,d\omega_1 \qquad (10.1.2.4)\end{aligned}$$

† Zaslavskiĭ (1970) has considered the basis for the statistical approach to various systems and has elucidated the criteria for stochastization.

where

$$\left.\begin{aligned} v(\boldsymbol{k}, \omega; \boldsymbol{k}_1, \omega_1) &= V(\boldsymbol{k}, \omega; \boldsymbol{k}_1, \omega_1)+V(\boldsymbol{k}, \omega; \boldsymbol{k}-\boldsymbol{k}_1, \omega-\omega_1), \\ U(\boldsymbol{k}, \omega; \boldsymbol{k}_1, \omega_1) &= V(\boldsymbol{k}, \omega; \boldsymbol{k}_1, \omega_1; -\boldsymbol{k}_1, -\omega_1)+V(\boldsymbol{k}, \omega; \boldsymbol{k}_1, \omega_1; \boldsymbol{k}, \omega), \end{aligned}\right\} \quad (10.1.2.5)$$

while the quantities V are defined by eqns. (10.1.1.12).

In the case of a weakly turbulent plasma in which we are interested the bilinear expression on the right-hand side of eqn. (10.1.2.4) turns out to be proportional to the small parameter γ/ω, where $|\gamma|$ is the growth rate of the oscillations given by the linear theory. Introducing the notation

$$\varepsilon'(\boldsymbol{k}, \omega) = \operatorname{Re} \varepsilon(\boldsymbol{k}, \omega), \quad \varepsilon''(\boldsymbol{k}, \omega) = \operatorname{Im} \varepsilon(\boldsymbol{k}, \omega),$$

and bearing in mind that the growth rate of the oscillations in the linear theory is proportional to the imaginary part of the dielectric permittivity,

$$\gamma(\boldsymbol{k}) = \varepsilon''(\boldsymbol{k}, \omega_k) \left[\frac{\partial \varepsilon'(\boldsymbol{k}, \omega)}{\partial \omega} \right]^{-1}_{\omega_k}, \tag{10.1.2.6}$$

where ω_k is the frequency of the plasma eigenoscillation, we get in zeroth approximation in the parameter γ/ω

$$\varepsilon'(\boldsymbol{k}, \omega)\, I(\boldsymbol{k}, \omega) = 0. \tag{10.1.2.7}$$

The solution of eqn. (10.1.2.7) clearly has the following structure:

$$I(\boldsymbol{k}, \omega) = I_+(\boldsymbol{k})\, \delta(\omega-\omega_k)+I_-(\boldsymbol{k})\, \delta(\omega+\omega_{-k}),$$

where I_+ and I_- are functions of the wavevector $\boldsymbol{k}$ (in order not to complicate the formulae we restrict our discussion in the present subsection to considering only one kind of collective oscillations; the generalization to the case of several branches of oscillations will be given in the next subsection). Bearing in mind that

$$I(\boldsymbol{k}, \omega) = I(-\boldsymbol{k}, -\omega),$$

we see that the functions I_+ and I_- are not independent, but connected through the relation

$$I_+(\boldsymbol{k}) = I_-(-\boldsymbol{k}).$$

Therefore, if we agree to understand by ω_{-k} the quantity $-\omega_k$, we can write the correlator $I(\boldsymbol{k}, \omega)$ in the more compact form

$$I(\boldsymbol{k}, \omega) = I(\boldsymbol{k})\, \delta(\omega-\omega_k). \tag{10.1.2.8}$$

The next-order equation clearly determines the additional damping and the shift of their eigenfrequency of the oscillations caused by non-linear effects. The correction $\Delta\omega$ to the eigenfrequency is here small compared to the frequency itself, $\Delta\omega \approx \gamma$. As far as the additional damping of the oscillations is concerned, the corresponding damping rate may turn out to be of the same order of magnitude as the growth rate given by the linear theory, or even exceed it. We shall therefore restrict ourselves to determining the non-linear damping rate of the oscillations and shall not be interested in the correction to their frequency caused by non-linear effects.

Substituting (10.1.2.8) into eqn. (10.1.2.4) we can write the imaginary part of the resulting expression in the form

$$\gamma\{I(\boldsymbol{k})\} = 0, \tag{10.1.2.9}$$

where

$$\gamma I(\boldsymbol{k}) = \gamma(\boldsymbol{k})\, I(\boldsymbol{k}) - I(\boldsymbol{k}) \left[\frac{\partial \varepsilon'(\boldsymbol{k}, \omega)}{\partial \omega}\right]_{\omega_k} \int \operatorname{Im}\left\{U(\boldsymbol{k}, \omega; \boldsymbol{k}_1, \omega_{k_1})\right.$$

$$\left. + \frac{v(\boldsymbol{k}, \omega_k; \boldsymbol{k}_1, \omega_{k_1})\, v(\boldsymbol{k}-\boldsymbol{k}_1, \omega_k-\omega_{k_1}; \boldsymbol{k}, \omega_k)}{\varepsilon(\boldsymbol{k}-\boldsymbol{k}_1, \omega_k-\omega_{k_1}+io)}\right\} I(\boldsymbol{k}_1)\, d^3\boldsymbol{k}_1 - \frac{\pi}{2}\left[\frac{\partial \varepsilon'(\boldsymbol{k}, \omega)}{\partial \omega}\right]_{\omega_k}$$

$$\times \int |v(\boldsymbol{k}, \omega_k; \boldsymbol{k}_1, \omega_{k_1})|^2\, I(\boldsymbol{k}_1)\, I(\boldsymbol{k}-\boldsymbol{k}_1)\, \delta(\omega_k-\omega_{k_1}-\omega_{k-k_1})\, d^3\boldsymbol{k}_1. \tag{10.1.2.10}$$

Equation (10.1.2.9) determines the correlator $I(\boldsymbol{k})$ in the case of a stationary and uniform distribution of random waves.

This equation can be physically interpreted in a simple way. Indeed, the quantity $2\gamma\{I\}$ determines clearly the decrease in the function $I(\boldsymbol{k})$ per unit time due to the simultaneous effect of the non-linear damping and the "build-up" of the waves, determined by the linear theory. Equation (10.1.2.9) can therefore be interpreted as the condition that the total damping rate of the oscillations—which is the (algebraic) sum of the growth rate of the oscillations given by the linear theory and the damping rate caused by non-linear effects—vanish.

We can easily generalize eqn. (10.1.2.9) to the case of non-stationary and non-uniform wave distributions. To do this we note that if there were no non-linear effects at all, the function $I(\boldsymbol{k})$ would in the case of weak inhomogeneities ($k\lambda \gg 1$, where λ is the characteristic size of the inhomogeneities) satisfy the equation

$$\frac{\partial I}{\partial t} + \left(\frac{\partial \omega_k}{\partial \boldsymbol{k}} \cdot \nabla\right) I + 2\gamma(\boldsymbol{k})\, I = 0.$$

It is clearly necessary in this equation to replace the damping rate $\gamma(\boldsymbol{k})$ of the linear theory by the non-linear damping rate $\gamma\{I\}/I$ given by eqn. (10.1.2.10), in order to take the non-linear wave–particle and wave–wave interactions into account. As a result we get the equation

$$\frac{\partial I}{\partial t} + \left(\frac{\partial \omega_k}{\partial \boldsymbol{k}} \cdot \nabla\right) I + 2\gamma\{I\} = 0 \tag{10.1.2.11}$$

This equation is usually called the *kinetic equation for the waves.*

Using the function $I(\boldsymbol{k})$ we can write eqn. (10.1.1.5) for the averaged particle distribution functions in the form

$$\frac{\partial F_\alpha^{(0)}}{\partial t} + (\boldsymbol{v}\cdot\nabla)\, F_\alpha^{(0)} = \frac{\pi e_\alpha^2}{m_\alpha^2}\left(\frac{\partial}{\partial \boldsymbol{v}} \cdot \int \boldsymbol{k} I(\boldsymbol{k})\, \delta[\omega-(\boldsymbol{k}\cdot\boldsymbol{v})] \left(\boldsymbol{k}\cdot\frac{\partial F_\alpha^{(0)}}{\partial \boldsymbol{v}}\right) d^3\boldsymbol{k}\right). \tag{10.1.2.12}$$

These equations form, together with eqn. (10.1.2.11) for $I(\boldsymbol{k})$, a complete set of equations to describe weak-turbulence states of the plasma.

We note that eqn. (10.1.2.12) which describes the slow change in the particle distribution functions under the influence of waves is, of course, nothing but the equation for the particle distribution functions of the quasi-linear theory (see Chapter 9).

We can put the kinetic equation for the waves in a more familiar form if we introduce the number $N(\boldsymbol{k})$ of plasma waves per unit volume instead of the correlator $I(\boldsymbol{k})$. To do this we write the energy density of the plasma oscillations in the form

$$W = \frac{1}{8\pi}\int k^2 \left.\frac{\partial}{\partial\omega}[\omega\varepsilon'(\boldsymbol{k}, \omega)]\right|_{\omega_{\boldsymbol{k}}} I(\boldsymbol{k})\, d^3\boldsymbol{k}. \tag{10.1.2.13}$$

Noting that the energy density of the oscillations must be connected with the number of waves by the relation

$$W = \int \hbar\omega_{\boldsymbol{k}} N(\boldsymbol{k}) \frac{d^3\boldsymbol{k}}{(2\pi)^3}, \tag{10.1.2.14}$$

we have

$$N(\boldsymbol{k}) = \frac{\pi^2 k^2}{\hbar}\left[\frac{\partial\varepsilon'(\boldsymbol{k}, \omega)}{\partial\omega}\right]_{\omega_{\boldsymbol{k}}} I(\boldsymbol{k}). \tag{10.1.2.15}$$

Substituting (10.1.2.15) into (10.1.2.11) and introducing instead of the function $\gamma\{I\}$ the collision integral

$$\left(\frac{\partial N}{\partial t}\right)_{\mathrm{c}} = 2\frac{\pi^2 k^2}{\hbar}\left[\frac{\partial\varepsilon'(\boldsymbol{k}, \omega)}{\partial\omega}\right]_{\omega_{\boldsymbol{k}}} \gamma\{I\},$$

we get the following kinetic equation for the plasmons:

$$\frac{\partial N}{\partial t} + \left(\frac{\partial\omega_{\boldsymbol{k}}}{\partial\boldsymbol{k}}\cdot\nabla\right)N + \left(\frac{\partial N}{\partial t}\right)_{\mathrm{c}} = 0. \tag{10.1.2.16}$$

We emphasize that the introduction of the distribution function $N(\boldsymbol{k}, \boldsymbol{r}, t)$ of the quasi-particles—plasmons—which depends both on the plasmon momentum $\hbar\boldsymbol{k}$ and on the coordinates $\boldsymbol{r}$ and the time t has a meaning only if the number N of plasmons changes slowly in space and time. This means that the change in the function N over distances of the order of the wavelength $\lambda = 2\pi/k$ and during times of the order of the period of the wave $T = 2\pi/\omega_{\boldsymbol{k}}$ must be much smaller than the function N itself.

We note that eqn. (10.1.2.16) is similar to the well-known kinetic equation for phonons in solid (see, for instance, Peierls, 1955).

10.1.3. THREE-WAVE PROCESSES AND NON-LINEAR LANDAU DAMPING

Let us now dwell upon the physical interpretation of the different terms on the right-hand side of eqn. (10.1.2.10). Bearing in mind that in the transparency region (when $\varepsilon'' \ll \varepsilon'$)

$$\frac{1}{\varepsilon(\boldsymbol{k}, \omega+io)} = \mathcal{P}\frac{1}{\varepsilon'(\boldsymbol{k}, \omega)} + \pi i\left|\frac{\partial\varepsilon'(\boldsymbol{k}, \omega)}{\partial\omega}\right|^{-1}_{\omega_{\boldsymbol{k}}} \{\delta(\omega-\omega_{\boldsymbol{k}})+\delta(\omega+\omega_{\boldsymbol{k}})\}\,\mathrm{sgn}\,\omega,$$

we can write the quantity $\gamma\{I\}$ in the form

$$\gamma\{I(\boldsymbol{k})\} = \gamma(\boldsymbol{k})\, I(\boldsymbol{k})+\Gamma(\boldsymbol{k})\, I(\boldsymbol{k})+S\{I(\boldsymbol{k})\}, \tag{10.1.3.1}$$

where $|\gamma(\boldsymbol{k})|$ is as before the growth rate of the oscillations given by the linear theory (see Chapter 4) while the quantities $\Gamma(\boldsymbol{k})$ and $S\{I(\boldsymbol{k})\}$ are given by the formulae

$$\Gamma(\boldsymbol{k}) = -\left[\frac{\partial\varepsilon'(\boldsymbol{k},\omega)}{\partial\omega}\right]^{-1}_{\omega_{\boldsymbol{k}}} \int \operatorname{Im}\left\{U(\boldsymbol{k},\omega_{\boldsymbol{k}};\boldsymbol{k}_1,\omega_{\boldsymbol{k}_1})+ \mathcal{P}\frac{v(\boldsymbol{k},\omega_{\boldsymbol{k}};\boldsymbol{k}_1,\omega_{\boldsymbol{k}_1})\, v(\boldsymbol{k}-\boldsymbol{k}_1,\omega_{\boldsymbol{k}}-\omega_{\boldsymbol{k}_1};\boldsymbol{k},\omega_{\boldsymbol{k}})}{\varepsilon(\boldsymbol{k}-\boldsymbol{k}_1,\omega_{\boldsymbol{k}}-\omega_{\boldsymbol{k}_1})}\right\} I(\boldsymbol{k}_1)\, d^3\boldsymbol{k}_1, \tag{10.1.3.2}$$

$$S\{I(\boldsymbol{k})\} = \pi I(\boldsymbol{k}) \left[\frac{\partial\varepsilon'(\boldsymbol{k},\omega)}{\partial\omega}\right]^{-1}_{\omega_{\boldsymbol{k}}} \int \left[\frac{\partial\varepsilon'(\boldsymbol{k}-\boldsymbol{k}_1,\omega)}{\partial\omega}\right]^{-1}_{\omega_{\boldsymbol{k}-\boldsymbol{k}_1}} \operatorname{Re}\{v(\boldsymbol{k},\omega_{\boldsymbol{k}};\boldsymbol{k}_1,\omega_{\boldsymbol{k}_1})$$
$$\times\, v(\boldsymbol{k}-\boldsymbol{k}_1,\omega_{\boldsymbol{k}-\boldsymbol{k}_1};\boldsymbol{k},\omega_{\boldsymbol{k}})\}\, I(\boldsymbol{k}_1)\,\delta(\omega_{\boldsymbol{k}}-\omega_{\boldsymbol{k}_1}-\omega_{\boldsymbol{k}-\boldsymbol{k}_1})\, d^3\boldsymbol{k}_1 - \frac{\pi}{2}\left[\frac{\partial\varepsilon'(\boldsymbol{k},\omega)}{\partial\omega}\right]^{-2}_{\omega_{\boldsymbol{k}}}$$
$$\times \int |v(\boldsymbol{k},\omega_{\boldsymbol{k}};\boldsymbol{k}_1,\omega_{\boldsymbol{k}_1})|^2\, I(\boldsymbol{k}_1)\, I(\boldsymbol{k}-\boldsymbol{k}_1)\,\delta(\omega_{\boldsymbol{k}}-\omega_{\boldsymbol{k}_1}-\omega_{\boldsymbol{k}-\boldsymbol{k}_1})\, d^3\boldsymbol{k}, \tag{10.1.3.3}$$

where the symbol $\mathcal{P}$ means that we are dealing with a principal value integral.

Equation (10.1.3.3) for the quantity S contains δ-functions which can be interpreted as giving the energy conservation law for wave–wave interactions,

$$\omega_{\boldsymbol{k}} = \omega_{\boldsymbol{k}_1}+\omega_{\boldsymbol{k}-\boldsymbol{k}_1}. \tag{10.1.3.4}$$

The quantity S therefore clearly characterizes the non-linear damping of waves due to three-wave—or three-plasmon—processes: the decay of one wave into two or the fusion of two waves into one. One sees easily that not for all dispersion laws $\omega = \omega_{\boldsymbol{k}}$ is this quantity non-vanishing, so that three-wave processes would be possible. For instance, in an isotropic medium condition (10.1.3.4) is satisfied if the phase velocity increases with increasing wave-number, $(\partial/\partial k)(\omega_k/k) > 0$—in such cases one says that the spectrum of the oscillations is *decayable*. If, on the other hand, the phase velocity decreases with increasing k, $(\partial/\partial k)(\omega_k/k) < 0$, condition (10.1.3.4) cannot be satisfied (non-decayable spectrum).

The term $\Gamma(\boldsymbol{k})\, I(\boldsymbol{k})$ in eqn. (10.1.3.1) does not contain $\delta(\omega_{\boldsymbol{k}}-\omega_{\boldsymbol{k}_1}-\omega_{\boldsymbol{k}-\boldsymbol{k}_1})$; this term characterizes clearly the damping of waves caused by the scattering of these waves by the plasma particles. This damping has the same nature as the collisionless Landau damping—and is therefore sometimes called the non-linear Landau damping. Indeed, looking at the definitions (10.1.1.12), (10.1.2.5), and (10.1.1.8) of the functions V, v, and $\boldsymbol{g}$, we see that the imaginary parts of these functions appear when one goes round the pole of expressions such as $\{(\omega_1-\omega_2)-([\boldsymbol{k}_1-\boldsymbol{k}_2]\cdot\boldsymbol{v})\}^{-1}$ and thus contain a factor $\delta(\{\omega_1-\omega_2\}-(\{\boldsymbol{k}_1-\boldsymbol{k}_2\}\cdot\boldsymbol{v}))$. Therefore, while the usual (linear) Landau damping occurs due to the absorption (or emission) of a wave with frequency ω and wavevector $\boldsymbol{k}$ by the plasma particles, the non-linear Landau damping is connected with the simultaneous absorption of a wave with frequency ω_1 and wavevector $\boldsymbol{k}_1$ and the emission of a wave with frequency ω_2 and wavevector $\boldsymbol{k}_2$.

Turning to wave–wave interaction processes we note that in the case of non-decayable spectra such processes are, of course, also possible; an example is the four-plasmon wave–wave scattering process. However, these processes occur only in higher approximations in the wave amplitude and contribute to the function $\gamma\{I\}$ a term, cubic in $I(\boldsymbol{k})$. Noting that the non-linear Landau damping, which contributes to the function $\gamma\{I\}$ a term quadratic in the correlator $I(\boldsymbol{k})$, is non-vanishing (for a non-zero particle temperature) both for decayable and for non-decayable spectra, we see that it must, in general, be unnecessary to take four-plasmon processes into account, even in the case of non-decayable spectra.

We shall now show how one can write the kinetic equation for the waves in the case when not one, but several modes of weakly damped oscillations can propagate in the plasma. One sees easily that the solution of eqn. (10.1.2.7) in that case has the form

$$I(\boldsymbol{k}, \omega) = \sum_{\mu} I_{\mu}(\boldsymbol{k})\, \delta(\omega - \omega_{\mu}(\boldsymbol{k})), \tag{10.1.3.5}$$

where the index μ numbers the oscillations branches. The kinetic eqn. (10.1.2.11) will therefore be valid for each of the functions $I_{\mu}(\boldsymbol{k})$, if we replaced in eqn. (10.1.2.10) for the functional γ the quantities $I(\boldsymbol{k})$, $I(\boldsymbol{k}_1)$ and $I(\boldsymbol{k}-\boldsymbol{k}_1)$ by $I_{\mu}(\boldsymbol{k})$, $I_{\mu_1}(\boldsymbol{k}_1)$ and $I_{\mu_2}(\boldsymbol{k}-\boldsymbol{k}_1)$ respectively, and sum over μ_1, μ_2 as well as integrate over $\boldsymbol{k}_1$.

10.2. Turbulent Processes in which Langmuir Waves Take Part

10.2.1. INTERACTION BETWEEN LANGMUIR WAVES AND ION-SOUND WAVES

It is well known that in a completely equilibrium plasma without an external magnetic field, only one kind of longitudinal oscillations—Langmuir oscillations with a non-decayable spectrum—is possible. If the electron temperature is much higher than the ion temperature, another branch of longitudinal oscillations appears—ion-sound waves also characterized by a non-decayable spectrum. Forgetting about possible processes involving transverse electromagnetic waves, we see that in an isotropic two-temperature plasma three-plasmon processes are possible involving one ion-sound and two Langmuir waves—the absorption (or emission) of an ion-sound wave by a Langmuir wave, the decay of an ion-sound wave into two Langmuir waves, and the inverse process of the fusion of two Langmuir waves with the formation of an ion-sound wave. We shall now start a study of these processes.

We shall characterize the intensities of the Langmuir and ion-sound waves by the functions $I_l(\boldsymbol{k})$ and $I_s(\boldsymbol{k})$, which are connected with the correlator $I(\boldsymbol{k}, \omega)$ by the relation

$$I(\boldsymbol{k}, \omega) = I_l(\boldsymbol{k})\, \delta(\omega - \omega_l(\boldsymbol{k})) + I_s(\boldsymbol{k})\, \delta(\omega - \omega_s(\boldsymbol{k})), \tag{10.2.1.1}$$

$$|\,\omega_l(\boldsymbol{k})\,| = \omega_{pe} + \tfrac{3}{2}\omega_{pe}(kr_D)^2, \qquad |\,\omega_s(k)\,| = v_s k, \tag{10.2.1.2}$$

with ω_{pe} the electron plasma frequency, $v_s = \sqrt{(T_e/m_i)}$ the ion-sound velocity, T_e the electron temperature, and r_D the electron Debye radius; we remind ourselves that we use a representation in which ω_l and ω_s can be both positive and negative, while $\omega_{\mu}(-\boldsymbol{k}) = -\omega_{\mu}(\boldsymbol{k})$, $\mu = 1$, s.

The kinetic equations for the interacting Langmuir and ion-sound oscillations have, according to (10.1.2.11) and (10.1.3.1), the form

$$\frac{\partial I_l}{\partial t}+(\boldsymbol{U}_l\cdot\nabla)\, I_l+2S\{I_l\}=0, \tag{10.2.1.3}$$

$$\frac{\partial I_s}{\partial t}+(\boldsymbol{U}_s\cdot\nabla)\, I_s+2S\{I_s\}=0, \tag{10.2.1.4}$$

where $\boldsymbol{U}_l$ and $\boldsymbol{U}_s$ are the respective group velocities,

$$\boldsymbol{U}_l(\boldsymbol{k})=\pm 3\omega_{pe}r_D^2\boldsymbol{k}, \quad \boldsymbol{U}_s(\boldsymbol{k})=\pm v_s\boldsymbol{k}/k. \tag{10.2.1.5}$$

To determine the functions $V(\boldsymbol{k}, \omega; \boldsymbol{k}_1, \omega_1)$ which occur in the expressions for $S\{I\}$ we shall start from the general formulae (10.1.1.12). Noting that the contribution from the ions to the function V is small both for Langmuir and for ion-sound waves, we have

$$V(\boldsymbol{k}, \omega_\mu(\boldsymbol{k}); \boldsymbol{k}_1, \omega_\nu(\boldsymbol{k}_1))=-\frac{4\pi e^3}{k^2m_e^2}\int d^3v[\omega_\mu(\boldsymbol{k})-(\boldsymbol{k}\cdot\boldsymbol{v})+io]^{-1}$$

$$\times\left(\boldsymbol{k}_1\cdot\frac{\partial}{\partial \boldsymbol{v}}\right)\left\{[\omega_\mu(\boldsymbol{k})-\omega_\nu(\boldsymbol{k}_1)-([\boldsymbol{k}-\boldsymbol{k}_1]\cdot\boldsymbol{v})+io]^{-1}\left([\boldsymbol{k}-\boldsymbol{k}_1]\cdot\frac{\partial}{\partial \boldsymbol{v}}F_e^{(0)}\right)\right\}, \quad \mu, \nu = l, s. \tag{10.2.1.6}$$

One can easily evaluate the integrals over the velocities which occur in these formulae if one takes into account that the velocity range for which $\omega_s \ll (\boldsymbol{k}\cdot\boldsymbol{v}) \ll \omega_l$ gives the main contribution to these integrals:

$$V(\boldsymbol{k}, \omega_l(\boldsymbol{k}); \boldsymbol{k}_1, \omega(\boldsymbol{k}_1))=-\frac{(\boldsymbol{k}\cdot\boldsymbol{k}_1)}{k^2}\frac{e}{T_e}, \quad V(\boldsymbol{k}, \omega_s(\boldsymbol{k}); \boldsymbol{k}_1, \omega_l(\boldsymbol{k}_1))=\frac{(\boldsymbol{k}\cdot\boldsymbol{k}_1)(\boldsymbol{k}\cdot\{\boldsymbol{k}-\boldsymbol{k}_1\})^2}{k^6}\frac{e}{T_e}; \tag{10.2.1.7}$$

we have not written down the expression for the function $V(\boldsymbol{k}, \omega_l(\boldsymbol{k}); \boldsymbol{k}_1, \omega_s(\boldsymbol{k}_1))$, as that function contains, when compared to the other two functions, an extra small factor $(kr_D)^2$. Substituting (10.2.1.7) into (10.1.2.5), we have

$$v(\boldsymbol{k}, \omega_l(\boldsymbol{k}); \boldsymbol{k}_1, \omega_l(\boldsymbol{k}_1))=-\frac{(\boldsymbol{k}\cdot\boldsymbol{k}_1)}{k^2}\frac{e}{T_e}, \quad v(\boldsymbol{k}, \omega_s(\boldsymbol{k}); \boldsymbol{k}_1, \omega_l(\boldsymbol{k}_1))=\frac{(\boldsymbol{k}\cdot\boldsymbol{k}_1)(\boldsymbol{k}\cdot\{\boldsymbol{k}-\boldsymbol{k}_1\})}{k^4}\frac{e}{T_e}. \tag{10.2.1.8}$$

Using then (10.1.3.3) we get finally the following kinetic equations which describe interacting Langmuir and ion-sound oscillations

$$\left[\frac{\partial}{\partial t}+(\boldsymbol{U}_l(\boldsymbol{k})\cdot\nabla)\right]I_l(\boldsymbol{k})+\frac{\pi}{4}\frac{e^2\omega_{pe}^2}{T_e^2k^2}\int k_1^2\cos^2\theta\{I_l(\boldsymbol{k})\,I_s(\boldsymbol{k}-\boldsymbol{k}_1)-I_l(\boldsymbol{k}_1)\,I_s(\boldsymbol{k}-\boldsymbol{k}_1)\}$$

$$\times\delta\{\omega_l(\boldsymbol{k})-\omega_l(\boldsymbol{k}_1)-\omega_s(\boldsymbol{k}-\boldsymbol{k}_1)\}\,d^3\boldsymbol{k}_1=0, \tag{10.2.1.9}$$

$$\left[\frac{\partial}{\partial t}+(\boldsymbol{U}_s(\boldsymbol{k})\cdot\nabla)\right]I_s(\boldsymbol{k})-\frac{\pi}{4}\frac{e^2k^2v_s^2}{m_e^2\omega_{pe}^4}\int k_1^2\cos^2\theta(k-k_1\cos\theta)^2\,I_l(\boldsymbol{k}_1)\,I_l(\boldsymbol{k}-\boldsymbol{k}_1)$$

$$\times\,\delta\{\omega_s(\boldsymbol{k})-\omega_l(\boldsymbol{k}_1)-\omega_l(\boldsymbol{k}-\boldsymbol{k}_1)\}\,d^3\boldsymbol{k}_1=0, \tag{10.2.1.10}$$

where θ is the angle between the vectors $\boldsymbol{k}$ and $\boldsymbol{k}_1$.

We can write eqns. (10.2.1.9) and (10.2.1.10) for the functions I_l and I_s in the form of kinetic equations for quasi-particles—the quanta of the plasma oscillations—if we use (10.1.2.15) to introduce instead of these functions the plasmon distribution functions

$$N_l(\boldsymbol{k}) = \frac{2\pi^2 k^2}{\hbar\omega_{pe}} I_l(\boldsymbol{k}), \quad N_s(\boldsymbol{k}) = \frac{2\pi^2}{\hbar k v_s r_D^2} I_s(\boldsymbol{k}). \tag{10.2.1.11}$$

Substituting these expressions into (10.2.1.9) and (10.2.1.10) we get

$$\left[\frac{\partial}{\partial t} + (\boldsymbol{U}_\mu(\boldsymbol{k})\cdot\nabla)\right] N_\mu(\boldsymbol{k}) + \left(\frac{\partial N_\mu}{\partial t}\right)_c = 0, \quad \mu = l, s, \tag{10.2.1.12}$$

where the collision integrals $(\partial N_\mu/\partial t)_c$ have the form

$$\left(\frac{\partial N_l(\boldsymbol{k})}{\partial t}\right)_c = \frac{e^2}{8\pi}\frac{\hbar v_s}{m_e T_e k^2}\int \cos^2\theta \{k_1^2 N_l(\boldsymbol{k}) - k^2 N_l(\boldsymbol{k}_1)\} \, |\boldsymbol{k}-\boldsymbol{k}_1| \, N_s(\boldsymbol{k}-\boldsymbol{k}_1)$$
$$\times \delta\{\omega_l(\boldsymbol{k}) - \omega_l(\boldsymbol{k}_1) - \omega_s(\boldsymbol{k}-\boldsymbol{k}_1)\} \, d^3\boldsymbol{k}_1, \tag{10.2.1.13}$$
$$\left(\frac{\partial N_s(\boldsymbol{k})}{\partial t}\right)_c = -\frac{e^2}{8\pi}\frac{\hbar k v_s}{m_e T_e}\int \cos^2\theta \, \frac{(k-k_1\cos\theta)^2}{|\boldsymbol{k}-\boldsymbol{k}_1|^2} N_l(\boldsymbol{k}) N_l(\boldsymbol{k}-\boldsymbol{k}_1)$$
$$\times \delta\{\omega_s(\boldsymbol{k}) - \omega_l(\boldsymbol{k}_1) - \omega_l(\boldsymbol{k}-\boldsymbol{k}_1)\} \, d^3\boldsymbol{k}_1.$$

We draw attention to the fact that the collision integral in the kinetic equation for the ion-sound waves is always negative, $(\partial N_s/\partial t)_c < 0$. The number of these waves and hence also their energy therefore always increase with time.

The increase with time of the energy of the ion-sound oscillations leads to the result that any distribution of Langmuir waves in a two-temperature plasma with $T_e \gg T_i$ turns out to be unstable. The physical mechanism of this instability consists in that low-frequency beats occurring when two Langmuir waves interact are transformed into ion-sound oscillations, leading thereby to the excitation of ion sound. Such an instability mechanism is possible because for two Langmuir waves and one ion-sound wave the decay conditions (10.1.3.4),

$$|\omega_l(\boldsymbol{k}_0)| = |\omega_l(\boldsymbol{k}_l)| + |\omega_s(\boldsymbol{k}_s)|, \qquad \boldsymbol{k}_0 = \boldsymbol{k}_l + \boldsymbol{k}_s, \tag{10.2.1.14}$$

can be satisfied; for this reason the instability itself is usually called a decay instability. We shall now turn to a study of the decay instability of Langmuir waves.

10.2.2. DECAY INSTABILITY OF LANGMUIR WAVES

We shall consider the problem of the instability of a separate finite-amplitude Langmuir wave propagating in a two-temperature plasma. We first of all note that equations (10.2.1.9) and (10.2.1.10)—as the original kinetic equation for the waves (10.1.2.11)—are intended for a study of the interaction between wavepackets which, in wavevector space, are rather broad and need some modifications if we want to consider processes involving monochromatic waves. We shall therefore start by not using these equations but eqn. (10.1.1.11) for the wave amplitude; afterwards we shall show how one can obtain the same results from the kinetic eqns. (10.2.1.9) and (10.2.1.10).

Turning therefore to eqn. (10.1.1.11) we can write this equation for the case of interacting Langmuir and ion-sound waves in the form

$$\left[\frac{\partial \varepsilon'(\boldsymbol{k},\omega)}{\partial \omega}\right]_{\omega_l(\boldsymbol{k})} \times [\omega-\omega_l(\boldsymbol{k})]\,\varphi_l(\boldsymbol{k},\omega) = \int V(\boldsymbol{k},\omega_l(\boldsymbol{k});\boldsymbol{k}_1,\omega_l(\boldsymbol{k}_1))\,\varphi_l(\boldsymbol{k}_1,\omega_1) \times \varphi_s(\boldsymbol{k}-\boldsymbol{k}_1,\omega-\omega_1)\,d^3\boldsymbol{k}_1\,d\omega_1, \tag{10.2.2.1}$$

$$\left[\frac{\partial \varepsilon'(\boldsymbol{k},\omega)}{\partial \omega}\right]_{\omega_s(\boldsymbol{k})} \times [\omega-\omega_s(\boldsymbol{k})]\,\varphi_s(\boldsymbol{k},\omega) = \int v(\boldsymbol{k},\omega_s(\boldsymbol{k});\boldsymbol{k}_1,\omega_l(\boldsymbol{k}_1))\,\varphi_l(\boldsymbol{k}_1,\omega_1) \times \varphi_l(\boldsymbol{k}-\boldsymbol{k}_1,\omega-\omega_1)\,d^3\boldsymbol{k}_1\,d\omega_1, \tag{10.2.2.2}$$

where the φ_μ ($\mu = \text{l, s}$) are the respective parts of the electrostatic potential while the functions V and v are again determined by formulae (10.2.1.7) and (10.2.1.8); we have symmetrized the equation for the quantity φ_s and neglected in both equations terms containing the potential in powers higher than the second.

We shall assume that in the initial state a single Langmuir wave with amplitude φ_0 and wavevector $\boldsymbol{k}_0$ propagates through the plasma,

$$\varphi_0(\boldsymbol{k},\omega) = \varphi_0\delta(\boldsymbol{k}-\boldsymbol{k}_0)\,\delta(\omega-\omega_l(\boldsymbol{k}_0)). \tag{10.2.2.3}$$

As we are interested in the instability of this wave against decay into an ion-sound wave with wavevector $\boldsymbol{k}_s$ and a Langmuir wave with wavevector $\boldsymbol{k}_l$ we write the perturbation of the electrostatic potential in the form

$$\varphi'(\boldsymbol{k},\omega) = 2\int \{\varphi_s \cos[(\boldsymbol{k}_s\cdot\boldsymbol{r})-\omega_s(\boldsymbol{k}_s)\,t]+\varphi_l' \cos[(\boldsymbol{k}_l\cdot\boldsymbol{r})-|\omega_l(\boldsymbol{k}_l)|\,t]\}\frac{d^3\boldsymbol{r}\,dt}{(2\pi)^4}, \tag{10.2.2.4}$$

where φ_s and φ_l' are slowly varying functions of the time.

In order that the above-mentioned decay be possible the vectors $\boldsymbol{k}_l$ and $\boldsymbol{k}_s$ must satisfy eqns. (10.2.1.14). Solving those equations and, to fix the ideas, assuming that $k_0 r_D \gg \sqrt{(m_e/m_i)}$ we get

$$k_l = k_0, \quad k_s = 2k_0 \sin\tfrac{1}{2}\alpha,$$

where α is the angle between the vectors $\boldsymbol{k}_0$ and $\boldsymbol{k}_l$.

Now substituting (10.2.2.3) and (10.2.2.4) into eqns. (10.2.2.1) and (10.2.2.2) and linearizing, we have

$$[\omega-\omega_l(\boldsymbol{k}_l)]\,\varphi_l' = -\frac{\omega_{pe}}{2}\frac{e\varphi_0}{T_e}\varphi_s\cos\alpha, \quad [\omega-\omega_s(\boldsymbol{k}_s)]\,\varphi_s = -\frac{k_0^3 v_s^2}{m_e\omega_{pe}^2}e\varphi_0\varphi_l'\sin^3\frac{\alpha}{2}. \tag{10.2.2.5}$$

Putting for each of the excited waves $\omega = \omega_\mu(\boldsymbol{k}_\mu)+i\nu(\alpha)$, we find the square of the growth rate of the oscillations, $\nu^2(\alpha)$:

$$\nu^2(\alpha) = \frac{\omega_{pe}^2}{2}(k_0 r_D)^3\sqrt{\frac{m_e}{m_i}}\frac{(e\varphi_0)^2}{T_e^2}\left[-\cos\alpha\times\sin^3\frac{\alpha}{2}\right]. \tag{10.2.2.6}$$

We see that perturbations for which $\alpha > \pi/2$ grow, while perturbations for which $\alpha = \pi$, that is, $\boldsymbol{k}_l = -\boldsymbol{k}_0$, $\boldsymbol{k}_s = 2\boldsymbol{k}_0$, are characterized by the largest growth rate. The growth rate of a decay instability, $\nu = \text{Max}\{\nu(\alpha)\}$, is thus determined by the formula (Oraevskiĭ and Sagdeev, 1963)

$$\nu = \omega_{pe}(k_0 r_D)^{3/2} \frac{|e\varphi_0|}{T_e} \left(\frac{m_e}{4m_i}\right)^{1/4}. \tag{10.2.2.7}$$

Let us now show how we can determine the growth rate of the decay instability, if we start from the kinetic eqns. (10.2.1.9) and (10.2.1.10) for interacting Langmuir and ion-sound waves rather than from eqn. (10.1.1.11). To do this we note that, if we take the growth (or damping) of waves into account in expression (10.1.3.3) for the function $S\{I\}$ we must perform the substitution

$$\delta(\omega) \to \frac{1}{\pi} \frac{|\nu|}{\omega^2+\nu^2}.$$

If we now take as the unperturbed state of the system of waves the state characterized by the correlator

$$I(\boldsymbol{k}, \omega) = I_0\{\delta(\boldsymbol{k}-\boldsymbol{k}_0)\,\delta(\omega-\omega_l(\boldsymbol{k}_0))+\delta(\boldsymbol{k}+\boldsymbol{k}_0)\,\delta(\omega+\omega_l(\boldsymbol{k}_0))\},$$

and impose a perturbation of the form

$$I'(\boldsymbol{k}, \omega) = I_l'(0)\,e^{\nu t}\{\delta(\boldsymbol{k}+\boldsymbol{k}_0)\,\delta[\omega-\omega_l(\boldsymbol{k}_0)]+\delta(\boldsymbol{k}-\boldsymbol{k}_0)\,\delta[\omega+\omega_l(\boldsymbol{k}_0)]\}$$
$$+I_s(0)\,e^{\nu t}\{\delta(\boldsymbol{k}-2\boldsymbol{k}_0)\,\delta[\omega-\omega_s(2\boldsymbol{k}_0)]+\delta(\boldsymbol{k}+2\boldsymbol{k}_0)\,\delta[\omega+\omega_s(2\boldsymbol{k}_0)]\},$$

where as before we have assumed that $kr_D \gg \sqrt{(m_e/m_i)}$, we get, after linearizing the kinetic equations,

$$\nu I_l' = \frac{e^2\omega_{pe}^2 I_0}{4|\nu|T_e^2} I_s, \qquad \nu I_s = \frac{e^2 v_s^2 k_0^6 I_0}{m_e^2 \omega_{pe}^4 |\nu|} I_l'. \tag{10.2.2.8}$$

Solving these equations we get for the growth rate an expression which is similar to (10.2.2.7) (Kadomtsev, 1965)

$$\nu = \omega_{pe}(k_0 r_D)^{3/2} \frac{(e^2 I_0)^{1/2}}{T_e} \left(\frac{m_e}{4m_i}\right)^{1/4}. \tag{10.2.2.9}$$

We note that eqns. (10.2.2.7) and (10.2.2.9) for the growth rate of the decay instability of a Langmuir wave are valid provided the growth rate of this instability exceeds the damping rate of all three waves which take part in the decay process. As far as the possibility itself of the decay instability is concerned, it is necessary for the occurrence of this instability that the condition $\nu \gg \gamma$ is realized for at least two—but not necessarily for all three—waves which take part in the process (Karplyuk, Oraevskiĭ, and Pavlenko, 1969).

When studying the decay instability we started from the assumption that all three waves taking part in the process are monochromatic and that their frequencies satisfy the decay condition (10.2.1.14). Simultaneously with the decay process of the waves the inverse processes—the processes in which waves fuse leading to a decrease in the growth rate of the decay instability—must, in principle, also take place. For monochromatic waves, satisfying the decay condition the condition for the fusion of waves $|\omega_l(\boldsymbol{k}_0)| = |\omega_l(\boldsymbol{k}_0+\boldsymbol{k}_s)| - |\omega_s(\boldsymbol{k}_s)|$, cannot be satisfied. However, in an actual physical situation we are, of course, not dealing with monochromatic waves, but with wavepackets, existing for a finite time. The first of conditions (10.2.1.14) must thus be replaced by the weaker condition

$$|\omega_l(\boldsymbol{k}_0)| = |\omega_l(\boldsymbol{k}_1)| \pm |\omega_s(\boldsymbol{k}_s)| + \Delta,$$

where $\boldsymbol{k}_1 = \boldsymbol{k}_0 \mp \boldsymbol{k}_s$ while Δ is a quantity characterizing the non-monochromatic packet —this non-monochromaticity can be caused by the damping or growth of the waves taking part in the process, by non-linear effects leading to a smearing-out of the wavepacket in frequency, or, finally, by the conditions under which the Langmuir wave whose decay we follow was excited. When the "frequency detuning" Δ is taken into account it is necessary to consider perturbations of a more general form than a single Langmuir and a single sound wave (Bakaĭ, 1970, 1971). In fact, it is necessary to take into account perturbations which are a superposition of three waves: an ion-sound wave with frequency $|\omega_s(\boldsymbol{k}_s)|$ and two Langmuir waves with frequencies $|\omega_l(\boldsymbol{k}_0 \pm \boldsymbol{k}_s)|$ which are close to, respectively, $|\omega_l(\boldsymbol{k}_0)|+|\omega_s(\boldsymbol{k}_s)|$ (fusion of waves) or $|\omega_l(\boldsymbol{k}_0)|-|\omega_s(\boldsymbol{k}_s)|$(decay of the wave). As a result the square of the growth rate of the decay instability becomes, in general, smaller than the quantity ν^2 given by formula (10.2.2.6). For sufficiently large values of the "detuning" Δ the quantity ν^2 may even become negative; in that case, of course, the quantity $|\nu|$ is no longer the growth rate of the oscillations, but a correction to their frequency.

10.2.3. NON-LINEAR DAMPING OF LANGMUIR WAVES

We showed in Chapter 4 that the damping rate of the usual (linear) Landau damping for Langmuir waves is exponentially small, $\gamma/\omega \propto \exp\{-\frac{1}{2}(kr_D)^{-2}\}$. This is connected with the fact that there is an exponentially small number of resonance particles—electrons with velocities close to the phase velocity of the Langmuir wave which therefore can strongly interact with that wave.

When two Langmuir waves interact non-linearly low-frequency beats with phase velocities of the order $kr_D v_e$, with $v_e = \sqrt{(2T_e/m_e)}$ the average thermal electron velocity, must occur. These beats, whose absorption by particles leads to the non-linear Landau damping, interact much more strongly with the electrons than the original Langmuir waves (when $T_e = T_i$, when the decay conditions for Langmuir oscillations are not satisfied; see subsection 10.2.2). Therefore, the decrease in the energy of Langmuir waves must, roughly speaking, be determined by their non-linear damping even for not very large amplitudes; we shall now turn to a study of this non-linear damping.

To determine the rate of the non-linear damping of Langmuir waves we shall use the general formula (10.1.3.2). To determine the functions U and v which occur in that formula we bear in mind that when we sum over the different kinds of particles in eqns. (10.1.1.12) it is sufficient to limit ourselves to the electron terms. Furthermore, the main contribution to the integrals on the right-hand sides of these equations comes from the range of velocities $v \sim v_e$ and an expansion of the integrands in power series in $(\boldsymbol{k}\cdot\boldsymbol{v})/\omega_l(\boldsymbol{k})$ therefore corresponds to an expansion of the functions U and v in power series in the small parameter kr_D.

To first order in that parameter we have

$$\left.\begin{aligned} v(\boldsymbol{k}, \omega_l(\boldsymbol{k}); \boldsymbol{k}_1, \omega_l(\boldsymbol{k}_1)) &= -\frac{e}{m_e}\frac{(\boldsymbol{k}\cdot\boldsymbol{k}_1)}{k^2}\frac{|\boldsymbol{k}-\boldsymbol{k}_1|^2}{\omega_l^2(\boldsymbol{k})}\varepsilon(\boldsymbol{k}-\boldsymbol{k}_1, \omega_l(\boldsymbol{k})-\omega_l(\boldsymbol{k}_1)), \\ U(\boldsymbol{k}, \omega_l(\boldsymbol{k}); \boldsymbol{k}_1, \omega_l(\boldsymbol{k}_1)) &= \frac{e}{m_e}\frac{(\boldsymbol{k}\cdot\boldsymbol{k}_1)}{k^2}\frac{|\boldsymbol{k}-\boldsymbol{k}_1|^2}{\omega_l^2(\boldsymbol{k})}v(\boldsymbol{k}-\boldsymbol{k}_1, \omega_l(\boldsymbol{k})-\omega_l(\boldsymbol{k}_1); \boldsymbol{k}, \omega_l(\boldsymbol{k})). \end{aligned}\right\} \tag{10.2.3.1}$$

We see that the first-order terms cancel one another and we must therefore go to the next approximation. In the next approximation in the parameter kr_D we have

$$\mathrm{Im}\left\{U(\boldsymbol{k},\,\omega_l(\boldsymbol{k});\,\boldsymbol{k}_1,\,\omega_l(\boldsymbol{k}_1))+\frac{v(\boldsymbol{k},\,\omega_l(\boldsymbol{k});\,\boldsymbol{k}_1,\,\omega_l(\boldsymbol{k}_1))\,v(\boldsymbol{k}-\boldsymbol{k}_1,\,\omega_l(\boldsymbol{k})-\omega_l(\boldsymbol{k}_1);\,\boldsymbol{k},\,\omega_l(\boldsymbol{k}))}{\varepsilon(\boldsymbol{k}-\boldsymbol{k}_1,\,\omega_l(\boldsymbol{k})-\omega_l(\boldsymbol{k}_1))}\right\}$$

$$=\frac{8\pi^2e^4(\boldsymbol{k}\cdot\boldsymbol{k}_1)^2}{m_e^3\omega_{pe}^6k^2}\int(\boldsymbol{k}\cdot\boldsymbol{v})^2\left(\{\boldsymbol{k}-\boldsymbol{k}_1\}\cdot\frac{\partial F_e^{(0)}}{\partial \boldsymbol{v}}\right)\delta\{\omega_l(\boldsymbol{k})-\omega_l(\boldsymbol{k}_1)-([\boldsymbol{k}-\boldsymbol{k}_1]\cdot\boldsymbol{v})\}\,d^3v. \tag{10.2.3.2}$$

Noting that, as to order of magnitude, $|\,\omega_l(\boldsymbol{k})-\omega_l(\boldsymbol{k}_1)\,|\sim(kr_D)^2\omega_{pe}$, we can write the integral occurring in this equation in a form symmetric in the vectors $\boldsymbol{k}$ and $\boldsymbol{k}_1$:

$$\int(\boldsymbol{k}\cdot\boldsymbol{v})^2\left([\boldsymbol{k}-\boldsymbol{k}_1]\cdot\frac{\partial F_e^{(0)}}{\partial \boldsymbol{v}}\right)\delta\{\omega_l(\boldsymbol{k})-\omega_l(\boldsymbol{k}_1)-([\boldsymbol{k}-\boldsymbol{k}_1]\cdot\cdot\boldsymbol{v})\}\,d^3v=-\frac{T_e\,|\,[\boldsymbol{k}\wedge\boldsymbol{k}_1]\,|^2}{4\pi^2e^2}$$
$$\times\mathrm{Im}\,\varepsilon(\boldsymbol{k}-\boldsymbol{k}_1,\,\omega_l(\boldsymbol{k})-\omega_l(\boldsymbol{k}_1)). \tag{10.2.3.3}$$

Substituting (10.2.3.2) and (10.2.3.3) into (10.1.3.2) we get the final expression for the non-linear damping rate of Langmuir waves (see Kadomtsev, 1965)

$$\Gamma(\boldsymbol{k})=\frac{e^2T_e}{m_e^3\omega_{pe}^5k^2}\int(\boldsymbol{k}\cdot\boldsymbol{k}_1)^2\,|\,[\boldsymbol{k}\wedge\boldsymbol{k}_1]\,|^2\,\mathrm{Im}\,\varepsilon(\boldsymbol{k}-\boldsymbol{k}_1,\,\omega_l(\boldsymbol{k})-\omega_l(\boldsymbol{k}_1))\,I(\boldsymbol{k}_1)\,d^3\boldsymbol{k}_1. \tag{10.2.3.4}$$

According to this formula the non-linear damping rate of Langmuir waves is, as to order of magnitude, equal to

$$\Gamma\sim\omega_{pe}(kr_D)^3\frac{W}{n_0T_e}, \tag{10.2.3.5}$$

where n_0 is the equilibrium electron density and W the energy density of the waves,

$$W=\frac{1}{4\pi}\int\frac{k^2}{1+\frac{3}{2}k^2r_D^2}\,I(\boldsymbol{k})\,d^3\boldsymbol{k}. \tag{10.2.3.6}$$

Bearing in mind that the linear damping rate for Langmuir waves is equal to

$$\gamma=\sqrt{\frac{\pi}{8}}\,\frac{\omega_{pe}}{(kr_D)^3}\exp\left[-\frac{3}{2}-\frac{1}{2(kr_D)^2}\right],$$

we see that the non-linear damping starts to play a decisive role for rather small wave amplitudes, namely, when $E>E_m$, where

$$E_m^2\sim\frac{n_0T_e}{(kr_D)^6}\exp\left[-\frac{1}{2(kr_D)^2}\right]. \tag{10.2.3.7}$$

Substituting (10.2.3.4) into (10.1.3.1) and (10.1.2.11) and neglecting the linear damping of the waves, we have

$$\frac{\partial I(\boldsymbol{k})}{\partial t}=-\frac{2e^2T_e}{m_e^3\omega_{pe}^5k^2}\,I(\boldsymbol{k})\int(\boldsymbol{k}\cdot\boldsymbol{k}_1)^2\,|\,[\boldsymbol{k}\wedge\boldsymbol{k}_1]\,|^2\,\mathrm{Im}\,\varepsilon((\boldsymbol{k}-\boldsymbol{k}_1,\,\omega_l(\boldsymbol{k})-\omega_l(\boldsymbol{k}_1))\,I(\boldsymbol{k}_1)\,d^3\boldsymbol{k}_1. \tag{10.2.3.8}$$

We limit ourselves to the case of spatially uniform distributions and assume that the electron temperature is not too different from the ion temperature so that three-wave processes involving Langmuir waves are impossible. Using (10.1.2.11) to introduce the number of Langmuir waves $N_l(\boldsymbol{k})$ we can write the kinetic eqn. (10.2.3.8) in the form

$$\frac{\partial N_l(\boldsymbol{k})}{\partial t}+\left(\frac{\partial N_l(\boldsymbol{k})}{\partial t}\right)_c = 0,$$

$$\left(\frac{\partial N_l(\boldsymbol{k})}{\partial t}\right)_c = \omega_{pe}\frac{e^2\hbar\omega_{pe}}{\pi^2 T_e^2} r_D^6 N_l(\boldsymbol{k})\int\frac{(\boldsymbol{k}\cdot\boldsymbol{k}_1)^2}{k^2k_1^2}$$
$$\times|[\boldsymbol{k}\wedge\boldsymbol{k}_1]|^2\,\mathrm{Im}\,\varepsilon((\boldsymbol{k}-\boldsymbol{k}_1,\,\omega_l(\boldsymbol{k})-\omega_l(\boldsymbol{k}_1))\,N_l(\boldsymbol{k}_1)\,d^3\boldsymbol{k}_1. \quad (10.2.3.9)$$

Integrating eqn. (10.2.3.9) over $\boldsymbol{k}$ and using the fact that $\mathrm{Im}\,\varepsilon(\boldsymbol{k},\omega) = -\mathrm{Im}\,\varepsilon(-\boldsymbol{k},-\omega)$, we easily check that the number of Langmuir waves per unit volume,

$$N_l = \int N_l(\boldsymbol{k})\,d^3\boldsymbol{k},$$

is conserved when there is non-linear damping,

$$\frac{\partial}{\partial t}N_l = 0. \quad (10.2.3.10)$$

It further follows from eqn. (10.2.3.9) that the total energy density of the Langmuir waves which in the approximation considered is connected with the number of waves through the relation $W = \hbar\omega_{pe}N_l$, is conserved (Drummond and Pines, 1962). Due to the non-linear damping the momentum of the waves,

$$\boldsymbol{P} = \int \hbar\boldsymbol{k}N_l(\boldsymbol{k})\,d^3\boldsymbol{k},$$

decreases. The damping rate Γ therefore in practice determines the inverse of the damping time of the momentum rather than of the energy of the waves.

If we want to determine the rate at which the energy of the Langmuir waves changes we must in our evaluation of the damping rate Γ consider terms of even higher order in the parameter kr_D. The characteristic time ν^{-1} for changes in the energy of the wave turns out to be much larger than Γ^{-1},

$$\nu \sim (kr_D)^2\Gamma. \quad (10.2.3.11)$$

The relation (10.2.3.10) which expresses the conservation of the total number of waves turns out to be valid not only in the first non-vanishing order in the parameter kr_D, but generally apart from exponentially small terms (Al'tshul' and Karpman, 1965). Indeed, integrating the first of equations (10.1.1.12) by parts and using (10.1.2.5) we have

$$v(\boldsymbol{k},\omega;\boldsymbol{k}_1,\omega_1) = \frac{|\boldsymbol{k}-\boldsymbol{k}_1|^2}{k^2}v(\boldsymbol{k}-\boldsymbol{k}_1,\omega-\omega_1;\boldsymbol{k},\omega).$$

One further easily verifies that if we neglect the exponentially small residues in the points $\omega = (\boldsymbol{k}\cdot\boldsymbol{v})$ and $\omega_1 = (\boldsymbol{k}_1\cdot\boldsymbol{v})$ we find

$$v(\boldsymbol{k}-\boldsymbol{k}_1,\omega-\omega_1;\boldsymbol{k},\omega) = v^*(-\boldsymbol{k}+\boldsymbol{k}_1,-\omega+\omega_1;-\boldsymbol{k},-\omega).$$

Finally we can easily prove for the function U the relation

$$\text{Im}\, U(\boldsymbol{k}, \omega; \boldsymbol{k}_1, \omega_1) = -\frac{k_1^2}{k^2}\, \text{Im}\, U(\boldsymbol{k}_1, \omega_1; \boldsymbol{k}, \omega).$$

If we, therefore, write the kinetic equation for the waves in the form

$$\frac{\partial N_l(\boldsymbol{k})}{\partial t} + \int K(\boldsymbol{k}, \boldsymbol{k}_1) N_l(\boldsymbol{k}) N_l(\boldsymbol{k}_1)\, d^3\boldsymbol{k}_1 = 0, \tag{10.2.3.12}$$

it turns out that the kernel $K(\boldsymbol{k}, \boldsymbol{k}_1)$ of the integral equation is antisymmetric in $\boldsymbol{k}$ and $\boldsymbol{k}_1$,

$$K(\boldsymbol{k}, \boldsymbol{k}_1) = -K(\boldsymbol{k}_1, \boldsymbol{k}).$$

Integrating equation (10.2.3.12) over $\boldsymbol{k}$ we get eqn. (10.2.3.10).

Bearing in mind the conservation of the number of Langmuir waves we easily understand why the characteristic time for the diminution of their energy is appreciably longer than the time of the damping of their momentum. Indeed, the non-linear damping of waves is caused by the absorption of one wave by a particle with the simultaneous emission of another wave; two other processes involving two waves leading to a change in the total number of waves—the simultaneous absorption or emission of two waves by a particle—make exponentially small contributions to the quantity Γ. The energy conservation law is satisfied,

$$\Delta\varepsilon = (\Delta\boldsymbol{p}\cdot\boldsymbol{v}),$$

where $\Delta\varepsilon = \hbar|\,\omega_l(\boldsymbol{k})\,| - \hbar|\,\omega_l(\boldsymbol{k}_1)\,|$ and $\Delta\boldsymbol{p} = \hbar\boldsymbol{k} - \hbar\boldsymbol{k}_1$ are the changes in the energy and the momentum of the wave. Noting that the energy of the Langmuir wave weakly depends on its momentum, we see that in each scattering process the relative change in the energy of the wave is much less than the relative change in its momentum,

$$\frac{\Delta\varepsilon}{\hbar\omega} \sim (kr_D)^2 \frac{|\Delta\boldsymbol{p}|}{\hbar k}.$$

Multiplying this relation by the number of scattering processes per unit time, we get eqn. (10.2.3.11).

From a quantum-mechanical point of view the non-linear damping of waves—in contrast to the linear damping—is thus not the absorption of oscillation quanta (plasmons) by plasma particles, but the diffusion of these quanta in momentum space due to their collisions with particles. The mixing of the plasmons with respect to the directions of their momenta is then characterized by the largest rate; the characteristic time for this process is equal to Γ^{-1}. The diffusion of plasmons in wavenumber space from the region of larger to that of smaller k-values, which leads to a slow change in the total plasmon energy, proceeds much more slowly, over a characteristic time of the order of $\nu^{-1} \sim (kr_D)^{-2}\Gamma^{-1}$.

We draw attention to the essential difference between plasma turbulence and ordinary hydrodynamic turbulence. It is well known that in a turbulent fluid energy is transferred from large to small vortices so that the energy flux in momentum space is directed towards larger k. In the larger k region the linear damping of the oscillations then begins to play a role; this is caused by the viscosity and thermal conductivity of the fluid. In a turbulent plasma the energy flux is in the opposite direction from larger to smaller k-values, that is, into the region where the linear collisionless damping is exponentially small.

One sees easily that only a single non-linear damping process cannot lead to the total dissipation of the energy of the Langmuir waves. Indeed, if initially N_l waves were excited in the plasma, their energy can not become less than $\hbar\omega_{pe}N_l$ as a result of non-linear damping.

In concluding this subsection we note that the non-linear wave-particle interaction leads to a damping of waves in the case when the equilibrium electron distribution function decreases with increasing energy. If, however, in the velocity region where $v \sim v_e k r_D$ the number of electrons increases with increasing energy, $\partial F_e^{(0)}/\partial v > 0$, the number of scattering processes in which $\Delta\varepsilon < 0$ will exceed the number of scattering processes with $\Delta\varepsilon > 0$ so that the non-linear wave-particle interaction leads to an increase in the wave energy rather than to a decrease (Silin, 1964a).

10.3. Ion-sound Turbulence

10.3.1. NON-LINEAR DAMPING OF ION SOUND

It is well known that in a two-temperature plasma with hot electrons and cold ions, $T_e \gg T_i$, apart from the high-frequency Langmuir waves, also low-frequency longitudinal oscillations—ion sound—can propagate. The usual (linear) Landau damping for ion-sound waves is mainly caused by the absorption of these waves by electrons and is characterized by a damping rate

$$\gamma_s^{e(0)}(\boldsymbol{k}) = \sqrt{\frac{\pi m_e}{8m_i}}\,\omega_s(\boldsymbol{k}); \tag{10.3.1.1}$$

the contribution of the ions to the damping rate is exponentially small,

$$\gamma_s^{i}(\boldsymbol{k}) \sim \omega_s(\boldsymbol{k}) \exp\left[-\frac{T_e}{2T_i}\right].$$

This small value of the ion damping is connected with the fact that the number of resonance ions, that is, ions with a velocity close to the phase velocity of the ion-sound wave which therefore can interact strongly with that wave is exponentially small, while the number of resonance electrons is only linear in the small parameter $\sqrt{(m_e/m_i)}$.

The relatively large value of the linear damping rate of ion-sound oscillations due to their interaction with electrons leads to the result that in the case of a plasma with an isotropic electron velocity distribution it is unnecessary to consider the non-linear damping of these oscillations together with the linear damping. We remind ourselves that in the case of Langmuir waves we are in a completely different situation—one needs to take the non-linear damping into account even for comparatively small wave amplitudes.

The situation changes in the case of a plasma with a directed electron motion. We have shown in Chapter 6 that in such a plasma

$$\gamma_s^e(\boldsymbol{k}) = \gamma_s^{e(0)}(\boldsymbol{k})\left[1-\frac{(\boldsymbol{k}\cdot\boldsymbol{u})}{kv_s}\right], \tag{10.3.1.2}$$

where $\boldsymbol{u}$ is the average directed electron velocity (the velocity of the electron stream) and v_s the ion-sound velocity. For electron stream velocities larger than the sound velocity the interaction between ion-sound waves and electrons leads thus to a growth of the waves, rather than a damping. Under these conditions it is necessary to take the non-linear damping of ion-sound into account.

Turning to a determination of the function $\Gamma(\boldsymbol{k})$ which characterizes the non-linear interaction of ion-sound waves with plasma particles we note first of all that three different processes contribute to it: the scattering of an ion-sound wave by an ion, the absorption (or emission) of two waves by an electron, and the scattering of an ion-sound wave by an electron (the contribution from a fourth possible process—the simultaneous absorption or emission of two waves by an ion—is proportional to the number of ions with velocities $v \approx v_s$ and therefore exponentially small).

The contribution to Γ of the first of these processes is clearly determined by the ion terms in equations (10.1.1.12) and we shall show that it is proportional to the small parameter T_i/T_e. One can easily verify that the contribution of the two other processes is proportional to the number of resonance electrons and thus proportional to the small parameter $\sqrt{(m_e/m_i)}$. Assuming that the inequalities

$$1 \gg \frac{T_i}{T_e} \gg \sqrt{\frac{m_e}{m_i}}$$

are satisfied, we can limit ourselves to considering the interaction of ion-sound waves with the ions and neglect the contribution of the electrons to the function Γ.

The non-linear damping of ion-sound waves as a result of their interaction with ions has the same physical nature as the non-linear damping of Langmuir waves due to their interaction with electrons, which we considered in the preceding section. In fact, when two ion-sound oscillations with wavevectors $\boldsymbol{k}$ and $\boldsymbol{k}_1$ interact beats occur with a phase velocity

$$V_{ph} = \left[1+\frac{4kk_1}{(k-k_1)^2}\sin^2\frac{1}{2}\vartheta\right]^{-1/2} v_s,$$

where ϑ is the angle between the vectors $\boldsymbol{k}$ and $\boldsymbol{k}_1$. If the frequencies of the interacting waves lie close to one another,

$$\frac{k-k_1}{\sqrt{(kk_1)}} \sim \sqrt{\frac{T_i}{T_e}}\sin\frac{1}{2}\vartheta,$$

the velocity V_{ph} becomes of the same order of magnitude as the thermal ion velocity, $v_i = \sqrt{(2T_i/m_i)}$. Such beats are strongly absorbed by the ions, leading thereby to a damping of the original ion-sound waves.

Turning to eqn. (10.1.3.2) for the non-linear damping rate of waves, we note that in determining the functions U, v, and V which occur there we can limit ourselves in the summation over the different kinds of particles to the ion terms. Moreover, we bear in mind that the main contribution to the integrals on the right-hand sides of eqns. (10.1.1.12) comes from the velocity region $v \sim v_i$ so that the expansion of the integrands in powers of $(\boldsymbol{k}\cdot\boldsymbol{v})/\omega_s(\boldsymbol{k})$ corresponds to expanding the functions U and v in power series in the small parameter $\sqrt{(T_i/T_e)}$.

One checks easily that—as in the earlier considered case of the interaction of Langmuir waves with electrons—the leading terms in eqn. (10.1.3.2) cancel and that it is therefore necessary to take into account terms proportional to $\sqrt{(T_i/T_e)}$ and T_i/T_e. As a result we get

$$\operatorname{Im}\left\{U(\boldsymbol{k},\omega_s(\boldsymbol{k});\boldsymbol{k}_1,\omega_s(\boldsymbol{k}_1))+\frac{v(\boldsymbol{k},\omega_s(\boldsymbol{k});\boldsymbol{k}_1,\omega_s(\boldsymbol{k}_1))\,v(\boldsymbol{k}_2,\omega_2;\boldsymbol{k},\omega_s(\boldsymbol{k}))}{\varepsilon(\boldsymbol{k}_2,\omega_2)}\right\}$$

$$=\frac{16\pi^2e^4(\boldsymbol{k}\cdot\boldsymbol{k}_1)^2}{m_i^3\omega^6k^2}\int\left\{(\boldsymbol{k}\cdot\boldsymbol{v})^2\left[1-3\frac{\omega_2}{\omega}\frac{(\boldsymbol{k}\cdot\boldsymbol{k}_2)}{k_2^2}\right]+\frac{(\boldsymbol{k}\cdot\boldsymbol{v})^3}{\omega}\left[3-4\frac{\omega_2}{\omega}\frac{(\boldsymbol{k}\cdot\boldsymbol{k}_2)}{k_2^2}\right]\right.$$

$$\left.-\frac{(\boldsymbol{k}\cdot\boldsymbol{k}_2)^2}{k_2^4}\omega_2^2\right\}\left(\boldsymbol{k}_2\cdot\frac{\partial F_i^{(0)}}{\partial\boldsymbol{v}}\right)\delta\{\omega_2-(\boldsymbol{k}_2\cdot\boldsymbol{v})\}\,d^3\boldsymbol{v}, \qquad (10.3.1.3)$$

where $\omega=\omega_s(\boldsymbol{k})$, $\omega_2=\omega_s(\boldsymbol{k})-\omega(_s\boldsymbol{k}_1)$, and $\boldsymbol{k}_2=\boldsymbol{k}-\boldsymbol{k}_1$.

Substituting this expression into (10.1.3.2) we get after a few straightforward transformations

$$\Gamma(\boldsymbol{k})=-\frac{8\pi^2e^4T_ir_D^2}{m_i^4\omega^5}\int d^3\boldsymbol{k}_1(\boldsymbol{k}\cdot\boldsymbol{k}_1)^2\frac{|[\boldsymbol{k}\wedge\boldsymbol{k}_1]|^2}{|\boldsymbol{k}-\boldsymbol{k}_1|^2}\left[1+3\frac{k-k_1}{k}\right]I(\boldsymbol{k}_1)\int d^3\boldsymbol{v}\left([\boldsymbol{k}-\boldsymbol{k}_1]\cdot\frac{\partial F_i^{(0)}}{\partial\boldsymbol{v}}\right)$$

$$\times\delta\{\omega_s(\boldsymbol{k})-\omega_s(\boldsymbol{k}_1)-([\boldsymbol{k}-\boldsymbol{k}_1]\cdot\boldsymbol{v})\}. \qquad (10.3.1.4)$$

We can considerably simplify expression (10.3.1.4) for Γ if we take into account that the main contribution to the integral over $\boldsymbol{k}_1$ comes from the region of wavenumbers where k_1 is close to k, $|k_1-k|\sim\sqrt{(T_i/T_e)}$. Writing $I(k,\boldsymbol{n})\equiv I(\boldsymbol{k})$, expanding $I(\boldsymbol{k}_1)$ in a series,

$$I(k_1,\boldsymbol{n}_1)=I(k,\boldsymbol{n}_1)+(k_1-k)\frac{\partial I(k,\boldsymbol{n}_1)}{\partial k}+\ldots,$$

and integrating over k_1 on the right-hand side of relation (10.3.1.4) we get the following expression for the non-linear damping rate of ion-sound waves (Kadomtsev and Petviashvili, 1963; Petviashvili, 1964)

$$\Gamma(\boldsymbol{k})=-\omega_s(\boldsymbol{k})\frac{T_i}{T_e^3}2\pi e^2k\frac{\partial}{\partial k}\int\sin^2\vartheta\cos^2\vartheta\,k^3\,I(k,\boldsymbol{n}_1)\,d^2\,\boldsymbol{\omega}_1, \qquad (10.3.1.5)$$

where $\boldsymbol{n}=\boldsymbol{k}/k$, $\boldsymbol{n}_1=\boldsymbol{k}_1/k_1$, and ϑ is the angle between the vectors $\boldsymbol{k}$ and $\boldsymbol{k}_1$.

We see that the quantity $\Gamma(\boldsymbol{k})$ will be positive, if the function $k^3I(\boldsymbol{k})$ decreases for increasing k; in that case the non-linear interaction between the ion-sound oscillations and the ions in the plasma leads to a damping of the oscillation. If, however, the function $I(\boldsymbol{k})$ decreases more slowly than k^{-3} when k increases, the quantity $\Gamma(\boldsymbol{k})$ becomes negative, and the non-linear interaction between ion-sound waves and ions leads to an increase in the wave amplitude.

The non-linear damping (or growth) rate of ion-sound waves is, according to (10.3.1.5), as to order of magnitude, equal to

$$\Gamma \sim \omega \frac{T_i}{T_e} \frac{\omega \partial W/\partial \omega}{n_0 T_e}, \tag{10.3.1.6}$$

where $\partial W/\partial \omega$ is the spectral density of the wave energy per unit volume,

$$\frac{\partial W}{\partial \omega} = \frac{1}{8\pi v_s} \int \frac{\partial}{\partial \omega} [\omega \varepsilon'(\boldsymbol{k}, \omega)]\, I(\boldsymbol{k})\, k^4\, d^2\omega = \frac{k^2}{4\pi v_3 r_D^2} \int I(k, \boldsymbol{n})\, d^2\omega.$$

Substituting (10.3.1.5) into (10.1.3.1) and (10.1.2.11) and restricting ourselves to the case of spatially uniform distributions, we get

$$\frac{\partial I(k, \boldsymbol{n})}{\partial t} = 2\nu(\boldsymbol{k})\, I(k, \boldsymbol{n}) + \omega_s(\boldsymbol{k}) \frac{T_i}{T_e^3} 4\pi e^2 I(k, \boldsymbol{n})\, k \frac{\partial}{\partial k} \int \sin^2 \vartheta \cos^2 \vartheta\, k^3 I(k, \boldsymbol{n}_1)\, d^2\omega_1, \tag{10.3.1.7}$$

where $\nu(\boldsymbol{k}) \equiv -\gamma_s(\boldsymbol{k})$ is the growth rate of the oscillations given by the linear theory.

Using (10.2.1.11) to introduce the number $N_s(\boldsymbol{k})$ of the ion-sound waves, we can write the kinetic eqn. (10.3.1.7) in the form

$$\left.\begin{aligned} \frac{\partial N_s(\boldsymbol{k})}{\partial t} + \left[\frac{\partial N_s(\boldsymbol{k})}{\partial t}\right]_c^{(i)} &= 2\nu(\boldsymbol{k})\, N_s(\boldsymbol{k}), \\ \left[\frac{\partial N_s(\boldsymbol{k})}{\partial t}\right]_c^{(i)} &= -\frac{\hbar T_i}{8\pi^2 n_0 m_i T_e} k^2 N_s(\boldsymbol{k}) \frac{\partial}{\partial k} \int \sin^2 2\vartheta\, k^4 N_s(k, \boldsymbol{n}_1)\, d^2\omega_1. \end{aligned}\right\} \tag{10.3.1.8}$$

Integrating this equation over $\boldsymbol{k}$ we easily see that the non-linear damping conserves the number of ion-sound oscillations per unit volume. Strictly speaking, if the term $2\nu(\boldsymbol{k})\, N_s(\boldsymbol{k})$ which describes the growth of the waves and which is determined by the linear theory were absent, we would get the continuity equation in wavenumber space,

$$\frac{\partial N_s(k)}{\partial t} + \frac{\partial}{\partial k} J(k) = 0, \tag{10.3.1.9}$$

where $N_s(k) = \int k^2 N_s(k, \boldsymbol{n})\, d^2\omega$, and

$$J(k) = -\frac{\hbar T_i k^8}{4\pi^2 n_0 m_i T_e} \left\{ \left| \int \boldsymbol{n} N_s(k, \boldsymbol{n})\, d^2\omega \right|^2 - \left| \int (\boldsymbol{n} - \boldsymbol{n}')^2 N_s(k, \boldsymbol{n})\, N_s^*(k, \boldsymbol{n}')\, d^2\omega\, d^2\omega' \right|^2 \right\}. \tag{10.3.1.10}$$

From the quantum-mechanical point of view the non-linear damping of ion-sound oscillations—like the non-linear damping of Langmuir oscillations—is thus the diffusion of the quanta of the oscillations (plasmons) in momentum space due to their collisions with particles, rather than the absorption of these quanta by the plasma particles. As a result of the collisions packets of ion-sound waves are displaced in momentum space; the energy of the waves then changes proportional to the frequency.

We note that if there were no linear interaction between the ion-sound and the particles, the energy of the ion-sound waves could, in principle, be completely transferred to the ions in the plasma due to the non-linear damping. (This possibility is connected with the fact that the frequency of the ion-sound wave tends to zero as $k \to 0$; we remember that in the case of Langmuir waves for which $\omega_l(0) = \omega_{pe}$ the energy of the waves can not become less than $\hbar\omega_{pe}N_l$ as a result of the non-linear damping.) This feature of the non-linear damping of ion sound leads to the result that the non-linear damping can completely compensate the increase in the energy of the waves caused by the directed motion of the electrons. A stationary distribution of turbulent ion-sound oscillations can then be established in the plasma; we shall now turn to a study of those distributions.

10.3.2. STATIONARY DISTRIBUTIONS OF TURBULENT WAVES

To determine the stationary distributions of turbulent ion-sound waves we shall start from the kinetic equation for the waves (10.3.1.7). Putting $\partial I/\partial t = 0$ in that equation, we get

$$\nu(k,\theta)+\frac{(4\pi e)^2\, v_s T_i}{T_e^3}\,k^2\frac{\partial}{\partial k}\int \varkappa(\theta,\theta')\,k^3 I(k,\theta')\,d\cos\theta' = 0, \tag{10.3.2.1}$$

where θ (θ') is the angle between the vector $\boldsymbol{k}(\boldsymbol{k}_1)$ and the direction of the electron stream,

$$\varkappa(\theta,\theta') = \frac{1}{4\pi}\int \sin^2\vartheta\,\cos^2\vartheta\,d\varphi, \tag{10.3.2.2}$$

and φ is the angle between the $\boldsymbol{k}$, $\boldsymbol{u}$- and the $\boldsymbol{k}_1$, $\boldsymbol{u}$-planes.

If we substitute in this equation for the growth rate $\nu(\boldsymbol{k})$ the quantity $-\gamma_s^e(\boldsymbol{k})$ determined by eqn. (10.3.1.2), we get, as to order of magnitude, for the function I

$$I \sim \sqrt{\frac{m_e}{m_i}}\,\frac{T_e^3}{e^2 T_i k^3}\,f(kr_D),$$

where $f(x)$ is a function of x which changes more slowly than a power.

It is very obvious that when there are such strong oscillations present, the electron distribution function cannot be Maxwellian. We must therefore take into account the reaction of the oscillations on the averaged electron distribution function which leads to the formation of a plateau on that function and thereby decreases the growth rate of the ion-sound waves.

The presence of turbulent ion-sound waves in the plasma clearly strongly changes the velocity distribution only of those electrons which can strongly interact with these waves, that is, which have velocities satisfying the relation

$$(\boldsymbol{k}\cdot\boldsymbol{v}) = kv_s < (\boldsymbol{k}\cdot\boldsymbol{u}).$$

The function $\gamma_s^e(\boldsymbol{k})$ which describes the linear wave–electron interaction shall therefore change appreciably only in the wavevector region for which the linear theory forecasts a growth of the oscillations. In other words, the damping rate $\gamma_s^e(\boldsymbol{k})$ must for damped oscillations ($kv_s > (\boldsymbol{k}\cdot\boldsymbol{u})$) differ little from that given by eqn. (10.3.1.2). However, for growing oscillations ($kv_s < (\boldsymbol{k}\cdot\boldsymbol{u})$) the expression for the growth rate, when we take into account the reaction of the oscillations on the particle distribution function, must have the structure

$$\nu(\boldsymbol{k}) = -\varepsilon(\theta)\,\gamma_s^e(\boldsymbol{k}), \tag{10.3.2.3}$$

where $\gamma_s^e(\boldsymbol{k})$ is given by formula (10.3.1.2) and $\varepsilon(\theta)$ is a small factor connected with the appearance of a plateau on the electron distribution function.

To estimate the order of magnitude of the function $\varepsilon(\theta)$ we use the kinetic eqn. (10.1.2.12) in the quasi-linear approximation. We introduce into that equation the relaxation term $S\{F_e^{(0)}\}$ (for example, the electron-neutral particles collision integral),

$$\frac{\pi e^2}{m_e^2}\left(\frac{\partial}{\partial \boldsymbol{v}}\cdot \int \boldsymbol{k}I(\boldsymbol{k})\,\delta\{\omega_s(\boldsymbol{k})-(\boldsymbol{k}\cdot\boldsymbol{v})\}\left(\cdot\boldsymbol{k}\frac{\partial F_e^{(0)}}{\partial \boldsymbol{v}}\right)d^3\boldsymbol{k}\right) = S\{F_e^{(0)}\}. \qquad (10.3.2.4)$$

Bearing in mind that, as to order of magnitude, $S\{F_e^{(0)}\} \sim \tau_e^{-1}F_e^{(0)}$, where τ_e^{-1} is the electron collision frequency, and

$$\int k_i k_j I(\boldsymbol{k})\,\delta\{\omega_s(\boldsymbol{k})-(\boldsymbol{k}\cdot\boldsymbol{v})\}\,d^3\boldsymbol{k} \sim v_s^{-1}\int kI(\boldsymbol{k})\,d^3\boldsymbol{k},$$

we get for the derivative of the electron distribution function in the region of the plateau

$$\frac{\partial F_e^{(0)}}{\partial v_z} \sim \frac{m_e^2 v_s v_z F_e^{(0)}}{e^2\tau_e \int kI(\boldsymbol{k})\,d^3\boldsymbol{k}}, \qquad v_s < v_z < u,$$

where we have chosen the z-axis in the direction of the vector $\boldsymbol{u}$. Noting that in the case of a Maxwellian electron velocity distribution,

$$\frac{\partial F_e^{(0)}}{\partial v_z} = \frac{m_e v_z}{T_e}F_e^{(0)},$$

we see that as a result of the reaction of the waves on the particle distribution function the derivative $\partial F_e^{(0)}/\partial v_z$ and, hence, also the growth rate $\nu(\boldsymbol{k})$ decreases approximately by a factor ε^{-1}, where

$$\varepsilon \sim \frac{m_e T_e v_s}{e^2\tau_e \int kI(\boldsymbol{k})\,d^3\boldsymbol{k}}. \qquad (10.3.2.5)$$

If we now substitute expression (10.3.2.3) for the growth rate into eqn. (10.3.2.1) we get

$$I \sim \varepsilon\sqrt{\frac{m_e}{m_i}}\,\frac{T_e^3}{e^2 T_i k^3}f(kr_D).$$

Comparing this expression with eqn. (10.3.2.5) we find finally (Kadomtsev and Petviashvili 1963; Petviashvili, 1964)

$$\varepsilon \sim \left[\frac{T_i}{\omega_{pe}\tau_e T_e}\right]^{1/2}, \qquad (10.3.2.6)$$

$$I(\boldsymbol{k}) \sim \frac{T_e^2}{e^2k^3} f(kr_D) \left[\frac{m_e T_e}{m_i T_i \omega_{pe} \tau_e}\right]^{1/2}. \tag{10.3.2.7}$$

Equation (10.3.2.7) determines the order of magnitude of the intensity of the turbulent ion-sound waves.† It follows, in particular, from this formula that the correlator of the potential increases steeply with decreasing wavenumber so that the energy of the turbulent waves turns out to be concentrated, roughly speaking, in the long-wavelength part of the spectrum.

To determine the order of magnitude of the energy of the turbulent waves we note that eqn. (10.3.2.1)—like the original kinetic equation for the waves (10.1.2.11)—does not take into account the damping of the waves due to the collisions between the particles. The contribution from the collisions to the growth rate of the waves can qualitatively be taken into account by replacing formula (10.3.2.3) by the relation

$$\nu(\boldsymbol{k}) = -\varepsilon(\theta)\, \gamma_s^e(\boldsymbol{k}) + \frac{1}{\tau_i},$$

where τ_i^{-1} is the frequency of the ion collisions. We see that for very high frequencies, $\omega \sim \omega_m$, where

$$\omega_m \sim \frac{1}{\tau_i \varepsilon} \sqrt{\frac{m_i}{m_e}}, \tag{10.3.2.8}$$

the growth rate of the oscillations becomes negative. In the frequency range $\omega \lesssim \omega_m$ ion-sound waves therefore cease to be turbulent so that the function $I(\boldsymbol{k})$ is small in that frequency range. Using that fact we get for the energy density of the ion-sound waves, as to order of magnitude,

$$W \sim n_0 T_e \left[\frac{m_e T_e}{m_i T_i \omega_{pe} \tau_e}\right]^{1/2} \ln\left[\frac{r_D}{v_s \tau_i \varepsilon} \sqrt{\frac{m_i}{m_e}}\right]. \tag{10.3.2.9}$$

We note that in the short-wavelength region eqns. (10.3.1.7) and (10.3.2.1) are valid down to wavelength $k^{-1} \sim r_D$. One can easily generalize these equations to the case of oscillations of even shorter wavelengths down to the upper limit of the transparency region $k^{-1} \sim r_{Di}$ ($r_{Di} = r_D \sqrt{(T_i/T_e)}$ is the ion Debye radius). To do this, it is sufficient to take in eqn. (10.1.3.2), and also when determining the growth rate $\nu(\boldsymbol{k})$, the dispersion of the ion-sound into account. We shall, however, not be interested in the function $I(\boldsymbol{k})$ in the region $kr_D > 1$, bearing in mind that, roughly speaking, the energy of the turbulent waves is concentrated in the long-wavelength part or the spectrum.

† The angular dependence of $I(\boldsymbol{k})$ and a detailed analysis of the function $f(kr_D)$ can be found in the work Akhiezer (1964b, 1965c).

10.4. Interaction Between Magneto-sound and Alfvén Waves

10.4.1. THE COLLISION INTEGRAL AND THE H-THEOREM FOR THE GAS OF PLASMONS

In the preceding subsections we have considered the non-linear wave–particle interaction in a plasma without external fields. Using the same method for constructing the kinetic equation for the waves we could, in principle, generalize the discussion to the case of a magneto-active plasma. In that case we should start, rather than from eqns. (10.1.1.1), from more general kinetic equations which take into account the presence of a magnetic field. As the plasma oscillations in a magnetic field are not purely electrostatic we should use the complete set of Maxwell equations instead of the Poisson equation. It is clear that then the turbulent fluctuations can no longer be characterized by a single scalar function $I(\boldsymbol{k}, \omega)$, but that we must introduce a tensor correlation function of the form $\langle E_i(\boldsymbol{k}, \omega)\, E_j^*(\boldsymbol{k}, \omega)\rangle$.

Although all these complications do not change the principles, they nonetheless make the derivation of the kinetic equation for the waves in a magneto-active plasma much more difficult, if we want to take into account both the non-linear wave–wave and the non-linear wave–particle interactions. We shall therefore not obtain here the general kinetic equation for the waves, which would generalize eqns. (10.1.2.10) and (10.1.2.11) for the case of a magneto-active plasma, but we shall restrict ourselves to a study of only one group of non-linear processes—three-wave processes, involving low-frequency waves (Oraevskiĭ and Sagdeev, 1963; Galeev and Oraevskiĭ, 1963). We shall use for the study of these processes a somewhat different method, which is close to the method applied to the study of wave–wave interactions in solid state theory.

If the intensity of the plasma waves is large and there is an appreciable probability that wave–wave interactions occur, the phases of the waves will, in general, be random functions of time. We need therefore not be interested in the phases and can average over them. Under those conditions the vibrational state of the plasma can be described in terms of the numbers of plasmons, that is, quasi-particles, which are plasma waves. Of course, the concept of a plasmon has a meaning only when the plasmon frequency ω_μ (μ: kind of plasmon) is appreciably larger than the inverse of its life time τ_μ. This time is, in general, determined both by the plasmon–plasmon interaction processes and by the processes of the interaction between plasmons and the electrons and ions in the plasma.

If the condition $\omega_\mu \tau_\mu \gg 1$ is satisfied, we can introduce the number $N_\mu(\boldsymbol{k}, t)$ of plasmons of different kinds with well-defined wavevectors $\boldsymbol{k}$ and frequencies $\omega_\mu(\boldsymbol{k})$ and study how these numbers change due to the processes of interaction of plasmons with other plasmons or with plasma particles. If the plasma wave intensity is sufficiently large the plasmon–plasmon interaction processes can become more likely than the plasmon–plasma particle interaction processes. Under these conditions—which we shall assume to be realized—we may assume the plasma to consist of two weakly interacting subsystems: the particle subsystem and the wave subsystem, which slowly exchange energy between each other. In other words, the relaxation in the plasma is then a two-stage process—firstly statistical equilibrium is established in the plasmon subsystem—corresponding to a temperature which, in general, will differ from the particle temperature, and after that the slower process of the equalization of the plasmon and particle temperatures will take place.

The energy and momentum conservation laws hold for the plasmon–plasmon interaction. In particular, in processes involving only three plasmons, that is, in the processes in which two plasmons fuse into one or in which one plasmon splits into two ($\mu_1 \rightleftarrows \mu_2+\mu_3$) the following conservation laws hold:

$$\omega_{\mu 1}(\boldsymbol{k}_1) = \omega_{\mu 2}(\boldsymbol{k}_2)+\omega_{\mu 3}(\boldsymbol{k}_3), \qquad \boldsymbol{k}_1 = \boldsymbol{k}_2+\boldsymbol{k}_3, \tag{10.4.1.1}$$

where the indices 1 and 2, 3 indicate the plasmons in the initial and final states.

One can easily write down a general expression for the change in the plasmon numbers in each given plasmon interaction process. To do this it is sufficient to take into account that the plasma wave intensity, which is proportional to the number of plasmons, can be arbitrarily large so that the plasmons must satisfy Bose–Einstein statistics. For instance, the change in the plasmon number $N_1 \equiv N_{\mu_1}(\boldsymbol{k}_1, t)$ per unit time due to three-plasmon processes ($1 \rightleftarrows 2+3$) can be written in the following general form:

$$(\dot{N}_1)_c = \sum_{2,\,3} \mathscr{L}_1\{N\},$$

$$\mathscr{L}_1\{N\} = \frac{2\pi}{\hbar}\,|V(1, 2, 3)|^2\,\{[(N_1+1)\,N_2N_3-N_1(N_2+1)\,(N_3+1)]\,\delta^{(4)}(k_1-k_2-k_3)$$
$$+2[(N_1+1)\,N_2(N_3+1)-N_1(N_2+1)\,N_3]\,\delta^{(4)}(k_1-k_2+k_3)\}, \tag{10.4.1.2}$$

where $V(1, 2, 3)$ is the matrix element for the plasmon decay (fusion) process $1 \rightleftarrows 2+3$ (the indices 1, 2, 3 indicate both the kind of plasmon and their wavevector: for instance, $1 \equiv \mu_1, \boldsymbol{k}_1$; the summation is over the kinds μ_2 and μ_3 and over the wavevectors $\boldsymbol{k}_2$ and $\boldsymbol{k}_3$ of the plasmons involved in the processes $1 \rightleftarrows 2+3$). As the energy and momentum conservation laws (10.4.1.1) are satisfied in the processes $1 \rightleftarrows 2+3$, there appears a product of δ-functions in eqn. (10.4.1.2):

$$\delta^{(4)}(k_1-k_2\mp k_3) = \delta(\hbar\omega_1-\hbar\omega_2\mp\hbar\omega_3)\,\Delta(\boldsymbol{k}_1-\boldsymbol{k}_2\mp\boldsymbol{k}_3), \qquad \Delta(\boldsymbol{k}) = \begin{cases} 1, & \text{if} \quad \boldsymbol{k} = 0 \\ 0, & \text{if} \quad \boldsymbol{k} \neq 0 \end{cases},$$

where $\omega_1 \equiv \omega_{\mu_1}(\boldsymbol{k}_1)$; we have assumed that the plasma is in a box of finite volume so that the wavevectors take on discrete values. It is natural to call expression (10.4.1.2) the collision integral for the plasmons.

One checks easily that the collision integral $(\dot{N}_1)_c$ vanishes for the Planck distribution function

$$N_\mu^{(0)} = \left[\exp\frac{\hbar\omega_\mu}{T^*}-1\right]^{-1}, \tag{10.4.1.3}$$

where T^* is a constant which can be interpreted as the temperature of the plasmon gas in the statistical equilibrium state. In the case of low-frequency oscillations for which $\hbar\omega_\mu \ll T^*$ the plasmon equilibrium distribution is given by the Rayleigh–Jeans formula,

$$N_\mu^{(0)} = \frac{T^*}{\hbar\omega_\mu}. \tag{10.4.1.4}$$

To conclude this subsection we show that one has a H-theorem for the plasmon collision integral (10.4.1.2). To do this we use the general expression for the entropy of a boson gas,

$$S^* = \sum_1 \{(N_1+1)\ln(N_1+1) - N_1 \ln N_1\}. \tag{10.4.1.5}$$

Taking the time-derivative of that expression, we get

$$\dot{S}^* = \sum_1 (\dot{N}_1)_c \ln \frac{N_1+1}{N_1}. \tag{10.4.1.6}$$

If we substitute here for $(\dot{N}_1)_c$ expression (10.4.1.2), we see easily that

$$\dot{S}^* = \frac{2\pi}{\hbar} \sum_{1,2,3} |V(1, 2, 3)|^2 \delta^{(4)}(k_1-k_2-k_3)\{(N_1+1)N_2N_3 - N_1(N_2+1)(N_3+1)\}$$
$$\times \ln \frac{(N_1+1)N_2N_3}{N_1(N_2+1)(N_3+1)}, \tag{10.4.1.7}$$

whence follows the H-theorem.

10.4.2. THE HAMILTONIAN OF A SYSTEM OF PLASMONS

We have written down the plasmon collision integral corresponding to processes involving three plasmons. We can similarly write down the general form of collision integrals corresponding to processes involving four or more plasmons. Apart from the numbers of plasmons in different states and δ-functions guaranteeing energy and momentum conservation in the expressions for these collision integrals there will appear the matrix elements for the various plasmon interaction processes, that is, the quantities which are the analogue of $V(1, 2, 3)$.

The matrix elements are determined by the dynamical laws of the wave interactions and can, in principle, be found using the kinetic equations for the particles and the Maxwell equations for the fields. We shall restrict ourselves to the calculation of the matrix elements $V(1, 2, 3)$ corresponding to three-plasmon processes involving low-frequency—Alfvén and magneto-sound—waves which, we know, can propagate not only in a well-conducting fluid, but also in a collisionless plasma with hot electrons and cold ions in the frequency range $\omega \ll \omega_{Bi}$, where ω_{Bi} is the ion cyclotron frequency (Akhiezer, Aleksin, and Khodusov, 1971; Aleksin and Khodusov, 1970).

To describe these waves we can use—both in the case of a conducting fluid and in the case of a collisionless plasma—the equations

$$\varrho_m \left[\frac{\partial}{\partial t} + (\boldsymbol{v} \cdot \nabla) \right] \boldsymbol{v} = -v_s^2 \nabla \varrho_m + \frac{1}{c} [\boldsymbol{j} \wedge \boldsymbol{B}],$$

$$\frac{\partial \varrho_m}{\partial t} + \operatorname{div} \varrho_m \boldsymbol{v} = 0, \qquad \operatorname{curl} \boldsymbol{B} = \frac{4\pi}{c} \boldsymbol{j}, \tag{10.4.2.1}$$

$$\boldsymbol{E} = -\frac{1}{c} [\boldsymbol{v} \wedge \boldsymbol{B}], \qquad \operatorname{div} \boldsymbol{B} = 0,$$

where ϱ_m and $\boldsymbol{v}$ are the mass density and the hydrodynamical velocity of the ions, $\boldsymbol{j}$ the current density, and v_s the sound velocity, which in the case of a non-isothermal collisionless plasma is equal to $v_s = \sqrt{(T_e/m_i)}$, where T_e is the electron temperature and m_i the ion mass.

To find the probabilities for the different interaction processes for magneto-hydrodynamical waves, it is convenient to put the equations of motion (10.4.2.1) in Hamiltonian form and to separate explicitly the Hamiltonian of the free magneto-hydrodynamical waves and the Hamiltonian of their interactions. To do this we change to the Lagrangian way to describe the motion. Let $\boldsymbol{r}_0$ and $\boldsymbol{r}$ be the coordinates of a plasma element at $t = 0$, corresponding to an equilibrium state, and at time t. Introducing the displacement vector $\boldsymbol{\xi} = \boldsymbol{r} - \boldsymbol{r}_0$, which we shall assume to be small, and noting that

$$\frac{\partial}{\partial x_i} = \frac{\partial}{\partial x_{0i}} - \sum_j \frac{\partial \xi_j}{\partial x_{0i}} \frac{\partial}{\partial x_{0j}} + \ldots,$$

we can write eqn. (10.4.2.1), up to terms of second order in $\boldsymbol{\xi}$ in the form

$$\varrho_m \frac{d^2 \xi}{dt^2} = F, \tag{10.4.2.2}$$

where

$$F_i = \frac{1}{4\pi} \Bigg\{ (4\pi \varrho_m v_s^2 + B^2) \frac{\partial}{\partial x_i} \operatorname{div} \boldsymbol{\xi} - \sum_{j,l} B_j B_l \frac{\partial^2 \xi_j}{\partial x_i \partial x_l} - \sum_j B_j B_i \frac{\partial}{\partial x_j} \operatorname{div} \boldsymbol{\xi} + \sum_{j,l} B_j B_l \frac{\partial^2 \xi_i}{\partial x_j \partial x_l}$$

$$+ 2\pi \varrho_m v_s^2 \frac{\partial}{\partial x_i} (\operatorname{div} \boldsymbol{\xi})^2 - \sum_{j,l} \left(2\pi \varrho_m v_s^2 + \frac{1}{2} B^2 \right) \frac{\partial}{\partial x_i} \left(\frac{\partial \xi_j}{\partial x_l} \frac{\partial \xi_l}{\partial x_j} \right) - \sum_j (4\pi \varrho_m v_s^2 + B^2) \frac{\partial}{\partial x_j} \left(\frac{\partial \xi_j}{\partial x_i} \operatorname{div} \boldsymbol{\xi} \right)$$

$$+ \sum_j \frac{1}{2} B_i B_j \frac{\partial}{\partial x_j} (\operatorname{div} \boldsymbol{\xi})^2 + \sum_{j,l} B_j B_l \frac{\partial}{\partial x_i} \left(\operatorname{div} \boldsymbol{\xi} \frac{\partial \xi_j}{\partial x_l} \right) - \sum_{j,l,m} \frac{1}{2} B_j B_l \frac{\partial}{\partial x_i} \left(\frac{\partial \xi_m}{\partial x_j} \frac{\partial \xi_m}{\partial x_l} \right)$$

$$- \sum_{j,l} B_j B_l \frac{\partial}{\partial x_l} \left(\operatorname{div} \boldsymbol{\xi} \frac{\partial \xi_i}{\partial x_j} \right) + \sum_{j,l,m} \frac{1}{2} B_i B_j \frac{\partial}{\partial x_j} \left(\frac{\partial \xi_l}{\partial x_m} \frac{\partial \xi_m}{\partial x_l} \right) + \sum_{j,m,l} B_j B_l \frac{\partial}{\partial x_m} \left(\frac{\partial \xi_m}{\partial x_i} \frac{\partial \xi_j}{\partial x_l} \right) \Bigg\}, \tag{10.4.2.3}$$

where we have here and henceforth dropped the index 0 of $\boldsymbol{r}_0$ and all equilibrium quantities.

Multiplying eqn. (10.4.2.2) by $\dot{\boldsymbol{\xi}}$ and integrating over space and time, we find the Hamiltonian, corresponding to the hydrodynamical system of equations:

$$H = H_0 + H_{\text{int}},$$

$$H_0 = \frac{1}{2}\int d^3r \left\{ \varrho_m(\dot{\boldsymbol{\xi}}\cdot\dot{\boldsymbol{\xi}}) + \left[\varrho_m v_s^2 + \frac{B^2}{4\pi}\right](\operatorname{div}\boldsymbol{\xi})^2 - \sum_{i,j}\frac{B_iB_j}{2\pi}\frac{\partial\xi_i}{\partial x_j}\operatorname{div}\boldsymbol{\xi} + \sum_{i,j,l}\frac{B_iB_j}{4\pi}\frac{\partial\xi_l}{\partial x_i}\frac{\partial\xi_l}{\partial x_j}\right\}, \tag{10.4.2.4}$$

$$H_{\text{int}} = \frac{1}{2}\int d^3r \left\{\frac{1}{2}\varrho_m v_s^2(\operatorname{div}\boldsymbol{\xi})^3 - \left(\varrho_m v_s^2 + \frac{B^2}{4\pi}\right)(\operatorname{div}\boldsymbol{\xi})\sum_{i,j}\frac{\partial\xi_i}{\partial x_j}\frac{\partial\xi_j}{\partial x_i} - \sum_{i,j,l}\frac{B_iB_j}{4\pi}(\operatorname{div}\boldsymbol{\xi})\frac{\partial\xi_l}{\partial x_i}\frac{\partial\xi_l}{\partial x_j} \right.$$
$$\left. + \sum_{i,j}\frac{B_iB_j}{4\pi}\frac{\partial\xi_i}{\partial x_j}(\operatorname{div}\boldsymbol{\xi})^2 + \sum_{i,j,l,m}\frac{B_iB_j}{4\pi}\frac{\partial\xi_i}{\partial x_j}\frac{\partial\xi_l}{\partial x_m}\frac{\partial\xi_m}{\partial x_l}\right\}. \tag{10.4.2.5}$$

Fourier transforming the displacement vector,

$$\boldsymbol{\xi} = \frac{1}{\sqrt{V}}\sum_{\boldsymbol{k}} \boldsymbol{\xi}_{\boldsymbol{k}} e^{i(\boldsymbol{k}\cdot\boldsymbol{r})},$$

where V is the plasma volume, we can write H_0 and H_{int} in the form

$$H_0 = \frac{1}{8\pi}\sum_{\boldsymbol{k}}\{4\pi\varrho_m\dot{\boldsymbol{\xi}}_{\boldsymbol{k}}\dot{\boldsymbol{\xi}}_{-\boldsymbol{k}} + (4\pi\varrho_m v_s^2 + B^2)(\boldsymbol{k}\cdot\boldsymbol{\xi}_{\boldsymbol{k}})(\boldsymbol{k}\cdot\boldsymbol{\xi}_{-\boldsymbol{k}}) + (\boldsymbol{\xi}_{\boldsymbol{k}}\cdot\boldsymbol{\xi}_{-\boldsymbol{k}})(\boldsymbol{k}\cdot\boldsymbol{B})^2$$
$$-(\boldsymbol{k}\cdot\boldsymbol{B})(\boldsymbol{B}\cdot\boldsymbol{\xi}_{-\boldsymbol{k}})(\boldsymbol{k}\cdot\boldsymbol{\xi}_{\boldsymbol{k}}) - (\boldsymbol{k}\cdot\boldsymbol{B})(\boldsymbol{B}\cdot\boldsymbol{\xi})(\boldsymbol{k}\cdot\boldsymbol{\xi}_{-\boldsymbol{k}})\}, \tag{10.4.2.6}$$

$$H_{\text{int}} = \frac{i}{24\pi\sqrt{V}}\sum_{1,2,3}\Delta(\boldsymbol{k}_1+\boldsymbol{k}_2+\boldsymbol{k}_3)\{(4\pi\varrho_m v_s^2 + B^2)[(\boldsymbol{k}_1\cdot\boldsymbol{\xi}_1)(\boldsymbol{k}_2\cdot\boldsymbol{\xi}_3)(\boldsymbol{k}_3\cdot\boldsymbol{\xi}_2) + (\boldsymbol{k}_2\cdot\boldsymbol{\xi}_2)(\boldsymbol{k}_1\cdot\boldsymbol{\xi}_3)$$
$$\times(\boldsymbol{k}_3\cdot\boldsymbol{\xi}_1) + (\boldsymbol{k}_3\cdot\boldsymbol{\xi}_3)(\boldsymbol{k}_1\cdot\boldsymbol{\xi}_2)(\boldsymbol{k}_2\cdot\boldsymbol{\xi}_1)] - 4\pi\varrho_m v_s^2(\boldsymbol{k}_1\cdot\boldsymbol{\xi}_1)(\boldsymbol{k}_2\cdot\boldsymbol{\xi}_2)(\boldsymbol{k}_3\cdot\boldsymbol{\xi}_3) + [(\boldsymbol{k}_1\cdot\boldsymbol{\xi}_1)(\boldsymbol{\xi}_2\cdot\boldsymbol{\xi}_3)$$
$$\times(\boldsymbol{k}_2\cdot\boldsymbol{B})(\boldsymbol{k}_3\cdot\boldsymbol{B}) + (\boldsymbol{k}_2\cdot\boldsymbol{\xi}_2)(\boldsymbol{\xi}_1\cdot\boldsymbol{\xi}_3)(\boldsymbol{k}_1\cdot\boldsymbol{B})(\boldsymbol{k}_3\cdot\boldsymbol{B}) + (\boldsymbol{k}_3\cdot\boldsymbol{\xi}_3)(\boldsymbol{\xi}_1\cdot\boldsymbol{\xi}_2)(\boldsymbol{k}_1\cdot\boldsymbol{B})(\boldsymbol{k}_2\cdot\boldsymbol{B})]$$
$$-(\boldsymbol{k}_1\cdot\boldsymbol{B})(\boldsymbol{\xi}_1\cdot\boldsymbol{B})[(\boldsymbol{k}_2\cdot\boldsymbol{\xi}_2)(\boldsymbol{k}_3\cdot\boldsymbol{\xi}_3) + (\boldsymbol{k}_2\cdot\boldsymbol{\xi}_3)(\boldsymbol{k}_3\cdot\boldsymbol{\xi}_2)] - (\boldsymbol{k}_2\cdot\boldsymbol{B})(\xi_2\cdot\boldsymbol{B})[(\boldsymbol{k}_1\cdot\boldsymbol{\xi}_1)(\boldsymbol{k}_3\cdot\boldsymbol{\xi}_3)$$
$$+(\boldsymbol{k}_1\cdot\boldsymbol{\xi}_3)(\boldsymbol{k}_3\cdot\boldsymbol{\xi}_1)] - (\boldsymbol{k}_3\cdot\boldsymbol{B})(\boldsymbol{\xi}_3\cdot\boldsymbol{B})[(\boldsymbol{k}_1\cdot\boldsymbol{\xi}_1)(\boldsymbol{k}_2\cdot\boldsymbol{\xi}_2) + (\boldsymbol{k}_1\cdot\boldsymbol{\xi}_2)(\boldsymbol{k}_2\cdot\boldsymbol{\xi}_1)]\}, \tag{10.4.2.7}$$

where the indices 1, 2, 3 of the displacements $\boldsymbol{\xi}$ correspond to the wavevectors $\boldsymbol{k}_1$, $\boldsymbol{k}_2$, $\boldsymbol{k}_3$.

One verifies easily that the Hamiltonian H_0 is the Hamiltonian of the free magneto-hydrodynamical waves—Alfvén, fast, and slow magneto-sound waves. Indeed, we shall diagonalize the expression (10.4.2.6) which is quadratic in the $\boldsymbol{\xi}$. To do this we write the vector $\boldsymbol{\xi}_k$ in the form

$$\boldsymbol{\xi}_k = \xi_k^{(a)}\boldsymbol{e}_k^{(a)} + \xi_k^{(f)}\boldsymbol{e}_k^{(f)} + \xi_k^{(s)}\boldsymbol{e}_k^{(s)}, \tag{10.4.2.8}$$

where $\boldsymbol{e}_k^{(a)}$, $\boldsymbol{e}_k^{(f)}$, and $\boldsymbol{e}_k^{(s)}$ are three mutually orthogonal unit vectors,

$$\boldsymbol{e}_k^{(a)} = \frac{[\boldsymbol{k}\wedge\boldsymbol{B}]}{\varkappa_\perp kB}, \quad \boldsymbol{e}_k^{(f)} = \frac{\beta}{\varkappa_\perp}\frac{\boldsymbol{k}}{k} + \left(\alpha - \beta\frac{\varkappa_\parallel}{\varkappa_\perp}\right)\frac{\boldsymbol{B}}{B}, \quad \boldsymbol{e}_k^{(s)} = \frac{\alpha}{\varkappa_\perp}\frac{\boldsymbol{k}}{k} - \left(\beta + \alpha\frac{\varkappa_\parallel}{\varkappa_\perp}\right)\frac{\boldsymbol{B}}{B}, \tag{10.4.2.9}$$

$$\alpha = \frac{1}{\sqrt{2}}[1-Q]^{1/2}, \quad \beta = \frac{1}{\sqrt{2}}[1+Q]^{1/2}, \quad \varkappa_\parallel = \frac{(\boldsymbol{k}\cdot\boldsymbol{B})}{kB}, \quad \varkappa_\perp = [1-\varkappa_\parallel]^{1/2},$$

$$Q = [\varrho_m v_s^2(\varkappa_\perp^2 - \varkappa_\parallel^2) + (B^2/4\pi)]\{[\varrho_m v_s^2(\varkappa_\perp^2 - \varkappa_\parallel^2) + (B^2/4\pi)]^2 + 4(\varrho_m v_s^2\varkappa_\parallel\varkappa_\perp)^2\}^{-1/2}.$$

The Hamiltonian (10.4.2.6) expressed in the variables $\xi_k^{(\mu)}$ has the form

$$H_0 = \tfrac{1}{2}\varrho_m \sum_{k,\,\mu} \{\dot{\xi}_k^{(\mu)}\dot{\xi}_{-k}^{(\mu)} + \omega_\mu^2(k)\,\xi_k^{(\mu)}\xi_{-k}^{(\mu)}\}, \tag{10.4.2.10}$$

where $\mu = \text{a, f, s}$; ω_a is the Alfvén wave frequency,

$$\omega_a(k) = kv_A\varkappa_{||}, \quad v_A = \frac{B}{\sqrt{(4\pi\varrho_m)}}, \tag{10.4.2.11}$$

and ω_f and ω_s are the frequencies of the fast and the slow magneto-sound waves,

$$\omega_f = kv_+, \quad \omega_s = kv_-, \quad v_\pm = \tfrac{1}{2}[v_A^2+v_s^2+2v_Av_s\varkappa_{||}]^{1/2} \pm \tfrac{1}{2}[v_A^2+v_s^2-2v_Av_s\varkappa_{||}]^{1/2}. \tag{10.4.2.12}$$

The variables $\xi_k^{(\mu)}$ are thus the amplitudes of the Alfvén ($\mu = \text{a}$), the fast ($\mu = \text{f}$), and the slow ($\mu = \text{s}$) magneto-sound waves, while the vectors $e_k^{(\mu)}$ are the corresponding polarization vectors of these waves.

The part H_{int} of the total Hamiltonian which is cubic in the wave amplitudes must clearly be considered as the Hamiltonian of the interaction between the magneto-hydrodynamical waves. The interaction Hamiltonian (10.4.2.7) has in the normal coordinates $\xi_k^{(\mu)}$ the form

$$H_{int} = \frac{i}{8\pi\sqrt{V}} \sum_{1,2,3} \xi_1\xi_2\xi_3 F(1,2,3)\, B^2k_1k_2k_3\, \Delta(k_1+k_2+k_3), \tag{10.4.2.13}$$

where

$$\begin{aligned} F(1,2,3) = {} & \left[1+\frac{v_s^2}{v_A^2}\right][(\varkappa_1\cdot e_1)(\varkappa_2\cdot e_3)(\varkappa_3\cdot e_2)+(\varkappa_2\cdot e_2)(\varkappa_1\cdot e_3)(\varkappa_3\cdot e_1)\\ & +(\varkappa_3\cdot e_3)(\varkappa_1\cdot e_2)(\varkappa_2\cdot e_1)] - \frac{v_s^2}{v_A^2}(\varkappa_1\cdot e_1)(\varkappa_2\cdot e_2)(\varkappa_3\cdot e_3)+[(\varkappa_1\cdot e_1)(\varkappa_2\cdot n)(\varkappa_3\cdot n)(e_2\cdot e_3)\\ & +(\varkappa_2\cdot e_2)(\varkappa_1\cdot n)(\varkappa_3\cdot n)(e_1\cdot e_3)+(\varkappa_3\cdot e_3)(\varkappa_1\cdot n)(\varkappa_2\cdot n)(e_1\cdot e_2)]-(\varkappa_1\cdot n)(e_1\cdot n)\\ & \times[(\varkappa_2\cdot e_2)(\varkappa_3\cdot e_3)+(\varkappa_2\cdot e_3)(\varkappa_3\cdot e_2)]-(\varkappa_2\cdot n)(e_2\cdot n)[(\varkappa_1\cdot e_1)(\varkappa_3\cdot e_3)+(\varkappa_1\cdot e_3)(\varkappa_3\cdot e_1)]\\ & -(\varkappa_3\cdot n)(e_3\cdot n)[(\varkappa_1\cdot e_1)(\varkappa_2\cdot e_2)+(\varkappa_1\cdot e_2)(\varkappa_2\cdot e_1)], \end{aligned} \tag{10.4.2.14}$$

where $n = B/B$, $\varkappa = k/k$, and the indices 1, 2, 3 are used to indicate both the wavevector and the kind of plasmon; for instance, $e_1 = e_{k_1}^{(\mu_1)}$, $\xi_1 = \xi_{k_1}^{(\mu_1)}$. We note that if all three waves involved in a process are Alfvén waves, we have $F(1, 2, 3) = 0$ according to (10.4.2.14). This corresponds to Oraevskiĭ's result (1963) that an Alfvén wave cannot split up into two Alfvén waves.

10.4.3. PROBABILITIES FOR THREE-PLASMON PROCESSES

We can treat the interaction Hamiltonian H_{int} as a perturbation causing transitions in the system of magneto-hydrodynamic waves. To find the probabilities for such processes it is simplest to use quantum field theory methods and to change from the classical free wave Hamiltonian (10.4.2.10) to the corresponding quantal Hamiltonian. To do this we must introduce operators for the creation $c_k^{(\mu)+}$ and the annihilation $c_k^{(\mu)}$ of a plasmon of kind

μ (μ = a, f, s) with wavevector $\boldsymbol{k}$,

$$c_{\boldsymbol{k}}^{(\mu)} = \left[\frac{\varrho_{\mathrm{m}}}{2\hbar\omega_\mu(\boldsymbol{k})}\right]^{1/2} \{\omega_\mu(\boldsymbol{k})\,\xi_{\boldsymbol{k}}^{(\mu)} + i\dot{\xi}_{-\boldsymbol{k}}^{(\mu)}\}, \quad c_{\boldsymbol{k}}^{(\mu)+} = \left[\frac{\varrho_{\mathrm{m}}}{2\hbar\omega_\mu(\boldsymbol{k})}\right]^{1/2} \{\omega_\mu(\boldsymbol{k})\,\xi_{-\boldsymbol{k}}^{(\mu)} - i\dot{\xi}_{\boldsymbol{k}}^{(\mu)}\}, \quad (10.4.3.1)$$

which satisfy the usual boson operator commutation relations.

We can express the displacement operators $\xi_{\boldsymbol{k}}^{(\mu)}$ in terms of the plasmon creation and annihilation operators:

$$\xi_{\boldsymbol{k}}^{(\mu)} = \left[\frac{\hbar}{2\varrho_{\mathrm{m}}\omega_\mu(\boldsymbol{k})}\right]^{1/2} [c_{\boldsymbol{k}}^{(\mu)} + c_{-\boldsymbol{k}}^{(\mu)+}]. \quad (10.4.3.2)$$

Substituting this expression into (10.4.2.10), we can write the Hamiltonian H_0 as a sum of independent oscillators corresponding to different kinds of plasmons with different values of the wavevector,

$$H_0 = \sum_{\boldsymbol{k},\,\mu} \hbar\omega_\mu(\boldsymbol{k})[N_\mu(\boldsymbol{k}) + \tfrac{1}{2}], \quad N_\mu(\boldsymbol{k}) = c_{\boldsymbol{k}}^{(\mu)+} c_{\boldsymbol{k}}^{(\mu)}. \quad (10.4.3.3)$$

The interaction Hamiltonian (10.4.2.13) in terms of the plasmon creation and annihilation operators takes the form

$$H_{\mathrm{int}} = i \sum_{1,\,2,\,3} V(1, 2, 3)\{\Delta(\boldsymbol{k}_1 - \boldsymbol{k}_2 - \boldsymbol{k}_3)\, c_1 c_2^+ c_3^+ + \Delta(\boldsymbol{k}_1 - \boldsymbol{k}_2 + \boldsymbol{k}_3)\, c_1^+ c_2 c_3^+$$
$$+ \Delta(\boldsymbol{k}_1 + \boldsymbol{k}_2 - \boldsymbol{k}_3)\, c_1^+ c_2^+ c_3\} + \text{h.c.}, \quad (10.4.3.4)$$

where

$$V(1, 2, 3) = \frac{v_{\mathrm{A}}^2 \hbar^{3/2}}{4\sqrt{(2\varrho_{\mathrm{m}} V)}} \frac{k_1 k_2 k_3}{\sqrt{(\omega_1\omega_2\omega_3)}} F(1, 2, 3); \quad (10.4.3.5)$$

$F(1, 2, 3)$ is the function given by formula (10.4.2.14), $c_1 = c_{\boldsymbol{k}_1}^{(\mu_1)}$, $\omega_1 = \omega_{\mu_1}(\boldsymbol{k}_1)$, while h.c. indicates the Hermitean conjugate terms. Bearing in mind that the only non-vanishing matrix elements of the operators $c_{\boldsymbol{k}}^{(\mu)}$ and $c_{\boldsymbol{k}}^{(\mu)+}$ are those for which the number of plasmons $N_\mu(\boldsymbol{k})$ changes by unity,

$$\langle N_\mu(\boldsymbol{k}) - 1\,|\,c_{\boldsymbol{k}}^{(\mu)}\,|\,N_\mu(\boldsymbol{k})\rangle = \sqrt{N_\mu(\boldsymbol{k})}\exp[-i\omega_\mu(\boldsymbol{k})\,t],$$
$$\langle N_\mu(\boldsymbol{k}) + 1\,|\,c_{\boldsymbol{k}}^{(\mu)+}\,|\,N_\mu(\boldsymbol{k})\rangle = \sqrt{[N_\mu(\boldsymbol{k}) + 1]}\exp[i\omega_\mu(\boldsymbol{k})\,t],$$

we see that the function $V(1, 2, 3)$ is exactly the amplitude of the three-plasmon processes and hence is the same as the matrix element occurring in eqn. (10.4.1.2) for the collision integral.

We shall now immediately write down the final expressions for the plasmon collision integrals, restricting ourselves to the case the gas-kinetic pressure p of the plasma is small compared to the magnetic pressure, $p \ll B^2/8\pi$ (Aleksin and Khodusov, 1970; Akhiezer, Aleksin and Khodusov, 1971). In that case the most important plasmon interaction processes are the following ones:

$$\mathrm{a} \rightleftarrows \mathrm{a+s}, \quad \mathrm{a} \rightleftarrows \mathrm{f+s}, \quad \mathrm{f} \rightleftarrows \mathrm{f+s}, \quad \mathrm{f} \rightleftarrows \mathrm{a+s},$$

and the collision integrals have the form

$$\left.\begin{aligned}
(\dot N_{a1})_c &= 4\pi \sum_{2,3} \{w_{a\to a+s}(N_{a2}N_{s3}-N_{a1}N_{s3}-N_{a1}N_{a2})+w_{a+s\to a}(N_{a1}N_{a2}+N_{a2}N_{s3}-N_{a1}N_{s3})\\
&\quad + w_{a\to f+s}(N_{f2}N_{s3}-N_{a1}N_{s3}-N_{a1}N_{f2})+w_{s+a\to f}(N_{a1}N_{f2}+N_{f2}N_{s3}-N_{a1}N_{s3})\},\\
(\dot N_{f2})_c &= 4\pi \sum_{1,3} \{w_{f\to a+s}(N_{a1}N_{s3}-N_{a1}N_{f2}-N_{f2}N_{s3})+w_{s+f\to a}(N_{a1}N_{s3}+N_{a1}N_{f2}-N_{f2}N_{s3})\\
&\quad + w_{f\to s+f}(N_{f1}N_{s2}-N_{f2}N_{f1}-N_{f2}N_{s1})+w_{f+s\to f}(N_{f1}N_{s3}+N_{f1}N_{f2}-N_{f2}N_{s3})\},\\
(\dot N_{s3})_c &= 4\pi \sum_{1,2} \{2w_{a+s\to a}(N_{a1}N_{a2}+N_{s3}N_{a2}-N_{a1}N_{s3})+2w_{f+s\to f}(N_{f1}N_{f2}+N_{s3}N_{f2}-N_{s3}N_{f1})\},
\end{aligned}\right\} \tag{10.4.3.6}$$

where

$$\left.\begin{aligned}
w_{a\to a+s} &= M_1\delta^{(4)}(k_1^a-k_2^a-k_3^s), & w_{a+s\to a} &= M_1\delta^{(4)}(k_1^a-k_2^a+k_3^s),\\
w_{a\rightleftarrows f+s} &= M_2\delta^{(4)}(k_1^a-k_2^f-k_3^s), & w_{a+s\rightleftarrows f} &= M_2\delta^{(4)}(k_1^a-k_2^f+k_3^s),\\
w_{f\to f+s} &= M_3\delta^{(4)}(k_1^f-k_2^f-k_3^s), & w_{f+s\to f} &= M_3\delta^{(4)}(k_1^f-k_2^f+k_3^s),
\end{aligned}\right\} \tag{10.4.3.7}$$

$N_{\mu_1} = N_\mu(\boldsymbol{k}_1)$, while k^μ stands for both the wavevector and the energy of a plasmon of kind μ. Furthermore,

$$\left.\begin{aligned}
M_1 &= A\varkappa_{1||}^2\varkappa_{2||}^2\varkappa_{3||}^2\,\frac{([\boldsymbol{\varkappa}_1\wedge\boldsymbol{n}]\cdot[\boldsymbol{\varkappa}_2\wedge\boldsymbol{n}])^2}{\varkappa_{1\perp}^2\varkappa_{2\perp}^2},\\
M_2 &= A\varkappa_{1||}^2\,\frac{\varkappa_{2\perp}^2}{\varkappa_{1\perp}^2}\left\{(\boldsymbol{\varkappa}_3\cdot[\boldsymbol{\varkappa}_1\wedge\boldsymbol{n}])+\frac{\varkappa_{2||}\varkappa_{3||}}{\varkappa_{2\perp}^2}(\boldsymbol{\varkappa}_2\cdot[\boldsymbol{\varkappa}_1\wedge\boldsymbol{n}])\right\}^2,\\
M_3 &= A\,\frac{\varkappa_{3||}^2}{\varkappa_{1\perp}^2\varkappa_{2\perp}^2}\{1-\varkappa_{1\perp}^2\varkappa_{2||}(\boldsymbol{\varkappa}_2\cdot\boldsymbol{\varkappa}_3)-\varkappa_{2\perp}^2\varkappa_{1||}(\boldsymbol{\varkappa}_1\cdot\boldsymbol{\varkappa}_3)-\varkappa_{1||}\varkappa_{2||}\varkappa_{3||}\\
&\quad\times([\boldsymbol{\varkappa}_1\wedge\boldsymbol{n}]\cdot[\boldsymbol{\varkappa}_2\wedge\boldsymbol{n}])\}^2,\\
A &= \frac{v_A^4}{32V\varrho_m}\,\hbar^2\,\frac{k_1^2k_2^2k_3^2}{\omega_a(\boldsymbol{k}_1)\,\omega_f(\boldsymbol{k}_2)\,\omega_s(\boldsymbol{k}_3)}.
\end{aligned}\right\} \tag{10.4.3.8}$$

10.4.4. PLASMON LIFETIMES

Let us now determine the lifetimes of the plasmons due to their interactions with one another. To do this we consider the collision integral for plasmons of a given kind μ with a given wavevector $\boldsymbol{k}$, and we shall assume that all plasmons of other kinds and also plasmons of the given kind but with other values of $\boldsymbol{k}$ are in a state of statistical equilibrium. As to the plasmon number $N_\mu(\boldsymbol{k})$ we shall assume that it differs little from its equilibrium value, $N_\mu(\boldsymbol{k}) = N_\mu^{(0)}(\boldsymbol{k})+\Delta N_\mu(\boldsymbol{k})$, where $\Delta N_\mu \ll N_\mu^{(0)}$. Bearing in mind that the collision integral vanishes for an equilibrium plasmon distribution, we see that in the case considered the collision integral will be proportional to ΔN_μ,

$$[\dot N_\mu(\boldsymbol{k})]_c = \left[\frac{\delta\mathcal{L}_\mu}{\delta N_\mu(\boldsymbol{k})}\right]_0 \Delta N_\mu(\boldsymbol{k}), \tag{10.4.4.1}$$

where $(\delta\mathcal{L}_\mu/\delta N_\mu)_0$ is the value of the functional derivative of the collision integral $\mathcal{L}_\mu$ (compare equation (10.4.1.2)) with respect to N_μ for $N_\mu = N_\mu^{(0)}$.

It is clear that the quantity $-(\delta\mathcal{L}_\mu/\delta N_\mu(\boldsymbol{k}))_0$ will be the larger the faster plasmons of kind μ approach their equilibrium distribution thanks to plasmon–plasmon interaction processes. It is therefore obvious to consider the quantity

$$\tau_\mu(\boldsymbol{k}) = -\left[\frac{\delta\mathcal{L}_\mu}{\delta N_\mu(\boldsymbol{k})}\right]_0^{-1} \tag{10.4.4.2}$$

to be the lifetime of plasmons of kind μ and wavevector $\boldsymbol{k}$ with regard to three-plasmon processes.

If we average the quantity $\tau_\mu^{-1}(\boldsymbol{k})$ over the intensity of the equilibrium plasmon distribution, we get the average reciprocal of the lifetime of plasmons of kind μ thanks to plasmon–plasmon interactions,

$$\frac{1}{\tau_\mu} = \frac{\int [\tau_\mu(\boldsymbol{k})]^{-1}\, \hbar\omega N_\mu^{(0)}(\boldsymbol{k})\, d^3\boldsymbol{k}}{\int N_\mu^{(0)}(\boldsymbol{k})\, \hbar\omega\, d^3\boldsymbol{k}}. \tag{10.4.4.3}$$

The integration here must be taken over volumes in wavevector space where the plasmons can exist and can interact with one another. Those are the volumes determined by the condition that the plasmon damping is small and by the energy and momentum conservation laws for the plasmon interactions; they turn out to be equal to

$$\Gamma \sim \frac{\omega_{Bi}^3}{6\pi^2 v_A^3}.$$

One can show that, as to order of magnitude, the average values of the reciprocal lifetimes of magneto-hydrodynamic waves are equal to

$$\frac{1}{\tau_a} \sim \frac{1}{\tau_f} \sim \frac{W}{n_i T_e}\,\omega_{Bi}, \qquad \frac{1}{\tau_s} \sim \frac{W}{n_i T_e}\left(\frac{v_s}{v_A}\right)^2 \omega_{Bi}, \tag{10.4.4.4}$$

where $W = T^*\Gamma$ is the wave energy and n_i the ion density. We note that

$$\tau_a \sim \tau_f \ll \tau_s;$$

it therefore takes the slow magneto-sound plasmons much longer to reach equilibrium than the Alfvén and the fast magneto-sound plasmons.

We emphasize that the stationary plasmon distribution, established thanks to the plasmon–plasmon interactions is, according to (10.4.1.2) a Rayleigh–Jeans distribution. In contrast to this, the stationary ion-sound distribution, studied in Section 10.3, differs appreciably from a Rayleigh distribution. This is connected with the totally different nature of the system considered here (a closed system—the gas of mutually interacting plasmons) and the system considered in Section 10.3 (an open system consisting of ion-sound waves interacting both with the electrons which are injected from outside and with the ions in the plasma).

Concluding this subsection, we note that the times τ_μ have a real meaning of the plasmon lifetimes only in the case when they are appreciably less than the plasmon lifetimes caused by the interaction of the plasmons with the plasma particles. On the other hand, these times must be large in comparison with the inverse of the plasmon frequency.

If we start from the linear theory in which the damping rate of the plasmons caused by the interaction with the plasma particles is given by the formula

$$\gamma_\mu^{(0)} \sim \sqrt{\frac{m_e}{m_i}\frac{v_s}{v_A}}\,\omega_{Bi}, \tag{10.4.4.5}$$

we get the following condition for the applicability of the treatment of the magneto-hydrodynamical waves as a gas of quasi-particles (plasmons) which interact only with themselves:

$$\sqrt{\frac{m_e}{m_i}\frac{v_A}{v_s}} \ll \frac{W}{n_i T_e} \ll 1. \tag{10.4.4.6}$$

One can, however, only use the linear theory of the damping of the waves, if the inequality

$$\frac{W}{n_i T_e} \ll \frac{v_s}{v_A}\frac{\nu}{|\omega_{Be}|} \tag{10.4.4.7}$$

is satisfied, where ω_{Be} is the electron cyclotron frequency and $\nu = 2\pi e^4 n_e \Lambda / m_e^2 v_e^3$ the frequency of Coulomb collisions, with Λ the Coulomb logarithm.

As the level of the plasmon energy, obtained from (10.4.4.6), exceeds the level (10.4.4.7) we must use the non-linear theory for the damping of the waves; in it the damping is less than in the linear theory and the Cherenkov damping rate is given by the formula (Rowlands, Sizonenko, and Stepanov, 1966)

$$\gamma_\mu \sim \gamma_\mu^{(0)}\frac{\nu}{|\omega_{Be}|}\left(\frac{W}{n_i T_e}\right)^{-1}, \tag{10.4.4.8}$$

rather than by (10.4.4.5). Using that formula we find, instead of (10.4.4.6), the following condition for the applicability of the approach used:

$$\sqrt{\frac{m_e}{m_i}}\frac{\nu}{|\omega_{Be}|} \ll \left(\frac{W}{n_i T_e}\right)^2 \ll 1. \tag{10.4.4.9}$$

CHAPTER 11

Fluctuations in a Plasma

11.1. Fluctuation–Dissipation Relation

11.1.1. SPACE–TIME CORRELATION FUNCTIONS

It is well known that any physical quantity characterizing a macroscopic system which is in equilibrium can experience deviations from its average value. These deviations from the average value, called the *fluctuations* of the physical quantity, are determined by the temperature and the macroscopic properties of the system (we refer to the literature (for instance, Landau and Lifshitz, 1960, 1969; Rytov, 1953; Levin and Rytov, 1967) for a discussion of the general theory of fluctuations).

For a description of fluctuations we introduce *correlation functions* defined as the average values of products of fluctuations in a single or in several quantities at different points in space and at different times. The average is here both over the quantum-mechanical state of the system and over the statistical distribution of various quantum-mechanical states of the system. If the medium is spatial homogeneous and if we consider stationary states of the system, the second-order space–time correlation functions will depend only on the relative distances and the absolute value of the time interval between the points in which the fluctuations are considered. If, to fix the ideas, we consider a vector quantity $\boldsymbol{j}(\boldsymbol{r}, t)$ which is distributed continuously in space—for instance, the current density in the system—and assume that its average vanishes, we define the space–time correlation functions of the components $j_i(\boldsymbol{r}, t)$ and $j_j(\boldsymbol{r}, t)$ of the vector $\boldsymbol{j}$ by

$$\langle j_i(\boldsymbol{r}_1, t_1)\, j_j(\boldsymbol{r}_2, t_2)\rangle = \langle j_i j_j\rangle_{\boldsymbol{r}t}, \tag{11.1.1.1}$$

where $\boldsymbol{r} = \boldsymbol{r}_2 - \boldsymbol{r}_1$, $t = t_2 - t_1$, while the brackets $\langle \ldots \rangle$ indicate the above-mentioned averaging.

We introduce the *spectral representation* or *spectral density* of the correlation function, that is, the space–time Fourier components of $\langle j_i j_j\rangle_{\boldsymbol{r}t}$:

$$\langle j_i j_j\rangle_{\boldsymbol{k}\omega} = \int \langle j_i j_j\rangle_{\boldsymbol{r}t} e^{-i(\boldsymbol{k}\cdot\boldsymbol{r})+i\omega t}\, d^3\boldsymbol{r}\, dt. \tag{11.1.1.2}$$

In what follows we shall call this quantity the spectral density of the fluctuations.

Introducing Fourier transforms through the relations

$$A(\boldsymbol{k}, \omega) = \int A(\boldsymbol{r}, t)\, e^{-i(\boldsymbol{k}\cdot\boldsymbol{r})+i\omega t}\, d^3\boldsymbol{r}\, dt, \quad A(\boldsymbol{r}, t) = \frac{1}{(2\pi)^4}\int A(\boldsymbol{k}, \omega)\, e^{i(\boldsymbol{k}\cdot\boldsymbol{r})-i\omega t}\, d^3\boldsymbol{k}\, d\omega, \tag{11.1.1.3}$$

we see easily that the average of the products of the Fourier components of fluctuating quantities will be connected with the spectral density of the correlation function through the equation

$$\langle j_i^{\dagger}(\boldsymbol{k}, \omega)\, j_j(\boldsymbol{k}', \omega')\rangle = (2\pi)^4 \langle j_i j_j\rangle_{\boldsymbol{k}\omega}\, \delta(\boldsymbol{k}-\boldsymbol{k}')\, \delta(\omega-\omega'), \tag{11.1.1.4}$$

where the symbol † indicates the Hermitean conjugate.

The average of a product of fluctuations of any quantities at different points in space but at the same time is called a *spatial correlation function*,

$$\langle j_i(\boldsymbol{r}_1, t)\, j_j(\boldsymbol{r}_2, t)\rangle \equiv \langle j_i j_j\rangle_{\boldsymbol{r}}. \tag{11.1.1.5}$$

One checks easily that the Fourier component of the spatial correlation function is the integral over all frequencies of the spectral representation of the space–time correlation function:

$$\langle j_i j_j\rangle_{\boldsymbol{k}} = \frac{1}{2\pi} \int_{-\infty}^{\infty} \langle j_i j_j\rangle_{\boldsymbol{k}\omega}\, d\omega. \tag{11.1.1.6}$$

We note that the spatial correlation function for the particle density $\langle \delta n(\boldsymbol{r}_1)\, \delta n(\boldsymbol{r}_2)\rangle$ is connected with the two-particle distribution function $n_2(\boldsymbol{r}_1, \boldsymbol{r}_2)$ introduced in Section 1.2 through the formula

$$\langle \delta n(\boldsymbol{r}_1)\, \delta n(\boldsymbol{r}_2)\rangle = n_2(\boldsymbol{r}_1, \boldsymbol{r}_2) - n(\boldsymbol{r}_1)\, n(\boldsymbol{r}_2) + n(\boldsymbol{r}_1)\, \delta(\boldsymbol{r}_1 - \boldsymbol{r}_2). \tag{11.1.1.7}$$

The average of a product of fluctuations of any quantities in the same point in space but at different times is called the *time correlation function* or the *autocorrelation* function,

$$\langle j_i(\boldsymbol{r}, t_1)\, j_j(\boldsymbol{r}, t_2)\rangle = \langle j_i j_j\rangle_t. \tag{11.1.1.8}$$

We clearly have the relation

$$\langle j_i j_j\rangle_{\omega} = \frac{1}{(2\pi)^3} \int \langle j_i j_j\rangle_{\boldsymbol{k}\omega}\, d^3\boldsymbol{k}. \tag{11.1.1.9}$$

We can similarly introduce higher-order correlation functions, but we shall restrict ourselves here to considering merely the second-order correlation functions.

11.1.2. THE SPECTRAL DENSITY OF FLUCTUATIONS AND THE ENERGY DISSIPATION IN A MEDIUM

We shall now show that the spectral representation of a correlation function is determined by the dissipative properties of the medium. To do this we shall evaluate the average of the product of $j_i^{\dagger}(\boldsymbol{k}, \omega)$ and $j_j(\boldsymbol{k}, \omega)$. If the system is in a well-defined stationary state n, the quantum-mechanical average is defined as the diagonal matrix element of the operator $j_i^{\dagger}(\boldsymbol{k}, \omega)\, j_j(\boldsymbol{k}', \omega')$:

$$[j_i^{\dagger}(\boldsymbol{k}, \omega)\, j_j(\boldsymbol{k}', \omega')]_{nn} = \sum_m [j_i^{\dagger}(\boldsymbol{k}, \omega)]_{nm}\, [j_j(\boldsymbol{k}', \omega')]_{mn}, \tag{11.1.2.1}$$

where the summation is over all states of the system. The matrix elements of the operator $j_{k\omega}$ between stationary states, which occur here, have the following structure:

$$(\boldsymbol{j}_{k\omega})_{nm} = 2\pi(j_k)_{nm}\,\delta(\omega+\omega_{nm}), \tag{11.1.2.2}$$

where $\omega_{nm} = (E_n - E_m)/\hbar$ is the frequency of the transition between the states n and m and $(j_k)_{nm}$ are quantities which do not contain δ-functions. Substituting this expression and a similar one for $(j^{\dagger}_{k\omega})_{nm}$ into (11.1.2.1) and averaging, we get

$$\langle j_i^{\dagger}(\boldsymbol{k}, \omega)\, j_j(\boldsymbol{k}', \omega')\rangle = 2\pi\langle j_i^{\dagger}(\boldsymbol{k})\, j_j(\boldsymbol{k}')\rangle_{\omega}\,\delta(\omega-\omega'), \tag{11.1.2.3}$$

$$\langle j_i^{\dagger}(\boldsymbol{k})\, j_j(\boldsymbol{k}')\rangle_{\omega} = 2\pi \sum_{m,\,n} f(E_n)[j_i^{\dagger}(\boldsymbol{k})]_{nm}\,[j_j(\boldsymbol{k}')]_{mn}\,\delta(\omega-\omega_{nm}), \tag{11.1.2.4}$$

where $f(E_n)$ is the statistical distribution function for the different quantum-mechanical states of the system, which in the case of statistical equilibrium is given by Gibbs' formula

$$f(E_n) = e^{(F-E_n)/T}, \tag{11.1.2.5}$$

where F is the free energy and T the temperature of the system.

Let us now connect the correlation function (11.1.2.4) with the energy absorbed by the system due to dissipation. We shall assume that there is a periodic perturbation, with energy V proportional to $\boldsymbol{j}$, acting upon the system. If, to fix our ideas, we assume $\boldsymbol{j}$ to be the electrical current density, V has the form

$$V = -\int (\boldsymbol{A}(\boldsymbol{r}, t)\cdot\boldsymbol{j}(\boldsymbol{r}, t))\, d^3\boldsymbol{r}, \tag{11.1.2.6}$$

where $\boldsymbol{A}$ is the vector potential of the perturbing field.

Taking the spatial Fourier transforms of $\boldsymbol{A}$ and $\boldsymbol{j}$ we can write the perturbation energy in the form

$$V = -\tfrac{1}{2}\,\mathrm{Re}\sum_{\boldsymbol{k}} (\boldsymbol{A}_k(t)\cdot\boldsymbol{j}_k^{\dagger}(t)), \tag{11.1.2.7}$$

where $\boldsymbol{A}_k(t)$ is, by definition, a harmonic function of the time,

$$\boldsymbol{A}_k(t) = \boldsymbol{A}_{k\omega}\, e^{-i\omega t}. \tag{11.1.2.8}$$

Transitions between different states of the system become possible under the action of the perturbation (11.1.2.6). Using (11.1.2.7) we can easily evaluate the matrix element of the perturbation energy corresponding to the transition $n \to m$:

$$V_{nm} = -\pi \sum_{\boldsymbol{k}} \left\{(\boldsymbol{A}_{k\omega}\cdot\{\boldsymbol{j}_k^{\dagger}\}_{nm})\,\delta(\omega-\omega_{nm}) + (\boldsymbol{A}^{*}_{k\omega}\cdot\{\boldsymbol{j}_k\}_{nm})\,\delta(\omega+\omega_{nm})\right\}. \tag{11.1.2.9}$$

Hence it follows that the probability for a transition in the system per unit time is equal to

$$w_{nm} = \frac{\pi}{2\hbar^2} \sum_{\boldsymbol{k},\,\boldsymbol{k}',\,i,\,j} A_i(\boldsymbol{k}, \omega)\, A_j^{*}(\boldsymbol{k}', \omega)\{[j_i^{\dagger}(\boldsymbol{k})]_{nm}\,[j_j(\boldsymbol{k}')]_{mn}\,\delta(\omega-\omega_{nm})$$
$$+\,[j_i^{\dagger}(\boldsymbol{k})]_{mn}\,[j_j(\boldsymbol{k}')]_{nm}\,\delta(\omega+\omega_{nm}). \tag{11.1.2.10}$$

In each transition $n \to m$ the system absorb an energy $\hbar\omega_{mn}$; the source for this energy is the external perturbation. The energy absorbed by the system per unit time is thus equal to

$$Q_n = \sum_m w_{nm}\hbar\omega_{nm}. \tag{11.1.2.11}$$

We find the average absorbed energy by averaging (11.1.2.11) over all stationary states n:

$$Q = \sum_{m,\,n} f(E_n)\, w_{nm}\, \hbar\omega_{nm}. \tag{11.1.2.12}$$

If we restrict ourselves in what follows to considering equilibrium states we must substitute for $f(E_n)$ the canonical distribution (11.1.2.5). Substituting formula (11.1.2.10) into (11.1.2.12) and interchanging in one of the terms the summation indices n and m, we get

$$Q = \frac{\pi\omega}{2\hbar}\left[\exp(\hbar\omega/T)-1\right] \sum_{\mathbf{k},\,\mathbf{k}'\,i,\,j} A_i(\mathbf{k},\omega)\, A_j^*(\mathbf{k}',\omega)$$
$$\times \sum_{m,\,n} \exp\{(F-E_n)/T\}[j_i^\dagger(\mathbf{k})]_{nm}\,[j_j(\mathbf{k}')]_{mn}\,\delta(\omega-\omega_{nm}). \tag{11.1.2.13}$$

Comparing this expression with eqn. (11.1.2.4) we find the following relation between the average energy absorbed by the system in unit time and the correlation function of fluctuating quantities:

$$Q = \frac{\omega}{4\hbar}\left[\exp(\hbar\omega/T)-1\right] \sum_{\mathbf{k},\,\mathbf{k}',\,i,\,j} A_i(\mathbf{k},\omega)\, A_j(\mathbf{k}',\omega)\,\langle j_i^\dagger(\mathbf{k})\, j_j(\mathbf{k}')\rangle_\omega. \tag{11.1.2.14}$$

On the other hand, the absorbed energy Q can be connected with the macroscopic parameters characterizing the dissipative properties of the system. When there is no external perturbation the average of $\mathbf{j}$ vanishes, $\langle \mathbf{j}\rangle = 0$. The action of the perturbation (11.1.2.6) leads to a non-vanishing average of $\mathbf{j}$, which is proportional to the magnitude of the perturbing potential $\mathbf{A}$:

$$j_i = \sum_j \hat{\alpha}_{ij} A_j, \tag{11.1.2.15}$$

where $\hat{\alpha}_{ij}$ is a linear space–time integral operator. For the Fourier components of the quantities $\mathbf{j}$ and $\mathbf{A}$ we can write this connection in the form

$$j_i(\mathbf{k},\omega) = \sum_j \alpha_{ij}(\mathbf{k},\omega)\, A_j(\mathbf{k},\omega), \tag{11.1.2.16}$$

where $\alpha_{ij}(\mathbf{k},\omega)$ is a tensor characterizing the dissipative properties of the system, which we shall call the *response tensor*.

Processes for which eqn. (11.1.2.15) holds are usually called linear dissipative processes. In the case of a linear dissipative process the absorbed energy Q can be expressed directly in terms of the coefficients α_{ij}. Indeed, the change in the average internal energy of the system is equal to the average of the partial time derivative of the Hamiltonian of the system. As in the Hamiltonian only the perturbation V depends explicitly on the time, we find for the change in the internal energy of the system

$$\frac{\partial U}{\partial t} = -\int (\dot{\mathbf{A}}(\mathbf{r},t)\cdot\mathbf{j}(\mathbf{r},t))\, d^3r. \tag{11.1.2.17}$$

Fourier transforming, using (11.1.2.15), and time-averaging, we get the following expression for the average energy absorbed per unit time:

$$Q = \tfrac{1}{4} i\omega \sum_{k,\, i,\, j} (\alpha_{ij}^* - \alpha_{ji})\, A_i(\boldsymbol{k}, \omega)\, A_j^*(\boldsymbol{k}, \omega). \tag{11.1.2.18}$$

Comparing this equation with formula (11.1.2.14), we find that

$$\langle j_i^\dagger(\boldsymbol{k})\, j_j(\boldsymbol{k}')\rangle_\omega = \frac{8\pi^3 \hbar}{\exp(\hbar\omega/T) - 1}\, i(\alpha_{ij}^* - \alpha_{ji})\, \delta(\boldsymbol{k} - \boldsymbol{k}'). \tag{11.1.2.19}$$

This gives us the connection between the correlation function of fluctuating quantities and the dissipative properties of the system, characterized by the coefficients α_{ij}.

Using (11.1.2.3) and (11.1.1.5) we get from this the following formula for the spectral density of the current fluctuations:

$$\langle j_i j_j \rangle_{\boldsymbol{k}\omega} = \frac{\hbar}{\exp(\hbar\omega/T) - 1}\, i\{\alpha_{ij}^*(\boldsymbol{k}, \omega) - \alpha_{ji}(\boldsymbol{k}, \omega)\}. \tag{11.1.2.20}$$

This important relation—called the *fluctuation–dissipation relation*—completely determines the fluctuations in equilibrium systems; it was obtained by Callen and Welton (1951).

If we use the fluctuation–dissipation relation to find the fluctuations of a quantity $\boldsymbol{j}$, we must first of all use the expression for the change in the energy of the system to determine from (11.1.2.17) the quantity $\boldsymbol{A}$ corresponding to $\boldsymbol{j}$, and then we must use (11.1.2.15) to find the quantities α_{ji} which directly determine the spectral representation of the fluctuations.

Nyquist's well-known formula (Nyquist, 1928),

$$\langle I^2 \rangle_\omega = \frac{\hbar\omega}{2\pi\, |Z|^2}\, R \coth \frac{\hbar\omega}{2T}, \tag{11.1.2.21}$$

for the fluctuations of the current $\boldsymbol{I}$ in a system with impedance Z ($R = \operatorname{Re} Z$) is a particular example of eqn. (11.1.2.20).

If we remember that the fluctuations in the quantity $\boldsymbol{j}$ are the effect of the action of a random potential $\boldsymbol{A}$ on the system, we can write eqn. (11.1.2.10) in a different form. Noting that

$$A_i = \sum_j \alpha_{ij}^{-1} j_j, \tag{11.1.2.22}$$

and using (11.1.2.20) we find the spectral representation of the random potential:

$$\langle A_i A_j \rangle_{\boldsymbol{k}\omega} = \frac{\hbar}{\exp(\hbar\omega/T) - 1}\, i\{\alpha_{ji}^{-1}(\boldsymbol{k}, \omega) - \alpha_{ij}^{-1*}(\boldsymbol{k}, \omega)\}. \tag{11.1.2.23}$$

Let us also give an expression for the spectral representation of the symmetrized space time correlation function,

$$\langle \tfrac{1}{2}\{ j_i(\boldsymbol{r}_1, t_1)\, j_j(\boldsymbol{r}_2, t_2) + j_j(\boldsymbol{r}_2, t_2)\, j_i(\boldsymbol{r}_1, t_1)\}\rangle \equiv \langle j_i j_j \rangle^{\mathrm{s}}_{\boldsymbol{r}t}.$$

One sees easily that

$$\langle j_i j_j \rangle^{\mathrm{s}}_{\boldsymbol{k}\omega} = \frac{1}{2}\, \hbar i \{\alpha_{ij}^*(\boldsymbol{k}, \omega) - \alpha_{ji}(\boldsymbol{k}, \omega)\} \coth \frac{\hbar\omega}{2T}. \tag{11.1.2.24}$$

At sufficiently high temperatures, when $T \gg \hbar\omega$, eqn. (11.1.2.20) becomes

$$\langle j_i j_j \rangle_{k\omega} = \frac{T}{\omega} i\{\alpha_{ij}^*(\boldsymbol{k}, \omega) - \alpha_{ji}(\boldsymbol{k}, \omega)\}, \tag{11.1.2.25}$$

that is, the spectral representation is, as should be expected, independent of the quantum of action, $\hbar$. We shall in what follows restrict ourselves to considering only that case.

11.1.3. SYMMETRY OF THE RESPONSE TENSOR

The response tensor $\alpha_{ij}(\boldsymbol{k}, \omega)$ which connects the quantities $j_i(\boldsymbol{k}, \omega)$ and $A_j(\boldsymbol{k}, \omega)$ has a number of symmetry properties. First of all, it follows immediately from the fact that the quantities $\boldsymbol{j}$ and $\boldsymbol{A}$ are real that

$$\alpha_{ij}(\boldsymbol{k}, \omega) = \alpha_{ij}^*(-\boldsymbol{k}, -\omega). \tag{11.1.3.1}$$

Denoting the real and imaginary parts of α_{ij} by α_{ij}' and α_{ij}'', we have thus

$$\alpha_{ij}'(\boldsymbol{k}, \omega) = \alpha_{ij}'(-\boldsymbol{k}, -\omega), \quad \alpha_{ij}''(\boldsymbol{k}, \omega) = -\alpha_{ij}''(-\boldsymbol{k}, -\omega). \tag{11.1.3.2}$$

Then we have the symmetry relations

$$\alpha_{ij}(\boldsymbol{k}, \omega) = \alpha_{ji}(-\boldsymbol{k}, \omega). \tag{11.1.3.3}$$

To see that these relations are valid we must use the fact that the symmetrized correlation functions are invariant under time reversal. As time reversal is equivalent to the substitution $\omega \to -\omega$ in the Fourier components, we have

$$\langle j_i j_j \rangle_{\boldsymbol{k}\omega}^{s} = \langle j_i j_j \rangle_{\boldsymbol{k}-\omega}^{s}.$$

On the other hand, one easily checks that

$$\langle j_i j_j \rangle_{\boldsymbol{k}-\omega}^{s} = \langle j_j j_i \rangle_{-\boldsymbol{k}\omega}^{s},$$

whence the symmetry relations follow.

If the plasma is in an external magnetic field $\boldsymbol{B}$, we must in (11.1.3.3) together with the change $\boldsymbol{k} \to -\boldsymbol{k}$ change the sign of $\boldsymbol{B}$:

$$\alpha_{ij}(\boldsymbol{B}, \boldsymbol{k}, \omega) = \alpha_{ji}(-\boldsymbol{B}, -\boldsymbol{k}, \omega). \tag{11.1.3.4}$$

Equations (11.1.3.1) and (11.1.3.3) simplify considerably if the medium is isotropic and if there is no external field. In that case we have only one vector—the wavevector— at our disposal to construct the tensor $\alpha_{ij}(\boldsymbol{k}, \omega)$. The tensor α_{ij} therefore must have the following structure:

$$\alpha_{ij}(\boldsymbol{k}, \omega) = A(\boldsymbol{k}, \omega)\, \delta_{ij} + B(\boldsymbol{k}, \omega)\, k_i k_j, \tag{11.1.3.5}$$

where A and B are functions of the absolute magnitude of the wavevector. From this it follows at once that the tensor α_{ij} in the isotropic case is symmetric in the indices i and j:

$$\alpha_{ij}(\boldsymbol{k}, \omega) = \alpha_{ji}(\boldsymbol{k}, \omega), \tag{11.1.3.6}$$

and is an even function of $\boldsymbol{k}$:

$$\alpha_{ij}(\boldsymbol{k}, \omega) = \alpha_{ij}(-\boldsymbol{k}, \omega). \tag{11.1.3.7}$$

According to (11.1.3.2) the real part of the tensor α_{ij} is in that case an even, and the imaginary part of α_{ij} an odd function of the frequency,

$$\alpha'_{ij}(\boldsymbol{k}, \omega) = \alpha'_{ij}(\boldsymbol{k}, -\omega), \quad \alpha''_{ij}(\boldsymbol{k}, \omega) = -\alpha''_{ij}(\boldsymbol{k}, -\omega). \tag{11.1.3.8}$$

Considered as a function of the complex variable ω the tensor $\alpha_{ij}(\boldsymbol{k}, \omega)$ is an analytical function in the upper half-plane of ω. This is connected with the fact that the value of the quantity $\boldsymbol{j}$ at time t can depend only on values of the external action $\boldsymbol{A}$ at earlier times t' ($t' \leq t$). From eqn. (11.1.2.18) which connects the imaginary part of α_{ij} with the energy dissipation it follows that the function $\alpha''_{ij}(\boldsymbol{k}, \omega)$ has no zeroes for real, non-vanishing values of ω.

From these properties we find, in turn, the following relations:

$$\alpha'_{ij}(\boldsymbol{k}, \omega) - \alpha'_{ij}(\boldsymbol{k}, \infty) = \frac{1}{\pi} \mathcal{P} \int_{-\infty}^{+\infty} \frac{\alpha''_{ij}(\boldsymbol{k}, \omega')}{\omega' - \omega} d\omega', \tag{11.1.3.9}$$

$$\alpha''_{ij}(\boldsymbol{k}, \omega) = \frac{1}{\pi} \mathcal{P} \int_{-\infty}^{+\infty} \frac{\alpha'_{ij}(\boldsymbol{k}, \infty) - \alpha'_{ij}(\boldsymbol{k}, \omega')}{\omega' - \omega} d\omega', \tag{11.1.3.10}$$

which connect the real and imaginary parts of the tensor α_{ij}. Using these relations—the so-called *Kramers–Kronig relations*—we can easily find the correlation function of the current density at equal times:

$$\langle j_i j_j \rangle_{\boldsymbol{k}} = \tfrac{1}{2} T \{ \alpha^*_{ij}(\boldsymbol{k}, 0) - \alpha^*_{ij}(\boldsymbol{k}, \infty) + \alpha_{ji}(\boldsymbol{k}, 0) - \alpha_{ji}(\boldsymbol{k}, \infty) \}, \tag{11.1.3.11}$$

where we have assumed that $T \gg \hbar\omega$.

11.2. Electromagnetic Fluctuations in an Equilibrium Plasma

11.2.1. ELECTROMAGNETIC FLUCTUATIONS IN MEDIA WITH SPACE–TIME DISPERSION

Let us now apply the general theory of fluctuations discussed in the preceding section to a study of electromagnetic fluctuations in media with space–time dispersion, in particular, in an equilibrium plasma.† Let $E_i(\boldsymbol{k}, \omega)$ and $j_i(\boldsymbol{k}, \omega)$ denote the Fourier components of the electrical field and of the current density. These quantities are related to one another through the Maxwell equations for a vacuum:

$$\sum_j \Lambda^{(0)}_{ij} E_j(\boldsymbol{k}, \omega) = -\frac{4\pi i}{\omega} j_i(\boldsymbol{k}, \omega), \tag{11.2.1.1}$$

† Leontovich and Rytov (1952) developed the theory of electromagnetic fluctuations. Silin (1959b) generalized this to the case of a medium with spatial dispersion. Akhiezer, Akhiezer, and Sitenko (1962; Sitenko, 1967) studied fluctuations in an equilibrium plasma; in the present section we follow their work.

where

$$\Lambda_{ij}^{(0)} = n^2\left(\frac{k_i k_j}{k^2} - \delta_{ij}\right) + \delta_{ij}, \quad n = \frac{kc}{\omega}. \tag{11.2.1.2}$$

To find the fluctuating fields and currents we must introduce into the material equations connecting the current and the field the external field $\mathscr{E}(\boldsymbol{k}, \omega)$:

$$j_i = \frac{1}{4\pi}\sum_j (\varepsilon_{ij} - \delta_{ij})(\dot{E}_j + \dot{\mathscr{E}}_j), \tag{11.2.1.3}$$

where $\varepsilon_{ij} = \varepsilon_{ij}(\boldsymbol{k}, \omega)$ is the dielectric permittivity tensor of the plasma, and we must determine the change in the energy of the medium caused by the action of the external field,

$$\frac{\partial U}{\partial t} = \int (\mathscr{E}(\boldsymbol{r}, t) \cdot \boldsymbol{j}(\boldsymbol{r}, t))\, d^3\boldsymbol{r} = -\int (\dot{\boldsymbol{A}}(\boldsymbol{r}, t) \cdot \boldsymbol{j}(\boldsymbol{r}, t))\, d^3\boldsymbol{r}, \tag{11.2.1.4}$$

where

$$\mathscr{E}(\boldsymbol{r}, t) = -\dot{\boldsymbol{A}}(\boldsymbol{r}, t).$$

According to (11.1.2.25), the spectral representation of the current density fluctuations is determined by the coefficients of proportionality in the relation connecting j_i and $A_j = -(i/\omega)\,\mathscr{E}_j$. To find those we must use (11.2.1.3) to eliminate the field $\boldsymbol{E}(\boldsymbol{k}, \omega)$ from (11.2.1.3):

$$j_i(\boldsymbol{k}, \omega) = \sum_j \alpha_{ij}(\boldsymbol{k}, \omega)\, A_j(\boldsymbol{k}, \omega), \tag{11.2.1.5}$$

where

$$\alpha_{ij}(\boldsymbol{k}, \omega) = \frac{\omega^2}{4\pi}\left\{\Lambda_{ij}^{(0)} - \sum_{k,l} \Lambda_{ik}^{(0)} \Lambda_{kl}^{-1} \Lambda_{lj}^{(0)}\right\}, \tag{11.2.1.6}$$

while Λ_{ij}^{-1} is the tensor which is the inverse of the tensor Λ_{ij},

$$\Lambda_{ij} = n^2\left(\frac{k_i k_j}{k^2} - \delta_{ij}\right) + \varepsilon_{ij}. \tag{11.2.1.7}$$

According to the fluctuation–dissipation relation (11.1.2.25) the current fluctuations in an equilibrium system are determined by the anti-Hermitean part of the tensor α_{ij} and we have seen that this, in turn, is determined by the dielectric permittivity tensor of the medium. Substituting (11.2.1.6) into (11.1.2.25) we get the following expression for the spectral representation of the current density fluctuations in an equilibrium system:

$$\langle j_i j_j \rangle_{k\omega} = \frac{i}{4\pi}\, \omega T \sum_{k,l} \Lambda_{ik}^{(0)} \{\Lambda_{kl}^{-1} - \Lambda_{lk}^{-1*}\}\, \Lambda_{lj}^{(0)}. \tag{11.2.1.8}$$

This formula determines the fluctuations of the electrical current in any system with space–time dispersion in a state of statistical equilibrium. In particular, the spectral representation (11.2.1.8) is valid both for an unmagnetized plasma and for a plasma in an external magnetic field; it is only necessary that the plasma is in equilibrium.

One easily concludes from the form of the tensor α_{ij} that its singularities are the same as the zeroes of the determinant of the Λ_{ij} (the tensor Λ_{ij}^{-1} is inversely proportional to

Det $\Lambda_{ij} \equiv \Lambda$). On the other hand, we saw in Chapter 4 that the equation $\Lambda(\boldsymbol{k}, \omega) = 0$ determines the spectrum of the plasma eigen oscillations. The spectral density of the fluctuations of the plasma thus has steep maxima at frequencies which are the same as the frequencies of its eigenoscillations.

In the transparency region (Im $\Lambda \ll$ Re Λ) we can neglect damping of the waves and the spectral density of the current fluctuations will have the form

$$\langle j_i j_j \rangle_{\boldsymbol{k}\omega} = \tfrac{1}{2}\,\omega T \sum_{k,l} \Lambda_{ik}^{(0)} \lambda_{kl} \Lambda_{lj}^{(0)}\, \delta\{\Lambda(\boldsymbol{k}, \omega)\}, \tag{11.2.1.9}$$

where λ_{kl} is the cofactor of Λ_{kl} in the determinant Λ ($\Sigma_k \lambda_{ik} \Lambda_{kj} = \Lambda \delta_{ij}$).

We see that in the transparency region the spectral representation of the current fluctuations have δ-function-like maxima near the eigenfrequencies, that is, the frequency spectrum of the fluctuations contains solely the eigenfrequencies of the oscillations in the plasma.

Using the connection between the charge densities and the current,

$$\omega \varrho_{\boldsymbol{k}\omega} = (\boldsymbol{k} \cdot \boldsymbol{j}_{\boldsymbol{k}\omega}), \tag{11.2.1.10}$$

which follows from the equation of continuity, we easily find, using (11.2.1.8), the spectral density of the charge density fluctuations in the plasma:

$$\langle \varrho^2 \rangle_{\boldsymbol{k}\omega} = \frac{1}{2\pi} \frac{T}{\omega} \operatorname{Im} \sum_{i,j} k_i \Lambda_{ij}^{-1*} k_j. \tag{11.2.1.11}$$

Finally, expressing the field $E_i(\boldsymbol{k}, \omega)$ in terms of $j_j(\boldsymbol{k}, \omega)$ and using (11.2.1.8), we find the spectral density of the electrical field fluctuations in the plasma:

$$\langle E_i E_j \rangle_{\boldsymbol{k}\omega} = 4\pi i \frac{T}{\omega} \{\Lambda_{ij}^{-1} - \Lambda_{ji}^{-1*}\} \tag{11.2.1.12}$$

In the transparency region (Im $\Lambda \ll$ Re Λ),

$$\Lambda_{ij}^{-1} - \Lambda_{ji}^{-1*} \to \frac{\pi}{|\operatorname{Im} \Lambda|} (\Lambda^* \lambda_{ij} - \Lambda \lambda_{ji}^*)\, \delta\{\Lambda(\boldsymbol{k}, \omega)\},$$

that is, the spectral representation of the electrical field fluctuations, like that of the current fluctuations, has δ-function-like maxima at the frequencies of the plasma eigen oscillations.

Noting further that, if ω and $\boldsymbol{k}$ satisfy the dispersion relation, we have the relation

$$\lambda_{ij} = e_i e_j^* \operatorname{Tr} \lambda, \tag{11.2.1.13}$$

where the $\boldsymbol{e}$ are the polarization vectors of the appropriate oscillations, and bearing in mind that (Tr λ/ω) Im $\Lambda > 0$, we can write the spectral representation of the electrical field fluctuations in the form

$$\langle E_i E_j \rangle_{\boldsymbol{k}\omega} = 8\pi^2 e_i^* e_j T \frac{|\operatorname{Tr} \lambda|}{|\omega|}\, \delta\{\Lambda(\boldsymbol{k}, \omega)\}. \tag{11.2.1.14}$$

This formula is valid not only for a plasma, but also for any system with space–time dispersion.

11.2.2. ELECTROMAGNETIC FLUCTUATIONS IN AN ISOTROPIC PLASMA

The formulae for the spectral densities of fluctuations become much simpler for an isotropic plasma when there are no external fields, as in that case the tensor α_{ij} has a simple structure: namely, like the dielectric permittivity tensor ε_{ij} we can split it into a longitudinal and a transverse part:

$$\alpha_{ij} = \frac{k_i k_j}{k^2}\alpha_l + \left(\delta_{ij} - \frac{k_i k_j}{k^2}\right)\alpha_t, \tag{11.2.2.1}$$

where the quantities α_l and α_t which are the coefficients giving the proportionality between the longitudinal and transverse components of the current $\boldsymbol{j}$ and the potential $\boldsymbol{A}$ depend only on the absolute magnitude of the wavevector, k, and on the frequency ω and are connected with the longitudinal and transverse components of the dielectric permittivity ε_l and ε_t by the relations

$$\alpha_l = \frac{\omega^2}{4\pi}\frac{\varepsilon_l - 1}{\varepsilon_l}, \qquad \alpha_t = \frac{\omega^2}{4\pi}(1-n^2)\frac{\varepsilon_t - 1}{\varepsilon_t - n^2}. \tag{11.2.2.2}$$

The real parts α_l' and α_t' of the quantities α_l and α_t are even, and the imaginary parts α_l'' and α_t'' odd functions of the frequency. They are related through Kramers–Kronig relations

$$\alpha_l'(\boldsymbol{k}, \omega) - \alpha_l'(\boldsymbol{k}, \infty) = \frac{1}{\pi}\mathcal{P}\int_{-\infty}^{+\infty}\frac{\alpha_l''(\boldsymbol{k}, \omega')}{\omega' - \omega}\,d\omega',$$

$$\alpha_t'(\boldsymbol{k}, \omega) - \alpha_t'(\boldsymbol{k}, \infty) = \frac{1}{\pi}\mathcal{P}\int_{-\infty}^{+\infty}\frac{\alpha_t''(\boldsymbol{k}, \omega')}{\omega' - \omega}\,d\omega'. \tag{11.2.2.3}$$

Substituting expressions (11.2.2.2) into (11.1.2.25) we get

$$\langle j_i j_j \rangle_{k\omega} = \frac{\omega}{2\pi}T\left\{\frac{k_i k_j}{k^2}\frac{\operatorname{Im}\varepsilon_l}{|\varepsilon_l|^2} + \left(\delta_{ij} - \frac{k_i k_j}{k^2}\right)(1-n^2)\frac{\operatorname{Im}\varepsilon_t}{|\varepsilon_t - n^2|^2}\right\}. \tag{11.2.2.4}$$

The first term determines here the fluctuations of the longitudinal current, and the second term the fluctuations of the transverse current.

We note that the current density fluctuations are connected with the longitudinal current fluctuations because of the continuity equation. The spectral density of the charge density fluctuations in an isotropic plasma is determined by the formula

$$\langle \varrho^2 \rangle_{k\omega} = \frac{k^2}{2\pi}\frac{T}{\omega}\frac{\operatorname{Im}\varepsilon_l}{|\varepsilon_l|^2}. \tag{11.2.2.5}$$

Using the connection between the field $\boldsymbol{E}$ and the current $\boldsymbol{j}$, as well as eqn. (11.2.2.4), we can easily obtain a formula for the spectral density of the electrical field strength fluctuations:

$$\langle E_i E_j \rangle_{k\omega} = 8\pi\frac{T}{\omega}\left\{\frac{k_i k_j}{k^2}\frac{\operatorname{Im}\varepsilon_l}{|\varepsilon_l|^2} + \left(\delta_{ij} - \frac{k_i k_j}{k^2}\right)\frac{\operatorname{Im}\varepsilon_t}{|\varepsilon_t - n^2|^2}\right\}. \tag{11.2.2.6}$$

The magnetic field is connected with the electrical field through the relation $\boldsymbol{B} = (c/\omega)[\boldsymbol{k} \wedge \boldsymbol{E}]$; the spectral density of the magnetic field fluctuations is thus given by the equation

$$\langle B_i B_j \rangle = 8\pi \frac{T}{\omega}\left(\delta_{ij} - \frac{k_i k_j}{k^2}\right) n^2 \frac{\operatorname{Im} \varepsilon_t}{|\varepsilon_t - n^2|^2}. \tag{11.2.2.7}$$

Using the Kramers–Kronig relations (11.2.2.3) we can integrate the spectral density over the frequencies, in its general form. Integrating (11.2.2.4) we get in this way the correlation function of the current fluctuations at one time:

$$\langle j_i j_j \rangle_k = T\left\{\frac{k_i k_j}{k^2}[\alpha_l(\boldsymbol{k}, 0) - \alpha_l(\boldsymbol{k}, \infty)] + \left(\delta_{ij} - \frac{k_i k_j}{k^2}\right)[\alpha_t(\boldsymbol{k}, 0) - \alpha_t(\boldsymbol{k}, \infty)]\right\}. \tag{11.2.2.8}$$

In deriving this equation we used the fact that the imaginary parts of α_l and α_t vanish as $\omega \to 0$ and as $\omega \to \infty$.

We can similarly find the correlation function for the charge density fluctuations at the same time:

$$\langle \varrho^2 \rangle_k = \frac{k^2}{4\pi} T\left\{1 - \frac{1}{\varepsilon_l(\boldsymbol{k}, 0)}\right\}. \tag{11.2.2.9}$$

Using the integral relations (11.2.2.3) we can also in a general form find the quantity

$$\langle \omega^2 \rangle = \int_{-\infty}^{+\infty} \omega^2 \langle \varrho^2 \rangle_{k\omega}\, d\omega \Big/ \int_{-\infty}^{+\infty} \langle \varrho^2 \rangle_{k\omega}\, d\omega,$$

which can be considered to be the mean square frequency of the charge density fluctuations. Using (11.2.2.3) we easily check that this frequency equals

$$\langle \omega^2 \rangle = \frac{\omega_{pe}^2}{1 - \varepsilon_l(\boldsymbol{k}, 0)^{-1}}, \tag{11.2.2.10}$$

where $\omega_{pe}^2 = \lim_{\omega \to \infty} \omega^2\{1 - \varepsilon_l(\boldsymbol{k}, \omega)\} = 4\pi n_0 e^2/m_e$. We can find the mean square frequency of the transverse fluctuations in a similar way.

The fluctuations of the different electromagnetic quantities are thus in an equilibrium isotropic plasma completely determined by the longitudinal and the transverse components of the plasma dielectric permittivity.

11.2.3. CHARGE DENSITY FLUCTUATIONS

Let us now use the explicit expressions for the components of the dielectric permittivity of an equilibrium isotropic plasma:

$$\left.\begin{aligned} \varepsilon_l(\boldsymbol{k}, \omega) &= 1 + \frac{1}{k^2 r_D^2}\{2 - \varphi(z) - \varphi(\mu z) + i\sqrt{\pi}\, z[e^{-z^2} + \mu e^{-\mu^2 z^2}]\}, \\ \varepsilon_t(\boldsymbol{k}, \omega) &= 1 - \frac{\omega_{pe}^2}{\omega^2}\{\varphi(z) - i\sqrt{\pi}\, z\, e^{-z^2}\}, \end{aligned}\right\} \tag{11.2.3.1}$$

twhere $z = \omega/\sqrt{2}\, kv_e$, $\mu = \sqrt{(m_i/m_e)}$, and $\varphi(z) = 2z \exp(-z^2) \int\limits_0^z \exp(x^2)\, dx$, and study in more detail the nature of the spectral representations of the fluctuations for various quanities.

Let us first of all consider the charge density fluctuations of an equilibrium isotropic plasma. Substituting (11.2.3.1) into (11.2.2.9) we get the following general formula for the spectral density of the charge density fluctuations in a plasma where the ion motion is taken into account:

$$\langle \varrho^2 \rangle_{k\omega} = \sqrt{(2\pi)}\, e^2 n_0 \frac{k^3 r_D^3}{\omega_{pe}} \frac{\exp(-z^2) + \mu \exp(-\mu^2 z^2)}{[k^2 r_D^2 + 2 - \varphi(z) - \varphi(\mu z)]^2 + \pi z^2 [\exp(-z^2) + \mu \exp(-\mu^2 z^2)]^2}. \tag{11.2.3.2}$$

In the region of sufficiently low frequencies, $z \ll \mu^{-1}$, this formula gives

$$\langle \varrho^2 \rangle_{k\omega} = \frac{\tilde{\omega}^2}{\tilde{\omega}^2 + \omega^2} \langle \varrho^2 \rangle_{k0},$$

where

$$\tilde{\omega}^2 = \frac{2}{\pi} \frac{(2 + k^2 r_D^2)^2}{(1+\mu)^2} k^2 r_D^2, \quad \langle \varrho^2 \rangle_{k_0} = \sqrt{(2\pi)}\, e^2 n_0 \frac{k^3 r_D^3}{\omega_{pe}} \frac{1+\mu}{(2 + k^2 r_D^2)^2}.$$

In the high-frequency region, $z \gg \mu^{-1}\sqrt{(\ln \mu)}$, we can neglect the ion motion and the expression for the spectral density of the fluctuations simplifies considerably

$$\langle \varrho^2 \rangle_{k\omega} = \sqrt{(2\pi)}\, e_0 n^2 \frac{k^3 r_D^3}{\omega_{pe}} \frac{\exp(-z^2)}{[k^2 r_D^2 + 1 - \varphi(z)]^2 + \pi z^2 \exp(-2z^2)}. \tag{11.2.3.3}$$

In the limiting case $z \gg 1$, that is, $\omega \gg kv_e$, we get then

$$\langle \varrho^2 \rangle_{k\omega} = 2\pi e^2 n_0 \frac{k^2 v_e^2}{\omega} \delta(\omega^2 - \omega_{pe}^2 - 3k^2 v_e^2). \tag{11.2.3.4}$$

In the high-frequency region there is thus only the plasma frequency present in the spectrum of the charge density fluctuations in a plasma.

We have given in Fig. 11.2.1 the spectral representation of the charge density fluctuations as function of the dimensionless frequency $z = \omega/\sqrt{2}\, kv_e$ for different values of the parameter $k^2 r_D^2$. At large values of the parameter $k^2 r_D^2$ the low-frequency fluctuations play the main role. When the parameter $k^2 r_D^2$ decreases the effective frequency of the fluctuations increases. When $k^2 r_D^2 \ll 1$, there remain in the fluctuation spectrum only frequencies close to the eigenfrequencies of the oscillations of the plasma density.

If $k^2 r_D^2 \gg 1$, one can neglect the Coulomb interactions and the spectral density is determined by the equation

$$\langle \varrho^2 \rangle_{k\omega} = \sqrt{(2\pi)} \frac{e^2 n_0}{kv_e} [\exp(-z^2) + \mu \exp(-\mu^2 z^2)]. \tag{11.2.3.5}$$

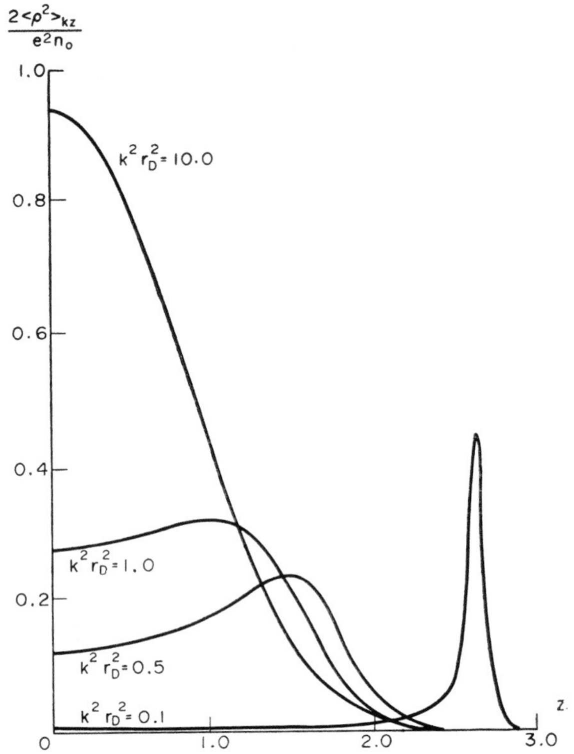

FIG. 11.2.1. The spectral representation of the charge density fluctuations in a plasma, $2\langle\varrho^2\rangle_{kz}/e^2n_0$, as function of the dimensionless frequency $z = \omega/\sqrt{2}kv_e$ for different values of the parameter $k^2r_D^2$.

According to (11.2.2.9) and (11.2.3.1) the spectral density of the charge density fluctuations, integrated over the frequency, is determined by the formula

$$\langle\varrho^2\rangle_k = \frac{2e^2n_0k^2}{k^2+(2/r_D^2)}. \tag{11.2.3.6}$$

If $k^2r_D^2 \gg 1$, we get from this the well-known result for a gas of non-interacting particles:

$$\langle\varrho^2\rangle_k = 2e^2n_0, \quad kr_D \gg 1. \tag{11.2.3.7}$$

We find the integrated contribution of the low-frequency fluctuations to the total intensity (11.2.3.6) by integrating (11.2.3.4) over the frequency:

$$\langle\tilde{\varrho}^2\rangle_k^2 = \frac{e^2n_0k^2r_D^2}{1+3k^2r_D^2}.$$

The ratio of the low-frequency to the total intensity equals

$$\frac{\langle\tilde{\varrho}^2\rangle_k}{\langle\varrho^2\rangle_k} = \frac{1}{2}\,\frac{2+k^2r_D^2}{1+3k^2r_D^2}.$$

This ratio equals 1/6 when $k^2r_D^2 \gg 1$ and equals 1 when $k^2r_D^2 \ll 1$.

We have taken both the contribution from the electrons and that from the ions into account in eqn. (11.2.3.6); the latter is especially important when $kr_D > 1$. If we neglect the ion motion, we get, instead of (11.2.3.6)

$$\langle \varrho^2 \rangle_k = \frac{e^2 n_0 k^2 r_D^2}{1+k^2 r_D^2}. \qquad (11.2.3.8)$$

We see that the quantity $\langle \varrho^2 \rangle_k$ is increased by a factor 2 by the effect of the ions when $k^2 r_D^2 \gg 1$. When kr_D decreases, so does the contribution from the ions.

Using (11.2.3.6) we can easily find the spatial correlation function for the charge density fluctuations:

$$\langle \varrho^2 \rangle_r = 2e^2 n_0 \left[\delta(\boldsymbol{r}) - \frac{1}{4\pi r_D^2} \frac{e^{-r/\bar{r}_D}}{r} \right], \qquad (11.2.3.9)$$

where $\bar{r}_D = r_D/\sqrt{2}$. We see that the correlation between the charge density fluctuations in a plasma occurs mainly at distances of the order of the Debye radius.

According to (11.2.2.10) and (11.2.3.1) the root mean square frequency of the charge density fluctuations is equal to

$$\sqrt{\langle \omega^2 \rangle} = \sqrt{[\omega_{pe}^2 + k^2 v_e^2]}. \qquad (11.2.3.10)$$

We note that the root mean square frequency of the fluctuations $\sqrt{\langle \omega^2 \rangle}$ is less than the eigenfrequency $\omega_p(\boldsymbol{k})$ of the plasma oscillations.

Let us also give the spectral representation of the charge density fluctuations for a relativistic electron plasma. The longitudinal component of the dielectric permittivity tensor of a plasma is for $T \gg m_e c^2$ determined by the formula

$$\varepsilon_l(\boldsymbol{k}, \omega) = 1 + \frac{1}{k^2 r_D^2} \left\{ 1 + \frac{\omega}{2kc} \ln \frac{|\omega - kc|}{\omega + kc} + \frac{\pi}{4} i \frac{\omega}{kc} \left[1 - \frac{\omega - kc}{|\omega - kc|} \right] \right\}, \qquad (11.2.3.11)$$

and substituting this into (11.2.2.5), we get

$$\langle \varrho^2 \rangle_{k\omega} = \frac{\pi e^2 n_0}{\omega} \left\{ \frac{\omega}{kc} \frac{\theta(kc-\omega)}{\left[1 + \frac{1}{k^2 r_D^2} \left(1 + \frac{\omega}{2kc} \ln \frac{|\omega - kc|}{\omega + kc} \right) \right]^2 + \frac{\pi^2}{4} \frac{\omega^2}{k^2 c^2} \frac{1}{k^4 r_D^4}} \right.$$
$$\left. + \frac{1}{2} k^2 r_D^2 \, \delta \left(1 + \frac{1}{k^2 r_D^2} \left[1 + \frac{\omega}{2kc} \ln \frac{|\omega - kc|}{\omega + kc} \right] \right) \right\}, \qquad (11.2.3.12)$$

where $\theta(x)$ is the Heaviside step-function.

We note that if we integrate (11.2.3.12) over the frequency we get, as in the non-relativistic case, eqn. (11.2.3.8).

11.2.4. CURRENT DENSITY FLUCTUATIONS

Let us now consider the current density fluctuations in an equilibrium, isotropic plasma. Using eqn. (11.2.3.1) we get the following formula for the spectral representation of the current density fluctuations:

$$\langle j_i j_j \rangle_{k\omega} = \frac{1}{\sqrt{(8\pi)}} \frac{\omega^2}{kv_e} T \left\{ \frac{k_i k_j}{k^2} \frac{\exp(-z^2)+\mu \exp(-\mu^2 z^2)}{[k^2 r_D^2 + 2 - \varphi(z) - \varphi(\mu z)]^2 + \pi z^2 [\exp(-z^2) + \mu \exp(-\mu^2 z^2)]^2} \right.$$

$$\left. + \left(\delta_{ij} - \frac{k_i k_j}{k^2}\right) \frac{\omega_{pe}^2}{\omega^2} (1-n^2)^2 \frac{\exp(-z^2)}{[n^2 - 1 + (\omega_{pe}^2/\omega^2)\varphi(z)]^2 + \pi(\omega_{pe}^4/\omega^4)\, z^2 \exp(-2z^2)} \right\}. \quad (11.2.4.1)$$

The first term in (11.2.4.1) determines the fluctuations in the longitudinal current and the second one those in the transverse current. As the corrections to the transverse dielectric permittivity, connected with the ion motion, are negligibly small, the motion of the ions practically does not affect the fluctuations of the transverse current density. However, the ion motion greatly affects the fluctuations of the longitudinal current density as it does the fluctuations charge density.

We shall give approximate expressions for the spectral representation of the transverse current density in the low- and high-frequency regions. If $\omega \ll kc$, we have

$$\langle j_t^2 \rangle k\omega = \frac{1}{\sqrt{(2\pi)}} \frac{\omega_{pe}^2}{kv_e} T \exp\left(-\frac{1}{2} \frac{\omega^2}{k^2 v_e^2}\right). \quad (11.2.4.2)$$

When $\omega \gg kv_e$ $(\omega \sim kc)$, we have

$$\langle j_t^2 \rangle_{k\omega} = \frac{\omega_{pe}^4}{\omega} T\delta(\omega^2 - \omega_{pe}^2 - k^2c^2), \quad (11.2.4.3)$$

that is, only the eigenfrequency of the transverse plasma oscillations occurs in the high-frequency region in the fluctuation spectrum.

If we integrate the spectral representation (11.2.4.1) over the frequency we can find the Fourier component of the spatial correlation function of the current density fluctuations in a plasma, $\langle j_i j_j \rangle_{\boldsymbol{k}}$. Noting that for an equilibrium plasma $\alpha_l(\boldsymbol{k}, 0) = \alpha_t(\boldsymbol{k}, 0) = 0$ and $\alpha_l(\boldsymbol{k}, \infty) = \alpha_t(\boldsymbol{k}, \infty) = -\omega_{pe}^2/4\pi$, we find from (11.2.2.8)

$$\langle j^2 \rangle_{\boldsymbol{k}} = \frac{3e^2 n_0 T}{m_e}. \quad (11.2.4.4)$$

The root mean square frequency of the transverse current fluctuations is equal to the Langmuir frequency,

$$\sqrt{\langle \omega^2 \rangle} = \omega_{pe}.$$

Using (11.2.4.4) we check easily that

$$\langle j^2 \rangle = \frac{3e^2 n_0 T}{m_e} \delta(r),$$

that is, there is no spatial correlation between the current density fluctuations in the plasma.

11.2.5. ELECTROMAGNETIC FIELD FLUCTUATIONS

The spectral representations of the electrical and magnetic fields in an isotropic equilibrium plasma are determined by the general formulae (11.2.2.6) and (11.2.2.7). Using eqn. (11.2.3.1) for the dielectric permittivity we can write the spectral representations for the field fluctuations in the form:

$$\langle \boldsymbol{E}^2\rangle_{k\omega} = 2(2\pi)^{3/2}\frac{kr_{\rm D}}{\omega_{\rm pe}}T\left\{\frac{\exp(-z^2)+\mu\exp(-\mu^2z^2)}{[k^2r_{\rm D}^2+2-\varphi(z)-\varphi(\mu z)]^2+\pi z^2[\exp(-z^2)+\mu\exp(-\mu^2z^2)]^2}\right.$$
$$\left.+\frac{2\omega_{\rm pe}^2}{\omega^2}\frac{\exp(-z^2)}{[n^2-1+(\omega_{\rm pe}^2/\omega^2)\varphi(z)]^2+\pi(\omega_{\rm pe}/\omega)^4z^2\exp(-2z^2)}\right\}, \tag{11.2.5.1}$$

$$\langle \boldsymbol{B}^2\rangle = 4(2\pi)^{3/2}\frac{kr_{\rm D}}{\omega}\frac{\omega_{\rm pe}}{\omega}n^2T\frac{\exp(-z^2)}{[n^2-1+(\omega_{\rm pe}/\omega)^2\varphi(z)]^2+\pi(\omega_{\rm pe}/\omega)^4z^2\exp(-2z^2)}. \tag{11.2.5.2}$$

The ion motion contributes only to the fluctuations of the longitudinal electrical field.

In the high-frequency region, $\omega \gg kv_{\rm e}$ ($\omega \approx kc$) the spectral representations of the transverse electrical field and of the magnetic field fluctuations have the form

$$\begin{aligned}\langle \boldsymbol{E}_{\rm t}^2\rangle_{k\omega} &= 16\pi^2T\omega\,\delta(\omega^2-\omega_{\rm pe}^2-k^2c^2),\\ \langle \boldsymbol{B}_{\rm t}^2\rangle_{k\omega} &= 16\pi^2T\omega\,\delta(\omega^2-\omega_{\rm pe}^2-k^2c^2).\end{aligned} \tag{11.2.5.3}$$

Integrating (11.2.5.1) and (11.2.5.2) over the frequency, we get

$$\langle \boldsymbol{E}^2\rangle_k = 8\pi T\left[1+\frac{1}{2+k^2r_{\rm D}^2}\right], \quad \langle \boldsymbol{B}^2\rangle_k = 8\pi T. \tag{11.2.5.4}$$

Using (11.2.5.4) we find easily the spatial correlation functions for the electrical and magnetic fields in an equilibrium plasma:

$$\langle \boldsymbol{E}^2\rangle_{\boldsymbol{r}} = 8\pi T\left[\delta(\boldsymbol{r})+\frac{1}{4\pi r_{\rm D}^2}\frac{\exp(-r/\bar{r}_{\rm D})}{r}\right], \quad \langle \boldsymbol{B}^2\rangle_{\boldsymbol{r}} = 8\pi T\,\delta(\boldsymbol{r}). \tag{11.2.5.5}$$

11.2.6. ELECTRON AND ION DENSITY FLUCTUATIONS

Let us now turn to a determination of the electron and ion density fluctuations. To do this we must introduce into the material eqns. (11.2.1.3) random fields which act independently on the electrons and the ions and then use the Maxwell equations and the relations

$$\varrho = e(\delta n_{\rm i}-\delta n_{\rm e}), \quad \boldsymbol{j} = \boldsymbol{j}_{\rm e}+\boldsymbol{j}_{\rm i}, \tag{11.2.6.1}$$

where $\delta n_{\rm i}$ and $\delta n_{\rm e}$ are the deviations of the ion and electron densities from their equilibrium value n_0 and $\boldsymbol{j}_{\rm e}$ and $\boldsymbol{j}_{\rm i}$ are the electron and ion components of the current, to determine the proportionality coefficients in the relation between the electron and ion current densities and the corresponding random potentials. These coefficients will be functions not only of the plasma dielectric permittivities $\varepsilon_{\rm l}$ and $\varepsilon_{\rm t}$, but also of the electron and ion plasma susceptibilities $\varkappa_l^e$, $\varkappa_l^i$, $\varkappa_t^e$, and $\varkappa_t^i$. Using the fluctuation–dissipation relation we get the following formulae for the spectral densities of the electron and ion current density fluctuations:

$$\left.\begin{aligned}\langle j_i^e j_j^e\rangle_{k\omega} &= 2T\omega\left\{\frac{k_ik_j}{k^2}\,\mathrm{Im}\,\frac{\varkappa_l^e(1+4\pi\varkappa_l^i)}{\varepsilon_l}+\left(\delta_{ij}-\frac{k_ik_j}{k^2}\right)\mathrm{Im}\,\frac{\varkappa_t^e(1+4\pi\varkappa_t^i-n^2)}{\varepsilon_t-n^2}\right\},\\ \langle j_i^i j_j^i\rangle_{k\omega} &= 2T\omega\left\{\frac{k_ik_j}{k^2}\,\mathrm{Im}\,\frac{\varkappa_l^i(1+4\pi\varkappa_l^e)}{\varepsilon_l}+\left(\delta_{ij}-\frac{k_ik_j}{k^2}\right)\mathrm{Im}\,\frac{\varkappa_t^i(1+4\pi\varkappa_t^e-n^2)}{\varepsilon_t-n^2}\right\}.\end{aligned}\right\} \tag{11.2.6.2}$$

Using the continuity equations for the electrons and the ions we easily get from these equations the spectral densities of the electron and ion density fluctuations:

$$\langle\delta n_e^2\rangle_{k\omega} = 8\pi n_0\,\frac{k^2r_D^2}{\omega}\,\mathrm{Im}\,\frac{\varkappa_l^e(1+4\pi\varkappa_l^i)}{\varepsilon_l},\quad \langle\delta n_i^2\rangle_{k\omega} = 8\pi n_0\,\frac{k^2r_D^2}{\omega}\,\mathrm{Im}\,\frac{\varkappa_l^i(1+4\pi\varkappa_l^e)}{\varepsilon_l}. \tag{11.2.6.3}$$

Integrating these equations over the frequency and using eqns. (11.2.2.3) we get

$$\langle\delta n_e^2\rangle_k = n_0\,\frac{1+k^2r_D^2-\varphi(\mu z)}{2+k^2r_D^2-\varphi(\mu z)}\bigg|_{z\to 0},\quad \langle\delta n_i^2\rangle_k = n_0\,\frac{(1+k^2r_D^2)\,[1-\varphi(\mu z)]}{2+k^2r_D^2-\varphi(\mu z)}\bigg|_{z\to 0}. \tag{11.2.6.4}$$

If we assume the ions to have infinite inertia ($\mu\to\infty$, $\varphi(\mu z)\to 1$), we get

$$\langle\delta n_e^2\rangle_k = n_0\,\frac{k^2r_D^2}{1+k^2r_D^2},\quad \langle\delta n_i^2\rangle_k = 0. \tag{11.2.6.5}$$

In actual fact, the ions are characterized by a finite mass so that $\varphi(\mu z)|_{z\to 0} = 0$, and hence

$$\langle\delta n_e^2\rangle_k = \langle\delta n_i^2\rangle_k = n_0\,\frac{1+k^2r_D^2}{2+k^2r_D^2}. \tag{11.2.6.6}$$

If $k^2r_D^2 \gg 1$, we can neglect the Coulomb interaction between the particles, and eqn. (11.2.6.6) leads to the well-known result for the mean square of the fluctuations in the density of neutral particles:

$$\langle\delta n_e^2\rangle_k = \langle\delta n_i^2\rangle_k = n_0,\quad k^2r_D^2 \gg 1. \tag{11.2.6.7}$$

In the limiting case $k^2r_D^2 \ll 1$ the mean square of the fluctuations turns out to be smaller by a factor two:

$$\langle\delta n_e^2\rangle_k = \langle\delta n_i^2\rangle = \tfrac{1}{2}n_0,\quad k^2r_D^2 \ll 1. \tag{11.2.6.8}$$

11.2.7. FLUCTUATIONS IN A PLASMA IN A MAGNETIC FIELD

Let us turn to a determination of fluctuations in an equilibrium plasma in a constant uniform magnetic field $\boldsymbol{B}_0$. In that case the tensor Λ_{ij} has the form

$$\Lambda_{ij} = \begin{bmatrix} -n^2\cos^2\vartheta+\varepsilon_{11} & \varepsilon_{12} & n^2\sin\vartheta\cos\vartheta+\varepsilon_{13}\\ -\varepsilon_{12} & -n^2+\varepsilon_{22} & \varepsilon_{23}\\ n^2\sin\vartheta\cos\vartheta+\varepsilon_{13} & -\varepsilon_{23} & -n^2\sin^2\vartheta+\varepsilon_{33}\end{bmatrix}, \tag{11.2.7.1}$$

where the z-axis is taken along $\boldsymbol{B}_0$, and the x-axis in the plane through $\boldsymbol{k}$ and $\boldsymbol{B}_0$, and ϑ is the angle between $\boldsymbol{k}$ and $\boldsymbol{B}_0$. The inverse tensor Λ^{-1} is given by the equation

$$\Lambda_{ij}^{-1} = \frac{\lambda_{ij}}{\Lambda},$$

where λ_{ij} is the cofactor and Λ the determinant of the matrix (11.2.7.1). The determinant Λ is equal to

$$\Lambda(\boldsymbol{k}, \omega) = An^4 + Bn^2 + C, \tag{11.2.7.2}$$

where $n = kc/\omega$ while the coefficients A, B, and C can, according to formulae (5.2.2.6), be expressed in terms of the components of the tensor ε_{ij}.

The charge density fluctuations in a plasma in a magnetic field are connected with the current fluctuations (11.2.1.8) by the relation

$$\langle \varrho^2 \rangle_{k\omega} = \frac{1}{\omega^2} \sum_{i,j} k_i k_j \langle j_i j_j \rangle_{k\omega}. \tag{11.2.7.3}$$

Using (11.2.1.8) we find

$$\langle \varrho^2 \rangle_{k\omega} = \frac{k^2}{2\pi} \frac{T}{\omega} \operatorname{Im} \left[1 - \frac{\Lambda(\boldsymbol{k}, \omega)}{\Delta(\boldsymbol{k}, \omega)} \right], \tag{11.2.7.4}$$

where

$$\Delta(\boldsymbol{k}, \omega) = n^4 - (\varepsilon_{11} \cos^2 \vartheta + \varepsilon_{22} + \varepsilon_{33} \sin^2 \vartheta - 2\varepsilon_{13} \sin \vartheta \cos \vartheta)\, n^2 \\ + (\varepsilon_{11}\varepsilon_{22} + \varepsilon_{12}^2) \cos^2 \vartheta + (\varepsilon_{22}\varepsilon_{33} + \varepsilon_{23}^2) \sin^2 \vartheta + 2(\varepsilon_{12}\varepsilon_{23} - \varepsilon_{22}\varepsilon_{13}) \sin \vartheta \cos \vartheta. \tag{11.2.7.5}$$

We shall integrate the spectral representation (11.2.7.4) over the frequency, using the Kramers–Kronig relations. Noting that $\Lambda(\boldsymbol{k}, 0)/\Delta(\boldsymbol{k}, 0) = 1 + 2(kr_{\mathrm{D}})^{-2}$, we find

$$\langle \varrho^2 \rangle_k = \frac{2n_0 e^2 k^2 r_{\mathrm{D}}^2}{2 + k^2 r_{\mathrm{D}}^2}. \tag{11.2.7.6}$$

We see thus that the magnetic field does not affect the spatial correlation function of the density fluctuations in the plasma.

However, the magnetic field does considerably affect the spectral density of the fluctuations of various quantities in the plasma. In the region where the plasma is transparent the spectral density of the current fluctuations has the form

$$\langle j_i j_j \rangle_{k\omega} = \tfrac{1}{2} \omega T \sum_{k,l} \Lambda_{jk}^{(0)} \lambda_{kl} \Lambda_{li}^{(0)}\, \delta(An^4 + Bn^2 + C). \tag{11.2.7.7}$$

The argument of the δ-function is the left-hand side of the dispersion equation of the plasma in a magnetic field. In the case of weak damping there occur thus only the eigenfrequencies of the plasma oscillations in a magnetic field in the fluctuation spectrum.

The spectral density of the density fluctuations which are connected with the Langmuir oscillations is given by the formula:

$$\langle \varrho^2 \rangle_{k\omega} = \frac{1}{4} k^2 T \frac{\omega^2}{\omega_{\mathrm{pe}}^2} \frac{(\omega^2 - \omega_B^2)^2}{\omega^4 \sin^2 \vartheta + (\omega^2 - \omega_B^2)^2 \cos^2 \vartheta} \{\delta(\omega - \omega_+) + \delta(\omega - \omega_-) + \delta(\omega + \omega_+) + \delta(\omega + \omega_-)\}, \tag{11.2.7.8}$$

where ω_+ and ω_- are the eigenfrequencies of the plasma oscillations in a magnetic field.

The spectral density of the electrical field fluctuations in a plasma in a magnetic field is given by the general formula (11.2.1.14). Noting that

$$\delta\{\Lambda(\boldsymbol{k},\omega)\} = \frac{1}{|\mathrm{Tr}\,\lambda|}\left\{\frac{1}{|\boldsymbol{e}|^2-|(\boldsymbol{k}\cdot\boldsymbol{e})|^2k^{-2}}\,[\delta(n^2-n_+^2)+\delta(n^2-n_-^2)]+\delta(\Lambda)\right\},$$

we can split off in (11.2.1.14) the contributions connected with separate eigen oscillations in the plasma. For instance, we have for the spectral density of the fluctuations connected with the ordinary and the extra-ordinary waves the formula

$$\langle E_iE_j\rangle_{k\omega} = 8\pi^2\frac{e_i^*e_j}{|\boldsymbol{e}|^2-|(\boldsymbol{k}\cdot\boldsymbol{e})|^2\,k^{-2}}\,\frac{T}{\omega}\,\delta(n^2-n_\pm^2), \tag{11.2.7.9}$$

where the polarization vectors are given by eqn. (5.1.1.10). The spectral density for the Langmuir fluctuations of the electrical field is given by the equation

$$\langle E_iE_j\rangle_{k\omega} = 4\pi^2\,\frac{k_ik_j}{k^2}\,T\,\frac{|\omega^2-\omega_B^2|}{\omega_+^2-\omega_-^2}\,\{\delta(\omega-\omega_+)+\delta(\omega+\omega_+)+\delta(\omega-\omega_-)+\delta(\omega+\omega_-)\}, \tag{11.2.7.10}$$

or

$$\langle E_iE_j\rangle_{k\omega} = 8\pi^2\frac{k_ik_j}{k^2}\,\frac{T}{\omega}\,\frac{n}{A_0}\,\delta(n^2-n_{\mathrm{pe}}^2), \tag{11.2.7.11}$$

if we take the dispersion of the Langmuir waves into account. Equations (11.2.7.9) to (11.2.7.11) describe the fluctuations in the high-frequency part of the plasma spectrum where the effect of the ions is unimportant.

11.3. Inversion of the Fluctuation–Dissipation Relation

11.3.1. CONNECTION BETWEEN THE DIELECTRIC PERMITTIVITY OF THE PLASMA AND THE CORRELATION FUNCTION OF THE FLUCTUATIONS FOR A SYSTEM OF NON-INTERACTING PARTICLES

In the preceding section we formulated the fluctuation–dissipation relation which connects the fluctuations in the electromagnetic quantities with the dielectric permittivity tensor of the medium. Using this relation and starting from an expression for the dielectric permittivity tensor of the plasma we found the correlation functions for the fluctuations in the fields and the charge and current densities in the plasma. However, we can use the fluctuation–dissipation relation also for something else. In fact, if we first of all use the microscopic theory to determine directly the charge and current density fluctuations, we can find the dielectric permittivity tensor using the fluctuation–dissipation relation.† A characteristic

† Shafranov (1958a), Kubo (1957), and Nakano (1954, 1957) suggested this approach. Sitenko (1966, 1967) has given a determination of the dielectric permittivity tensor, based on a conversion of the fluctuation–dissipation theorem for a non-equilibrium plasma.

feature of this method to determine the dielectric permittivity tensor is that it is possible to take thermal effects into account without recourse to a kinetic equation.

To explain this method we shall first of all consider an isotropic medium and turn to eqns. (11.2.2.5) and (11.2.2.4) which connect the spectral representations of the current density and the transverse current fluctuations with the longitudinal and transverse dielectric permittivities ε_l and ε_t and expand the correlation functions and the components of the dielectric permittivity which are functions of the elementary charge e in a power series in e—or rather in e^2. As the electrical conductivity contains a factor e^2, we get, retaining the first terms of the expansion

$$\operatorname{Im} \varepsilon_l^{(2)} = \frac{2\pi}{k^2}\frac{\omega}{T}\langle\varrho^2\rangle_{k\omega}^{(0)}, \quad \operatorname{Im} \varepsilon_t^{(2)} = \frac{\pi}{\omega T}\langle j_t^2\rangle_{k\omega}^0, \tag{11.3.1.1}$$

where $\langle\varrho^2\rangle_{k\omega}^{(0)}$ and $\langle j_t^2\rangle_{k\omega}^{(0)}$ are the spectral densities of the charge and current densities for a gas of non-interacting charged particles and $\varepsilon_{l,t}^{(2)}$ the value of the quantities $\varepsilon_{l,t}$ up to terms of order e^4.

Up to those terms we can now find the quantities $\operatorname{Re} \varepsilon_{l,t}$. To do that we use the Kramers–Kronig relations (11.2.2.3):

$$\left.\begin{aligned}\operatorname{Re} \varepsilon_l &= 1+\frac{2}{k^2T}\mathcal{P}\int_{-\infty}^{+\infty}\frac{\langle\varrho^2\rangle_{k\omega'}^{(0)'}}{\omega'-\omega}\,\omega'\,d\omega',\\ \operatorname{Re} \varepsilon_t &= 1-\frac{\omega_{pe}^2}{\omega^2}+\frac{1}{\omega^2T}\mathcal{P}\int_{-\infty}^{+\infty}\frac{\langle j_t^2\rangle_{k\omega'}^{(0)}}{\omega'-\omega}\,\omega'\,d\omega';\end{aligned}\right\} \tag{11.3.1.2}$$

we drop here and henceforth the index (2) of the $\varepsilon_{l,t}$. We can clearly combine these formula with eqns. (11.3.1.1) to write the longitudinal and transverse permittivities in the form (Sitenko, 1967):

$$\begin{aligned}\varepsilon_l(\boldsymbol{k}, \omega) &= 1+\frac{2}{k^2T}\int_{-\infty}^{+\infty}\frac{\langle\varrho^2\rangle_{k\omega'}^{(0)}}{\omega'-\omega-io}\,\omega'\,d\omega',\\ \varepsilon_t(\boldsymbol{k}, \omega) &= 1-\frac{\omega_{pe}^2}{\omega^2}+\frac{1}{\omega^2T}\int_{-\infty}^{+\infty}\frac{\langle j_t^2\rangle_{k\omega'}^{(0)}}{\omega'-\omega-io}\,\omega'\,d\omega',\end{aligned} \tag{11.3.1.3}$$

where the term $-io$ in the denominator of the integrand indicates that the integration over ω' is along a contour in the complex ω'-plane along the real axis, going round the point $\omega' = \omega$ from below.

Equations (11.3.1.3) determine the quantities $\varepsilon_l(\boldsymbol{k}, \omega)$ and $\varepsilon_t(\boldsymbol{k}, \omega)$ up to terms of order e^4 in the spectral densities of the charge density and transverse current fluctuations, $\langle\varrho^2\rangle_{k\omega}$ and $\langle j_t^2\rangle_{k\omega}$, for a gas of non-interacting charged particles.

We can similarly consider an anisotropic medium characterized by the dielectric permittivity tensor ε_{ij}. Turning in that case to eqn. (11.2.1.8) and noting that up to terms of order e^4 we have the expansion

$$\Lambda_{kl}^{-1} = \Lambda_{kl}^{(0)-1}-\sum_{n,m}\Lambda_{km}^{(0)-1}(\varepsilon_{mn}-\delta_{mn})\,\Lambda_{nl}^{(0)-1},$$

we get

$$\varepsilon_{ij}^{(2)} - \varepsilon_{ji}^{(2)*} = \frac{4\pi i}{\omega T} \langle j_i j_j \rangle_{k\omega}^{(0)}, \tag{11.3.1.4}$$

where $\langle j_i j_j \rangle_{k\omega}^{(0)}$ is the spectral density of the current fluctuations in the medium, if the electromagnetic interactions between the particles are neglected. Hence, proceeding as for the derivation of (11.3.1.3) we can find ε_{ij} (Sitenko, 1967):

$$\varepsilon_{ij}(\boldsymbol{k}, \omega) = \delta_{ij} + \frac{2}{\omega T} \int_{-\infty}^{+\infty} \frac{\langle j_j j_i \rangle_{k\omega'}^{(0}}{\omega' - \omega - io} \, d\omega'. \tag{11.3.1.5}$$

In deriving this formula we have used the relation $\int_{-\infty}^{+\infty} \langle j_i j_j \rangle_{k\omega}^{(0)} \, d\omega = \frac{1}{2} T \omega_{pe}^2 \, \delta_i$.

11.3.2. DIELECTRIC PERMITTIVITY OF AN ISOTROPIC PLASMA

Let us now evaluate the quantity $\langle \varrho^2 \rangle_{k\omega}^{(0)}$ for the case of an isotropic plasma, neglecting for the sake of simplicity the ion motion. We introduce the electron density $n(\boldsymbol{r}, t)$ in the point $\boldsymbol{r}$ at time t:

$$n(\boldsymbol{r}, t) = \sum_{\alpha} \delta(\boldsymbol{r} - \boldsymbol{r}_\alpha(t)), \tag{11.3.2.1}$$

where $\boldsymbol{r}_\alpha(t)$ is the radius vector of the αth electron at time t and where the summation is over all electrons in unit volume. As we are interested in the correlation function $\langle \varrho^2 \rangle_{k\omega}^{(0)}$ for the case when there are no interactions between the electrons, we may assume that

$$\boldsymbol{r}_\alpha(t) = \boldsymbol{r}_\alpha + \boldsymbol{v}_\alpha t,$$

where $\boldsymbol{v}_\alpha$ is the electron velocity. The charge density fluctuation in the plasma is connected with $n(\boldsymbol{r}, t)$ by the relation

$$\varrho(\boldsymbol{r}, t) = -e\{n(\boldsymbol{r}, t) - n_0\}, \tag{11.3.2.2}$$

where $n_0 = \langle n(\boldsymbol{r}, t) \rangle$ is the average electron density. Substituting this expression into the definition of the correlation function of the charge density fluctuations,

$$\langle \varrho^2 \rangle_{rt}^{(0)} \equiv \langle \varrho(\boldsymbol{r}_1, t_1) \, \varrho(\boldsymbol{r}_2, t_2) \rangle,$$

and using (11.3.2.2) and (11.3.2.1), we get

$$\langle \varrho^2 \rangle_{rt}^{(0)} = e^2 \left\langle \sum_{\alpha} \delta(\boldsymbol{r}_1 - \boldsymbol{r}_\alpha - \boldsymbol{v}_\alpha t_1) \, \delta(\boldsymbol{r}_2 - \boldsymbol{r}_\alpha - \boldsymbol{v}_\alpha t_2) \right\rangle, \tag{11.3.2.3}$$

where the brackets $\langle \ldots \rangle$ indicate a statistical average.

We introduce the single-particle electron distribution function $f_0(v)$, normalized by $\int f_0(v) \, d^3\boldsymbol{v} = n_0$. We then have, clearly,

$$\langle \varrho^2 \rangle_{rt}^{(0)} = e^2 \int f_0(v) \, \delta(\boldsymbol{r} - \boldsymbol{v}t) \, d^3\boldsymbol{v}. \tag{11.3.2.4}$$

To find the spectral density of the charge density fluctuations, $\langle \varrho^2 \rangle^{(0)}_{k\omega}$, in the case when we neglect the interactions between the charged particles, we must evaluate the Fourier component of $\langle \varrho^2 \rangle^{(0)}_{rt}$:

$$\langle \varrho^2 \rangle^{(0)}_{k\omega} = 2\pi e^2 \int f_0(v)\, \delta[\omega - (\boldsymbol{k}\cdot\boldsymbol{v})]\, d^3\boldsymbol{v}. \tag{11.3.2.5}$$

If we, finally, substitute this expression into (11.3.1.3) we find the longitudinal dielectric permittivity of the plasma

$$\varepsilon_l(\boldsymbol{k}, \omega) = 1 + \frac{1}{k^2 r_D^2} \int \frac{(\boldsymbol{k}\cdot\boldsymbol{v})\, f_0(v)}{(\boldsymbol{k}\cdot\boldsymbol{v}) - \omega - io}\, d^3\boldsymbol{v}, \tag{11.3.2.6}$$

where $r_D^2 = T/4\pi n_0 e^2$. This expression is the same as expression (4.3.4.3) for $\varepsilon_l(\boldsymbol{k}, \omega)$ which we obtained, using the Vlasov equation. On the other hand, we showed earlier that eqn. (11.3.2.6) is valid up to terms of order e^4. We can thus state that eqn. (4.3.4.3), based upon a kinetic discussion involving self-consistent fields is valid with the same accuracy. Proceeding along the same lines we can also obtain eqn. (4.3.4.3) for the transverse dielectric permittivity of an isotropic plasma.

11.3.3. DIELECTRIC PERMITTIVITY TENSOR OF A PLASMA IN A MAGNETIC FIELD

Let us now find by the same method the dielectric permittivity tensor of a plasma in a constant, uniform magnetic field. Neglecting the particle–particle interactions, the particles move along helices:

$$\boldsymbol{r}(t) = \boldsymbol{r}_0 + \boldsymbol{R}(t), \tag{11.3.3.1}$$

where $\boldsymbol{R}(t)$ is a vector with components $-(v_\perp/\omega_B)\cos\omega_B t$, $(v_\perp/\omega_B)\sin\omega_B t$, $v_{||}t$, where $v_{||}$ and $v_\perp$ are the longitudinal and transverse components of the electron velocity, with respect to the magnetic field $\boldsymbol{B}_0$ which is along the z-axis.

The correlation function for the charge density fluctuations is, as before, given by the equation

$$\langle \varrho^2 \rangle^{(0)}_{rt} = e^2 \left\langle \sum_\alpha \delta(\boldsymbol{r}_1 - \boldsymbol{r}_\alpha(t_1))\, \delta(\boldsymbol{r}_2 - \boldsymbol{r}_\alpha(t_2)) \right\rangle, \tag{11.3.3.2}$$

where $\boldsymbol{r} = \boldsymbol{r}_2 - \boldsymbol{r}_1$ and $t = t_2 - t_1$, and where the summation is over all electrons in unit volume. Introducing the single-particle distribution function $f_0(v)$, we get for the correlation function $\langle \varrho^2 \rangle_{rt}$ the expression

$$\langle \varrho^2 \rangle^{(0)}_{rt} = e^2 \int f_0(v)\, \delta(\boldsymbol{r} - \boldsymbol{r}(t) + \boldsymbol{r}(0))\, d^3\boldsymbol{v}. \tag{11.3.3.3}$$

The spectral density of the correlation function of the charge density fluctuations is given by the equation

$$\langle \varrho^2 \rangle^0_{k\omega} = e^2 \int f_0(v) \int_{-\infty}^{+\infty} e^{-i(\boldsymbol{k}\cdot\{\boldsymbol{r}(t)-\boldsymbol{r}(0)\}) + i\omega t}\, dt\, d^3\boldsymbol{v}. \tag{11.3.3.4}$$

Using the relations

$$e^{-ia\sin\psi} = \sum_{l=-\infty}^{+\infty} J_l(a)\, e^{-il\psi}, \qquad \int_0^{2\pi} e^{ia\sin\psi - il\psi} = 2\pi J_l(a),$$

where $J_l(a)$ is a Bessel function, we get for the spectral density of the correlation function of the charge density fluctuations the following formula:

$$\langle \varrho^2 \rangle^{(0)}_{k\omega} = 2\pi e^2 \sum_{l=-\infty}^{+\infty} \int f_0(v)\, J_l^2\left(\frac{k_\perp v_\perp}{\omega_B}\right) \delta(\omega - l\omega_B - k_{||}v_{||})\, d^3v. \quad (11.3.3.5)$$

We can similarly find the general form of the spectral density of the correlation function of the electron current fluctuations when there is an external magnetic field present:

$$\langle j_i j_j \rangle^{(0)}_{k\omega} = 2\pi e^2 \sum_{l=-\infty}^{+\infty} \int f_0(v)\, \Pi_{ji}(l, v)\, \delta(\omega - l\omega_B - k_{||}v_{||})\, d^3v, \quad (11.3.3.6)$$

where the tensor Π_{ij} is given by eqn. (5.2.1.15).

If the electron distribution function is Maxwellian, we can integrate in (11.3.3.3)

$$\langle \varrho^2 \rangle^{(0)}_{rt} = e^2 n_0 \left[\frac{m_e}{2\pi T}\right]^{3/2} \frac{\omega_B^2}{4t \sin^2 \frac{1}{2}\omega_B t} \exp\left[-\frac{m_e}{2T}\left\{\frac{\omega_B^2}{4\sin^2\frac{1}{2}\omega_B t}(x^2+y^2)+\frac{z^2}{t^2}\right\}\right]. \quad (11.3.3.7)$$

Hence we find by Fourier transforming the spectral density of the correlation function

$$\langle \varrho^2 \rangle^{(0)}_{k\omega} = e^2 n_0 \int_{-\infty}^{+\infty} \exp\left\{-\frac{T}{2m_e}\left[2\frac{k_\perp^2}{\omega_B^2}(1-\cos\omega_B t)+k_{||}^2 t^2\right]+i\omega t\right\} dt. \quad (11.3.3.8)$$

We can similarly find the spectral density of the current fluctuations in a plasma in an external magnetic field

$$\langle j_i j_j \rangle^{(0)}_{k\omega} = e^2 n_0 \frac{T}{m_e} \int_{-\infty}^{+\infty} \Pi_{ji}(t) \exp\left\{-\frac{T}{2m_e}\left[2\frac{k_\perp^2}{\omega_B^2}(1-\cos\omega_B t)+k_{||}^2 t^2\right]+i\omega t\right\} dt, \quad (11.3.3.9)$$

where

$$\Pi_{ij}(t) = \begin{bmatrix} \cos\omega_B t - \frac{T}{m_e}\frac{k_\perp^2}{\omega_B^2}\sin^2\omega_B t & -\left[1-\frac{T}{m_e}\frac{k_\perp^2}{\omega_B^2}(1-\cos\omega_B t)\right]\sin\omega_B t & -\frac{T}{m_e}\frac{k_\perp k_{||}}{\omega_B} t\sin\omega_B t \\ \left[1-\frac{T}{m_e}\frac{k_\perp^2}{\omega_B^2}(1-\cos\omega_B t)\right]\sin\omega_B t & \cos\omega_B t+\frac{T}{m_e}\frac{k_\perp^2}{\omega_B^2}(1-\cos\omega_B t)^2 & \frac{T}{m_e}\frac{k_\perp k_{||}}{\omega_B} t(1-\cos\omega_B t) \\ -\frac{T}{m_e}\frac{k_\perp k_{||}}{\omega_B} t\sin\omega_B t & -\frac{T}{m_e}\frac{k_\perp k_{||}}{\omega_B} t(1-\cos\omega_B t) & 1-\frac{T}{m_e}\frac{k_{||}^2 t}{\omega_B} \end{bmatrix} \quad (11.3.3.10)$$

Using (11.3.1.15) we can hence obtain the following formula for the dielectric permittivity tensor of an equilibrium electron plasma in a magnetic field:

$$\varepsilon_{ij}(\boldsymbol{k}, \omega) = \delta_{ij} + i\frac{\omega_{pe}^2}{\omega}\int_0^{\infty} \Pi_{ij}(t)\exp\left[-\frac{T}{2m_e}\left\{2\frac{k_\perp^2}{\omega_B^2}(1-\cos\omega t)+k_{||}^2 t^2\right\}+i\omega t\right] dt. \quad (11.3.3.11)$$

Using the expansion

$$\exp(\beta \cos \omega_B t) = \sum_{l=-\infty}^{+\infty} I_l(\beta) \exp(il\omega_B t),$$

where $I_l(\beta)$ is a modified Bessel function, and noting that

$$\int_0^\infty e^{i\alpha t - q^2 t^2}\, dt = \frac{i}{\alpha} \{\varphi(z) - i\sqrt{\pi}\, z e^{-z^2}\}, \quad \varphi(z) = 2z e^{-z^2} \int_0^z e^{x^2}\, dx, \tag{11.3.3.12}$$

$z = \alpha/2q$, we can write ε_{ij} in the form (5.2.2.4).

11.4. Electromagnetic Fluctuations in a Non-isothermal Plasma

11.4.1. FLUCTUATIONS IN AN ISOTROPIC NON-ISOTHERMAL PLASMA

In the preceding sections we have determined the fluctuations of the electromagnetic quantities in an equilibrium plasma. To find these fluctuations we needed to know the dielectric permittivity tensor of the plasma. In the general case of a non-equilibrium plasma the knowledge of the dielectric permittivity tensor is not yet sufficient to determine the fluctuations in the plasma.

Turning now to a study of fluctuations in a non-equilibrium plasma we start with a consideration of fluctuations in a non-isothermal unmagnetized plasma where the electrons and ions are characterized by Maxwellian distributions with different temperatures T_e and T_i (Akhiezer, Akhiezer, and Sitenko, 1962; Sitenko, 1967; Salpeter, 1960a, 1961).

Because of the great difference in the masses of the electrons and the ions the exchange of energy between them is much slower than that between particles of the same kind, so that a non-isothermal plasma can be considered to be a quasi-equilibrium system and for a study of the fluctuations we can apply to it the fluctuation–dissipation relation. It is, however, necessary to bear in mind that the electrons and the ions are connected through the self-consistent fields so that the fluctuations in the charge and current densities of the electron and ion components are not independent and, in particular, will be determined by the two temperatures T_e and T_i.

We introduce into the material eqns. (11.2.1.3) the external random fields acting independently on the electrons and ions $\mathcal{E}_e$ and $\mathcal{E}_i$:

$$j_i^\alpha = -i\omega \sum_j \varkappa_{ij}^\alpha (E_j + \mathcal{E}_j^\alpha), \quad \alpha = \mathrm{e, i}, \tag{11.4.1.1}$$

where the $\varkappa_{ij}^\alpha$ are the components of the dielectric conductivity tensor of the plasma and $\boldsymbol{E}$ the self-consistent field, determined by the Maxwell equations,

$$\sum_j \Lambda_{ij}^{(0)} E_j = -\frac{4\pi i}{\omega}(j_i^e + j_i^i). \tag{11.4.1.2}$$

Using these equations to eliminate $\boldsymbol{E}$ from (11.4.1.1) we get the following relations between the electron and ion current densities $\boldsymbol{j}^\alpha$ and the random potentials $\boldsymbol{A}^\alpha$ ($\mathcal{E}^\alpha = i\omega A^\alpha$):

$$j_\text{l}^\text{e} = \omega^2 \frac{\varkappa_\text{l}^\text{e}}{\varepsilon_\text{l}}\{(1+4\pi\varkappa_\text{l}^\text{i})A_\text{l}^\text{e} - 4\pi\varkappa_\text{l}^\text{i}A_\text{l}^\text{i}\}, \quad j_\text{l}^\text{i} = \omega^2\frac{\varkappa_\text{l}^\text{i}}{\varepsilon_\text{l}}\{-4\pi\varkappa_\text{l}^\text{e}A_\text{l}^\text{e}+(1+4\pi\varkappa_\text{l}^\text{i})A_\text{l}^\text{i}\}; \quad (11.4.1.3)$$

$$\boldsymbol{j}_\text{t}^\text{e} = \omega^2\frac{\varkappa_\text{t}^\text{e}}{\varepsilon_\text{t}-n^2}\{(1+4\pi\varkappa_\text{t}^\text{i}-n^2)\boldsymbol{A}_\text{t}^\text{e} - 4\pi\varkappa_\text{t}^\text{i}\boldsymbol{A}_\text{t}^\text{i}\}, \quad \boldsymbol{j}_\text{t}^\text{i} = \omega^2\frac{\varkappa_\text{l}^\text{i}}{\varepsilon_\text{t}-n^2}\{4\pi\varkappa_\text{t}^\text{e}\boldsymbol{A}_\text{t}^\text{e}+(1+4\varkappa\pi_\text{t}^\text{i}-n^2)\boldsymbol{A}_\text{t}^\text{i}\}. \quad (11.4.1.4)$$

We draw attention to the fact that thanks to the self-consistent field the electron current turns out to depend on the ion external potential, and the ion current on the electron external current.

As the currents $\boldsymbol{j}^\text{e}$ and $\boldsymbol{j}^\text{i}$ are connected and as the system is not in a state of complete equilibrium, it is impossible to use directly eqn. (11.1.2.25) to determine the fluctuations from a knowledge of the proportionality coefficients connecting $\boldsymbol{j}^\alpha$ and $\boldsymbol{A}^\alpha$. However, we can use eqns. (11.4.1.3) and (11.4.1.4) to express the current correlators in terms of the correlators of the external potentials $\boldsymbol{A}^\alpha$. The external potentials are independent so that the fluctuations of each of the potentials will be determined solely by the temperature of the appropriate subsystem—the electron or the ion temperature.

Solving the set (11.4.1.3) and (11.4.1.4) for the potentials we find

$$\omega^2 A^\alpha = \frac{1}{\varkappa_\text{l}^\alpha} j_\text{l}^\alpha + 4\pi j_\text{l}, \quad \omega^2 \boldsymbol{A}_\text{t}^\alpha = \frac{1}{\varkappa_\text{t}^\alpha}\boldsymbol{j}_\text{t}^\alpha + \frac{4\pi}{1-n^2}\boldsymbol{j}_\text{t}, \quad \alpha = \text{e, i}. \quad (11.4.1.5)$$

Using then the fluctuation–dissipation relation in the form (11.1.2.23) we get the following formulae for the spectral density of the external potential fluctuations:

$$\langle A_i^\alpha A_j^\alpha \rangle_{k\omega} = 2\frac{T_\alpha}{\omega^3}\left\{\frac{k_ik_j}{k^2}\frac{\text{Im}\,\varkappa_\text{l}^\alpha}{|\varkappa_\text{l}^\alpha|^2} + \left(\delta_{ij} - \frac{k_ik_j}{k^2}\right)\frac{\text{Im}\,\varkappa_\text{t}^\alpha}{|\varkappa_\text{t}^\alpha|^2}\right\}, \quad \alpha = \text{e, i}. \quad (11.4.1.6)$$

$$\langle A_i^\text{e} A_j^\text{i} \rangle_{k\omega} = 0. \quad (11.4.1.7)$$

As we should have expected, the fluctuations of the electron potential are expressed solely in terms of the electron conductivity of the plasma and the fluctuations of the ion potential in terms of the ion plasma conductivity.

Using eqns. (11.4.1.3) and (11.4.1.2) we get then the following general formulae for the spectral densities of the current fluctuations in a non-isothermal plasma:

$$\langle j_i^\text{e} j_j^\text{e} \rangle_{k\omega} = \frac{2\omega}{|\varepsilon_\text{l}|^2}\frac{k_ik_j}{k^2}\{T_\text{e}|1+4\pi\varkappa_\text{l}^\text{i}|^2\,\text{Im}\,\varkappa_\text{l}^\text{e} + T_\text{i}16\pi^2|\varkappa_\text{l}^\text{e}|^2\,\text{Im}\,\varkappa_\text{l}^\text{i}\}$$
$$+\frac{2\omega}{|\varepsilon_\text{t}-n^2|^2}\left(\delta_{ij}-\frac{k_ik_j}{k^2}\right)\{T_\text{e}|1+4\pi\varkappa_\text{t}^\text{i}-n^2|^2\,\text{Im}\,\varkappa_\text{t}^\text{e} + T_\text{i}16\pi^2|\varkappa_\text{t}^\text{e}|^2\,\text{Im}\,\varkappa_\text{t}^\text{i}\}, \quad (11.4.1.8)$$

$$\langle j_i^\text{i} j_j^\text{i} \rangle_{k\omega} = \frac{2\omega}{|\varepsilon_\text{l}|^2}\frac{k_ik_j}{k^2}\{T_\text{e}16\pi^2|\varkappa_\text{l}^\text{i}|^2\,\text{Im}\,\varkappa_\text{l}^\text{e} + T_\text{i}|1+4\pi\varkappa_\text{l}^\text{e}|^2\,\text{Im}\,\varkappa_\text{t}^\text{i}\}$$
$$+\frac{2\omega}{|\varepsilon_\text{t}-n^2|^2}\left(\delta_{ij}-\frac{k_ik_j}{k^2}\right)\{T_\text{e}16\pi^2|\varkappa_\text{t}^\text{i}|^2\,\text{Im}\,\varkappa_\text{t} + T_\text{i}|1+4\pi\varkappa_\text{t}^\text{e}-n^2|\,\text{Im}\,\varkappa_\text{t}^\text{i}\}, \quad (11.4.1.9)$$

$$\langle j_i^e j_j^i\rangle_{k\omega} = \langle j_i^i j_j^e\rangle_{k\omega}^* = -\frac{2\omega}{|\varepsilon_l|^2}\frac{k_i k_j}{k^2}\{T_e(1+4\pi\varkappa_l^i)\,4\pi\varkappa_l^{i*}\,\mathrm{Im}\,\varkappa_l^e + T_i 4\pi\varkappa_l^e(1+4\pi\varkappa_l^e)^*\,\mathrm{Im}\,\varkappa_l^i\}$$

$$-\frac{2\omega}{|\varepsilon_t - n^2|^2}\left(\delta_{ij} - \frac{k_i k_j}{k^2}\right)\{T_e(1+4\pi\varkappa_t^i - n^2)\,4\pi\varkappa_t^{i*}\,\mathrm{Im}\,\varkappa_t^e + T_i 4\pi\varkappa_t^e(1+4\pi\varkappa_t^{e*} - n^2)\,\mathrm{Im}\,\varkappa_t^i\}. \quad (11.4.1.10)$$

These formulae enable us to find the spectral densities of any quantity in a non-isothermal plasma. For instance, the spectral densities of the electron and ion density fluctuations in a two-temperature plasma have the form:

$$\langle \delta n_e^2\rangle_{k\omega} = \frac{8\pi n_0 k^2 r_D^2}{\omega|\varepsilon_l|^2}\left\{|1+4\pi\varkappa_l^i|^2\,\mathrm{Im}\,\varkappa_l^e + 16\pi^2\frac{T_i}{T_e}|\varkappa_l^e|^2\,\mathrm{Im}\,\varkappa_l^i\right\}, \quad (11.4.1.11)$$

$$\langle \delta n_i^2\rangle_{k\omega} = \frac{8\pi n_0 k^2 r_D^2}{\omega|\varepsilon_l|^2}\left\{16\pi^2|\varkappa_l^i|^2\,\mathrm{Im}\,\varkappa_l^e + \frac{T_i}{T_e}|1+4\pi\varkappa_l^e|^2\,\mathrm{Im}\,\varkappa_l^i\right\}, \quad (11.4.1.12)$$

$$\langle \delta n_e\,\delta n_i\rangle_{k\omega} = \langle \delta n_i\,\delta n_e\rangle_{k\omega}^* = \frac{32\pi^2 n_0 k^2 r_D^2}{\omega|\varepsilon_l|^2}\left\{(1+4\pi\varkappa_l^i)\,\varkappa_l^{i*}\,\mathrm{Im}\,\varkappa_l^e + \frac{T_i}{T_e}\varkappa_l^e(1+4\pi\varkappa_l^{e*})\,\mathrm{Im}\,\varkappa_l^i\right\}. \quad (11.4.1.13)$$

Using (11.4.1.11) to (11.4.1.13) we can find the spectral density of the charge density fluctuations in a two-temperature plasma:

$$\langle \varrho^2\rangle_{k\omega} = \frac{8\pi e^2 r_0 k^2 r_D^2}{\omega|\varepsilon_l|^2}\,\mathrm{Im}\left(\varkappa_l^e + \frac{T_i}{T_e}\varkappa_l^i\right). \quad (11.4.1.14)$$

This density can be written in the form

$$\langle \varrho^2\rangle_{k\omega} = \sqrt{(2\pi)}\,e^2 n_0\frac{k^3 r_D^3}{\omega_{pe}}\,\frac{\exp(-z^2)+\mu\exp(-\mu^2 z^2)}{\{k^2 r_D^2 + 1 - \varphi(z) + p[1-\varphi(\mu z)]\}^2 + \pi z^2\{\exp(-z^2)+p\mu\exp(-\mu^2 z^2)\}^2}, \quad (11.4.1.15)$$

where $z = \omega/\sqrt{2}kv_e$, $\mu = \sqrt{(m_i/m_e)}$, and $p = T_e/T_i$.

We shall, finally, give approximate formulae for the spectral density of the charge density in various limiting cases when $k^2 r_D^2 \ll 1$. In the low-frequency region, $\omega \ll kv_i$ $(z \ll \mu^{-1})$, we have

$$\langle \varrho^2\rangle_{k\omega} = \sqrt{(2\pi)}\,e^2 n_0\frac{k^3 r_D^3}{\omega_{pe}}\sqrt{\frac{m_i}{m_e}}\,\frac{T_i^2}{(T_e+T_i)^2}. \quad (11.4.1.16)$$

When the frequency increases the quantity $\langle \varrho^2\rangle_{k\omega}$ decreases exponentially. If $z \lesssim \mu^{-1} \ll 1$, we have

$$\langle \varrho^2\rangle_{k\omega} = \sqrt{(2\pi)}\,e^2 n_0\frac{k^3 r_D^3}{\omega_{pe}}\,\frac{\mu\exp(-\mu^2 z^2)}{\{1+p[1-\varphi(\mu z)]\}^2 + \pi\mu^2 z^2\exp(-2\mu^2 z^2)}. \quad (11.4.1.17)$$

The width of the maximum is in the low-frequency region of the order of magnitude of $z \sim \mu^{-1}$. The height of the maximum depends according to (11.4.1.16) strongly on the degree of non-isothermy of the plasma. If the plasma is strongly non-isothermal $(p \gg 1)$

the height of the maximum (11.4.1.16) is much lower than in the isothermal case. If $\mu^{-1} \ll z \ll 1$, we have

$$\langle \varrho^2 \rangle_{k\omega} = \sqrt{(2\pi)}\, e^2 n_0 \frac{k^3 r_D^3}{\omega_{pe}} \frac{1+\mu \exp(-\mu^2 z^2)}{[k^2 r_D^2 + 1 - \frac{1}{2} p\mu^{-2} z^{-2}]^2 + \pi z^2 [1 + p\mu \exp(-\mu^2 z^2)]^2}. \quad (11.4.1.18)$$

In this frequency region the spectral density $\langle \varrho^2 \rangle_{k\omega}$ is very small for an isothermal plasma ($p = 1$), smaller by a factor μ than the value (11.4.1.16) at the maximum. In a strongly non-isothermal plasma ($p \gg 1$) the spectral density of the fluctuations has steep maxima at $\omega = \pm\, \omega_s(\boldsymbol{k})$ where $\omega_s(\boldsymbol{k}) = kv_s$ is the frequency of the non-isothermal sound oscillations:

$$\langle \varrho^2 \rangle_{\boldsymbol{k}\omega} = \pi e^2 n_0 k^4 r_D^4 \{\delta(\omega - kv_s) + \delta(\omega + kv_s)\}. \quad (11.4.1.19)$$

When the frequency increases further, $\omega \sim kv_e$ ($z \sim 1$), the spectral density of the fluctuations decreases exponentially, as $\exp(-z^2)$. At high frequencies, $\omega \gg kv_e$ ($z \gg 1$) the electrons play the main role in the fluctuations in the charge density and the quantity $\langle \varrho^2 \rangle_{k\omega}$ is given by formula (11.2.3.3). In particular, $\langle \varrho^2 \rangle_{k\omega}$ has maxima at frequencies corresponding to plasma oscillations.

11.4.2. FLUCTUATIONS IN AN ANISOTROPIC NON-ISOTHERMAL PLASMA

Let us now generalize the results obtained to the case of an anisotropic non-isothermal plasma (Akhiezer, Akhiezer, and Sitenko, 1962). In particular, this generalization will be suitable for a non-isothermal plasma in an external constant and uniform field $\boldsymbol{B}_0$.

Introducing external fields into the material eqns. (11.4.1.1) and using the Maxwell eqn. (11.4.1.2), we can write the self-consistent electrical field in the form

$$E_j = -4\pi i\omega \sum_{i,k} \Lambda_{ji}^{-1} \{\varkappa_{ik}^{e} A_k^{e} + \varkappa_{ik}^{i} A_k^{i}\}. \quad (11.4.2.1)$$

The dielectric permittivity tensor ε_{ij} of the plasma can in the anisotropic case not be reduced to a longitudinal and a transverse component, and neither can the conductivities $\varkappa_{ij}^{e}$ and $\varkappa_{ij}^{i}$. Substituting (11.4.2.1) into (11.4.1.1) we can express the currents in terms of the external potentials $\boldsymbol{A}^{\alpha}$:

$$j_i^{\alpha} = \omega^2 \sum_{\beta} \sum_{j,l} (\delta_{\alpha\beta}\delta_{ij} - 4\pi \sum_{k} \varkappa_{ik}^{\alpha} \Lambda_{kj}^{-1})\, \varkappa_{jl}^{\beta} A_l^{\beta}, \quad (11.4.2.2)$$

hence

$$\omega^2 A_i^{\alpha} = \sum_{j} \{(\varkappa^{\alpha-1})_{ij} j_j^{\alpha} + 4\pi (\Lambda^{(0)-1})_{ij}\, j_j\}. \quad (11.4.2.3)$$

Noting that $\Lambda_{ij}^{(0)}$ is a real symmetric tensor and using the fluctuation–dissipation relation (11.1.2.23) we find

$$\langle A_i^{\alpha} A_j^{\beta} \rangle_{k\omega} = \frac{T_{\alpha}}{\omega}\, i\{\varkappa_{ji}^{\alpha-1} - (\varkappa_{ij}^{\alpha-1})^*\}. \quad (11.4.2.4)$$

Using then (11.4.2.2) we easily get the general formulae which determine the spectral densities of the fluctuations in the electron and ion currents in an anisotropic non-isothermal plasma:

$$\langle j_i^\alpha j_j^\beta \rangle_{k\omega} = i\omega \sum_\gamma \sum_{m,n} \left(\delta_{\alpha\gamma}\delta_{im} - 4\pi \sum_k \varkappa_{ik}^\alpha \Lambda_{km}^{-1}\right)^* \left(\delta_{\beta\gamma}\delta_{jn} - 4\pi \sum_l \varkappa_{jl}^\beta \Lambda_{ln}^{-1}\right) \times T_\gamma(\varkappa_{mn}^{\gamma *} - \varkappa_{nm}^\gamma), \qquad \alpha, \beta = \text{e, i}. \tag{11.4.2.5}$$

Noting that the total current $\boldsymbol{j}$ is equal to the sum of the electron and the ion currents $\boldsymbol{j}_e$ and $\boldsymbol{j}_i$, we get the following formulae for the current correlators:

$$\langle j_i j_j \rangle_{k\omega} = i\omega \sum_{m,n} \left(\delta_{im} - 4\pi \sum_k \varkappa_{ik} \Lambda_{km}^{-1}\right)^* \left(\delta_{jn} - 4\pi \sum_l \varkappa_{jl} \Lambda_{ln}^{-1}\right) \sum_\gamma T_\gamma(\varkappa_{mn}^{\gamma *} - \varkappa_{\gamma m}^\gamma), \tag{11.4.2.6}$$

$$\langle j_i^\alpha j_j \rangle_{k\omega} = i\omega \sum_\gamma \sum_{m,n} \left(\delta_{\alpha\gamma}\delta_{im} - 4\pi \sum_k \varkappa_{ik}^\alpha \Lambda_{km}^{-1}\right)^* \left(\delta_{jn} - 4\pi \sum_l \varkappa_{jl} \Lambda_{ln}^{-1}\right) T_\gamma(\varkappa_{mn}^{\gamma *} - \varkappa_{nm}^\gamma). \tag{11.4.2.7}$$

Using the Maxwell equations we can now obtain by means of eqns. (11.4.2.5) and (11.4.2.7) the correlators of all quantities in which we are interested. In particular, the correlators of the fluctuations in the electron density, and in the electrical and magnetic fields have the form:

$$\langle \delta n_e^2 \rangle_{k\omega} = \sum_{i,j} \frac{k_i k_j}{e^2 \omega^2} \langle j_i^e j_j^e \rangle_{k\omega}, \tag{11.4.2.8}$$

$$\langle E_i E_j \rangle_{k\omega} = \frac{16\pi^2 i}{\omega^2} \sum_{k,l} \Lambda_{ik}^{*-1} \Lambda_{jl}^{-1} \sum_\gamma T_\gamma(\varkappa_{kl}^{\gamma *} - \varkappa_{lk}^\gamma), \tag{11.4.2.9}$$

$$\langle B_i B_j \rangle_{k\omega} = \frac{16\pi^2}{\omega^2} \frac{n^2}{n^2 - 1} \sum_{k,l,m,n} \varepsilon_{ikl} \varepsilon_{jmn} \frac{k_k k_m}{k^2} \langle j_l j_n \rangle_{k\omega}, \tag{11.4.2.10}$$

where ε_{ijk} is the completely antisymmetric third rank unit tensor.

We have already noted that the correlation functions have steep maxima near those values of ω and $\boldsymbol{k}$ which satisfy the dispersion equation $\Lambda(\boldsymbol{k}, \omega_r) = 0$, where the index r numbers the eigenoscillations. One can easily establish the form of the correlation functions near such maxima. For instance, we have for the quantity $\langle j_i j_j \rangle_{k\omega}$:

$$\langle j_i j_j \rangle_{k\omega} = \sum_r B_{ij}(\boldsymbol{k}, \omega)\, \delta\{\omega - \omega_r(\boldsymbol{k})\}, \tag{11.4.2.11}$$

where

$$B_{ij}(\boldsymbol{k}, \omega) = \pi i \omega \left[\frac{\partial \Lambda}{\partial \omega}\right]^{-1} \sum_{k,l,m,n} \frac{16\pi^2 \varkappa_{ik} \varkappa_{jl}^* \lambda_{km} \lambda_{ln}^* \sum_\gamma T_\gamma(\varkappa_{mn}^{\gamma *} - \varkappa_{nm}^\gamma)}{\operatorname{Im} \Lambda},$$

while the λ_{ij} are determined by the relation $\sum_j \lambda_{ij} \Lambda_{jk} = \Lambda \delta_{ik}$.

11.4.3. FLUCTUATIONS IN A NON-ISOTHERMAL PLASMA IN A MAGNETIC FIELD

As an example of an application of the general formulae which we have obtained we shall consider fluctuations in a non-isothermal two-temperature plasma in an external magnetic field, $\boldsymbol{B}_0$ (Sitenko, 1967). As the electrons play the major role in the high-frequency region for the collective motions of the plasma we can neglect the effect of the ions in the high-frequency region. The correlation functions for the fluctuations in the high-frequency region will therefore be the same as for the equilibrium case with a temperature equal to the electron temperature. In the low-frequency region, however, the fluctuations for a non-isothermal plasma will differ greatly from those in an equilibrium plasma.

The possibility that there may occur weakly damped magneto-hydrodynamical ocillations in a plasma in an external magnetic field leads to the appearance of additional maxima, connected with these oscillations, in the low-frequency part of the fluctuation spectrum. The fluctuations connected with Alfvén and magneto-sound oscillations in a magneto-active plasma are described by the general formulae (11.4.2.9) to (11.4.2.11), where we must use expressions (5.4.1.3) and (5.4.2.4) for the dielectric permittivity. Retaining the highest powers in the large parameter ω_{Bi}/ω we can obtain simple formulae for the correlation functions when $\omega \ll \omega_{Bi}$. For instance, the spectral density of the electrical field fluctuations near the Alfvén frequency can be written in the form

$$\langle E_i E_j \rangle_{k\omega} = 4\pi^2 e_i^* e_j T_e \frac{v_A^2}{c^2} \{\delta(\omega - kv_A \cos\vartheta) + \delta(\omega + kv_A \cos\vartheta)\}, \qquad (11.4.3.1)$$

where the polarization vector $\boldsymbol{e}$ is defined by the relation

$$\boldsymbol{e}_a = \left\{1, \; -i\frac{\omega}{\omega_{Bi}} \cot^2\vartheta, \; -\frac{v_s^2}{v_A^2}\frac{\omega^2}{\omega_{Bi}^2}(\sin\vartheta\cos\vartheta)^{-1}\right\}. \qquad (11.4.3.2)$$

The magneto-sound fluctuations of the electrical field are characterized by the spectral densities

$$\langle E_i E_j \rangle_{k\omega} = 4\pi^2 e_i^* e_j T_e \frac{v_A^2}{c^2} \{\delta(\omega - kv_A) + \delta(\omega + kv_A)\}, \qquad (11.4.3.3)$$

$$\langle E_i E_j \rangle_{k\omega} = 4\pi^2 e_i^* e_j T_e k^2 r_D^2 \{\delta(\omega - kv_s \cos\vartheta) + \delta(\omega + kv_s \cos\vartheta), \qquad (11.4.3.4)$$

where the polarization vectors for the fast and slow magneto-sound waves have the components

$$\boldsymbol{e}_f = \left\{-i\frac{\omega}{\omega_{Bi}} \sin^{-2}\vartheta, \; 1, \; -i\frac{v_s^2}{v_A^2}\frac{\omega}{\omega_{Bi}} \sin\vartheta\cos\vartheta\right\}, \qquad (11.4.3.5)$$

$$\boldsymbol{e}_s = \left\{\sin\vartheta, \; -i\frac{v_s^2}{v_A^2}\frac{\omega_{Bi}}{\omega} \sin\vartheta\cos\vartheta, \; \cos\vartheta\right\}. \qquad (11.4.3.6)$$

We note that the spectral densities for the low-frequency fluctuations are proportional to the square of the ratio of the phase velocity of the wave in question and the velocity of light *in vacuo*.

11.5. Electromagnetic Fluctuations in a Non-equilibrium Plasma

11.5.1. SPECTRAL DENSITIES OF FLUCTUATIONS IN A PLASMA WITH NON-EQUILIBRIUM, BUT STABLE DISTRIBUTION FUNCTIONS

We determined in the preceding sections the fluctuations of electromagnetic quantities in an equilibrium or a quasi-equilibrium plasma; in that case it was sufficient to know the dielectric permittivity of the plasma. We shall now study fluctuations in a non-equilibrium plasma with particle distributions which differ greatly from the Maxwell distribution.† Such a situation occurs in a collisionless plasma in which the paucity of collisions leads to very long times needed for the establishing of an equilibrium state. We shall therefore be dealing with times which are short compared to the relaxation times of the plasma, that is, we shall be dealing with fluctuations referring to such short time intervals. We shall start by considering an unmagnetized plasma with non-equilibrium, but stable particle distribution functions. We determined in Section 11.3 the spectral density of the charge density fluctuations in a plasma, neglecting the particle interactions:

$$\langle \varrho^2 \rangle^{(0)}_{k\omega} = 2\pi e^2 \int f_0(v)\, \delta\{\omega - (\boldsymbol{k}\cdot\boldsymbol{v})\}\, d^3v, \tag{11.5.1.1}$$

where $f_0(v)$ is the electron distribution function. When deriving this formula we did not require that the plasma had to be in an equilibrium state. We may thus assume that formula (11.5.1.1) determines the charge density correlation function also in the case of a non-equilibrium plasma described by the distribution function $f_0(v)$.

A similar formula is valid for the spectral densities of the electron and ion current fluctuations,

$$\langle j_i^\alpha j_j^\beta \rangle^{(0)}_{k\omega} = 2\pi e^2\, \delta_{\alpha\beta} \int f_0^\alpha(v)\, v_i v_j\, \delta\{\omega - (\boldsymbol{k}\cdot\boldsymbol{v})\}\, d^3v, \tag{11.5.1.2}$$

where $f_0^\alpha(v)$ is the non-equilibrium distribution function for the electrons or the ions ($\alpha = \mathrm{e, i}$).

If the plasma is in an external magnetic field $\boldsymbol{B}_0$ the spectral densities of the current fluctuations are determined by the formula

$$\langle j_i^\alpha j_j^\beta \rangle^{(0)}_{k\omega} = 2\pi e^2\, \delta_{\alpha\beta} \sum_l \int f_0^\alpha(v)\, \Pi_{ji}^\alpha(l, \boldsymbol{v})\, \delta(\omega - l\omega_{B\alpha} - k_{||}v_{||})\, d^3v, \tag{11.5.1.3}$$

where the tensor $\Pi_{ij}(l, \boldsymbol{v})$ is determined by eqn. (5.2.1.15).

Let us now take into account the self-consistent interaction between the particles. To do this we turn to the material eqns. (11.4.1.1) and introduce into them the random external currents $\boldsymbol{j}_0^\alpha$:

$$\boldsymbol{j}^\alpha = -i\omega\hat{\varkappa}^\alpha \boldsymbol{E} + \boldsymbol{j}_0^\alpha, \tag{11.5.1.4}$$

where the tensor $\hat{\varkappa}^\alpha$ is constructed using the non-equilibrium distribution functions $f_0^\alpha(v)$, as follows ($\boldsymbol{b} = \boldsymbol{B}_0/B_0$):

$$\varkappa_{ij}^\alpha(\boldsymbol{k}, \omega) = \frac{e^2}{m_\alpha\omega}\left\{\sum_l \int \left[\frac{\omega - k_{||}v_{||}}{v_\perp}\frac{\partial f_0^\alpha}{\partial v_\perp} + k_{||}\frac{\partial f_0^\alpha}{\partial v_{||}}\right]\frac{\Pi_{ij}(l, \boldsymbol{v})}{\omega - l\omega_{B\alpha} - k_{||}v_{||}}\, d^3v \right.$$
$$\left. - \left(n_0 + \int \frac{v_{||}^2}{v_\perp}\frac{\partial f_0^\alpha}{\partial v_\perp}\, d^3v\right) b_i b_j\right\}. \tag{11.5.1.5}$$

As the random external currents cannot depend on the self-consistent interaction, the correlation functions for them will be the same as for the currents of non-interacting particles, that is, they will be determined by eqns. (11.5.1.3). Using the Maxwell equations to eliminate the self-consistent field $\boldsymbol{E}$ from (11.5.1.4), we find

$$\langle j_i^\alpha j_j^\beta \rangle_{k\omega} = \sum_\gamma \sum_{m,n} \left(\delta_{\alpha\gamma}\delta_{im} - 4\pi \sum_k \varkappa_{ik}^\alpha \Lambda_{km}^{-1}\right)\left(\delta_{\beta\gamma}\delta_{jn} - 4\pi \sum_l \varkappa_{jl}^\beta \Lambda_{ln}^{-1}\right)^* \langle j_m^\gamma j_n^\gamma \rangle^{(0)}_{km}. \tag{11.5.1.6}$$

† Thompson and Hubbard (1960), Rostoker (1961), and Sitenko (1967) have studied fluctuations in a non-equilibrium plasma; we shall follow the last author.

Equation (11.5.1.6) establishes a general connection between the correlation function for the current fluctuations, with the self-consistent interaction between the charged particles in the plasma taken into account, and the correlation functions for the current fluctuations of independent particles.

The spectral densities of the fluctuations in the total current and in the electrical field in a non-equilibrium plasma are determined by the equations

$$\langle j_i j_j \rangle_{k\omega} = \sum_{m,\,n} \Big(\delta_{im} - 4\pi \sum_k \varkappa_{ik} \Lambda_{km}^{-1}\Big)^* \Big(\delta_{jn} - 4\pi \sum_l \varkappa_{jl} \Lambda_{ln}^{-1}\Big) \langle j_m j_n \rangle_{k\omega}^{(0)}, \tag{11.5.1.7}$$

$$\langle E_i E_j \rangle_{k\omega} = \sum_{k,\,l} \frac{16\pi^2}{\omega^2} \Lambda_{ik}^{*-1} \Lambda_{jl}^{-1} \langle j_k j_l \rangle_{k\omega}^{(0)}. \tag{11.5.1.8}$$

We can easily find in a similar way the correlation functions for other quantities in the plasma.

We saw in Section 11.4 that the correlation functions of the external electron and ion current fluctuations can in the case of an equilibrium or quasi-equilibrium two-temperature plasma be expressed in terms of the plasma conductivity $\varkappa_{ij}^{\alpha}(\boldsymbol{k}, \omega)$:

$$\langle j_i^{\alpha} j_j^{\beta} \rangle_{k\omega}^{(0)} = i T_{\alpha} (\varkappa_{ij}^{\alpha *} - \varkappa_{ji}^{\alpha})\, \delta_{\alpha\beta}. \tag{11.5.1.9}$$

In the case of a non-equilibrium plasma the current correlation functions can be expressed in terms of the quantities $\langle j_i^{\alpha} j_j^{\beta} \rangle_{k\omega}^{(0)}$ which, however, can no longer be expressed in terms of the plasma conductivity; knowledge of the $\varkappa_{ij}^{\alpha}$ is therefore in the case of a non-equilibrium plasma insufficient to describe the fluctuations, in contrast to the case of an equilibrium or even a quasi-equilibrium non-isothermal plasma.

The expressions for the correlation functions of the fluctuations in a non-equilibrium plasma simplify considerably in the particular case of an isotropic particle distribution in the plasma. Indeed, introducing the longitudinal and transverse permittivities and conductivities, ε_l and ε_t, $\varkappa_l^{\alpha}$, and $\varkappa_t^{\alpha}$, of the non-equilibrium plasma and noting that in the isotropic case

$$\Lambda_{ij}^{-1} = \frac{k_i k_j}{k^2} \varepsilon_l^{-1} + \Big(\delta_{ij} - \frac{k_i k_j}{k^2}\Big)(\varepsilon_t - n^2)^{-1}, \tag{11.5.1.10}$$

we can write the spectral density of the electron current fluctuations in a non-equilibrium plasma in the form:

$$\langle j_i^{e} j_j^{e} \rangle_{k\omega} = \frac{2\pi e^2 \omega^2}{k^2} \Big\{ \frac{k_i k_j}{k^2} [\,|1 + 4\pi \varkappa_l^{i}|^2 A^{e} + 16\pi^2 |\varkappa_l^{e}|^2 A^{i}] \frac{1}{|\varepsilon_l|^2}$$
$$+ \frac{1}{2} \Big(\delta_{ij} - \frac{k_i k_j}{k^2}\Big) [\,|1 + 4\pi \varkappa_t^{i}|^2 B^{e} + 16\pi^2 |\varkappa_t^{e}|^2 B^{i}] \frac{1}{|\varepsilon_t - n^2|^2}, \tag{11.5.1.11}$$

where

$$A^{\alpha} = \int f_0^{\alpha}(v)\, \delta\{\omega - (\boldsymbol{k}\cdot\boldsymbol{v})\}\, d^3v, \quad B^{\alpha} = \int f_0^{\alpha}(v) \frac{|[\boldsymbol{k} \wedge \boldsymbol{v}]|^2}{\omega^2}\, \delta\{\omega - (\boldsymbol{k}\cdot\boldsymbol{v})\}\, d^3v. \tag{11.5.1.12}$$

The spectral density of the electron density fluctuations in an isotropic non-equilibrium plasma is given by the formula

$$\langle \delta n_e^2 \rangle_{k\omega} = \frac{2\pi}{|\varepsilon_l|^2} \{|1+4\pi\varkappa_l^i|^2 A^e + 16\pi^2 |\varkappa_l^e|^2 A^i\}. \tag{11.5.1.13}$$

We see that in a non-equilibrium plasma the spectral density of the electron density fluctuations is determined not only by the electron and ion plasma conductivity, but also by the quantities $A^\alpha(\boldsymbol{k}, \omega) = A^\alpha$ which, in turn, are determined by the particle distribution functions with respect to the velocity components along the wavevector $\boldsymbol{k}$.

The fluctuations in the ion density and the ion current are determined by equations which are the analogues of eqns. (11.5.1.11) and (11.5.1.13) in which we need only change the index e to i. Thanks to the interaction between electrons and ions there exists a correlation between the fluctuations in the electron and ion densities. The spectral density of these fluctuations has the form

$$\langle \delta n_e \delta n_i \rangle_{k\omega} = \langle \delta n_i \delta n_e \rangle_{k\omega}^* = -\frac{8\pi^2}{|\varepsilon_l|^2} \{(1+4\pi\varkappa_l^i)\, \varkappa_l^{i*} A^e + \varkappa_l^e (1+4\varkappa_l^e)^* A^i\}. \tag{11.5.1.14}$$

Using eqn. (11.5.1.13) for the electron density fluctuations and the analogous expression for the ion density fluctuations we can obtain the spectral density for the charge density fluctuations:

$$\langle \varrho^2 \rangle_{k\omega} = \frac{2\pi e^2}{|\varepsilon_l|^2} (A^e + A^i). \tag{11.5.1.15}$$

The spectral densities of the fluctuations in the longitudinal electrival field and in the longitudinal current can be expressed in terms of $\langle \varrho^2 \rangle_{k\omega}$ using the Poisson equation and the equation of continuity. We note that in the case of a two-temperature plasma the quantities A^e and A^i can be directly expressed in terms of the imaginary parts of the electron and ion plasma conductivities:

$$A^e = \frac{k^2}{\pi e^2 n_0 \omega} T_e \operatorname{Im} \varkappa_l^e, \quad A^i = \frac{k^2}{\pi e^2 n_0 \omega} T_i \operatorname{Im} \varkappa_e^i.$$

Using these expressions we can write eqn. (11.5.1.15) in the form

$$\langle \varrho^2 \rangle_{k\omega} = \frac{2k^2}{\omega |\varepsilon_l|^2} \{T_e \operatorname{Im} \varkappa_l^e + T_i \operatorname{Im} \varkappa_e^i\}. \tag{11.5.1.16}$$

This formula is the same as eqn. (11.4.1.14).

11.5.2. COLLECTIVE FLUCTUATIONS AND EFFECTIVE TEMPERATURE

The spectral densities of the fluctuations have in the transparency region of the plasma sharp δ-function-like maxima at the frequencies of the eigenoscillations of the plasma which satisfy the dispersion equation $\Lambda(\boldsymbol{k}, \omega) = 0$. One can easily establish the form of the spectral densities near such maxima (Sitenko and Kirochkin, 1966). In this case we shall speak of resonance or coherent fluctuations in contrast to the incoherent fluctuations when $\boldsymbol{k}$ and ω are not connected by a well-defined relation.

As in the transparency region of the plasma $\operatorname{Im} \Lambda \ll \operatorname{Re} \Lambda$, we have

$$\Lambda_{ik}^{*-1}\Lambda_{jl}^{-1} \to \pi \frac{\lambda_{ik}^{*}\lambda_{jl}}{|\operatorname{Im} \Lambda|} \delta(\Lambda). \tag{11.5.2.1}$$

Noting further that we have for the eigenoscillations the relation $\lambda_{ij} = e_i e_j^* \operatorname{Tr} \lambda$, we can write the spectral density of the electrical field fluctuations in the form

$$\langle E_i E_j \rangle_{k\omega} = 8\pi^2 e_i^* e_j \tilde{T}(\boldsymbol{k}, \omega) \frac{|\operatorname{Tr} \lambda|}{|\omega|} \delta(\Lambda), \tag{11.5.2.2}$$

where

$$\tilde{T}(\boldsymbol{k}, \omega) = 2\pi \frac{\operatorname{Tr}\lambda}{\omega \operatorname{Im} \Lambda} \sum_{i,j} \langle j_i j_j \rangle_{k\omega}^{(0)} e_i e_j^* . \tag{11.5.2.3}$$

We note that $\tilde{T}$ satisfies the relation

$$\tilde{T}(\boldsymbol{k}, \omega) = \tilde{T}(-\boldsymbol{k}, -\omega).$$

One sees easily that in the case of an equilibrium plasma the quantity $\tilde{T}$ is the same as the plasma temperature. It is thus natural to call the quantity $\tilde{T}$ the effective temperature which characterizes the mean square amplitude of the fluctuating oscillations of the electrical field in the plasma. In a non-equilibrium plasma the effective temperature can take on large values. If the state of the plasma approaches the boundaries of the region of kinetic stability of the plasma, $\operatorname{Im} \Lambda \to 0$, and the effective temperature increases without bounds. Noting that

$$\operatorname{Im} \Lambda = \frac{i}{2} [\operatorname{Tr} \lambda] \sum_{i,j} (\varepsilon_{ji}^* - \varepsilon_{ij}) e_i^* e_j, \tag{11.5.2.4}$$

we can write the effective temperature in the form

$$\tilde{T}(\boldsymbol{k}, \omega) = \frac{4\pi}{i\omega} \frac{\sum_{i,j} \langle j_i j_j \rangle_{k\omega}^{(0)} e_i e_j^*}{\sum_{k,l} (\varepsilon_{kl}^* - \varepsilon_{lk}) e_k e_l^*} . \tag{11.5.2.5}$$

In particular, the effective temperature for a non-isothermal plasma equals

$$\tilde{T}(\boldsymbol{k}, \omega) = \frac{\sum_{\alpha} \sum_{i,j} T_\alpha (\varkappa_{ij}^{\alpha *} - \varkappa_{ji}^{\alpha}) e_i e_j^*}{\sum_{\alpha} \sum_{k,l} (\varkappa_{kl}^{\alpha *} - \varkappa_{lk}^{\alpha}) e_k e_l^*} . \tag{11.5.2.6}$$

It is immediately clear from this formula that in the case of an equilibrium plasma $\tilde{T} = T$.

We can directly express the spectral density of the fluctuations in the partial currents in the transparency region of the plasma in terms of the electrical field correlation function,

$$\langle j_i^\alpha j_j^\beta \rangle_{k\omega} = \omega^2 \sum_{k,l} \varkappa_{ik}^{\alpha} \varkappa_{jl}^{\beta} \langle E_k E_l \rangle_{k\omega}. \tag{11.5.2.7}$$

We draw attention to the fact that the same formula would appear if we had in (11.5.1.4) dropped the term $\boldsymbol{j}_0$. This is connected with the fact that in the transparency region the δ-function-like terms contribute to the spectral densities.

To conclude this subsection we shall give the formulae connecting the charge density and electron density fluctuations with the electrical field fluctuations:

$$\langle \varrho^2 \rangle_{k\omega} = \frac{k^2}{16\pi^2} \langle E_l^2 \rangle_{k\omega}, \tag{11.5.2.8}$$

$$\langle \delta n_e^2 \rangle_{k\omega} = \frac{k^2}{e^2} (\varkappa_l^e)^2 \langle E_l^2 \rangle_{k\omega}. \tag{11.5.2.9}$$

11.5.3. CRITICAL FLUCTUATIONS NEAR THE BOUNDARIES OF PLASMA INSTABILITY

Equations (11.5.1.8) and (11.5.2.2) determine the field fluctuations in a plasma with arbitrary non-equilibrium, but stable, particle distribution functions. The average directed particle velocities need not necessarily vanish so that one can also apply these equations for a study of fluctuations in a plasma through which a particle beam passes and also in a plasma with electrons moving relative to the ions, provided the beam velocity—or the electron velocity—does not exceed the critical value above which instability sets in.

Let us, as an example, consider fluctuations in a plasma consisting of cold ions and hot electrons which are moving relative to one another (Ichimaru, 1962; Ichimaru, Pines, and Rostoker, 1962; Bogdankevich, Rukhadze, and Silin, 1962). The charge density correlation function can in that case be determined using formula (11.5.1.16), provided we make the substitution $\omega \to \omega - (\boldsymbol{k} \cdot \boldsymbol{u})$ in the quantity $\varkappa_l^e(\boldsymbol{k}, \omega)$ which occurs in that equation:

$$\langle \varrho^2 \rangle_{k\omega} = \frac{2k^2}{|\varepsilon_l(\boldsymbol{k}, \omega)|^2} \left\{ \frac{T_e}{\omega - (\boldsymbol{k} \cdot \boldsymbol{u})} \operatorname{Im} \varkappa_l^e(\boldsymbol{k}, \omega - (\boldsymbol{k} \cdot \boldsymbol{u})) + \frac{T_i}{\omega} \operatorname{Im} \varkappa_l^i(\boldsymbol{k}, \omega) \right\}, \tag{11.5.3.1}$$

where

$$\varepsilon_l(\boldsymbol{k}, \omega) = 1 + 4\pi\varkappa_l^e(\boldsymbol{k}, \omega - (\boldsymbol{k} \cdot \boldsymbol{u})) + 4\pi\varkappa_l^i(\boldsymbol{k}, \omega). \tag{11.5.3.2}$$

We see that for velocities for which $(\boldsymbol{k} \cdot \boldsymbol{u})$ lies close to ω, that is, when we approach the boundaries of the stability region, the fluctuations increase strongly, becoming infinite on the boundary of that region—of course, as long as we restrict ourselves to the linear theory. This phenomenon is analogous to the growth of fluctuations near a critical point (phase transition point; see, for instance, Landau and Lifshitz, 1969) which is well known in statistical physics.

Let us consider in some detail the critical fluctuations in the case when a compensated beam of charged particles passes through the plasma. The spectral density of the longitudinal field fluctuations are, according to (11.5.1.8) determined by the equation

$$\langle E^2 \rangle_{k\omega} = \frac{16\pi^2}{k^2 |\varepsilon_l|^2} \langle \varrho^2 \rangle_{k\omega}^{(0)}, \tag{11.5.3.3}$$

where $\langle \varrho^2 \rangle_{k\omega}^{(0)}$ is the spectral density of the charge density fluctuations in the case when there are no interactions between the particles.

We shall assume that the particle distributions in the plasma, f_0, and in the beam, f_0', are Maxwellian,

$$f_0 = n_0 \left(\frac{m_e}{2\pi T}\right)^{3/2} \exp\left(-\frac{m_e v^2}{2T}\right), \quad f_0' = n_0' \left(\frac{m_e}{2\pi T'}\right)^{3/2} \exp\left(-\frac{m_e |\boldsymbol{v}-\boldsymbol{u}|^2}{2T'}\right),$$

where $\boldsymbol{u}$ is the beam velocity, so that we can start from the following expressions for $\langle \varrho^2 \rangle^{(0)}_{k\omega}$ and $\varepsilon_l(\boldsymbol{k}, \omega)$:

$$\langle \varrho^2 \rangle^{(0)}_{k\omega} = \sqrt{(2\pi)}\, e^2 \left\{ \sum_\alpha \frac{n_0^\alpha}{kv_\alpha} \exp(-z_\alpha^2) + \sum_\alpha \frac{n_0^{\alpha'}}{kv_\alpha'} \exp(-y_\alpha^2), \right. \tag{11.5.3.4}$$

$$\varepsilon_l(\boldsymbol{k}, \omega) = 1 + \sum_\alpha (kr_{D\alpha})^{-2} \{1 - \varphi(z_\alpha) + i\sqrt{\pi}\, z_\alpha \exp(-z_\alpha^2)\}$$
$$+ \sum_\alpha (kr_{D\alpha}')^{-2} \{1 - \varphi(y_\alpha) + i\sqrt{\pi}\, y_\alpha \exp(-y_\alpha^2)\}, \tag{11.5.3.5}$$

where $z_\alpha = \omega/\sqrt{2}\, kv_\alpha$, $y_\alpha = \{\omega - (\boldsymbol{k}\cdot\boldsymbol{u})\}/\sqrt{2}\, kv_\alpha'$, and the function φ is again given by equation (11.3.3.12); the summation is over all different kinds of particles in the plasma and in the beam. In the transparency region of the plasma we have Im $\varepsilon_l \ll$ Re ε_l, and eqn. (11.5.3.3) becomes

$$\langle E^2 \rangle_{k\omega} = 8\pi^2 \frac{\tilde{T}(\boldsymbol{k}, \omega)}{|\omega|} \delta\{\varepsilon_l(\boldsymbol{k}, \omega)\}, \tag{11.5.3.6}$$

where

$$\tilde{T}(\boldsymbol{k}, \omega) = \frac{2\pi\omega \langle \varrho^2 \rangle^{(0)}_{k\omega}}{k^2 \operatorname{Im} \varepsilon_l(\boldsymbol{k}, \omega)}.$$

When the state of the system approaches the boundary of the region of kinetic stability, Im $\varepsilon_l(\boldsymbol{k}, \omega) \to 0$, and the effective temperature increases without bounds.

Neglecting the thermal motion of the ions and assuming that the condition $\{\omega - (\boldsymbol{k}\cdot\boldsymbol{u})\}^2 \ll k^2 v_e'^2$ is satisfied, we can write the effective temperature in the form

$$\tilde{T} = \frac{a(z)}{1 - \{(\boldsymbol{k}\cdot\boldsymbol{u})/k\tilde{u}\}} T, \tag{11.5.3.7}$$

where

$$a(z) = \frac{1 + \dfrac{n_0'}{n_0}\left(\dfrac{T}{T'}\right)^{1/2} e^{z^2}}{1 + \dfrac{n_0'}{n_0}\left(\dfrac{T}{T'}\right)^{3/2} e^{z^2}},$$

and

$$\tilde{u} = \frac{\omega}{k}\left\{1 + \frac{n_0}{n_0'}\left(\frac{T'}{T}\right)\right\}^{3/2} e^{-z^2}.\Big\} \tag{11.5.3.8}$$

The quantity $\tilde{u}$ plays the role of a critical beam velocity; when it is reached, the state of the plasma-beam system turns out to be unstable.

If we assume that the beam density is sufficiently low, $n_0' \ll n_0$, we can neglect the effect of the beam on the dispersion of the waves—the effect of the beam on the effective temperature will be important also when $n_0' \ll n_0$. Equation (11.5.3.6) becomes in that case in the high-frequency region for Langmuir oscillations

$$\langle E^2\rangle_{k\omega} = 4\pi^2\tilde{T}(\boldsymbol{k}, \omega)\{\delta(\omega-\omega_{pe})+\delta(\omega+\omega_{pe})\}, \tag{11.5.3.9}$$

where the effective temperature is given by eqn. (11.5.3.7) with $z^2 = \frac{1}{2}(kr_D)^{-2}+\frac{3}{2}$. The critical velocity then turns out to be of the order of the electron thermal velocity in the plasma at rest,

$$\tilde{u} = \frac{\omega_{pe}}{k}\left[1+\frac{n_0}{n_0'}\left(\frac{T'}{T}\right)^{3/2}\exp\left\{-\frac{1}{2k^2r_D^2}-\frac{3}{2}\right\}\right]. \tag{11.5.3.10}$$

The average energy of the fluctuating Langmuir oscillations equals

$$\langle E^2\rangle_{k} = 2\pi\{\tilde{T}(\boldsymbol{k}, \omega_{pe})+\tilde{T}(\boldsymbol{k}, -\omega_{pe})\}. \tag{11.5.3.11}$$

For beam velocities close to $\tilde{u}$ this energy can greatly exceed the thermal value.

Using (11.5.2.8) we find easily the spectral density of the charge density fluctuations

$$\langle \varrho^2\rangle_{k\omega} = \tfrac{1}{4}k^2\tilde{T}(\boldsymbol{k}, \omega)\,\{\delta(\omega-\omega_{pe})+\delta(\omega+\omega_{pe})\}. \tag{11.5.3.12}$$

Similarly we find by using (11.5.2.9) the spectral density of the electron density fluctuations

$$\langle \delta n_e^2\rangle_{k\omega} = \frac{k^2}{4e^2}\tilde{T}(\boldsymbol{k}, \omega)\{\delta(\omega-\omega_{pe})+\delta(\omega+\omega_{pe})\}. \tag{11.5.3.13}$$

Let us now consider low-frequency oscillations in a two-temperature plasma with hot electrons and cold ions. Equation (11.5.3.6) becomes in that case when $n_0' \ll n_0$

$$\langle E^2\rangle_{k\omega} = 4\pi^2k^2r_D^2\tilde{T}(\boldsymbol{k}, \omega)\{\delta(\omega-kv_s)+\delta(\omega+kv_s)\}, \tag{11.5.3.14}$$

where v_s is the non-isothermal sound velocity and

$$\tilde{T}(\boldsymbol{k}, \omega) = \frac{T}{1-(\boldsymbol{k}\cdot\boldsymbol{u})/k\tilde{u}}, \qquad \tilde{u} = \frac{n_0}{n_0'}\left(\frac{T'}{T}\right)^{3/2} v_s. \tag{11.5.3.15}$$

In contrast to the high-frequency region, in the low-frequency region the quantity $\tilde{u}$ can be both larger and smaller than the electron thermal velocity. The average energy of the fluctuating sound oscillations is equal to

$$\langle E^2\rangle_{k} = 2\pi k^2r_D^2\{\tilde{T}(\boldsymbol{k}, kv_s)+\tilde{T}(\boldsymbol{k}, -kv_s)\}. \tag{11.5.3.16}$$

Using (11.5.2.8) we find the spectral density of the charge density fluctuations

$$\langle \varrho^2\rangle_{k\omega} = \tfrac{1}{4}k^4r_D^2\tilde{T}(\boldsymbol{k}, \omega)\{\delta(\omega-kv_s)+\delta(\omega+kv_s)\}. \tag{11.5.3.17}$$

The spectral density of the electron density fluctuations is given by the formula

$$\langle \delta n_e^2\rangle_{k\omega} = \pi n_0\frac{\tilde{T}(\boldsymbol{k}, \omega)}{T}\{\delta(\omega-kv_s)+\delta(\omega+kv_s)\}. \tag{11.5.3.18}$$

Equations (11.5.3.9) to (11.5.3.18) have been obtained by using the linear theory and are applicable only to fluctuations with wavevectors $\boldsymbol{k}$ satisfying the conditions $u < \tilde{u}(\boldsymbol{k})$. Fluctuating oscillations for which $u > \tilde{u}(\boldsymbol{k})$ lead to plasma instabilities.

Let us now consider fluctuations in a plasma consisting of cold ions and hot electrons which move relative to the ions with a velocity $\boldsymbol{u}$. In that case the spectral densities of the fluctuations will be described by the formulae for the plasma-beam case, provided we put in those formulae the electron density in the plasma equal to zero and take the electron density in the beam to be equal to the ion density. In that case $a(z) \rightarrow T_e/T$, and the effective temperature will be equal to

$$\tilde{T}(\boldsymbol{k}, \omega) = \frac{T_e}{1-(\boldsymbol{k}\cdot\boldsymbol{u})/k\tilde{u}}, \quad \tilde{u} = \frac{\omega}{k}, \tag{11.5.3.19}$$

where T_e is the electron temperature. In the low-frequency region the spectral density of the charge density and the electron density fluctuations are in this case given by the expressions.

$$\langle \varrho^2 \rangle_{k\omega} = \frac{\pi e^2 n_0 k^4 r_D^4}{1-(\boldsymbol{k}\cdot\boldsymbol{u})/k\tilde{u}} \{\delta(\omega - kv_s) + \delta(\omega + kv_s)\}, \tag{11.5.3.20}$$

$$\langle \delta n_e^2 \rangle_{k\omega} = \frac{\pi n_0}{1-(\boldsymbol{k}\cdot\boldsymbol{u})/k\tilde{u}} \{\delta(\omega - kv_s) + \delta(\omega + kv_s)\}. \tag{11.5.3.21}$$

11.5.4. FLUCTUATIONS IN A NON-EQUILIBRIUM PLASMA IN A MAGNETIC FIELD

Let us now consider field fluctuations in a non-equilibrium plasma in an external magnetic field $\boldsymbol{B}_0$ (Sitenko and Kirochkin, 1966). We shall assume that a compensated beam of charged particles passes through the plasma with a velocity $\boldsymbol{u}$ in the direction of the magnetic field $\boldsymbol{B}_0$. If the particle distributions in the plasma and in the beam are Maxwellian with temperatures T and T' the components of the dielectric permittivity of the plasma-beam system will have the form

$$\begin{aligned}\varepsilon_{ij}(\boldsymbol{k}, \omega) = \delta_{ij} - \sum_\alpha \frac{\omega_{p\alpha}^2}{\omega^2} \left\{ e^{-\beta_\alpha} \sum_l \frac{z_0^\alpha}{z_l^\alpha} \Pi_{ij}(z_l^\alpha) [\varphi(z_l^\alpha) - i\sqrt{\pi}\, z_l^\alpha \exp(-z_l^{\alpha 2})] - 2z_0^{\alpha 2} b_i b_j \right\} \\ - \sum_\alpha \frac{\omega_{p\alpha}'^2}{\omega^2} \left\{ e^{-\beta'_\alpha} \sum_l \frac{y_0^\alpha}{y_l^\alpha} \Pi_{ij}(z_l^{\alpha\prime}) [\varphi(y_l^\alpha) - i\sqrt{\pi}\, y_l^\alpha \exp(-y_l^{\alpha 2})] - 2z_0'^{\alpha 2} b_i b_j \right\},\end{aligned} \tag{11.5.4.1}$$

where the primes indicate quantities referring to the beam, $y_l^\alpha = \{\omega - l\omega_{B\alpha} - k_{||}u\}/\sqrt{2}\,|k_{||}|\,v'_\alpha$, while the remaining notation is as in eqn. (11.3.3.9), and $\beta = k_\perp^2 T/m_\alpha \omega_{B\alpha}^2$.

The spectral density of the fluctuations in the current of the non-interacting particles is given by the formula

$$\langle j_i j_j \rangle_{k\omega}^{(0)} = \sqrt{(2\pi)} \frac{e^2}{|k_{||}|} \left\{ \sum_\alpha n_0^\alpha v_\alpha e^{-\beta_\alpha} \sum_l \Pi_{ji}(z_l^\alpha) \exp(-z_l^{\alpha 2}) + \sum_\alpha n_0^{\alpha\prime} v'_\alpha e^{-\beta_\alpha'} \sum_l \Pi_{ji}(z_l'^\alpha) \exp(-y_l^{\alpha 2}) \right\}. \tag{11.5.4.2}$$

Using these equations we can from (11.5.2.2) find the spectral density of the field fluctuations in the plasma-beam system.

Assuming the beam density to be sufficiently low we get for the spectral density of the field fluctuations in the low-frequency region the following formula:

$$\langle E_i E_j \rangle_{k\omega} = 8\pi^2 \frac{\tilde{T}(\boldsymbol{k}, \omega)}{|\omega|} \left\{ \frac{e_i^* e_j}{|\boldsymbol{e}|^2 - |(\boldsymbol{k} \cdot \boldsymbol{e})|^2 k^{-2}} [\delta(n^2 - n_+^2) + \delta(n^2 - n_-^2)] + \frac{k_i k_j}{k^2} \delta(A) \right\}. \tag{11.5.4.3}$$

As there is no damping of the waves in a cold plasma, it is necessary when evaluating the effective temperature to take into account both the thermal motion of the electrons and the presence of the beam. We can neglect the thermal motion of the ions in the plasma and in the beam.

Using (11.5.4.1) and (11.5.4.2) we can write the effective temperature which occurs in (11.5.4.3) in the form

$$\tilde{T}(\boldsymbol{k}, \omega) = \frac{R(\boldsymbol{k}, \omega)}{1 - (\boldsymbol{k} \cdot \boldsymbol{u})/k\tilde{u}} T, \tag{11.5.4.4}$$

where

$$R(\boldsymbol{k}, \omega) = \frac{1 + \dfrac{n_0'}{n_0} \left(\dfrac{T}{T'} \right)^{1/2} K}{1 + \dfrac{n_0'}{n_0} \left(\dfrac{T}{T'} \right)^{3/2} K}, \quad \tilde{u} = \frac{\omega}{k} \left[1 + \frac{n_0}{n'} \left(\frac{T'}{T} \right)^{3/2} K^{-1} \right]. \tag{11.5.4.5}$$

The quantity K is in the case of fluctuating Langmuir oscillations given by the formula

$$K = e^{\beta - \beta'} \frac{\sum_l I_l(\beta') \exp(-y_l^2)}{\sum_l I_l(\beta) \exp(-z_l^2)}, \tag{11.5.4.6}$$

and for fluctuating ordinary and extra-ordinary electromagnetic waves by the formula

$$K = e^{\beta - \beta'} \frac{\sum_l \{[\varrho'^2 + (l^2 + \beta'^2)\, \varepsilon_2^2 (n^2 - \varepsilon_1)^{-2}]\, I_l(\beta') - 2\beta' \varrho' \varepsilon_2 (n^2 - \varepsilon_1)^{-1} I_l'(\beta')\} \exp(-y_l^2)}{\sum_l \{[\varrho^2 + (l^2 + \beta^2)\, \varepsilon_2^2 (n^2 - \varepsilon_1)^{-2}]\, I_l(\beta) - 2\beta \varrho \varepsilon_2 (n^2 - \varepsilon_1)^{-1} I_l'(\beta)\} \exp(-z_l^2)}, \tag{11.5.4.7}$$

$$\varrho = l + \beta \frac{\varepsilon_2}{n^2 - \varepsilon_1} + \sqrt{(2\beta)} \frac{k_{||}}{|k_{||}|} \frac{n^2 \sin\vartheta \cos\vartheta}{n^2 \sin^2\vartheta - \varepsilon_3} z_l.$$

The quantity $\tilde{u}$ plays the role of a critical velocity; when it is reached the fluctuations in the plasma grow without bounds, and the plasma becomes unstable.

In the limiting case $n^2 \gg 1$ eqns. (11.5.4.6) and (11.5.4.7) become the same. We note that when $n^2 \gg 1$ the spectral density of the fluctuations of the ordinary and the extra-ordinary waves is the same as that for fluctuating Langmuir waves.

The effective temperature for low-frequency fluctuating oscillations in a magneto-active plasma through which a beam of charged particles is passing is given by the formula

$$\tilde{T}(\boldsymbol{k}, \omega) = \frac{T}{1 - (u/\tilde{u}) \cos\vartheta}, \tag{11.5.4.8}$$

where the critical velocities for the Alfvén, the fast, and the slow magneto-sound fluctuations are, respectively, equal to

$$\tilde{u}_a = \frac{\omega}{k}\left\{1+\frac{n_0}{n_0'}\left(\frac{T}{T'}\right)^{1/2}\frac{\sin^4\vartheta+\cos^4\vartheta}{\cos^4\vartheta+(\cos^2\vartheta-[T/T'])^2}\exp y_0^2\right\}, \qquad (11.5.4.9)$$

$$\tilde{u}_f = \frac{\omega}{k}\left\{1+\frac{n_0}{n_0'}\left(\frac{T}{T'}\right)^{1/2}\frac{\exp y_0^2}{1+(1-[T/T'])^2}\right\}, \quad \tilde{u}_s = \frac{\omega}{k}\left\{1+\frac{n_0}{n_0'}\left(\frac{T'}{T}\right)^{3/2}\exp y_0^2\right\}. \qquad (11.5.4.10)$$

The spectral densities of the low-frequency fluctuations in a non-equilibrium plasma are given by eqns. (11.4.3.1), (11.4.3.3), and (11.4.3.4) in which we must substitute for T_e the effective temperature (11.5.4.8).

11.6. Kinetic Theory of Fluctuations

11.6.1. FLUCTUATIONS IN THE DISTRIBUTION FUNCTION

We have shown in the preceding sections that the fluctuations in the electromagnetic quantities in an equilibrium plasma—and also in any other medium in a state of statistical equilibrium—are completely determined by its dielectric permittivity tensor. The equations of motion of the plasma particles, that is, the kinetic equations for the electrons and the ions are not directly needed for a study of fluctuations—these equations are used only to determine the plasma dielectric permittivity tensor, but as soon as it is found we can use the fluctuation–dissipation relation which connects the spectral density of the fluctuations with the dielectric permittivity tensor.

The situation is considerably more complex in the case of a non-equilibrium plasma for which the spectral densities of the fluctuations are, according to (11.5.1.6), determined not solely by the plasma dielectric permittivity tensor but also directly by the distribution functions of its particles.

In the present section we shall reproduce the results obtained in Section 11.5 using now a kinetic theory of fluctuations, and we shall find the fluctuations not only in the macroscopic quantities, but also in the particle distribution functions.†

We shall start by considering fluctuations of an equilibrium plasma. It is well known that if we want to find the fluctuations of various physical quantities of some system we must introduce into the equations describing that system additional external quantities—the so-called random forces. As we are interested in the fluctuations in the particle distribution functions we must introduce random forces in the right-hand side of the kinetic equations which determine the particle distribution functions $F^{(\alpha)}(\boldsymbol{p}, \boldsymbol{r}, t)$, where the index α is used to indicate the kind of particle:

$$\left\{\frac{\partial}{\partial t}+(\boldsymbol{v}\cdot\nabla)+e_\alpha\left(\left\{\boldsymbol{E}+\frac{1}{c}[\boldsymbol{v}\wedge\boldsymbol{B}]\right\}\cdot\frac{\partial}{\partial \boldsymbol{p}}\right)\right\}F^{(\alpha)} = \left[\frac{\partial F^{(\alpha)}}{\partial t}\right]_c + y^{(\alpha)}, \qquad (11.6.1.1)$$

† A kinetic theory of fluctuations in a plasma was developed by Akhiezer, Akhiezer, and Sitenko (1962)

where $[\partial F^{(\alpha)}/\partial t]_c$ is a collision integral, $y^{(\alpha)}(\boldsymbol{p}, \boldsymbol{r}, t)$ a random force, and $\boldsymbol{E}$ and $\boldsymbol{B}$ are the self-consistent electrical and magnetic fields.†

Following the general rules of the theory of fluctuations, we choose as the generalized thermodynamic velocities $\dot{x}^{(\alpha)}$ which occur in that theory the quantities

$$\dot{x}^{(\alpha)} = \left[\frac{\partial F^{(\alpha)}}{\partial t}\right]_c + y^{(\alpha)}. \tag{11.6.1.2}$$

As suming for the sake of simplicity that the collision integral is of the form

$$\left[\frac{\partial F^{(\alpha)}}{\partial t}\right]_c = -\frac{1}{\tau_\alpha} f^{(\alpha)}, \tag{11.6.1.3}$$

where $f^{(\alpha)}$ is the deviation of the distribution function $F^{(\alpha)}$ from the Maxwell distribution $F_0^{(\alpha)}$, $f^{(\alpha)} = F^{(\alpha)} - F_0^{(\alpha)}$, we have

$$\dot{x}^{(\alpha)} = -\frac{1}{\tau_\alpha} f^{(\alpha)} + y^{(\alpha)}. \tag{11.6.1.4}$$

We now introduce the generalized thermodynamic force $X^{(\alpha)}$, as the functional derivative of $\dot{S}$ with respect to $\dot{x}^{(\alpha)}$:

$$X^{(\alpha)} = -\frac{\delta \dot{S}}{\delta \dot{x}^{(\alpha)}},$$

where $\dot{S}$ is the time-derivative of the entropy of the system,

$$\dot{S} = -\sum_\alpha \int \left\{\left[\frac{\partial F^{(\alpha)}}{\partial t}\right]_c + y^{(\alpha)}\right\} \ln F^{(\alpha)}\, d^3\boldsymbol{p}\, d^3\boldsymbol{r}. \tag{11.6.1.5}$$

Using (11.6.1.3) we can write $\dot{S}$ in the form

$$\dot{S} = \int \left\{\frac{f^{(e)}}{\tau_e} \ln F_0^{(e)} + \frac{f^{(i)}}{\tau_i} \ln F_0^{(i)}\right\} d^3\boldsymbol{p}\, d^3\boldsymbol{r} + \int \left\{\left(\frac{f^{(e)}}{\tau_e} - y^{(e)}\right)\frac{f^{(e)}}{F_0^{(e)}} + \left(\frac{f^{(i)}}{\tau_i} - y^{(i)}\right)\frac{f^{(i)}}{F_0^{(i)}}\right\} d^3\boldsymbol{p}\, d^3\boldsymbol{r}. \tag{11.6.1.6}$$

As the entropy must be given for fixed values of the energy and number of particles of each kind, the first term on the right-hand side vanishes, and hence

$$\dot{S} = \int \left\{\left(\frac{f^{(e)}}{\tau_e} - y^{(e)}\right)\frac{f^{(e)}}{F_0^{(e)}} + \left(\frac{f^{(i)}}{\tau_i} - y^{(i)}\right)\frac{f^{(i)}}{F_0^{(i)}}\right\} d^3\boldsymbol{p}\, d^3\boldsymbol{r}. \tag{11.6.1.7}$$

The thermodynamic force therefore has the form

$$X^{(\alpha)} = \frac{f^{(\alpha)}}{F_0^{(\alpha)}}. \tag{11.6.1.8}$$

‡ The present method is a generalization of the one used by Abrikosov and Khalatnikov (1958) to find the fluctuations in the distribution function in an equilibrium Fermi-liquid.

Comparing this equation with (11.6.1.4) we see that

$$\mathring{x}^{(\alpha)} = -\gamma^{(\alpha)} X^{(\alpha)} + y^{(\alpha)}, \tag{11.6.1.9}$$

where

$$\gamma^{(\alpha)}(p) = \frac{1}{\tau_\alpha} F_0^{(\alpha)}(p).$$

These quantities are called the kinetic coefficients. According to the general theory of fluctuations they determine the correlation functions of the random forces:

$$\langle y^{(\alpha)}(\boldsymbol{p}, \boldsymbol{r}, t)\, y^{(\alpha')}(\boldsymbol{p}', \boldsymbol{r}', t')\rangle = 2\delta_{\alpha\alpha'}\gamma^{(\alpha)}(p)\, \delta(\boldsymbol{p}-\boldsymbol{p}')\, \delta(\boldsymbol{r}-\boldsymbol{r}')\, \delta(t-t'). \tag{11.6.1.10}$$

From this we easily find the spectral densities of the correlation functions of the random forces

$$\langle y^{(\alpha)}(\boldsymbol{p})\, y^{(\alpha')}(\boldsymbol{p}')\rangle_{k\omega} = 2\delta_{\alpha\alpha'}\gamma^{(\alpha)}(p)\, \delta(\boldsymbol{p}-\boldsymbol{p}'). \tag{11.6.1.11}$$

Knowing the correlation functions for the random forces we can use the kinetic equations and the Maxwell equations to express the particle distribution functions in terms of the random forces and to find the correlation functions of the distribution functions.

We shall give the expression for the spectral density of the fluctuations in the electron distribution function in a collisionless plasma (Akhiezer, Akhiezer, and Sitenko, 1962):

$$\langle f(\boldsymbol{p}) f(\boldsymbol{p}')\rangle_{k\omega} = 2\pi\, \delta(\boldsymbol{p}-\boldsymbol{p}')\, \delta\{\omega-(\boldsymbol{k}\cdot\boldsymbol{v})\}\, F_0(p)$$
$$+\frac{8\pi^2 e^2}{k^2} F_0(p)\, F_0(p') \left\{S_l(\boldsymbol{v}, \boldsymbol{v}') - \frac{k^2(\boldsymbol{v}\cdot\boldsymbol{v}')-(\boldsymbol{k}\cdot\boldsymbol{v})(\boldsymbol{k}\cdot\boldsymbol{v}')}{\omega^2} S_t(\boldsymbol{v}, \boldsymbol{v}')\right\}, \tag{11.6.1.12}$$

where

$$S_l(\boldsymbol{v}, \boldsymbol{v}') = \frac{1}{T}\left\{\frac{(\boldsymbol{k}\cdot\boldsymbol{v})}{\omega-(\boldsymbol{k}\cdot\boldsymbol{v})-io}\, \varepsilon_l^{*-1}\, \delta\{\omega-(\boldsymbol{k}\cdot\boldsymbol{v}')\} + \frac{(\boldsymbol{k}\cdot\boldsymbol{v}')}{\omega-(\boldsymbol{k}\cdot\boldsymbol{v})+io}\, \varepsilon_l^{-1}\, \delta\{\omega-(\boldsymbol{k}\cdot\boldsymbol{v})\}\right.$$
$$\left. + \frac{1}{\pi}\frac{(\boldsymbol{k}\cdot\boldsymbol{v})}{\omega-(\boldsymbol{k}\cdot\boldsymbol{v})-io}\, \frac{(\boldsymbol{k}\cdot\boldsymbol{v}')}{\omega-(\boldsymbol{k}\cdot\boldsymbol{v}')+io}\, \frac{\operatorname{Im}\varepsilon_l}{\omega|\,\varepsilon_l\,|^2}\right\}, \tag{11.6.1.13}$$

$$S_t(\boldsymbol{v}, \boldsymbol{v}') = \frac{1}{\pi}\left\{\frac{\omega}{\omega-(\boldsymbol{k}\cdot\boldsymbol{v})-io}\, [\varepsilon_t^* - n^2]^{-1}\, \delta\{\omega-(\boldsymbol{k}\cdot\boldsymbol{v}')\} + \frac{\omega}{\omega-(\boldsymbol{k}\cdot\boldsymbol{v}')+io}\, [\varepsilon_t - n^2]^{-1}\, \delta\{\omega-(\boldsymbol{k}\cdot\boldsymbol{v})\}\right.$$
$$\left. + \frac{1}{\pi}\frac{\omega}{\omega-(\boldsymbol{k}\cdot\boldsymbol{v})-io}\, \frac{1}{\omega-(\boldsymbol{k}\cdot\boldsymbol{v}')+io}\, \frac{\operatorname{Im}\varepsilon_t}{|\,\varepsilon_t-n^2\,|^2}\right\}. \tag{11.6.1.14}$$

The first term on the right-hand side of (11.6.1.12) is the same as the correlation function of the fluctuations of the distribution function in a gas of non-interacting particles (Kadomtsev, 1957), while the second and third terms take into account the interaction between the particles caused by the self-consistent field. The second term is here connected with the longitudinal (electrostatic) part of the self-consistent field, while the third term takes into account the transverse part of the self-consistent field. We note that S_l has sharp maxima at the frequencies of the longitudinal plasma oscillations, and S_t at the frequencies of the transverse electromagnetic waves. One can neglect the contribution S_t under non-relativistic conditions.

Once we know the spectral distribution of the fluctuations in the particle distribution function we can easily determine the spectral densities of the various macroscopic quantities, such as the electron density. The fluctuations in the electron density are connected with f through the relation

$$\delta n = \int f \, d^3\mathbf{p}. \tag{11.6.1.15}$$

Using this relation and eqn. (11.6.1.12) we can easily obtain the following expression for the spectral density of the electron density fluctuations in a collisionless plasma:

$$\langle \delta n^2 \rangle_{k\omega} = \sqrt{(2\pi)} \frac{n_0}{\omega_{pe}} k^3 r_D^3 \frac{\exp(-z^2)}{[k^2 r_D^2 + 1 - \varphi(z)]^2 + \pi z^2 \exp(-2z^2)} . \tag{11.6.1.16}$$

We can similarly study the fluctuations in the electron temperature. This quantity is connected with f by the relation

$$\delta T = \frac{m_e}{3n_0} \int v^2 f \, d^3\mathbf{p}; \tag{11.6.1.17}$$

using this relation we can easily show that

$$\langle \delta T^2 \rangle_{k\omega} = \frac{\sqrt{(2\pi)}}{9} \frac{T_0^2}{n_0 \omega_{pe}} \frac{1}{k r_D} \frac{P(z) \exp(-z^2)}{[k^2 r_D^2 + 1 - \varphi(z)]^2 + \pi z^2 \exp(-2z^2)} , \tag{11.6.1.18}$$

where

$$P(z) = k^4 r_D^4 (4z^4 + 4z^2 + 5) - k^2 r_D^2 (8\varphi(z) + 4z^2 - 10) + 4\varphi(z)^2 - 8\varphi(z) + 5.$$

We also give the spectral density

$$\langle \delta n \, \delta T \rangle_{k\omega} = \frac{\sqrt{(2\pi)}}{3} \frac{T_0}{\omega_{pe}} k r_D \frac{[k^2 r_D^2 (2z^2 - 1) - 1] \exp(-z^2)}{[k^2 r_D^2 + 1 - \varphi(z)]^2 + \pi z^2 \exp(-2z^2)} . \tag{11.6.1.19}$$

In the case of short-wavelength waves ($k^2 r_D^2 \gg 1$) the spectral densities (11.6.1.16), (11.6.1.18), and (11.6.1.19) are characterized by broad maxima at zero frequency. In the long-wavelength region ($k^2 r_D^2 \ll 1$) the spectral densities are characterized by δ-function-

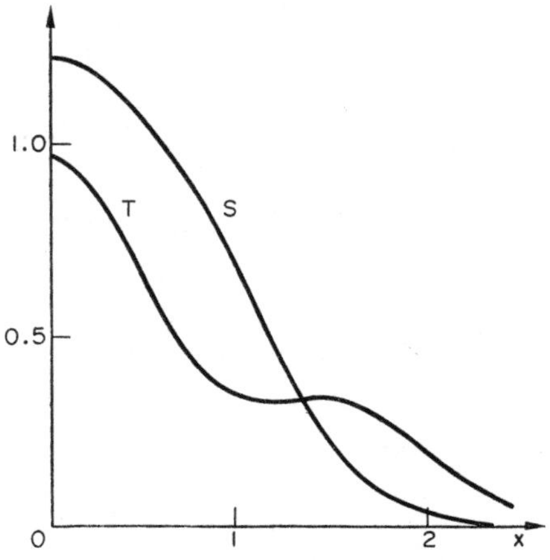

FIG. 11.6.1. Spectral densities of the fluctuations in the density $S(x) = \langle \delta n^2 \rangle_{kx} / 2n_0$, and in the temperature, $T(x) = n_0 \langle \delta T^2 \rangle_{kx} / 2T_0^2$, in a collisionless plasma for the case when $k^2 r_D^2 = 5$; $x = \omega/\sqrt{2}\, k v_e$ is the dimensionless frequency.

like maxima at frequencies corresponding to the frequency of the longitudinal plasma oscillations. The spectral density of the temperature fluctuations also has a maximum at zero frequency, which is connected with entropy waves in the plasma. We show in Fig. 11.6.1 the spectral densities of the density and temperature fluctuations in a collisionless plasma for the case where $k^2r_D^2 = 5$.

Integrating (11.6.1.16), (11.6.1.18), and (11.6.1.19) over the frequency we find the correlation functions for the momentaneous density and temperature fluctuations in the plasma:

$$\langle \delta n^2 \rangle_k = \frac{n_0 k^2}{k^2 + 4\pi e^2 n_0 / T_0}, \quad \langle \delta T^2 \rangle_k = \frac{2}{3} \frac{T_0^2}{n_0}, \quad \langle \delta n \, \delta T \rangle_k = 0. \tag{11.6.1.20}$$

It follows from the last of eqns. (11.6.1.20) that the density and temperature fluctuations in a plasma are statistically independent. In concluding this subsection we emphasize that our results were obtained assuming that $\tau \to \infty$, that is, they refer to a collisionless plasma.

11.6.2. FLUCTUATIONS IN THE DISTRIBUTION FUNCTIONS IN A NON-ISOTHERMAL PLASMA

One can easily generalize the method for studying fluctuations which was given in the preceding subsection to the case of a non-isothermal, two-temperature plasma (Akhiezer, Akhiezer and Sitenko, 1962). The fact that it is possible to introduce such a generalization is connected with the fact that due to the large difference in mass between the electrons and ions the exchange of energy between particles of the same kind takes place much faster than the exchange of energy between particles of different kinds. If, therefore, we neglect the exchange of energy between particles of different kinds, the state of the plasma with different electron and ion temperatures will correspond to a maximum of the entropy—for given values of the particle numbers and of the electron and ion energies, separately. This means that we can use the previous expression (11.6.1.6) for the time-derivative of the entropy of a two-temperature plasma:

$$\dot{S} = \int \left\{ \frac{f^{(e)}}{\tau_e} \ln F_0^{(e)} + \frac{f^{(i)}}{\tau_i} \ln F_0^{(i)} \right\} d^3\boldsymbol{p}\, d^3\boldsymbol{r} + \int \left\{ \left(\frac{f^{(e)}}{\tau_e} - y^{(e)} \right) \frac{f^{(e)}}{F_0^{(e)}} + \left(\frac{f^{(i)}}{\tau_i} - y^{(i)} \right) \frac{f^{(i)}}{F_0^{(i)}} \right\} d^3\boldsymbol{p}\, d^3\boldsymbol{r}, \tag{11.6.2.1}$$

where $F_0^{(e)}$ and $F_0^{(i)}$ are the Maxwell distributions for the electrons and ions with different temperatures, and $f^{(e)}$ and $f^{(i)}$ are small deviations from these distributions. As we must find the quantity $\dot{S}$ for given values of the electron and ion energies, the first term in the above expression for $\dot{S}$ vanishes, as in the case of an equilibrium plasma. In other words, the time-derivative of the entropy will, as in the case of an equilibrium plasma, be a bilinear expression in the quantities

$$\dot{x}^{(\alpha)} = -\frac{1}{\tau_\alpha} f^{(\alpha)} + y^{(\alpha)}, \tag{11.6.2.2}$$

which characterize the deviation of the state of the system from the quasi-equilibrium state with different electron and ion temperatures. Following the same road as in the case of a plasma in complete equilibrium we find the quantities

$$X^{(\alpha)} \equiv \frac{\delta\dot{S}}{\delta\dot{x}^{(\alpha)}} = \frac{f^{(\alpha)}}{F_0^{(\alpha)}}, \tag{11.6.2.3}$$

and we can write $\dot{x}^{(\alpha)}$ in the form

$$\dot{x}^{(\alpha)} = -\gamma^{(\alpha)}X^{(\alpha)} + y^{(\alpha)},$$

where

$$\gamma^{(\alpha)}(p) = \frac{1}{\tau_\alpha} F_0^{(\alpha)}(p). \tag{11.6.2.4}$$

Using (11.6.2.4) we get again the earlier expression for the correlation function of the random forces

$$\langle y^{(\alpha)}(\boldsymbol{p}, \boldsymbol{r}, t)\, y^{(\alpha')}(\boldsymbol{p}', \boldsymbol{r}', t')\rangle = 2\delta_{\alpha\alpha'}\gamma^{(\alpha)}(p)\,\delta(\boldsymbol{p}-\boldsymbol{p}')\,\delta(\boldsymbol{r}-\boldsymbol{r}')\,\delta(t-t'). \tag{11.6.2.5}$$

The expression for the spectral density of the correlation function of the random forces also remains the same:

$$\langle y^{(\alpha)}(\boldsymbol{p})\, y^{(\alpha')}(\boldsymbol{p}')\rangle_{\boldsymbol{k}\omega} = 2\delta_{\alpha\alpha'}\gamma^{(\alpha)}(p)\,\delta(\boldsymbol{p}-\boldsymbol{p}'). \tag{11.6.2.6}$$

As in the case of an equilibrium plasma we can use this expression to find the correlation functions of various physical quantities.

Let us give the general expression for the correlation function of the fluctuations in the particle distribution functions in a non-isothermal two-temperature plasma:

$$\langle f^{(\alpha)}(\boldsymbol{p})\, f^{(\alpha')}(\boldsymbol{p}')\rangle_{\boldsymbol{k}\omega} = 2\pi\,\delta_{\alpha\alpha'}F_0^{(\alpha)}(p)\,\delta(\boldsymbol{p}-\boldsymbol{p}')\,\delta\{\omega-(\boldsymbol{k}\cdot\boldsymbol{v})\}$$

$$+\frac{8\pi^2 e^2 Z_\alpha Z_{\alpha'}}{k^2} F_0^{(\alpha)}(p)\, F^{(\alpha')}(p')\left\{S_{\rm l}^{\alpha\alpha'}(\boldsymbol{v}, \boldsymbol{v}') + \frac{k^2(\boldsymbol{v}\cdot\boldsymbol{v}')-(\boldsymbol{k}\cdot\boldsymbol{v})(\boldsymbol{k}\cdot\boldsymbol{v}')}{\omega^2} S_{\rm t}^{\alpha\alpha'}(\boldsymbol{v}, \boldsymbol{v}')\right\}. \tag{11.6.2.7}$$

Here

$$S_{\rm l}^{\alpha\alpha'}(\boldsymbol{v}, \boldsymbol{v}') = \frac{1}{T_\alpha}\frac{(\boldsymbol{k}\cdot\boldsymbol{v})}{\omega-(\boldsymbol{k}\cdot\boldsymbol{v})-io}\,\varepsilon_{\rm l}^{*\,-1}\,\delta\{\omega-(\boldsymbol{k}\cdot\boldsymbol{v}')\} + \frac{1}{T_{\alpha'}}\frac{(\boldsymbol{k}\cdot\boldsymbol{v}')}{\omega-(\boldsymbol{k}\cdot\boldsymbol{v}')+io}\,\varepsilon_{\rm l}^{-1}\,\delta\{\omega-(\boldsymbol{k}\cdot\boldsymbol{v})\}$$

$$+\frac{4}{T_\alpha T_{\alpha'}}\frac{(\boldsymbol{k}\cdot\boldsymbol{v})}{\omega-(\boldsymbol{k}\cdot\boldsymbol{v})-io}\,\frac{(\boldsymbol{k}\cdot\boldsymbol{v}')}{\omega-(\boldsymbol{k}\cdot\boldsymbol{v}')+io}\,\frac{\operatorname{Im}(T_{\rm e}\varkappa_{\rm l}^{\rm e}+T_{\rm i}\varkappa_{\rm l}^{\rm i})}{\omega\,|\,\varepsilon_{\rm l}\,|^2}, \tag{11.6.2.8}$$

$$S_{\rm t}^{\alpha\alpha'}(\boldsymbol{v}, \boldsymbol{v}') = \frac{1}{T_\alpha}\frac{\omega}{\omega-(\boldsymbol{k}\cdot\boldsymbol{v})-io}\,\frac{\delta\{\omega-(\boldsymbol{k}\cdot\boldsymbol{v}')\}}{\varepsilon_{\rm t}^*-n^2} + \frac{1}{T_{\alpha'}}\frac{\omega}{\omega-(\boldsymbol{k}\cdot\boldsymbol{v}')+io}\,\frac{\delta\{\omega-(\boldsymbol{k}\cdot\boldsymbol{v})\}}{\varepsilon_{\rm t}-n^2}$$

$$+\frac{4}{T_\alpha T_{\alpha'}}\frac{\omega}{\omega-(\boldsymbol{k}\cdot\boldsymbol{v})-io}\,\frac{\omega}{\omega-(\boldsymbol{k}\cdot\boldsymbol{v}')+io}\,\frac{\operatorname{Im}(T_{\rm e}\varkappa_{\rm t}^{\rm e}+T_{\rm i}\varkappa_{\rm t}^{\rm i})}{\omega\,|\,\varepsilon_{\rm t}-n^2\,|^2}, \tag{11.6.2.9}$$

where T_α is the temperature of the αth component of the plasma, and Z_α its charge.

We can study in the same way the fluctuations in a two-temperature plasma in a constant, uniform magnetic field $\boldsymbol{B}_0$. When evaluating the correlation functions in that case we can, as before, start from the kinetic equations with random forces (11.6.1.1); it is only necessary to introduce in the left-hand sides of these equations the terms $(Z_\alpha e/c)([\boldsymbol{v} \wedge \boldsymbol{B}_0]\cdot\{\partial F_0^{(\alpha)}/\partial \boldsymbol{p}\})$ which take the effect of the magnetic field into account. Due to these additional terms the connection between the averaged products of various physical quantities and the averaged products of the random forces is changed; the normalization of the random forces is, as before, given by eqn. (11.6.2.6).

11.6.3. EVOLUTION IN TIME OF THE FLUCTUATIONS

In the preceding subsections we have determined the fluctuations of the particle distribution functions in the plasma and we have shown how from a knowledge of these fluctuations we can find the fluctuations in various macroscopic quantities such as the fields in an equilibrium and in a two-temperature plasma. We shall now turn to a calculation of fluctuations in a plasma with non-equilibrium, but stable distribution functions, using the kinetic description of the plasma (Akhiezer, 1962).

In the case of a plasma with non-equilibrium particle distribution functions the fluctuations are not necessarily determined by the averaged characteristics of the plasma, but may depend on its previous history. There is, however, a possibility that the process of establishing well-defined fluctuation distributions is much faster than the evolution of the distribution functions themselves. In such a case quasi-equilibrium fluctuation densities are set up in the plasma, and the correlation functions are fully determined by the averaged characteristics of the plasma, and especially by the non-equilibrium particle distributions, averaged over the fluctuations, and will be independent of the previous history. The slow change in the non-equilibrium particle distribution functions is accompanied by a slow change in the correlation functions of the quasi-equilibrium fluctuations, which are adjusted to these distribution functions.

To determine the distribution function $F^{(\alpha)}$ of particles of the αth kind we use the kinetic equation without the collision integral and with the initial condition

$$F^{(\alpha)}(\boldsymbol{p}, \boldsymbol{r}, t=0) - F_0^{(\alpha)}(p) = g^{(\alpha)}(\boldsymbol{p}, \boldsymbol{r}), \tag{11.6.3.1}$$

where $F_0^{(\alpha)}$ is the distribution function averaged over the fluctuations and $g^{(\alpha)}$ the fluctuation in the distribution function at time $t = 0$.

If we Laplace transform the kinetic equation and the Poisson equation with respect to time we can express the particle distribution functions and the various physical quantities determined by them at time t in terms of the initial values of the fluctuations $g^{(\alpha)}$ in the distribution functions. In particular, we find for the spatial Fourier component of the charge density

$$\varrho_k(t) = \frac{ie}{2\pi} \int\limits_{i\sigma-\infty}^{i\sigma+\infty} \frac{e^{-i\omega t}}{\varepsilon_l(\boldsymbol{k}, \omega)} \sum_\alpha Z_\alpha \int \frac{g^{(\alpha)}(\boldsymbol{p}, \boldsymbol{k})}{\omega - (\boldsymbol{k}\cdot\boldsymbol{v})} \, d^3\boldsymbol{p} \, d\omega, \tag{11.6.3.2}$$

where the integration over ω is along the straight line $\text{Im}\,\omega = \sigma$, which lies above all poles of the function ε_l^{-1}.

Let us now construct quadratic combinations of different physical quantities—referring to times t and t' which are not necessarily the same—and average these combinations over the random quantities $g^{(\alpha)}$. The quantities obtained in this way are the correlation functions of the system considered; once we know these we can, in particular, determine the squares of the amplitudes of the oscillations of physical quantities at time t (to do this it is sufficient to put $t = t'$ in the correlators).

As an example we give the expression for the charge density correlator

$$\langle \varrho^2 \rangle_{kt-t'} = \iint\limits_{i\sigma-\infty}^{i\sigma+\infty} e^{-i\omega t - i\omega' t'} \varepsilon_1^{-1}(\boldsymbol{k}, \omega)\, \varepsilon_1^{-1}(-\boldsymbol{k}, \omega')\, B(\boldsymbol{k}, \omega, \omega')\, d\omega\, d\omega', \qquad (11.6.3.3)$$

where

$$B(\boldsymbol{k}, \omega, \omega') = -e^2 \sum_{\alpha, \alpha'} Z_\alpha Z_{\alpha'} \iint \frac{\langle g^{(\alpha)}(\boldsymbol{p})\, g^{(\alpha')}(\boldsymbol{p}')\rangle}{[\omega - (\boldsymbol{k}\cdot\boldsymbol{v})]\{\omega' + (\boldsymbol{k}\cdot\boldsymbol{v}')\}}\, d^3\boldsymbol{p}\, d^3\boldsymbol{p}'. \qquad (11.6.3.4)$$

We can write the Fourier component of the averaged product of the initial values of the distribution function fluctuations which occurs in this equation in the following general form:

$$\langle g^{(\alpha)}(\boldsymbol{p})\, g^{(\alpha')}(\boldsymbol{p}')\rangle_k = \delta_{\alpha\alpha'} F_0^{(\alpha)}(p)\, \delta(\boldsymbol{p} - \boldsymbol{p}') + Y_{\alpha\alpha'}(\boldsymbol{k}, \boldsymbol{p}, \boldsymbol{p}'), \qquad (11.6.3.5)$$

where the first term corresponds to a perfect gas—each particle is correlated only with itself—while the second term is caused by the interaction between the particles. It is important that the second term is a smooth function of the velocities while the first one contains $\delta(\boldsymbol{p} - \boldsymbol{p}')$.

According to eqns. (11.6.1.12) and (11.6.1.13) the function Y has in the case of an equilibrium plasma the form

$$Y_{\alpha\alpha'} = -\frac{4\pi e^2 Z_\alpha Z_{\alpha'}}{T} F_0^{(\alpha)}(p)\, F_0^{(\alpha')}(p')\, (k^2 + \bar{r}_D^{-2})^{-1}, \qquad (11.6.3.6)$$

where $\bar{r}_D$ is the screening radius. Using the kinetic equation and the Poisson equation to express the fluctuations in the distribution function at time t in terms of the initial fluctuations and averaging over those, using eqns. (11.6.3.5) and (11.6.3.6), we arrive at the same eqns. (11.6.3.5) and (11.6.3.6) for the averaged values of the distribution function fluctuations. One can say that the equilibrium fluctuations reproduce themselves. Equation (11.6.3.3) then gives

$$\langle \varrho^2 \rangle_{kt} = \frac{k^2}{4\pi^2} T \int\limits_{-\infty}^{+\infty} \frac{e^{-i\omega t}}{\omega} \operatorname{Im} \varepsilon_1^{*-1}(\boldsymbol{k}, \omega)\, d\omega, \qquad (11.6.3.7)$$

which agrees with (11.2.2.5).

Let the averaged distribution function $F_0^{(\alpha)}$ as before be an equilibrium function but let the initial perturbations $g^{(\alpha)}$ be such that the function Y no longer is given by formula (11.6.3.6). We substitute (11.6.3.5) into (11.6.3.3) and use Cauchy's theorem to integrate over ω and ω'. We then get not only the term (11.6.3.7) which describes the equilibrium fluctuations but also other terms. These terms are, however, damped with a damping rate $\operatorname{Im} \omega_k$, where ω_k is a root of the equation $\varepsilon_1(\boldsymbol{k}, \omega) = 0$. The equilibrium fluctuations of the charge density and of other macroscopic quantities will be established therefore after a time $t \sim (\operatorname{Im} \omega_k)^{-1}$, notwithstanding the non-equilibrium nature of the initial fluctuations.

The quantity $\operatorname{Im} \omega_k$ is small for weakly damped eigenoscillations of the plasma so that states of the plasma for which the amplitudes of the eigenoscillations differ greatly from their equilibrium values may exist for a very long time and can be considered to be quasi-equilibrium states.

Let us consider the more general case of a plasma with non-equilibrium, albeit stable particle distribution functions. Substituting (11.6.3.5) into (11.6.3.4) we get

$$B(\boldsymbol{k},\omega,\omega') = -\sum_\alpha \frac{e^2Z_\alpha^2}{\omega+\omega'}\int F_0^{(\alpha)}(p)\left\{\frac{1}{\omega-(\boldsymbol{k}\cdot\boldsymbol{v})+io}+\frac{1}{\omega'+(\boldsymbol{k}\cdot\boldsymbol{v})+io}\right\}d^3\boldsymbol{p}+\delta B(\boldsymbol{k},\omega,\omega'). \tag{11.6.3.8}$$

The first term here arises from the δ-function-like term in eqn. (11.6.3.5) while δB indicates the contribution from the quantity Y to the function B. It is important that the first term in eqn. (11.6.3.8) has a pole at $\omega+\omega'=0$, while the quantity δB—like ε_1^{-1}—has no poles whatever in the upper ω- and ω'-halfplanes, nor for real vales of these variables. Due to this an undamped contribution to the correlator $\langle\varrho^2\rangle$, given by eqn. (11.6.3.3), is given only by the pole of the function B at $\omega+\omega'=0$; the remaining terms in the expression for $\langle\varrho^2\rangle$ will decrease with increasing time.

In a non-equilibrium, but stable plasma there are thus after a time $t\sim(\operatorname{Im}\omega_k)^{-1}$ established charge density fluctuations which are independent of the initial perturbations and which are determined by the correlation functions

$$\langle\varrho^2\rangle_{\boldsymbol{k}t}=\frac{1}{2\pi}\int e^{-i\omega t}\langle\varrho^2\rangle_{\boldsymbol{k}\omega}\,d\omega,\qquad \langle\varrho^2\rangle_{\boldsymbol{k}\omega}=\frac{2\pi}{|\varepsilon_1|^2}(A^{\mathrm{e}}+A^{\mathrm{i}}), \tag{11.6.3.9}$$

where the quantities A^α are given by the formula

$$A^\alpha=\int F_0^{(\alpha)}(p)\,\delta\{\omega-(\boldsymbol{k}\cdot\boldsymbol{v})\}\,d^3\boldsymbol{p}. \tag{11.6.3.10}$$

We note that, formally, eqn. (11.6.3.9) is the same as relation (11.4.1.14) for the charge density correlator in a quasi-equilibrium plasma—although, of course, the quantities A^α can no longer be expressed in terms of the imaginary parts of the electrical conductivities when the distribution functions $F_0^{(\alpha)}(p)$ are arbitrary.

As in the case of equilibrium distribution functions fluctuations with the frequencies of the plasma eigenoscillations will continue to depend on the initial perturbations for the longest periods, so that plasma states which are characterized by eigenoscillation amplitudes differing from those given by eqn. (11.6.3.9) can persist for very long periods.

We emphasize the characteristic peculiarities of fluctuations in a plasma with non-Maxwellian particle distribution functions. First of all, only the fluctuations of macroscopic quantities will get to a stage which is independent of the initial perturbations; the correlators of the distribution function fluctuations contain terms which are proportional to $\delta\{\omega-(\boldsymbol{k}\cdot\boldsymbol{v})\}$ which are determined by the initial distribution of the fluctuations. Secondly, the amplitudes of the random waves do not reach a stage independent of the initial perturbations if the phase velocity of these waves is larger than the velocities of all the particles in the plasma—in particular, this is true for transverse electromagnetic waves which are well known to have a phase velocity larger than the velocity of light. Indeed, for such waves the time to establish equilibrium distributions of the amplitudes is, when there are no collisions, $t\sim(\operatorname{Im}\omega_k)^{-1}\to\infty$.

11.6.4. FLUCTUATIONS IN A PLASMA-BEAM SYSTEM

If the particle distribution functions in the plasma are not only non-equilibrium, but also unstable, the situation becomes considerably more complicated (Akhiezer, 1962). The function ε_l^{-1} has in that case a pole in the upper ω-halfplane so that the contribution to the correlation functions from the quantities Y—which are determined by the previous history of the system and which in general are unknown—does not decrease, but increases with increasing time. Nevertheless, in the case of a system consisting of a plasma and a low density beam ($n_0' \ll n_0$, where n_0' is the beam density) the main terms in the correlator are independent of the random initial perturbations. We assume here that the perturbations did not increase so strongly during the period when the beam was introduced that the initial fluctuations, even though they differ from the equilibrium ones, are not of the same order of magnitude as the equilibrium fluctuations.

Indeed, substituting (11.6.3.5) and (11.6.3.8) into (11.6.3.3) and using Cauchy's theorem to integrate we see easily that the main terms in the correlator $\langle \varrho^2 \rangle$ turn out to be the terms given by the first term in (11.6.3.5) and, hence, the first term in (11.6.3.8) which contains $\omega+\omega'$ in the denominator. These terms are proportional to a power of n_0' less than the first power while the remaining terms contain n_0' at least to the first power.

In the general case the correlation functions and the squares of the amplitudes of the oscillations for a plasma-beam system contain terms which grow exponentially with a growth rate proportional to $\sqrt{n_0'}$ while the multiplying factor is also proportional to $\sqrt{n_0'}$. We have seen in Section 11.5 that if the beam velocity is large compared to the average thermal velocity of the electrons in the plasma, resonance occurs between the beam oscillations and the Langmuir oscillations in the plasma. The correlation functions then grow with a growth rate proportional to $n_0'^{1/3}$, while the multiplying factor is independent of the beam density.

Without wanting to go into the general expressions for the correlation functions we shall only give the formula for the charge density correlator with a resonance value of the wave-vector $|(\boldsymbol{k}\cdot\boldsymbol{u})| \cong \omega_{pe}$ for the case when t and $t' \gg (n_0/n_0')^{1/3}\omega_{pe}^{-1}$:

$$\langle \varrho(t)\,\varrho(t')\rangle_k = \frac{k^2}{72\pi}T\exp\left\{\frac{1}{2}\sqrt{3}\left(\frac{n_0'}{2n_0}\right)^{1/2}\omega_{pe}(t+t')-i\,\mathrm{sgn}\,(\boldsymbol{k}\cdot\boldsymbol{u})\left[1-\frac{1}{2}\left(\frac{n_0'}{2n_0}\right)^{1/3}\right]\omega_{pe}(t-t')\right\} \tag{11.6.4.1}$$

where T is the plasma temperature (we assume the beam to be cold) and $\boldsymbol{u}$ the beam velocity.

Putting $t = t'$ in that expression and using the Poisson equation we find the root mean square amplitude of the resonance oscillations of the electrical field in a plasma-beam system

$$\sqrt{\langle E^2(t)\rangle_k} = \frac{1}{3}\sqrt{(2\pi T)}\exp\left\{\frac{1}{2}\sqrt{3}\left(\frac{n_0'}{2n_0}\right)^{1/3}\omega_{pe}\,t\right\}. \tag{11.6.4.2}$$

11.6.5. EFFECT OF PARTICLE COLLISIONS ON FLUCTUATIONS IN A PLASMA

So far we have considered fluctuations in a collisionless plasma. We shall now turn to the problem of the influence of binary particle collisions on the fluctuations in a plasma (Sitenko and Gurin, 1966). The main difficulty is then connected with the complicated nature of the exact collision integral. To obtain a qualitative picture we shall therefore use the following model collision integral (Bhatnagar, Gross, and Krook, 1954):

$$\left[\frac{\partial F}{\partial t}\right]_c = \nu \frac{n}{n_0}[n\Phi - F], \quad \Phi = (2\pi m_e T)^{-3/2} \exp\left(-\frac{m_e}{2T}|\boldsymbol{v}-\boldsymbol{u}|^2\right). \tag{11.6.5.1}$$

Here $n(\boldsymbol{r}, t)$, $\boldsymbol{u}(\boldsymbol{r}, t)$, and $T(\boldsymbol{r}, t)$ are the density, macroscopic velocity, and temperature of the plasma electrons given by the equations

$$n = \int F d^3\boldsymbol{p}, \quad \boldsymbol{u} = \frac{1}{n}\int \boldsymbol{v} F\, d^3\boldsymbol{p}, \quad T = \frac{m_e}{3n}\int |\boldsymbol{v}-\boldsymbol{u}|^2 F\, d^3\boldsymbol{p}; \tag{11.6.5.2}$$

n_0 and T_0 are the equilibrium values of the electron density and temperature, and ν is the effective binary collision frequency which we shall assume to be independent of the relative velocity of the colliding particles.

One sees easily that this collision integral conserves the number of particles, the energy, and the momentum, and guarantees that the H-theorem is satisfied.

Introducing random forces into the kinetic equation and assuming that the deviation of the distribution function, f, from the equilibrium distribution function, F_0, is small, we find easily the following expression for the spectral density of the correlation function of the random forces:

$$\langle y(\boldsymbol{p})\, y(\boldsymbol{p}')\rangle_{k\omega} = 2\gamma(\boldsymbol{p}, \boldsymbol{p}'), \tag{11.6.5.3}$$

where

$$\gamma(\boldsymbol{p}, \boldsymbol{p}') = \nu \left\{F_0(p)\, \delta(\boldsymbol{p}-\boldsymbol{p}') - \frac{1}{n_0}\left[1 + \frac{m_e}{T_0}(\boldsymbol{v}\cdot\boldsymbol{v}') + \frac{3}{2}\left(1 - \frac{m_e v^2}{3T_0}\right)\left(1 - \frac{m_e v'^2}{3T_0}\right)\right] F_0(p)\, F_0(p')\right\}. \tag{11.6.5.4}$$

Using the kinetic equation we can express the fluctuations in the distribution function and the density and temperature fluctuations determined by them in terms of the random forces and then use eqn. (11.6.5.3) to find the correlation functions of these quantities. In this way we get for the Fourier components of the fluctuations in density $\delta n_{k\omega}$ and in temperature $\delta T_{k\omega}$ the relations

$$\alpha_{11}\delta n_{k\omega}/n_0 + \alpha_{12}\delta T_{k\omega}/T_0 = Y_{k\omega}, \quad \alpha_{21}\delta n_{k\omega}/n_0 + \alpha_{22}\delta T_{k\omega}/T_0 = \tilde{Y}_{k\omega}, \tag{11.6.5.5}$$

where the right-hand sides are determined by the random forces

$$Y_{k\omega} = \frac{i}{n_0}\int \frac{y_{k\omega}(\boldsymbol{p})}{\omega - (\boldsymbol{k}\cdot\boldsymbol{v}) + i\nu} d^3\boldsymbol{p}, \quad Y_{k\omega} = \frac{i m_e}{n_0 T_0}\int \frac{v^2 y_{k\omega}(\boldsymbol{p})}{\omega - (\boldsymbol{k}\cdot\boldsymbol{v}) + i\nu} d^3\boldsymbol{p}, \tag{11.6.5.6}$$

while the α_{ij} coefficients are equal to

$$\left.\begin{aligned}
\alpha_{11} &= 1 - i\frac{y}{z}I - (p + 2ixy)(I-1), \quad \alpha_{12} = \frac{i}{2}\frac{y}{z}\{1 + (1-2z^2)(I-1)\},\\
\alpha_{21} &= 3 - 2i\frac{y}{z}\{1 + (z^2+1)(I-1)\} + (p+2ixy)\{1 - 2(z^2+1)(I-1)\},\\
\alpha_{22} &= 3 + i\frac{y}{z}\{z^2 - 1 - (2z^4 + z^2 + 1)(I-1)\},\\
z &\equiv x + iy = \frac{\omega + i\nu}{\sqrt{2}\, k v_e}, \quad I \equiv I(z) = \frac{z}{\sqrt{\pi}}\int_{-\infty}^{+\infty} \frac{\exp(-\zeta^2)}{z - \zeta} d\zeta, \quad p = (k r_D)^{-2}.
\end{aligned}\right\} \tag{11.6.5.7}$$

Solving the set (11.6.5.5) for $\delta n_{k\omega}$ and $\delta T_{k\omega}$, and using (11.6.5.6) and (11.6.5.7) we get the following general formulae for the spectral densities of the density and temperature fluctuations in a plasma for which the binary particle collisions are taken into account:

$$\left.\begin{aligned}
\langle\delta n^2\rangle_{k\omega} &= \frac{|\alpha_{11}|^2\langle Y^*Y\rangle_{k\omega}-2\,\mathrm{Re}\,\alpha_{12}\alpha_{22}^*\langle Y^*\tilde{Y}\rangle_{k\omega}+|\alpha_{12}|^2\rangle\langle\tilde{Y}^*\tilde{Y}\rangle_{k\omega}}{|\alpha_{11}\alpha_{22}-\alpha_{12}\alpha_{21}|^2},\\
\langle\delta T^2\rangle_{k\omega} &= \frac{T_0^2}{n_0^2}\frac{|\alpha_{21}|^2\langle Y^*Y\rangle_{k\omega}-2\,\mathrm{Re}\,\alpha_{11}\alpha_{21}^*\langle Y^*\tilde{Y}\rangle_{k\omega}+|\alpha_{11}|^2\langle\tilde{Y}^*\tilde{Y}\rangle_{k\omega}}{|\alpha_{11}\alpha_{22}-\alpha_{12}\alpha_{21}|^2},\\
\langle\delta n\,\delta T\rangle_{k\omega} &= -\frac{T_0}{n_0}\frac{\alpha_{22}\alpha_{21}^*\langle Y^*Y\rangle_{k\omega}-(\alpha_{22}\alpha_{11}^*+\alpha_{12}\alpha_{21}^*)\langle Y^*\tilde{Y}\rangle_{k\omega}+\alpha_{12}\alpha_{11}\langle\tilde{Y}^*\tilde{Y}\rangle_{k\omega}}{|\alpha_{11}\alpha_{22}-\alpha_{12}\alpha_{21}|^2}.
\end{aligned}\right\}\quad(11.6.5.8)$$

The correlation functions $\langle Y^*Y\rangle_{k\omega}$ which occur in (11.6.5.8) and also the α_{ij} coefficients can be expressed in terms of the function $I(z)$:

$$\left.\begin{aligned}
\langle Y^*Y\rangle_{k\omega} &= \sqrt{2}\frac{n_0}{kv_e}|z|^{-2}\Big\{\mathrm{Re}\,iz^*I-y\Big[2|z|^2|I-1|^2+|I|^2\\
&\quad+\frac{1}{6}|1+(2z^2-1)(I-1)|^2\Big]\Big\},\\
\langle Y^*\tilde{Y}\rangle_{k\omega} &= \sqrt{2}\frac{n_0}{kv_e}|z|^{-2}\Big\{2\,\mathrm{Re}\,iz^*[1+(z^2+1)(I-1)]-y[2\{1+(z^2+1)(I-1)\}I^*\\
&\quad-2|z|^2\{1-2(z^2+1)(I-1)\}(I-1)^*\\
&\quad+\frac{1}{3}\{1-(2z^2-1)(I-1)\}^*\{z^2-1-(2z^4+z^2+1)(I-1)\}]\Big\},\\
\langle\tilde{Y}^*\tilde{Y}\rangle_{k\omega} &= \sqrt{2}\frac{n_0}{kv_e}|z|^{-2}\Big\{-\mathrm{Re}\,2iz^*[2z^4+z^2-4-2(z^4+2z^2+2)(I-1)]\\
&\quad-2y[2|1+(z^2+1)(I-1)|^2+|1-2(z^2+1)(I-1)|^2|z|^2\\
&\quad+\frac{1}{3}|z^2-1-(2z^4+z^2+1)(I-1)|^2]\Big\}.
\end{aligned}\right\}\quad(11.6.5.9)$$

The condition for the vanishing of the expression in the denominators of (11.6.5.8) is the dispersion equation for longitudinal oscillations in a plasma for which binary collisions are taken into account:

$$(3z-2iyI)\{z-iyI-(p+2ixy)\,z(I-1)\}-\tfrac{1}{2}i(p+1-2x^2)\,y\{1-(2z^2-1)(I-1)\}=0,$$
$$p=(kr_D)^{-2}. \qquad (11.6.5.10)$$

Equations (11.6.5.8) enable us to study the effect of interparticle collisions on the nature of the fluctuations in a plasma for all possible values of the effective collision frequency, starting from the collisionless case and ending up with the hydrodynamical limit. When collisions are neglected, the spectral densities of the density and temperature fluctuations in a plasma are given by formulae (11.6.1.16), (11.6.1.18), and (11.6.1.19).

If we may assume the effective binary collision frequency to be small we can expand (11.6.5.8) in a power series in ν and find the corrections to the spectral densities (11.6.1.16) to (11.6.1.19) caused by the binary collisions. The expressions for the spectral densities (11.6.5.8) can also be considerably simplified in the hydrodynamical limit for large values of the effective binary collision frequency. Expanding (11.6.5.8) in inverse powers of ν we find

$$\left.\begin{aligned}
\langle\delta n^2\rangle_{k\omega} &= \frac{4n_0}{\nu\Delta(\mathbf{k},\omega)}(6\omega^2+5k^2v_e^2)\,k^4v_e^4,\\
\langle\delta T^2\rangle_{k\omega} &= \frac{4T_0^2}{n_0\nu\Delta(\mathbf{k},\omega)}\left\{5(\omega^2-\omega_{pe}^2-k^2v_e^2)^2+\frac{8}{3}\,\omega^2k^2v_e^2\right\}k^2v_e^2,\\
\langle\delta n\,\delta T\rangle_{k\omega} &= \frac{4T_0}{\nu\Delta(\mathbf{k},\omega)}\{5(\omega^2-\omega_{pe}^2-k^2v_e^2)+4\omega^2\}\,k^4v_e^4,\\
\Delta(\mathbf{k},\omega) &= 9\omega^2\left(\omega^2-\omega_{pe}^2-\frac{5}{3}\,k^2v_e^2\right)^2+(9\omega^2-5\omega_{pe}^2-5k^2v_e^2)^2\,\frac{k^4v_e^4}{\nu^2}.
\end{aligned}\right\}\qquad(11.6.5.11)$$

If we let in (11.6.5.11) tend the effective collision frequency to infinity, $\nu\to\infty$, we get the formulae for the spectral densities of the fluctuations in ideal hydrodynamics:

$$\left.\begin{aligned}
\langle\delta n^2\rangle_{k\omega} &= 2\pi n_0\frac{k^2v_e^2}{\omega_s^2}\left\{\frac{2}{3}\,\frac{k^2v_e^2}{\omega_{pe}^2+k^2v_e^2}\,\delta(\omega)+\frac{1}{2}\,[\delta(\omega-\omega_s)+\delta(\omega+\omega_s)]\right\},\\
\langle\delta T^2\rangle_{k\omega} &= \frac{4\pi}{3}\,\frac{T_0^2}{n_0}\left\{\frac{\omega_{pe}^2+k^2v_e^2}{\omega_s^2}\,\delta(\omega)+\frac{1}{5}\,\frac{k^2s^2}{\omega_s^2}\,[\delta(\omega-\omega_s)+\delta(\omega+\omega_s)]\right\},\\
\langle\delta n\,\delta T\rangle_{k\omega} &= \frac{4\pi}{3}\,T_0\,\frac{k^2v_e^2}{\omega_s^2}\left\{-\delta(\omega)+\frac{1}{2}\,[\delta(\omega-\omega_s)+\delta(\omega+\omega_s)]\right\},\\
\omega_s^2 &= \omega_{pe}^2+k^2s^2,
\end{aligned}\right\}\qquad(11.6.5.12)$$

where $s=\sqrt{(5T_0/3m_e)}$ is the sound velocity.

The spectral densities of the fluctuations are according to (11.6.5.12) characterized by maxima at the frequencies corresponding to sound oscillations and also at zero frequency which is connected with entropy waves in the plasma. It is interesting to note that the spectral densities of the single-time density and temperature fluctuations in a plasma are the same for the case of large collision frequencies and for the collisionless case.

We have given in Fig. 11.6.2 the spectral densities of the density and temperature fluctuations in a plasma for intermediate values of the effective binary collision frequency and for values of the parameter p equal to 0, 1, 4, and 10. According to the curves given here taking collisions into account leads in the short-wavelength region to the appearance of a sound maximum, with a magnitude which increases with increasing collision frequency, in the particle density-fluctuation spectrum. Its width is determined by the plasma viscosity and thermal conductivity. In the long-wavelength case taking collisions into account leads to a shift to lower frequencies of the Langmuir maximum in the density fluctuation spectrum. This shift is connected with the change in the nature of the high-frequency oscillations in the plasma due to the interparticle collisions.

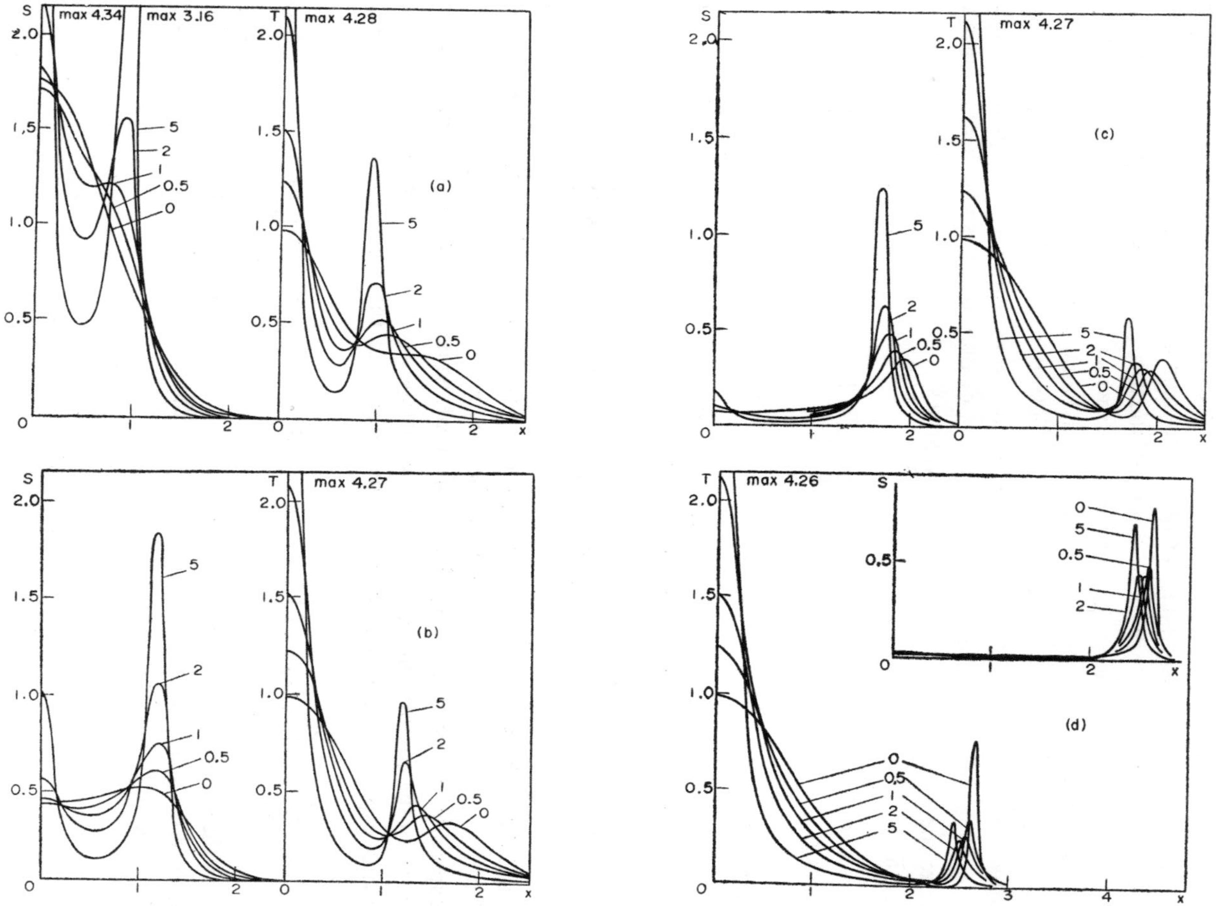

FIG. 11.6.2. Spectral densities of the fluctuations in the density, $S(x, y) = \langle \delta n^2 \rangle_{kx+iy}/2n_0$, and in the temperature, $T(x, y) = n_0 \langle \delta T^2 \rangle_{kx+iy}/2T_0^2$, where $x = \omega/\sqrt{2}kv_e$ is the dimensionless frequency, as functions of the collision frequency ν. The parameter $y = \nu/\sqrt{2}kv_e$ takes on the values 0, 0.5, 1, 2, and 5 for the cases a) $p = 0$, b) $p = 1$, c) $p = 4$, and d) $p = 10$.

11.6.6. TRANSITION TO THE HYDRODYNAMICAL THEORY OF FLUCTUATIONS

We can also obtain eqns. (11.6.5.12) in a different way, namely, by using the hydrodynamical theory of fluctuations (Landau and Lifshitz, 1957). To construct a hydrodynamical theory of fluctuations we must use the equations of hydrodynamics and introduce external forces into it. If we then find the change in the energy of the hydrodynamical medium caused by the external forces, we can obtain the response tensor which according to (11.1.2.25) determines the spectral densities of the fluctuating hydrodynamical quantities.

In the case of magneto-hydrodynamics we must introduce an external force $\boldsymbol{f}$ into the Navier–Stokes equation, an external current $\boldsymbol{g}$ into the heat transfer equation, and, finally, an external current $\boldsymbol{y}$ into Ohm's law. Restricting ourselves to small fluctuations we can linearize the equations of magneto-hydrodynamics and thus start from the following set of equations:

$$\left.\begin{aligned} &\varrho_0\frac{\partial \boldsymbol{v}}{\partial t} = -\nabla p+\frac{1}{c}[\boldsymbol{j}\wedge \boldsymbol{B}_0]+\eta\nabla^2\boldsymbol{v}+\left(\zeta+\frac{1}{3}\eta\right)\nabla(\nabla\cdot\boldsymbol{v})+\boldsymbol{f},\\ &\frac{\partial\varrho}{\partial t}+\varrho_0(\nabla\cdot\boldsymbol{v}) = 0,\\ &p = s^2(\varrho+\varrho_0\alpha T),\quad \varrho_0 c_V\frac{\partial T}{\partial t}-\alpha T_0 s^2\frac{\partial\varrho}{\partial t} = \varkappa\nabla^2 T-(\nabla\cdot\boldsymbol{g}),\\ &\operatorname{curl}\boldsymbol{E} = -\frac{1}{c}\frac{\partial \boldsymbol{B}}{\partial t},\quad \operatorname{curl}\boldsymbol{B} = \frac{4\pi}{c}\boldsymbol{j},\quad \boldsymbol{j} = \sigma\left\{\boldsymbol{E}+\frac{1}{c}[\boldsymbol{v}\wedge\boldsymbol{B}_0]\right\}+\boldsymbol{y}, \end{aligned}\right\} \tag{11.6.6.1}$$

where η and ζ are the viscosity coefficients, $\varkappa$ the thermal conductivity coefficient, σ the conductivity, $s^2 = (\partial p/\partial\varrho)_T$ the square of the sound velocity, $\alpha = -\varrho_0^{-1}(\partial\varrho/\partial T)_p$ the volume expansion coefficient, and c_V the specific heat, while ϱ and T are, respectively, the deviations from the equilibrium values of the density ϱ_0 and temperature T_0. We then get the following expression for the change in the energy of the magneto-hydrodynamical medium:

$$\frac{\partial U}{\partial t} = \int\left\{(\boldsymbol{v}\cdot\boldsymbol{f})+\left(\left\{\boldsymbol{E}+\frac{1}{c}[\boldsymbol{v}\wedge\boldsymbol{B}_0]\right\}\cdot\boldsymbol{y}\right)-\frac{T}{T_0}(\nabla\cdot\boldsymbol{g})\right\}d^3\boldsymbol{r}. \tag{11.6.6.2}$$

Assuming the external forces to vary with time as $e^{-i\omega t}$ and taking the Fourier transform with respect to the spatial coordinates, we can write the change in the energy of the system due to the external forces in the form

$$\frac{\partial U}{\partial t} = -\frac{1}{2}\operatorname{Re}\sum_{k}(\boldsymbol{v}_k\cdot\dot{\boldsymbol{P}}_k^*), \tag{11.6.6.3}$$

where

$$\boldsymbol{P}_{k\omega} = \frac{i}{\omega}\boldsymbol{f}_{k\omega}-\frac{\alpha s^2}{c_V}\frac{\boldsymbol{k}(\boldsymbol{k}\cdot\boldsymbol{g}_{k\omega})}{\omega(\omega+ik^2\varkappa)}+i\frac{c\{k^2[\boldsymbol{B}_0\wedge\boldsymbol{y}_{k\omega}]-[\boldsymbol{B}_0\wedge\boldsymbol{k}](\boldsymbol{k}\cdot\boldsymbol{y}_{k\omega})\}}{\omega(k^2c^2-4\pi i\sigma\omega)}. \tag{11.6.6.4}$$

We must then use the equations of motion to express the velocity Fourier components $\boldsymbol{v}_{k\omega}$

in terms of the $\boldsymbol{P}_{k\omega}$:

$$v_i(\boldsymbol{k}, \omega) = \sum_j \alpha_{ij}(\boldsymbol{k}, \omega)\, P_j(\boldsymbol{k}, \omega), \tag{11.6.6.5}$$

where the quantities α_{ij} form the response tensor. Using the coordinate system such that

$$\boldsymbol{e}_1 = \frac{[\boldsymbol{k} \wedge \boldsymbol{B}_0]}{|[\boldsymbol{k} \wedge \boldsymbol{B}_0]|}, \quad \boldsymbol{e}_2 = \frac{[\boldsymbol{B}_0 \wedge [\boldsymbol{k} \wedge \boldsymbol{B}_0]]}{|[\boldsymbol{B}_0 \wedge [\boldsymbol{k} \wedge \boldsymbol{B}_0]]|}, \quad \boldsymbol{e}_3 = \frac{\boldsymbol{B}_0}{B_0},$$

we obtain the following expressions for the components of the response tensor:

$$\alpha_{11} = D_1^{-1}, \quad \alpha_{22} = \left\{1 + i\eta \frac{k^2}{\omega\varrho_0} + (\gamma - i\omega\mu)\cos^2\theta\right\} D_2^{-1},$$

$$\alpha_{33} = \left\{1 + i\eta \frac{k^2}{\omega\varrho_0} - (\gamma - i\omega\mu)\sin^2\theta + \frac{4\pi i\sigma k^2 v_A^2}{\omega(k^2c^2 - 4\pi i\sigma\omega)}\right\} D_2^{-1}, \tag{11.6.6.6}$$

$$\alpha_{23} = (\gamma - i\omega\mu)\sin\theta\cos\theta\, D_2^{-1}, \quad \alpha_{32}(\boldsymbol{B}_0) = -\alpha_{23}(-\boldsymbol{B}_0), \quad \alpha_{12} = \alpha_{21} = \alpha_{13} = \alpha_{31} = 0,$$

$$\gamma = \frac{k^2 s^2}{\omega^2}\left\{1 + \frac{\alpha^2 s^2 T_0}{c_V}\left[1 + \frac{\varkappa^2 k^4}{\omega^2 \varrho_0{}^2 c_V^2}\right]^{-1}\right\}, \quad \mu = \left(\zeta + \frac{1}{3}\eta\right)\frac{k^2}{\omega^2\varrho_0} + \frac{\alpha^2 \varkappa T_0 k^4 s^4}{\omega^4 \varrho_0 c_V^2}\left[1 + \frac{\varkappa^2 k^4}{\omega^2 \varrho_0{}^2 c_V^2}\right]^{-1},$$

where

$$\left.\begin{aligned} D_1 &= -\varrho_0\left\{1 + i\eta\frac{k^2}{\omega\varrho_0} + i\frac{4\pi\sigma k^2 v_A^2\cos^2\theta}{\omega(k^2c^2 - 4\pi i\sigma\omega)}\right\}, \\ D_2 &= -\varrho_0\left\{\left(1 + i\eta\frac{k^2}{\omega\varrho_0}\right)\left[1 - \gamma + i\omega\left(\eta\frac{k^2}{\omega\varrho_0} + \mu\right)\right]\right. \\ &\quad \left. + i\frac{4\pi\sigma k^2 v_A^2}{\omega(k^2c^2 - 4\pi i\sigma\omega)}\left[1 - \gamma\cos^2\theta + i\omega\left(\eta\frac{k^2}{\omega^2\varrho_0} + \mu\cos^2\theta\right)\right]\right\}, \end{aligned}\right\} \tag{11.6.6.7}$$

and where θ is the angle between $\boldsymbol{k}$ and $\boldsymbol{B}_0$. Using equation (11.2.2.25) we can, once we know the quantities α_{ij} find the spectral density of the velocity fluctuations:

$$\langle v_i v_j \rangle_{k\omega} = \frac{T_0}{\omega} i(\alpha_{ji}^* - \alpha_{ij}) = 2\frac{T_0}{\omega}\,\mathrm{Im}\,\alpha_{ij}, \tag{11.6.6.8}$$

and also that of other quantities. In particular, the spectral density of the density fluctuations is given by the formula

$$\langle \varrho^2 \rangle_{k\omega} = 2\frac{T_0 \varrho_0^2 k^2}{\omega^3}(\sin^2\theta\,\mathrm{Im}\,\alpha_{22} + \cos^2\theta\,\mathrm{Im}\,\alpha_{33} + 2\sin\theta\cos\theta\,\mathrm{Im}\,\alpha_{23}). \tag{11.6.6.9}$$

The spectral density of the temperature fluctuations has the form ($\alpha'_{ij} = \mathrm{Re}\,\alpha_{ij}$, $\alpha''_{ij} = \mathrm{Im}\,\alpha_{ij}$)

$$\langle T^2 \rangle_{k\omega} = \frac{2T_0^2}{\omega^2\varrho_0^2 c_V^2 + \varkappa^2 k^4}\left\{\varkappa k^2 + \alpha^2 s^4 \varrho_0^2 T_0^2 \sum_{i,j} k_i k_j \left(\frac{\alpha''_{ij}}{\omega} - \frac{2\varrho_0 c_V \varkappa k^2 \alpha'_{ij}}{\omega^2\varrho_0^2 c_V^2 + \varkappa^2 k^4}\right)\right\}. \tag{11.6.6.10}$$

The spectral density of the magnetic field fluctuations is given by the formula

$$\langle B_i B_j \rangle_{k\omega} = \frac{32\pi^2 c^2 \sigma T_0}{k^4 c^4 + 16\pi^2\sigma^2\omega^2}\left\{k^2\delta_{ij} - k_i k_j - B_0^2\frac{\sigma}{c^2}\sum_{k,l}(k_3\delta_{ik} - \delta_{i3}k_k)(k_3\delta_{jl} - \delta_{j3}k_l)\right.$$
$$\left.\times\left[\frac{\alpha''_{kl}}{\omega} + \frac{8\pi\sigma k^2 c^2 \alpha'_{kl}}{k^4c^4 + 16\pi^2\sigma^2\omega^2}\right]\right\}. \tag{11.6.6.11}$$

If we let the thermal conductivity and the viscosity coefficients in eqn. (11.6.6.8) tend to zero, and after that let the electrical conductivity tend to infinity, we get the spectral densities of the velocity fluctuations in an ideal magneto-hydrodynamical medium

$$\langle v_1^2\rangle_{k\omega} = \frac{2\pi T_0}{\varrho_0|\omega|}\,\delta(\Delta_1), \quad \Delta_1 = 1-\frac{k^2v_A^2}{\omega^2}\cos^2\theta,$$

$$\langle v_2^2\rangle_{k\omega} = \frac{2\pi T_0}{\varrho_0|\omega|}\left\{1-\frac{k^2s^2}{\omega^2}\left(1+\frac{\alpha^2s^2T_0}{c_V}\right)\cos^2\theta\right\}\delta(\Delta_2),$$

$$\langle v_3^2\rangle_{k\omega} = \frac{2\pi T_0}{\varrho_0|\omega|}\frac{k^4s^4}{\omega^4}\sin^2\theta\cos^2\theta\frac{[1+(\alpha^2s^2T_0/c_V)]^2}{1-(k^2s^2/\omega^2)[1+(\alpha^2s^2T_0/c_V)]\cos^2\theta}\,\delta(\Delta_2), \tag{11.6.6.12}$$

$$\langle v_2v_3\rangle_{k\omega} = \frac{2\pi T_0}{\varrho_0|\omega|}\frac{k^2s^2}{\omega^2}\sin\theta\cos^2\theta\left(1+\frac{\alpha^2s^2T_0}{c_V}\right)\delta(\Delta_2),$$

$$\Delta_2 = 1-\frac{k^2v_A^2}{\omega^2}\cos^2\theta-\frac{k^2s^2}{\omega^2}\left(1+\frac{\alpha^2s^2T_0}{c_V}\right).$$

If we put the arguments of the δ-functions which appear in these formulae equal to zero, we get the dispersion equations for the Alfvén and the magneto-sound waves. The spectral densities of the velocity fluctuations therefore have steep maxima corresponding to the possibility of the propagation of weakly damped magneto-hydrodynamical waves. There are similar maxima in the spectral densities of the magnetic field fluctuations.

Integrating (11.6.6.12), (11.6.6.11), and (11.6.6.10) over the frequency we find the Fourier components for the spatial correlation functions for the velocity, magnetic field, and temperature fluctuations:

$$\langle v_iv_j\rangle_k = \frac{T_0}{\varrho_0}\,\delta_{ij}, \tag{11.6.6.13}$$

$$\langle B_1^2\rangle_k = 4\pi T_0, \quad \langle B_2^2\rangle_k = 4\pi T_0\cos^2\theta, \quad \langle B_3^2\rangle_k = 4\pi T_0\sin^2\theta,$$
$$\langle B_2B_3\rangle_k = -4\pi T_0\sin\theta\cos\theta, \tag{11.6.6.14}$$

$$\langle T^2\rangle_k = \frac{T_0^2}{\varrho_0c_V}. \tag{11.6.6.15}$$

11.7. Fluctuations in a Partially Ionized Plasma in an External Electrical Field

11.7.1. FLUCTUATIONS WHEN THERE IS NO MAGNETIC FIELD

We now turn to a study of fluctuations in a partially ionized plasma in a constant uniform electrical field (Angeleĭko and Akhiezer, 1968; Akhiezer and Angeleĭko, 1969b). We have shown in Chapter 7 that as a result of collisions between charged and neutral particles a stationary velocity distribution is set up in such a plasma which is characterized by a very high average energy of the random motion (effective temperature) of the electrons and a relatively small average velocity of their directed motion.

Of course, a plasma in an external electrical field is an essentially non-equilibrium system so that we cannot immediately apply to it the general theory of fluctuations. Nevertheless, we can clearly use the general formulae of Section 11.5 which describe fluctuation in a collisionless plasma when we study high-frequency fluctuations for which $\omega \gg \nu_{e,i}$, where ω is the frequency of the fluctuations and ν_α the effective collision frequency. The expressions for the correlation functions which we then obtain will, of course, depend on the form of the collision integral as the stationary electron distribution function $F_e^{(0)}$ depends on the form of the collision integral.

We can thus use eqns. (11.5.1.15) and (11.5.1.12) for the correlator of the high-frequency charge density fluctuations when there is no external magnetic field; in these equations we must substitute for the equilibrium electron distribution function the function $F_e^{(0)} = F_{e0}^{(0)} + ([\boldsymbol{v}/v]\cdot \boldsymbol{F}_{e1}^{(0)})$, given by formula (7.1.2.4), and for the equilibrium ion distribution function a Maxwellian distribution function with temperature T_i.

If we put the denominator of the expression for the correlator which we obtain in this way equal to zero and assume that

$$\frac{T_i}{m_i} \ll \frac{\omega^2}{k^2} \ll \frac{T_e}{m_e},$$

we get the frequency and damping rate, (7.2.1.7) and (7.2.1.10), of the ion-sound oscillations. Assuming that the damping rate of the ion-sound is small, $\gamma \ll \omega$, we can easily write eqn, (11.5.1.15) in the form

$$\langle \varrho^2 \rangle_{k\omega} = \frac{\sqrt{Z}\, k^5 r_D^2 v_s T_e}{2^{5/4}\sqrt{3}\,[1+k^2 r_D^2]^2 (R-\cos\chi)}\,\delta\left(\omega^2 - \frac{k^2 v_s^2}{1+k^2 r_D^2}\right). \tag{11.7.1.1}$$

The electron and ion densities correlators are, according to (11.5.1.13) and (11.5.1.14), given by the formulae

$$\begin{aligned} e^2 \langle n_e^2 \rangle_{k\omega} &= \frac{(Ze)^2}{(1+k^2 r_D^2)^2} \langle n_i^2 \rangle_{k\omega} = -\frac{Ze^2}{1+k^2 r_D^2} \langle n_e n_i \rangle_{k\omega} \\ &= \frac{\sqrt{Z}\, k v_s T_e}{2^{5/4}\sqrt{3}\, r_D^2 (R-\cos\chi)}\,\delta\left(\omega^2 - \frac{k^2 v_s^2}{1+k^2 r_D^2}\right), \end{aligned} \tag{11.7.1.2}$$

where R is given by (7.2.1.11), Ze is the ion charge, and χ is the angle between $\boldsymbol{k}$ and the field $\boldsymbol{E}_0$.

Let us now investigate how the nature of the ion-sound fluctuations depends on the external electrical field strength $\boldsymbol{E}_0$. At not too strong field strengths, when $R \gg 1$, the charge density and electron and ion densities fluctuation distributions are almost isotropic. As the field $\boldsymbol{E}_0$ increases in strength the fluctuation distribution becomes more anisotropic: the smaller the angle χ between the direction of propagation of the fluctuations and the direction of the external electrical field, the larger the intensity characteristic for the fluctuations.

If $E_0 \to E_c$, where E_c is the critical value of the electrical field, given by formula (7.2.1.12), the level of the fluctuations increases steeply; in the case of long wavelengths, $kr_D \ll 1$, expressions (11.7.1.1) and (11.7.1.2) then tend to infinity when $\chi = 0$. The steep increase of the level of the fluctuations when the electrical field approaches its critical value is connected with the fact that the plasma becomes unstable when $E = E_c$; this is caused by the growth of the ion-sound oscillations. Such an increase in the level of fluctuations is analogous to the appearance of critical fluctuations in a collisionless plasma, considered in Subsection 11.5.3.

We note that we showed in Subsection 7.2.1 that, if $Zm_0/m_i > 3\pi\, 2^{-3/2}$, the ion-sound oscillations do not grow for any value of the external electrical field. The correlation functions remain finite in that case for all values of E_0 and no critical fluctuations arise.

We emphasize that all expressions for the correlation functions given in the present and the next subsection are obtained in the framework of the linear theory of fluctuations and are therefore applicable in the case of damped oscillations. If the plasma oscillations grow, it is necessary to take non-linear effects into account when determining the correlation functions; these will limit the growth of the fluctuations.

In concluding this subsection we give the expression for the correlation function of the charge density fluctuations when $\omega^2 \gg k^2T_e/m_e$. Using (11.5.1.15) we have

$$\langle \varrho^2 \rangle_{k\omega} = \frac{k^2}{4|\omega|} T_L(k, \cos\chi)\, \delta\left\{1 - \frac{\omega_l(\boldsymbol{k})}{\omega}\right\} + T_L(k, -\cos\chi)\, \delta\left\{1 + \frac{\omega_l(-\boldsymbol{k})}{\omega}\right\}, \qquad (11.7.1.3)$$

where ω_l is the frequency of the Langmuir wave, given by eqn. (7.1.3.3) and T_L a function of the wavevector $\boldsymbol{k}$, which may be called the effective temperature of the Langmuir waves. The exact form of the function T_L depends in an essential way on the form of the unperturbed distribution function of the electron component of the plasma in the high-velocity region, $v \gg \sqrt{(T_e/m_e)}$. If the distribution function retains the form (7.1.2.4) in that region, we have (Angeleĭko, 1968) (for the sake of definiteness we assume that $kr_D \gg (m_e/m_0)^{1/4}$)

$$T(k, \cos\chi)_{k\omega} = \frac{\Gamma(\frac{1}{4})}{2\Gamma(\frac{3}{4})} (kr_D)^2 T_e \approx 1.51\, (kr_D)^2 T_e. \qquad (11.7.1.4)$$

We see that the effective temperature of the Langmuir oscillations is in that case appreciably lower than the average electron energy, $T_L \sim (kr_D)^2 T_e$. We note that the expression (11.2.3.4) for the charge density correlation function can also be written in the form (11.7.1.3) with $T_L = T_e$.

11.7.2. CRITICAL FLUCTUATIONS IN ELECTRICAL AND MAGNETIC FIELDS

Let us now consider high-frequency fluctuations in a partially ionized plasma in both an external electrical field and a constant and uniform magnetic field (Akhiezer and Angeleĭko, 1969 a, b). The correlation function of the charge density is, as before, given by formula (11.5.1.15) in which we must substitute expression (7.2.2.1) for the dielectric permittivity of the plasma.

If the magnetic field is not too strong so that only the electron component of the plasma is strongly magnetized ($\omega_{Bi} \ll \omega \ll |\omega_{Be}|$, where $\omega_{B\alpha} = e_\alpha B_0/m_\alpha c$ is the gyro-frequency of the αth kind of particles), the correlation function (11.5.1.15) has poles corresponding to the possibility of propagation through the plasma of ion-sound oscillations with a frequency

$$\omega_s(k) = \frac{kv_s}{[1 + k^2 r_D^2]^{1/2}};$$

their damping rate is given by formula (7.2.2.6). Bearing in mind that the damping rate of these oscillations is small compared to their frequency we can in the sound region ($T_i/m_i \ll \omega^2/k^2 \ll T_e/m_e$) write expression (11.5.1.15) in the form

$$\langle \varrho^2 \rangle_{k\omega} = \frac{\sqrt{Z}\, k^5 r_D^2 v_s T_e}{2^{5/4}\sqrt{3}\,(1+k^2 r_D^2)^2\, \mu^{\pm}(R^{\pm}-\cos\chi)}\, \delta(\omega^2-\omega_s^2(\boldsymbol{k})), \tag{11.7.2.1}$$

where χ is the angle between the direction of the wave propagation, that is, the vector $\boldsymbol{k}\,\mathrm{sgn}\,\omega$, and the direction of the average electron velocity $\boldsymbol{u}$, while the quantities $R^{\pm}$ and $\mu^{\pm}$ are given by formulae (7.2.2.7)—the upper sign refers to the case $\chi < \pi/2$ and the lower sign to the case $\chi > \pi/2$; we assume that the angle χ does not lie too close to $\pi/2$, $|\cos\chi| \gg \mathrm{Max}\{\sqrt{(Zm_e/m_i)}, \nu_e/\omega\}$.

The correlation functions of the electron and ion densities fluctuations, in the ion-sound wave, have the form

$$e^2\langle n_e^2\rangle_{k\omega} = \frac{(Ze)^2}{1+k^2r_D^2}\langle n_i^2\rangle_{k\omega} = -\frac{Ze^2}{1+k^2r_D^2}\langle n_e n_i\rangle_{k\omega}$$

$$= \frac{\sqrt{Z} k v_s T_e}{2^{5/4}\sqrt{3}\, r_D^2 \mu^{\pm}(R^{\pm}-\cos\chi)}\,\delta\left(\omega^2 - \frac{k^2v_s^2}{1+k^2r_D^2}\right). \tag{11.7.2.2}$$

Let us now consider the case of a very strong magnetic field ($\omega \ll \omega_{Bi}$) when not only the electron, but also the ion component of the plasma is strongly magnetized. The correlator (11.5.1.15) has in that case poles corresponding to the possibility that magneto-sound waves with frequency

$$\omega = \frac{k v_s}{[1+k^2r_D^2]^{1/2}}\,|\cos\chi|,$$

and damping rate γ, given by eqn. (7.2.2.9), can propagate in the plasma. Bearing in mind that $\gamma \ll \omega$ we can in the sound region write expression (11.5.1.15) for the case of strongly magnetized ions in the form

$$\langle \varrho^2\rangle_{k\omega} = \frac{\sqrt{Z}\, k^5 r_D^2 v_s T_e\,|\cos\chi|}{2^{5/4}\sqrt{3}\,(1+k^2r_D^2)^2\,(\xi - \mathrm{sgn}\cos\chi)}\,\delta\left(\omega^2 - \frac{k^2 v_s^2\cos^2\chi}{1+k^2r_D^2}\right), \tag{11.7.2.3}$$

where the quantity ξ is given by formula (7.2.2.10).

The correlation functions of the electron and ion densities fluctuations have, in the magneto-sound wave, the form

$$e^2\langle n_e^2\rangle_{k\omega} = \frac{(Ze)^2}{(1+k^2r_D^2)^2}\langle n_i^2\rangle_{k\omega} = -\frac{Ze^2}{1+k^2r_D^2}\langle n_e n_i\rangle_{k\omega}$$

$$= \frac{\sqrt{Z}\, k v_s T_e}{2^{5/4}\sqrt{3}\, r_D^2\,|\cos^3\chi|\,(\xi-\mathrm{sgn}\cos\chi)}\,\delta\left(\omega^2-\frac{k^2v_s^2\cos^2\chi}{1+k^2r_D^2}\right). \tag{11.7.2.4}$$

We emphasize that the original expression (11.5.1.15) for the correlator is valid in the high-frequency region, $\omega \gg \nu_\alpha$. Equations (11.7.2.3) and (11.7.2.4) are thus valid for angles

χ which are not too close to $\pi/2$,

$$\left|\frac{1}{2}\pi-\chi\right| \gg \frac{\nu_\alpha}{kv_s}\sqrt{(1+k^2r_D^2)}.$$

Let us now investigate how the nature of the fluctuations changes when the external fields $\boldsymbol{E}_0$ and $\boldsymbol{B}_0$ change. At not too large values of the external electrical field the spatial correlation function of the charge density fluctuations,

$$\langle\varrho^2\rangle_k = \int \langle\varrho^2\rangle_{k\omega}\frac{d\omega}{2\pi},$$

is almost isotropic. When the external magnetic field is not too strong ($\omega_{Bi} \ll kv_s(1+k^2r_D^2)^{-1/2} \ll |\omega_{Be}|$) the fluctuation density becomes more anisotropic as the external electrical field strength increases: the smaller the angle χ between the direction of propagation of the fluctuations and the direction of the electron velocity, the larger the intensity characterizing the fluctuations.

As $E_0 \to E_c(\boldsymbol{B}_0)$, where $E_c(\boldsymbol{B}_0)$ is the critical value of the electrical field given by formula (7.2.2.8), the level of the fluctuations increases steeply; expressions (11.7.2.1) and (11.7.2.2) then become infinite in the long-wavelength case ($kr_D \ll 1$) when $\chi = 0$.

In the case of a very strong magnetic field ($\omega_{Bi} \gg \omega_s(\boldsymbol{k})$) the function $\langle\varrho^2\rangle_k$ has, according to (11.7.2.3), the form

$$\langle\varrho^2\rangle_k = \frac{k^4r_D^2T_e}{2^{5/4}\sqrt{3}\,(1+k^2r_D^2)^{3/2}}\,\frac{\xi}{\xi^2-1}, \tag{11.7.2.5}$$

where ξ is again given by formula (7.2.2.10). In this case this function remains isotropic while increasing when the electrical field strength increases. If the electrical field E_0 approaches the critical value $E_c(\boldsymbol{B}_0)$, expression (11.7.2.5) tends to infinity when $kr_D \ll 1$ at the same time for all directions—and not only for $\chi = 0$ as in the case of weakly magnetized ions.

As in the case of a plasma when there is no magnetic field, considered in the preceding subsection, the steep increase in the level of the fluctuations when the electrical field approaches its critical value—the appearance of critical fluctuations—is connected with the fact that the plasma becomes unstable for $E_0 = E_c(\boldsymbol{B}_0)$; this instability is caused by the growth of the magneto-sound (or ion-sound) oscillations. Of course, if

$$\frac{Zm_0}{m_i} > 3\pi\, 2^{-3/2} \approx 3.3,$$

the magneto-sound (or ion-sound) oscillations do not grow for any value of the external electrical field and no critical fluctuations occur in that case.

CHAPTER 12

Scattering and Transformation of Waves in a Plasma

12.1. Scattering of Electromagnetic Waves in an Unmagnetized Plasma

12.1.1. SCATTERING CURRENT

When deriving the spectra of the eigen oscillations of the plasma we started from the linearized kinetic equations and neglected the non-linear terms $(e/m_e)(\{\boldsymbol{E}+[\boldsymbol{v} \wedge \boldsymbol{B}]/c\} \cdot \{\partial f/\partial \boldsymbol{v}\})$. The equation for the electrical field $\boldsymbol{E}$,

$$\text{curl curl}\,\boldsymbol{E}+\frac{\hat{\varepsilon}}{c^2}\frac{\partial^2 \boldsymbol{E}}{\partial t^2} = 0, \tag{12.1.1.1}$$

where $\hat{\varepsilon}$ is the plasma dielectric permittivity operator, which we obtain in this approximation, is linear, satisfies the superposition principle, and therefore corresponds to the possibility that different oscillations will propagate independently through the plasma.

In actual fact, the different oscillations will, however, not propagate independently of one another through the plasma, but interact with one another. This interaction is contained in the original kinetic equations and is described by the non-linear terms $(e/m_e)(\{\boldsymbol{E}+[\boldsymbol{v} \wedge \boldsymbol{B}]/c\} \cdot \{\partial f/\partial \boldsymbol{v}\})$ which were dropped when eqn. (12.1.1.1) was derived.

The interaction between oscillations leads to various scattering and transformation processes for waves in a plasma.† Let us, for instance, consider the propagation of a transverse electromagnetic wave in a plasma. Due to its interaction with fluctuating plasma oscillations the wave will be scattered and this may be accompanied by a change in frequency. The intensity of the scattered waves is determined by both the intensity of the incident wave and the level of fluctuations in the plasma.

As the fluctuation spectrum has steep maxima at the frequencies of the eigenoscillations of the plasma, there will also be steep maxima in the spectrum of the scattered waves at frequencies differing from the frequency of the incident wave by the frequencies of the plasma eigenoscillations (or by multiples of these frequencies).

The interaction of the waves propagating in the plasma with the fluctuating oscillations can also lead to a transformation of waves, for instance, to the transformation of a transverse wave into a longitudinal one, or of a longitudinal into a transverse wave. The probabilities

† Akhiezer, Prokhoda, and Sitenko (1958) predicted the occurrence of Raman scattering and wave transformation in a plasma.

for such processes, and also for the scattering processes, are determined by the level of fluctuations in the plasma.

In our study of fluctuations we now turn to an investigation of the scattering and transformation of waves in a plasma which, to begin with, we shall assume to be unmagnetized.‡ We note beforehand that the non-linear interaction between different plasma oscillations is smal. Thanks to that we can approximately separate off the field of the incident wave which, by delfinition, satisfies eqn. (12.1.1.1). We shall assume that the field of the incident wave **is** given and denote it by $\boldsymbol{E}^0(\boldsymbol{r}, t)$.

Due to the interaction of the incident wave $\boldsymbol{E}^0(\boldsymbol{r}, t)$ with the fluctuating field the wave is scattered; the total electrical field in the plasma when the wave is propagating can thus be written in the form

$$\boldsymbol{E}(\boldsymbol{r}, t) = \boldsymbol{E}^0(\boldsymbol{r}, t) + \delta\boldsymbol{E}(\boldsymbol{r}, t) + \boldsymbol{E}'(\boldsymbol{r}, t), \tag{12.1.1.2}$$

where $\delta\boldsymbol{E}$ is the fluctuating field and $\boldsymbol{E}'$ the field of the scattered wave. As the wave–wave interaction is weak, we may assume the field $\boldsymbol{E}'$ to be bilinear in the field of the incident wave and of the fluctuating field, $E' \propto E^0\, \delta E$.

Our problem lies thus in determining the field $\boldsymbol{E}'$ of the scattered wave. This field satisfies clearly the Maxwell equation

$$\text{curl curl}\, \boldsymbol{E}' + \frac{1}{c^2}\frac{\partial^2 \boldsymbol{E}'}{\partial t^2} = -\frac{4\pi}{c^2}\frac{\partial \boldsymbol{j}'}{\partial t}, \tag{12.1.1.3}$$

where $\boldsymbol{j}'$ is the current density producing the field $\boldsymbol{E}'$.

The current in the plasma is connected with the distribution functions through the relation

$$\boldsymbol{j} = \sum_\alpha e_\alpha \int \boldsymbol{v} f_\alpha \, d^3\boldsymbol{v}, \tag{12.1.1.4}$$

where f_α is the deviation of the distribution function $F^{(\alpha)}$ from the original distribution function $F_0^{(\alpha)}$. We must thus make clear what the form is of the distribution function when the incident wave propagates through the plasma. It is clear that we can write the function f_α in the form

$$f_\alpha = f_\alpha^0 + \delta f_\alpha + f_\alpha', \tag{12.1.1.5}$$

where f_α^0 is the deviation of the distribution function connected with the incident wave, δf_α the fluctuation in the distribution function, and f_α' the deviation of the distribution function connected with the scattered wave. The fluctuation of the distribution function is given by eqn. (11.6.1.1), while the functions f_α^0 and f_α' satisfy the equations

$$\frac{\partial f_\alpha^0}{\partial t} + (\boldsymbol{v}\cdot\nabla) f_\alpha^0 + \frac{e_\alpha}{m_\alpha}\left(\left\{\boldsymbol{E}^0 + \frac{1}{c}[\boldsymbol{v} \wedge \boldsymbol{B}^0]\right\}\cdot\frac{\partial F_0^{(\alpha)}}{\partial \boldsymbol{v}}\right) = 0, \tag{12.1.1.6}$$

$$\frac{\partial f_\alpha'}{\partial t} + (\boldsymbol{v}\cdot\nabla) f_\alpha' + \frac{e_\alpha}{m_\alpha}\left(\left\{\boldsymbol{E}' + \frac{1}{c}[\boldsymbol{v} \wedge \boldsymbol{B}']\right\}\cdot\frac{\partial F_0^{(\alpha)}}{\partial \boldsymbol{v}}\right)$$
$$+ \frac{e_\alpha}{m_\alpha}\left(\left\{\boldsymbol{E}^0 + \frac{1}{c}[\boldsymbol{v} \wedge \boldsymbol{B}^0]\right\}\cdot\frac{\partial \delta f_\alpha}{\partial \boldsymbol{v}}\right) + \frac{e_\alpha}{m_\alpha}\left(\left\{\delta\boldsymbol{E} + \frac{1}{c}[\boldsymbol{v} \wedge \delta\boldsymbol{B}]\right\}\cdot\frac{\partial f_\alpha^0}{\partial \boldsymbol{v}}\right) = 0, \tag{12.1.1.7}$$

‡ Akhiezer, Akhiezer, and Sitenko (1962; in what follows we follow this paper), Dougherty and Farley (1960), and Salpeter (1960 b, c) developed the theory of the scattering of electromagnetic waves in a plasma.

where $\boldsymbol{B}^0$ and $\boldsymbol{B}'$ are the magnetic fields of the incident and the scattered wave and $\delta\boldsymbol{B}$ the fluctuation in the magnetic field.

We shall assume that the incident wave is a plane monochromatic wave:

$$\boldsymbol{E}^0(\boldsymbol{r}, t) = \boldsymbol{E}^0\, e^{i(\boldsymbol{k}\cdot\boldsymbol{r})-i\omega t}.$$

The solution of equation (12.1.1.6) has in that case the form

$$(f_\alpha^0)_{\boldsymbol{k}\omega} = -i\frac{e_\alpha}{m_\alpha}\left(\left\{\boldsymbol{E}^0+\frac{1}{c}[\boldsymbol{v}\wedge\boldsymbol{B}^0]\right\}\cdot\left\{\frac{1}{\omega-(\boldsymbol{k}\cdot\boldsymbol{v})}\frac{\partial F_0^{(\alpha)}}{\partial\boldsymbol{v}}\right\}\right), \tag{12.1.1.8}$$

where we have omitted the factor $e^{i(\boldsymbol{k}\cdot\boldsymbol{r})-i\omega t}$. The current connected with this part of the distribution function equals

$$\boldsymbol{j}_{\boldsymbol{k}\omega} = -i\omega\frac{\varepsilon_l(\boldsymbol{k}, \omega)-1}{4\pi}\boldsymbol{E}^0. \tag{12.1.1.9}$$

Fourier transforming eqn. (12.1.1.7) we get

$$(f_\alpha')_{\boldsymbol{k}'\omega'} = -i\frac{e_\alpha}{m_\alpha}\frac{1}{\omega'-(\boldsymbol{k}'\cdot\boldsymbol{v})}\left\{\left(\left\{\boldsymbol{E}'_{\boldsymbol{k}'\omega'}+\frac{1}{c}[\boldsymbol{v}\wedge\boldsymbol{B}'_{\boldsymbol{k}'\omega'}]\right\}\cdot\frac{\partial F_0^{(\alpha)}}{\partial\boldsymbol{v}}\right)\right.$$
$$\left.+\left(\left\{\boldsymbol{E}^0+\frac{1}{c}[\boldsymbol{v}\wedge\boldsymbol{B}^0]\right\}\cdot\frac{\partial}{\partial\boldsymbol{v}}\right)(\delta f_\alpha)_{\boldsymbol{q}\Delta\omega}+\left(\left\{\delta\boldsymbol{E}_{\boldsymbol{q}\Delta\omega}+\frac{1}{c}[\boldsymbol{v}\wedge\delta\boldsymbol{B}_{\boldsymbol{q}\Delta\omega}]\right\}\cdot\frac{\partial}{\partial\boldsymbol{v}}\right)(f_\alpha^0)_{\boldsymbol{k}\omega}\right\}, \tag{12.1.1.10}$$

where

$$\Delta\omega = \omega'-\omega \quad \text{and} \quad \boldsymbol{q} = \boldsymbol{k}'-\boldsymbol{k}.$$

One can easily determine the current connected with the function f_α':

$$\boldsymbol{j}'_{\boldsymbol{k}'\omega'} = -i\omega'\frac{\varepsilon_l(\boldsymbol{k}', \omega')-1}{4\pi}\boldsymbol{E}'_{\boldsymbol{k}'\omega'}+\boldsymbol{I}_{\boldsymbol{k}'\omega'}, \tag{12.1.1.11}$$

where

$$\boldsymbol{I}_{\boldsymbol{k}'\omega'} = -i\sum_\alpha\frac{e_\alpha^2}{m_\alpha}\int\frac{\boldsymbol{v}}{\omega'-(\boldsymbol{k}'\cdot\boldsymbol{v})}\left\{\left(\left\{\boldsymbol{E}^0+\frac{1}{c}[\boldsymbol{v}\wedge\boldsymbol{B}^0]\right\}\cdot\frac{\partial}{\partial\boldsymbol{v}}\right)(\delta f_\alpha)_{\boldsymbol{q}\Delta\omega}\right.$$
$$\left.+\left(\left\{\delta\boldsymbol{E}_{\boldsymbol{q}\Delta\omega}+\frac{1}{c}[\boldsymbol{v}\wedge\delta\boldsymbol{B}_{\boldsymbol{q}\Delta\omega}]\right\}\cdot\frac{\partial}{\partial\boldsymbol{v}}\right)(f_\alpha^0)_{\boldsymbol{k}\omega}\right\}d^3\boldsymbol{v}. \tag{12.1.1.12}$$

Substituting expression (12.1.1.11) for the current $\boldsymbol{j}'$ into eqn. (12.1.1.3) for the field of the scattered wave we get

$$\text{curl curl}\,\boldsymbol{E}'+\frac{\hat{\varepsilon}}{c^2}\frac{\partial^2\boldsymbol{E}'}{\partial t^2} = -\frac{4\pi}{c^2}\frac{\partial\boldsymbol{I}}{\partial t}. \tag{12.1.1.13}$$

We see that we can consider the quantity $\boldsymbol{I}$ as the current which causes the scattered wave. We shall call this current the scattering current; it is proportional to the field of the incident wave and to a quantity characterizing the fluctuations in the plasma.

Equation (12.1.1.13) describes all scattering and wave transformation processes in an unmagnetized plasma. We shall in what follows consider only the scattering (and transformation) of high-frequency waves—transverse electromagnetic and longitudinal Langmuir waves. These processes are mainly produced by the electron component of the plasma. We need thus take for $\boldsymbol{I}$ in (12.1.1.13) merely the electron current density.

As the phase velocities of the waves considered are much larger than the electron thermal velocity, we can use the expansions

$$\{\omega-(\boldsymbol{k}\cdot\boldsymbol{v})\}^{-1}=\omega^{-1}\left\{1+\frac{(\boldsymbol{k}\cdot\boldsymbol{v})}{\omega}+\ldots\right\},\quad \{\omega'-(\boldsymbol{k}'\cdot\boldsymbol{v})\}^{-1}=\omega'^{-1}\left\{1+\frac{(\boldsymbol{k}'\cdot\boldsymbol{v})}{\omega'}+\ldots\right\}$$

when we evaluate the integrals occurring in the expression for the current $\boldsymbol{I}$. As a result we get the following expression for the Fourier components of the current $\boldsymbol{I}$:

$$\boldsymbol{I}_{k'\omega'}=-i\frac{e}{m_e\omega}\left\{\left[-e\delta n^e_{q\Delta\omega}+\frac{1}{\omega'}(\boldsymbol{k}'\cdot\boldsymbol{\delta j}^e_{q\Delta\omega})-\frac{1}{\omega}(\boldsymbol{k}\cdot\boldsymbol{\delta j}^e_{q\Delta\omega})\right]\boldsymbol{E}^o\right.$$
$$+\frac{\boldsymbol{k}}{\omega}(\boldsymbol{E}^o\cdot\boldsymbol{\delta j}^e_{q\Delta\omega})+\frac{1}{\omega'}(\boldsymbol{k}'\cdot\boldsymbol{E}^o)\,\boldsymbol{\delta j}^e_{q\Delta\omega}+\frac{i}{4\pi}\frac{\omega_{pe}^2}{\omega\omega'}(\boldsymbol{k}'\cdot\boldsymbol{\delta E}_{q\Delta\omega})\boldsymbol{E}^o$$
$$\left.+\frac{\omega'}{\Delta\omega}(\boldsymbol{E}^o\cdot\boldsymbol{\delta E}_{q\Delta\omega})\boldsymbol{q}+\left[(\boldsymbol{k}'\cdot\boldsymbol{E}^o)-\frac{\omega'}{\Delta\omega}(\boldsymbol{q}\cdot\boldsymbol{E}^o)+\frac{\omega'}{\omega}(\boldsymbol{k}\cdot\boldsymbol{E}^o)\right]\boldsymbol{\delta E}_{q\Delta\omega},\right. \qquad (12.1.1.14)$$

where δn^e and $\boldsymbol{\delta j}^e$ are the fluctuations in the electron density and the electron current density,

$$\delta n^e=\int\delta f_e\,d^3v,\quad \boldsymbol{\delta j}^e=-e\int\boldsymbol{v}\,\delta f_e\,d^3v.$$

We note that the fluctuation field $\boldsymbol{\delta E}$ which occurs in expression (12.1.1.14) is determined by the fluctuations in both the electron and ion currents.

Neglecting small relativistic corrections we can express the fluctuation current $\boldsymbol{\delta j}^e$ and field $\boldsymbol{\delta E}$ in terms of the fluctuations in the electron and ion densities, δn^e and δn^i:

$$\boldsymbol{\delta j}^e_{q\Delta\omega}=\frac{\Delta\omega}{q^2}\boldsymbol{q}\,\delta n^e_{q\Delta\omega},\quad \boldsymbol{\delta E}_{q\Delta\omega}=i\frac{4\pi e}{q^2}\boldsymbol{q}(\delta n^e_{q\Delta\omega}-\delta n^i_{q\Delta\omega}). \qquad (12.1.1.15)$$

Substituting these expressions into (12.1.1.14) we get

$$\boldsymbol{I}_{k'\omega'}=i\frac{e^2}{m_e\omega}\left\{\left[\boldsymbol{E}^o+\frac{\Delta\omega}{\omega'}\frac{\boldsymbol{k}'}{q^2}(\boldsymbol{q}\cdot\boldsymbol{E}^o)+\frac{\Delta\omega}{\omega}\frac{\boldsymbol{q}}{q^2}(\boldsymbol{k}\cdot\boldsymbol{E}^o)\right]\delta n^e_{q\Delta\omega}-\frac{\Delta\omega^2}{q^2\omega'^2}\left[(\boldsymbol{k}\cdot\boldsymbol{q})\boldsymbol{E}^o+(\boldsymbol{k}'\cdot\boldsymbol{E}^o)\boldsymbol{q}\right.\right.$$
$$\left.\left.+\frac{\omega'}{\omega}(\boldsymbol{k}\cdot\boldsymbol{E}^o)\boldsymbol{q}\right]\delta n^e_{q\Delta\omega}+\frac{\omega_{pe}^2}{q^2\omega'^2}\left[(\boldsymbol{k}'\cdot\boldsymbol{q})\boldsymbol{E}^o+(\boldsymbol{k}'\cdot\boldsymbol{E}^o)\boldsymbol{q}+\frac{\omega'}{\omega}(\boldsymbol{k}\cdot\boldsymbol{E}^o)\boldsymbol{q}\right](\delta n^e_{q\Delta\omega}-\delta n^i_{q\Delta\omega})\right\}. \qquad (12.1.1.16)$$

We shall show that we can neglect in this expression the two last terms. To do this we bear in mind that the longitudinal fluctuations in the plasma occur mainly with the Larmor frequency $\Delta\omega\approx\omega_{pe}$ and with frequencies much lower than the Langmuir frequency, $\Delta\omega\ll\omega_{pe}$. Let us first consider fluctuations with frequencies $\Delta\omega\ll\omega_{pe}$. For them we have

$$\langle(\delta n^e-\delta n^i)^2\rangle_{q\Delta\omega}\sim q^4r_D^4\langle\delta n_e^2\rangle_{q\Delta\omega}.$$

When $\Delta\omega\ll\omega_{pe}$ the ratio of the third term in (12.1.1.16) to the first one is thus of the order of $k^2v_e^2/\omega^2$, that is, much less than unity—as we have assumed that $\omega/k \gg v_e$. We can also neglect the second term in (12.1.1.16) when $\Delta\omega \ll \omega_{pe}$, as it contains the factor $\Delta\omega^2/\omega'^2$ ($\omega' > \omega_{pe}$).

Let us now assume that the fluctuation oscillation is a Langmuir wave, that is, $\Delta\omega \approx \omega_{pe}$. In that case $\delta n^i_{q\Delta\omega} \ll \delta n^e_{q\Delta\omega}$ and the second term in (12.1.1.16) cancels the third one. We can thus use the following expression for the Fourier component of the scattering current:

$$I_{k\omega'} = i\frac{e^2}{m\omega}\left\{E^\circ + \frac{\Delta\omega}{\omega'}\frac{k'}{q^2}(q\cdot E^\circ) + \frac{\Delta\omega}{\omega}\frac{q}{q^2}(k\cdot E^\circ)\right\}\delta n^e{}_{q\Delta\omega}. \tag{12.1.1.17}$$

We emphasize that this expression is valid provided the phase velocities of the incident and of the scattered waves are much larger than the electron thermal velocity.

Let us now prove that under those assumptions we can obtain eqn. (12.1.1.17) from a simple hydrodynamical picture if we introduce the hydrodynamical electron velocity $v(r, t)$ and the electron density $n(r, t)$. These quantities are connected with the electron current through the equation

$$j = -env, \tag{12.1.1.18}$$

and satisfy the equations

$$\frac{\partial v}{\partial t} + (v\cdot\nabla)\,v = -\frac{e}{m_e}\left(E + \frac{1}{c}[v \wedge B]\right), \quad \frac{\partial n}{\partial t} + \operatorname{div} nv = 0. \tag{12.1.1.19}$$

If there is a wave E° propagating through the plasma, we can write the electron density and hydrodynamical velocity in the form

$$n = n_0 + n^0 + \delta n + n', \quad v = v^0 + \delta v + v', \tag{12.1.1.20}$$

where n^0 is the change in the density and v^0 the electron velocity caused by the field of the incident wave; n' and v' the analogous quantities connected with the scattered wave E'; and δn and δv the fluctuations in the density and the hydrodynamical velocity.

The quantities n^0 and v^0 satisfy the equations

$$\frac{\partial v^0}{\partial t} = -\frac{e}{m_e}E^\circ, \quad \frac{\partial n^0}{\partial t} + n_0 \operatorname{div} v^0 = 0, \tag{12.1.1.21}$$

and the quantities n' and v' the equations

$$\frac{\partial v'}{\partial t} = -\frac{e}{m_e}E' - \frac{e}{m_e c}[\delta v \wedge B^0] - (v^0\cdot\nabla)\,\delta v - (\delta v\cdot\nabla)\,v^0,$$

$$\frac{\partial n'}{\partial t} + n_0 \operatorname{div} v' + n^0 \operatorname{div} \delta v + \delta n \operatorname{div} v^\circ = 0. \tag{12.1.1.22}$$

If we put in these equations $E^\circ(r, t) = E^\circ e^{i(k\cdot r) - i\omega t}$ and $E'(r, t) = E'\, e^{i(k'\cdot r) - i\omega' t}$, we find

$$v^0 = -i\frac{e}{m_e\omega}E^\circ, \quad n^0 = -i\frac{en_0}{m_e\omega^2}(k\cdot E^\circ),$$

$$v' = -i\frac{e}{m_e\omega'}\left(E' + \frac{1}{c}[\delta v_{q\Delta\omega} \wedge B^0]\right) + \frac{1}{\omega'}\{(q\cdot v^0)\,\delta v_{q\Delta\omega} + (k\cdot\delta v_{q\Delta\omega})\,v^0\}, \tag{12.1.1.23}$$

where $\boldsymbol{q} = \boldsymbol{k}' - \boldsymbol{k}$, $\Delta\omega = \omega' - \omega$, while $\delta \boldsymbol{v}_{\boldsymbol{q}\Delta\omega}$ is the Fourier component of the electron velocity fluctuation. The velocity fluctuations are connected with the electron density fluctuations through the continuity equation,

$$\frac{\partial \delta n}{\partial t} + n_0 \operatorname{div} \delta \boldsymbol{v} = 0. \tag{12.1.1.24}$$

For longitudinal fluctuations which are the only important ones we have

$$\delta \boldsymbol{v}_{\boldsymbol{q}\Delta\omega} = \frac{\Delta\omega}{n_0 q^2} \boldsymbol{q}\, \delta n_{\boldsymbol{q}\Delta\omega}.$$

Substituting this expression into (12.1.1.23), we find

$$\boldsymbol{v}' = -i \frac{e}{m_e \omega'} \left[\boldsymbol{E}' + \frac{\Delta\omega \delta n_{\boldsymbol{q}\Delta\omega}}{n_0 \omega q^2} \{ [\boldsymbol{q} \wedge [\boldsymbol{k} \wedge \boldsymbol{E}^{\circ}]] + (\boldsymbol{q} \cdot \boldsymbol{E}^{\circ}) \boldsymbol{q} + \boldsymbol{E}^{\circ} (\boldsymbol{k} \cdot \boldsymbol{q}) \} \right] \tag{12.1.1.25}$$

Let us now determine the current connected with the scattered wave:

$$\boldsymbol{j}' = -e(n_0 \boldsymbol{v}' + n^0 \delta \boldsymbol{v} + \boldsymbol{v}^0 \delta n). \tag{12.1.1.26}$$

If there were no fluctuations, the current $\boldsymbol{j}'$ would have the form

$$\boldsymbol{j}' |_{\delta n = 0} = -e n_0 \boldsymbol{v}' |_{\delta n = 0} = i \frac{e^2 n_0}{m_e \omega'} \boldsymbol{E}'.$$

Subtracting this expression from expression (12.1.1.26), we find the scattering current

$$\boldsymbol{I} = \boldsymbol{j}' - \boldsymbol{j}' |_{\delta n = 0}. \tag{12.1.1.27}$$

Using (12.1.1.25) we get from this the Fourier component $\boldsymbol{I}_{k,\omega}$ of the scattering current:

$$\boldsymbol{I}_{k'\omega'} = i \frac{e^2}{m_e \omega} \left\{ \boldsymbol{E}^{\circ} + \frac{\Delta\omega}{\omega'} \frac{\boldsymbol{k}'}{q^2} (\boldsymbol{q} \cdot \boldsymbol{E}^{\circ}) + \frac{\Delta\omega}{\omega} \frac{\boldsymbol{q}}{q^2} (\boldsymbol{k} \cdot \boldsymbol{E}^{\circ}) \right\} \delta n_{\boldsymbol{q}\Delta\omega}.$$

This expression is the same as expression (12.1.1.17) which was obtained using kinetic considerations.

12.1.2. SCATTERING CROSS-SECTION

Once we have an expression for the scattering current, we can use (12.1.1.13) to find the field of the scattered waves. It is clear that independent of the polarization properties of the incident wave the scattering current will contain both a transverse and a longitudinal part. The field of the scattered waves will therefore also contain transverse and longitudinal components. Separating these components we can study the scattering of transverse waves —that is, the transition of transverse into other transverse waves—the transformation of transverse waves into longitudinal ones, the scattering of longitudinal waves, and the transformation of longitudinal into transverse waves.

Let us first of all consider the scattering of transverse waves in a plasma, so that we can put $(\boldsymbol{k} \cdot \boldsymbol{E}^{\circ}) = 0$ and $(\boldsymbol{k}' \cdot \boldsymbol{E}') = 0$. Taking the transverse part of the current $\boldsymbol{I}$ we find the Fourier component of the field of the scattered transverse wave:

$$\boldsymbol{E}'_{k'\omega'} = -\frac{4\pi e^2\omega'}{m_e c^2\omega}\left\{k'^2 - \frac{\omega'^2}{c^2}\,\varepsilon(\omega')\right\}^{-1} \boldsymbol{E}^{\circ}_{\perp}\,\delta n_{q\Delta\omega}, \tag{12.1.2.1}$$

where $\boldsymbol{E}^{\circ}_{\perp}$ is the component of $\boldsymbol{E}^{\circ}$ at right angles to $\boldsymbol{k}'$ and $\varepsilon(\omega) = 1-(\omega_{pe}/\omega)^2$.

The average increase in the energy of the field of the scattered waves per unit time is clearly determined by the formula

$$I = -\tfrac{1}{2}\,\mathrm{Re}\int \langle(\boldsymbol{I}(\boldsymbol{r},t)\cdot\boldsymbol{E}'(\boldsymbol{r},t))\rangle\, d^3\boldsymbol{r}. \tag{12.1.2.2}$$

Substituting here expressions (12.1.2.1) and (12.1.1.17) for $\boldsymbol{E}'$ and $\boldsymbol{I}$, we find the increase in the energy of the field of the scattered transverse waves:

$$I_{t\to t'} = \frac{V}{(2\pi)^3}\,\frac{e^4}{m_e^2c^2\omega^2}\,\mathrm{Im}\int \frac{\omega' E^{\circ 2}_{\perp}\langle\delta n^2\rangle_{q\Delta\omega}}{k'^2-(\omega'^2/c^2)\,\varepsilon(\omega')}\,d^3\boldsymbol{k}'\,d\omega', \tag{12.1.2.3}$$

where V is the volume of a plasma.

Clearly, only the poles of the integrand contribute to the integral determining $I_{t\to t'}$. Integrating over the absolute magnitude of the vector $\boldsymbol{k}'$ we find the intensity of scattering in a frequency range $d\omega'$ and into an element of solid angle $d^2\boldsymbol{\omega}'$:

$$dI_{t\to t'} = \frac{Vc}{16\pi^2}\left(\frac{e^2}{m_ec^2}\right)^2 \frac{\omega'^2}{\omega^2}\sqrt{[\varepsilon(\omega')]}\,E^{\circ 2}_{\perp}\langle\delta n^2\rangle_{q\Delta\omega}\,d\omega'\,d^2\boldsymbol{\omega}', \tag{12.1.2.4}$$

where the frequency ω' and the wavevector $\boldsymbol{k}'$ of the scattered wave are connected through the relation $k'^2 = (\omega'^2/c^2)\,\varepsilon(\omega')$.

If the incident wave is unpolarized, we must average eqn. (12.1.2.4) over the different orientations of the vector $\boldsymbol{E}^{\circ}$. The mean square of the field, $\overline{E^{\circ 2}_{\perp}}$, is then equal to

$$\overline{E^{\circ 2}_{\perp}} = \tfrac{1}{2}(1+\cos^2\vartheta)\,E^{\circ 2}, \tag{12.1.2.5}$$

where ϑ is the angle between the vectors $\boldsymbol{k}$ and $\boldsymbol{k}'$, that is, the scattering angle.

Dividing the scattering intensity dI by the energy flux density of the incident wave,

$$S_0 = \frac{c}{8\pi}\sqrt{[\varepsilon(\omega)]}\,E^{\circ 2},$$

and by the magnitude of the scattering volume V, we find the differential scattering cross-section, or scattering coefficient, $d\Sigma$,

$$d\Sigma = \frac{dI}{VS_0}. \tag{12.1.2.6}$$

The differential scattering cross-section referred to an element of solid angle $d^2\boldsymbol{\omega}'$ and a frequency interval $d\omega'$ has the following form for an unpolarized wave (Akhiezer, Akhiezer, and Sitenko, 1962):

$$d\Sigma_{t\to t'} = \frac{1}{4\pi}\left(\frac{e^2}{m_ec^2}\right)^2\frac{\omega'^2}{\omega^2}\sqrt{\frac{\varepsilon(\omega')}{\varepsilon(\omega)}}\,(1+\cos^2\vartheta)\langle\delta n^2\rangle_{q\Delta\omega}\,d\omega'\,d^2\boldsymbol{\omega}'. \tag{12.1.2.7}$$

This formula is valid for any change in frequency; we have merely assumed that the frequencies ω and ω' exceed ω_{pe}. If $\Delta\omega \ll \omega$, the factor $(\omega'/\omega)^2 \sqrt{\{\varepsilon(\omega')/\varepsilon(\omega)\}}$ becomes unity and eqn. (12.1.2.7) becomes the well-known formula determining the cross-section for scattering by density fluctuations with small changes in frequency.

We note that although in the derivation of formula (12.1.2.7) we only took into account the scattering of electromagnetic waves by electron density fluctuations, it turns out that the scattering cross-section $d\Sigma_{t\to t'}$, also depends on the ion motion. This is connected with the fact that the spectral density of the electron density fluctuations $\langle\delta n^2\rangle_{q\Delta\omega}$ depends in an essential way also on the ion motion in the plasma, due to the self-consistent interaction between the electrons and the ions. The spectral distribution of the scattered waves is thus determined by the spectral density of the electron density fluctuations.

In an isothermal plasma the spectrum of the scattered radiation consists of a Doppler-broadened main line, $\Delta\omega \lesssim qv_i$, where v_i is the thermal ion velocity, and steep maxima occurring for $\Delta\omega = \pm\omega_{pe}$ (we assume that $qr_D \ll 1$). We can therefore assume the factor $(\omega'/\omega)^2\sqrt{\{\varepsilon(\omega')/\varepsilon(\omega)\}}$ to be equal to unity in the case of high frequencies, $\omega \gg \omega_{pe}$, which is of most interest to us. In that case we can integrate the scattering cross-section over the frequencies, using eqn. (11.2.2.3). As a result we obtain the following expression for the cross-section for the scattering of transverse waves in an isothermal plasma into a solid angle $d^2\boldsymbol{\omega}'$:

$$d\Sigma_{t\to t'} = \frac{1}{2} n_0 \left(\frac{e^2}{m_e c^2}\right)^2 \frac{1+q^2r_D^2}{2+q^2r_D^2} (1+\cos^2\vartheta)\, d^2\boldsymbol{\omega}', \qquad (12.1.2.8)$$

where $r_D^2 = T/4\pi n_0 e^2$, $q = (2\omega/c)\sin\frac{1}{2}\vartheta$, and ϑ the scattering angle.

Integrating (12.1.2.8) over angles we get the total scattering cross-section:

$$\Sigma_{t\to t'} = n_0\sigma_0 \left\{1 - \frac{3}{4k^2r_D^2} + \frac{3\ln(1+2k^2r_D^2)}{8k^2r_D^2} + \frac{3}{4\sqrt{2}}\,\frac{1-k^2r_D^2}{k^3r_D^3}\arctan(kr_D\sqrt{2})\right\}, \qquad (12.1.2.9)$$

where $\sigma_0 = (8\pi/3)(e^2/m_e c^2)^2$ is the Thomson cross-section for the scattering of electromagnetic waves by free electrons.

In the limiting cases of short and long wavelengths the scattering cross-section has the form

$$\Sigma = n_0\sigma_0, \qquad kr_D \gg 1; \quad \Sigma = \tfrac{1}{2} n_0\sigma_0, \qquad kr_D \ll 1. \qquad (12.1.2.10)$$

12.1.3. SPECTRAL DISTRIBUTION OF THE SCATTERED RADIATION

Let us study in some more detail the spectral distribution of the scattered radiation in an unmagnetized plasma.† In the case of short wavelengths, when $qr_D \gg 1$, the spectral density of the density fluctuations $\langle\delta n^2\rangle_{q\Delta\omega}$ is a Gaussian in the frequency $\Delta\omega$ and the spectral distribution of the scattered radiation is therefore also Gaussian in shape:

$$d\Sigma = \frac{1}{\sqrt{(8\pi)}} n_0 \left(\frac{e^2}{m_e c^2}\right)^2 \left(\frac{m_e}{q^2T_e}\right)^{1/2} (1+\cos^2\vartheta)\exp\left(-\frac{m_e\Delta\omega^2}{2q^2T_e}\right) d\omega'\, d^2\boldsymbol{\omega}'. \qquad (12.1.3.1)$$

This formula is valid, provided $\Delta\omega \neq \omega_{pi}$.

† Akhiezer, Akhiezer, and Sitenko (1962) and Dougherty and Farley (1960) have studied the spectral distribution of the scattered radiation in the case of an equilibrium plasma, and the first authors also in the case of a two-temperature plasma.

We see that the Doppler broadening of the line is given by the thermal electron velocity. The total scattering cross-section (12.1.2.10) is equal to the sum of the cross-sections for scattering by the separate electrons. The Coulomb interaction between the electrons and the ions is unimportant and the scattering is incoherent in character.

In the case of long wavelengths, when $kr_D \ll 1$, collective properties of the plasma come into play. This case is realized, in particular, in experiments on the scattering of radio-waves by density fluctuations in the upper layers of the ionosphere ($k^{-1} \sim 10$ cm, $r_D \sim 1$ cm). We shall give expressions for the spectral distribution of the scattered radiation in various ranges of frequencies, assuming that $kr_D \ll 1$.

If the change in frequency in the scattering process is small, $\Delta\omega \ll qv_i$, where v_i is the thermal ion velocity, the scattering cross-section is given by the formula

$$d\Sigma = \frac{1}{\sqrt{(8\pi)}} n_0 \left(\frac{e^2}{m_e c^2}\right)^2 \frac{\sqrt{m_e}\, T_e^{3/2} + \sqrt{m_i}\, T_i^{3/2}}{q(T_e+T_i)^2} (1+\cos^2\vartheta)\, d\omega'\, d^2\boldsymbol{\omega}', \qquad (12.1.3.2)$$

where T_e and T_i are the electron and ion temperatures. In the case of a strongly non-isothermal plasma the scattering cross-section for the case of small changes in the frequency is smaller than the corresponding cross-section for an isothermal plasma with temperature T_e by a factor $\frac{1}{4}\sqrt{(m_i/m_e)}$.

If $\Delta\omega \ll qv_i$, the scattering cross-section in an isothermal plasma has the form

$$d\Sigma = \frac{1}{\sqrt{(8\pi)}} n_0 \left(\frac{e^2}{m_e c^2}\right)^2 \frac{1}{qv_i} \frac{\exp(-z^2)}{[2-\varphi(z)]^2 + \pi z^2 \exp(-2z^2)} (1+\cos^2\vartheta)\, d\omega'\, d^2\boldsymbol{\omega}', \qquad (12.1.3.3)$$

where

$$z = \frac{1}{\sqrt{2}} \frac{\omega}{qv_i} \quad \text{and} \quad \varphi(z) = 2z\, e^{-z^2} \int_0^z e^{x^2}\, dx.$$

The scattering cross-section given by this formula decreases steeply when $\Delta\omega \sim qv_i$; the quantity $\Delta\omega \sim qv_i$ therefore characterizes the width of the spectral distribution of the scattered radiation in an isothermal plasma. We note that this quantity is determined by the thermal ion velocity, notwithstanding that the scattering is by the electrons.

If $qv_i \ll \Delta\omega \ll qv_e$, the scattering cross-section is very small for an isothermal plasma In a strongly non-isothermal plasma the scattering cross-section has steep maxima when the frequency shift $\Delta\omega$ coincides with the frequency of the low-frequency plasma oscillations with wavevector $\boldsymbol{q}$. In particular, if $qr_D \ll 1$, maxima occur when $\Delta\omega = \pm qv_s$, where v_s is the non-isothermal sound velocity. The scattering cross-section has near the maxima the form

$$d\Sigma = \frac{1}{4} n_0 \left(\frac{e^2}{m_e c^2}\right)^2 (1+\cos^2\vartheta)\, \delta(\Delta\omega - qv_s) + \delta(\Delta\omega + qv_s)\, d\omega'\, d^2\boldsymbol{\omega}'. \qquad (12.1.3.4)$$

Integrating (12.1.3.4) over the angles we find the total scattering cross-section in a non-isothermal plasma,

$$\Sigma = n_0 \sigma_0. \qquad (12.1.3.5)$$

This formula is valid, provided $(T_e/T_i)^3 \gg m_i/m_e$ and $kr_D \ll 1$. We note that the total cross-section is twice the scattering cross-section in an isothermal plasma, which was given by eqn. (12.1.2.10).

In the case of large changes in frequency, $\Delta\omega \gg qv_e$, the scattering cross-section has sharp maxima at $\Delta\omega = \pm\omega_{pe}$; these are connected with the scattering of the electromagnetic waves by the longitudinal electron oscillations. The differential scattering cross-section is in that range of frequencies for any ratio of the electron and ion temperatures given by the formula

$$d\Sigma = \frac{1}{4} n_0 \left(\frac{e^2}{m_e c^2}\right)^2 \frac{\omega'^2}{\omega^2} \sqrt{\left[\frac{\varepsilon(\omega')}{\varepsilon(\omega)}\right]}\, q^2 r_D^2 (1+\cos^2\vartheta)\{\delta(\Delta\omega-\omega_{pe})+\delta(\Delta\omega+\omega_{pe})\}\, d\omega'\, d^2\boldsymbol{\omega}'. \tag{12.1.3.6}$$

The total cross-section for the scattering of electromagnetic waves by Langmuir oscillations has for $\omega \gg \omega_{pe}$ the form

$$\Sigma = 2k^2 r_D^2 n_0 \sigma_0, \qquad kr_D \ll 1. \tag{12.1.3.7}$$

We see that the collective properties of the plasma affect the scattering particularly severely, when $kr_D \ll 1$. In that case the spectra of the scattered electromagnetic waves differ greatly for the two cases of an isothermal and a non-isothermal plasma.

In an isothermal plasma there is in the spectrum of the scattered radiation a central maximum caused by the incoherent scattering by the electron density fluctuations, with a width determined by the ion velocity, and satellites caused by the scattering by electron oscillations. The relative weight of the satellites—relative to the main maximum—is $\sim 2k^2 r_D^2$. In a strongly non-isothermal plasma there is no central maximum. There are two maxima, placed symmetrically with respect to $\Delta\omega = 0$, which are caused by scattering by sound oscillations, and satellites connected with the scattering by Langmuir oscillations. The relative weight of these satellites relative to the sound maxima is $\sim 2k^2 r_D^2$.

12.1.4. CRITICAL OPALESCENCE

Let us now consider the scattering of electromagnetic waves in a plasma in which the particles have directed velocities, for instance, because the electrons move relative to the ions or because a beam of charged particles moves through the plasma. Let us first of all dwell on the case when the directed electron (or beam) velocity is less than that critical velocity at which instability sets in. The scattering coefficient is then given by the general formula (12.1.2.7), in which we must substitute expressions (11.5.3.13) or (11.5.3.18) for the spectral density of the electron density fluctuations.

We shall show in Section 12.5 that as the directed velocity approaches its critical value, which defines the boundary of the stability region, the spectral density of the electron density fluctuations grows without bounds. The light scattering coefficient will then also grow, according to (12.1.2.7). In particular, if the electrons in a two-temperature plasma move relative to the ions with a velocity approaching the non-isothermal sound velocity v_s, the differential coefficient for the scattering of light by sound oscillations has, according to

(11.5.3.18) the form

$$d\Sigma = \frac{1}{2} n_0 \left(\frac{e^2}{m_e c^2}\right)^2 \frac{q^2 v_s^2}{|\Delta\omega - (\boldsymbol{q}\cdot\boldsymbol{u})|} (1+\cos^2\vartheta)\, \delta(\Delta\omega^2 - q^2 v_s^2)\, d\omega'\, d^2\boldsymbol{\omega}'. \quad (12.1.4.1)$$

If $|(\boldsymbol{q}\cdot\boldsymbol{u})| \to qv_s$ the coefficient in front of the δ-function tends to infinity.

Integrating eqn. (12.1.4.1) over ω' we get the cross-section for the scattering of light by sound oscillations into a solid angle $d^2\boldsymbol{\omega}'$:

$$d\Sigma = \frac{1}{2} n_0 \left(\frac{e^2}{m_e c^2}\right)^2 (1+\cos^2\vartheta) \left[1 - \frac{u^2(\cos\theta' - \cos\theta)^2}{2v_s^2 \sin^2 \frac{1}{2}\vartheta}\right]^{-1} d^2\boldsymbol{\omega}', \quad (12.1.4.2)$$

where θ and θ' are the angles between $\boldsymbol{u}$ and $\boldsymbol{k}$ or $\boldsymbol{k}'$ (to fix the ideas we assume that $\theta < \pi/2$).

One sees easily that when $u \gtrsim v_s$ there are directions of the wavevector $\boldsymbol{k}'$ for which $d\Sigma/d^2\boldsymbol{\omega}'$ is anomalously large. If $|1-(v_s/u)| \ll 1$, it is necessary for this that the vectors $\boldsymbol{k}$, $\boldsymbol{k}'$, and $\boldsymbol{u}$ lie almost in one plane and that the condition $\theta+\theta' \cong \pi$ is satisfied. In that case $d\Sigma/d^2\boldsymbol{\omega}' \to \infty$ when the angle φ between the $(\boldsymbol{k}', \boldsymbol{u})$- and $(\boldsymbol{k}, \boldsymbol{u})$-planes tends to $\pm\varphi_c$, where

$$\varphi_c^2 = \cot^2\theta \left\{4\left(\frac{u^2}{v_s^2} - 1\right) - (\pi - \theta - \theta')^2\right\}, \quad (12.1.4.3)$$

when the angles θ and θ' are not close to $\pi/2$, while

$$\varphi_c^2 = 4\left(\frac{u^2}{v_s^2} - 1\right)(\theta - \theta')^2, \quad (12.1.4.4)$$

when θ and $\theta' \approx \pi/2$.

The anomalous growth of the scattering coefficient near the boundary of the plasma instability is connected with the occurrence of critical fluctuations and may be called *critical opalescence* by analogy with the well-known critical opalescence phenomenon in condensed bodies near phase transition points.†

12.2. Transformation of Transverse and Longitudinal Waves in a Plasma

12.2.1. TRANSFORMATION OF A TRANSVERSE WAVE INTO A LONGITUDINAL ONE

Let us turn to the consideration of the transformation of a transverse wave into a longitudinal one. Separating in (12.1.1.13) the longitudinal part of the current $\boldsymbol{I}$ we find the Fourier component of the longitudinal component of the electrical field of the scattered wave

$$E^l_{k'\omega'} = \frac{4\pi e^2}{m_e \omega\omega'} \frac{\boldsymbol{k}'(\boldsymbol{k}'\cdot\boldsymbol{E}^0)}{k'^2 \varepsilon_l(\boldsymbol{k}', \omega')} \left[1 + \frac{\Delta\omega}{\omega'} \frac{k'^2}{q^2}\right] \delta n_{q\Delta\omega}, \quad (12.2.1.1)$$

where ε_l is the longitudinal dielectric permittivity of the plasma, given by eqn. (4.3.4.3).

† Ichimaru, Pines, and Rostoker (1962), Rosenbluth and Rostoker (1962), and Ichimaru (1962) predicted the occurrence of critical opalescence in a plasma.

Substituting the expression for the field $\boldsymbol{E}'$ into formula (12.1.2.2) and using expression (12.1.1.17) for the current $\boldsymbol{I}$ we find the intensity of the excitation of longitudinal waves:

$$I_{t\to l} = \frac{V}{(2\pi)^3}\frac{e^4}{m_e^2\omega^2}\,\mathrm{Im}\int \frac{(\boldsymbol{k}'\cdot\boldsymbol{E}^\circ)^2}{\omega' k'^2 \varepsilon_l^*(\boldsymbol{k}',\omega')}\left[1+\frac{\Delta\omega}{\omega'}\frac{k'^2}{q^2}\right]\langle\delta n^2\rangle_{q\Delta\omega}\,d\omega'\,d^3\boldsymbol{k}'. \quad (12.2.1.2)$$

Only those frequencies for which the denominator of the integrand vanishes, that is, for which $\varepsilon_l(\boldsymbol{k}',\omega') = 0$, contribute to this expression. This means that as a result of the absorption of a transverse electromagnetic wave eigenoscillations are excited.

Using (12.2.1.2) and assuming that the incident wave is unpolarized we find the *coefficient for the transformation* of transverse electromagnetic waves into longitudinal plasma waves (Sitenko, 1967):

$$d\Sigma_{t\to l} \equiv \frac{dI_{t\to l}}{VS_0}$$

$$= \frac{1}{4\pi}\left(\frac{e^2}{m_e c^2}\right)^2\left(\frac{m_e c^2}{3T}\right)^{3/2}\frac{\omega'^2}{\omega^2}\sqrt{\left[\frac{\varepsilon(\omega')}{\varepsilon(\omega)}\right]}\left(1+\frac{\Delta\omega}{\omega'}\frac{k'^2}{q^2}\right)^2\sin^2\vartheta\langle\delta n^2\rangle_{q\Delta\omega}\,d\omega'\,d^2\boldsymbol{\omega}'. \quad (12.2.1.3)$$

We emphasize that this formula is only applicable in the frequency region $\omega' \approx \omega_{pe}$, where the damping of the plasma waves is small.

The ratio of the transformation coefficient (12.2.1.3) to the scattering coefficient (12.1.2.7) is equal to

$$\frac{d\Sigma_{t\to l}}{d\Sigma_{t\to t}} = \left(\frac{m_e c^2}{3T}\right)^{3/2}\frac{\sin^2\vartheta}{1+\cos^2\vartheta}\left[1+\frac{\Delta\omega}{\omega'}\frac{k'^2}{q^2}\right]^2. \quad (12.2.1.4)$$

In the frequency region $\omega' \approx \omega_{pe}$ this quantity can be much greater than unity.

Let us now consider the transformation of a transverse wave into a longitudinal one in a two-temperature plasma, consisting of cold ions and hot electrons, which move with a velocity $\boldsymbol{u}$ relative to the ions (Akhiezer and Bolotin, 1963). Substituting eqn. (11.5.3.21) for the spectral density of the electron density fluctuations into formula (12.2.1.3), we get

$$d\Sigma_{t\to l} = \frac{1}{2}n_0\left(\frac{e^2}{m_e c^2}\right)^2\left(\frac{m_e c^2}{3T}\right)^{3/2}\frac{\omega'^2}{\omega^2}\sqrt{\left[\frac{\varepsilon(\omega')}{\varepsilon(\omega)}\right]}\sin^2\vartheta\,\frac{\Delta\omega^2}{|\Delta\omega-(\boldsymbol{q}\cdot\boldsymbol{u})|}$$
$$\times\,\delta(\Delta\omega^2-q^2v_s^2)\,d\omega'\,d^2\boldsymbol{\omega}'. \quad (12.2.1.5)$$

Bearing in mind that the frequency of the Langmuir oscillations in a plasma with a directed electron motion equals $\omega' = \omega_{pe}+(\boldsymbol{k}'\cdot\boldsymbol{u})$, we have when $k' \gg k$:

$$\frac{\Delta\omega^2}{|\Delta\omega-(\boldsymbol{q}\cdot\boldsymbol{u})|}\,\delta(\Delta\omega^2-q^2v_s^2) = \frac{k'v_s}{|(\boldsymbol{k}\cdot\boldsymbol{u})-(k^2c^2/2\omega_{pe})|}\left[\delta\left\{(\boldsymbol{k}'\cdot\boldsymbol{u})-k'v_s-\frac{k^2c^2}{2\omega_{pe}}+kv_s\cos\vartheta\right\}\right.$$
$$\left.+\,\delta\left\{(\boldsymbol{k}'\cdot\boldsymbol{u})+k'v_s-\frac{k^2c^2}{2\omega_{pe}}-kv_s\cos\vartheta\right\}\right]. \quad (12.2.1.6)$$

Using (12.2.1.6) to integrate (12.2.1.5) over the absolute magnitude of the vector $\boldsymbol{k}'$ we get the coefficient for the transformation of a transverse wave into a longitudinal one per unit solid angle:

$$\frac{d\Sigma_{t\to l}}{d^2\boldsymbol{\omega}'} = \frac{1}{2} n_0 \left(\frac{e^2}{m_e c^2}\right)^2 \frac{\varepsilon(\omega)}{\varepsilon(\omega')} \sin^2\vartheta \frac{\omega}{|(\boldsymbol{k}\cdot\boldsymbol{u})-(k^2c^2/2\omega_{pe})|} \{y(\theta, \theta', \varphi)+y(\theta, \pi-\theta', \varphi)\}, \tag{12.2.1.7}$$

where φ is the angle between the $(\boldsymbol{k}, \boldsymbol{u})$- and $(\boldsymbol{k}', \boldsymbol{u})$-planes, and

$$y(\theta, \theta', \varphi) = \left(\frac{1}{kv_s}\left|(\boldsymbol{k}\cdot\boldsymbol{u})-\frac{k^2c^2}{2\omega_{pe}}\right|+\theta' \sin\theta\cos\varphi\right)^3 \left(1-\frac{u}{v_s}+\frac{1}{2}\theta'^2\right)^{-4}, \tag{12.2.1.8}$$

where we have assumed that $\sin^2\theta' \ll 1$ as only in that case can the quantity $d\Sigma/d^2\boldsymbol{\omega}'$ be anomalously large. We see that as $\cos\theta \to kc^2/2\omega_{pe}u$ the coefficients in the expressions (12.2.1.5) and (12.2.1.7) grow without bounds.

12.2.2. TRANSFORMATION AND SCATTERING OF LONGITUDINAL WAVES

Let us now consider the transformation and scattering of longitudinal waves by density fluctuations in the plasma. The field of the scattered waves is determined as before by eqns. (12.1.1.13) and (12.1.1.17) in which we must take the field $\boldsymbol{E}^0$ to be parallel to $\boldsymbol{k}$. Splitting off the part of the current $\boldsymbol{I}$ which is transverse as far as $\boldsymbol{k}'$ is concerned we can find the intensity of the excited transverse waves. Dividing the intensity by the energy flux density in the incident waves,

$$S_0 = \frac{v_g}{16\pi}\frac{d}{d\omega}[\omega\varepsilon(\omega)] E^{0^2},$$

where $v_g = d\omega/dk$ is the group velocity of the longitudinal waves, we find the coefficient for the transformation of a longitudinal wave into a transverse one (Sitenko, 1967):

$$d\Sigma_{l\to t} = \frac{1}{2\pi}\left(\frac{e^2}{m_e c^2}\right)^2 \left(\frac{m_e c^2}{3T}\right)^{1/2} \frac{\omega'^2}{\omega^2} \sqrt{\left[\frac{\varepsilon(\omega')}{\varepsilon(\omega)}\right]} \left(1-\frac{\Delta\omega}{\omega}\frac{k^2}{q^2}\right)^2 \sin^2\vartheta \langle\delta n^2\rangle_{q\Delta\omega}\, d\omega'\, d^2\boldsymbol{\omega}'. \tag{12.2.2.1}$$

As the spectral density of the density fluctuations is characterized by maxima at $\Delta\omega = 0$ and $\Delta\omega = \pm\omega_{pe}$, the transverse electromagnetic waves will mainly be emitted at frequencies close to ω_{pe} and $2\omega_{pe}$.

Splitting off in (12.1.1.17) that part of the current which is longitudinal relative to $\boldsymbol{k}'$ we can determine the coefficient for the scattering of longitudinal waves by density fluctuations (Sitenko, 1967):

$$d\Sigma_{l\to l} = \frac{1}{2\pi}\left(\frac{e^2}{m_e c^2}\right)^2 \left(\frac{m_e c^2}{3T}\right)^2 \frac{\omega'^2}{\omega^2} \sqrt{\left[\frac{\varepsilon(\omega')}{\varepsilon(\omega)}\right]}$$
$$\times \left[\cos\vartheta + \frac{\Delta\omega}{\omega'}\frac{k'^2\cos\vartheta - kk'}{q^2} + \frac{\Delta\omega}{\omega}\frac{kk'-k^2\cos\vartheta}{q^2}\right]^2 \langle\delta n^2\rangle_{q\Delta\omega}\, d\omega'\, d^2\boldsymbol{\omega}', \tag{12.2.2.2}$$

where ϑ is the angle between the vectors $\boldsymbol{k}$ and $\boldsymbol{k}'$. The ratio of the coefficient of scattering of Langmuir waves to that for their transformation into transverse ones is of the order of $(m_e c^2/T)^{3/2} \cot^2 \vartheta$.

Let us consider in some detail the transformation and scattering of Langmuir waves in a two-temperature plasma consisting of cold ions and hot electrons moving relative to the ions (Akhiezer, Daneliya, and Tsintsadze, 1964). We shall here be interested in the case of critical fluctuations when the directed electron velocity u is close to the sound velocity v_s.

Substituting expression (11.5.3.21) into eqn. (12.2.2.1) we get the following expression for the coefficient for the transformation of a Langmuir wave into a transverse electromagnetic wave:

$$d\Sigma_{1\to t} = n_0 \left(\frac{e^2}{m_e c^2}\right)^2 \left(\frac{m_e c^2}{3T}\right)^{1/2} \frac{\omega'^2}{\omega^2} \sqrt{\left[\frac{\varepsilon(\omega')}{\varepsilon(\omega)}\right]} \sin^2\vartheta \frac{\Delta\omega^2}{|\Delta\omega - (\boldsymbol{q}\cdot\boldsymbol{u})|} \delta(\Delta\omega^2 - q^2 v_s^2)\, d\omega'\, d^2\boldsymbol{\omega}', \tag{12.2.2.3}$$

where $\Delta\omega = \sqrt{(\omega_{pe}^2 + k^2c^2)} - \omega_{pe} - (\boldsymbol{k}\cdot\boldsymbol{u})$ is the change in frequency. One sees easily that as $|(\boldsymbol{q}\cdot\boldsymbol{u})| \to qv_s$ the coefficient in front of the δ-function in the expression for $d\Sigma$ becomes anomalously large.

Integrating (12.2.2.3) over the absolute magnitude of the vector $\boldsymbol{k}'$ we get the intensity of the emission of transverse waves into a given element of solid angle. If $u \cong v_s$ and $\sin\theta \ll 1$, we have

$$\frac{d\Sigma_{1\to t}}{d^2\boldsymbol{\omega}'} = n_0 \left(\frac{e^2}{m_e c^2}\right)^2 \frac{m_e c^2}{3T} \frac{f(\theta, \theta')}{\varepsilon(\omega')}, \tag{12.2.2.4}$$

where

$$f(\theta, \theta') = \frac{\sin^2\theta'}{|\cos\theta'|} \left\{ \left(\sqrt{\left[1 + 2\frac{k^2c^2}{\omega_{pe} v_s \cos^2\theta'}\left(1 + \frac{u}{v_s}\cos\theta\right)\right]} - 1 \right)^{-1} + \left(1 - \sqrt{\left[1 - 2\frac{k^2c^2}{\omega_{pe} v_s \cos^2\theta'}\left(1 - \frac{u}{v_s}\cos\theta\right)\right]}\right)^{-1} \boldsymbol{\Theta}(\cos\theta') \right\}, \tag{12.2.2.5}$$

where $\boldsymbol{\Theta}(\cos\theta')$ is the step function. Here and henceforth θ and θ' are the angles which the vectors $\boldsymbol{k}$ and $\boldsymbol{k}'$ make with the direction of $\boldsymbol{u}$.

One sees easily that the quantity $d\Sigma/d^2\boldsymbol{\omega}'$ is anomalously large when the condition

$$1 \pm \frac{u}{v_s}\cos\theta \ll \frac{\omega_{pe} v_s}{kc^2}\cos^2\theta'$$

is satisfied. We note that this condition imposes very rigid limitations on the angle θ and the quantity u/v_s.

The coefficient for the scattering of Langmuir waves by anomalous sound fluctuations is according to (12.2.2.2) and (11.5.3.21) equal to

$$d\Sigma_{1\to 1'} = n_0 \left(\frac{e^2}{m_e c^2}\right)^2 \left(\frac{m_e c^2}{3T}\right)^2 \frac{\omega'^2}{\omega^2} \sqrt{\left[\frac{\varepsilon(\omega')}{\varepsilon(\omega)}\right]} \cos^2\vartheta \frac{\Delta\omega^2}{|\Delta\omega - (\boldsymbol{q}\cdot\boldsymbol{u})|} \delta(\Delta\omega^2 - q^2 v_s^2)\, d\omega'\, d^2\boldsymbol{\omega}', \tag{12.2.2.6}$$

where the change in frequency equals

$$\Delta\omega = (\boldsymbol{q}\cdot\boldsymbol{u}) - \tfrac{1}{2}(k'^2 - k^2)\frac{v_s^2}{\omega_{pe}}.$$

Integrating this relation over the absolute magnitude of the vector $\boldsymbol{k}'$ we can determine the angular distribution of the scattered Langmuir waves. Of most interest is the case of scattering over angles $\theta' \cong \pi - \theta$ and $\varphi \ll 1$, where φ is the angle between the $(\boldsymbol{k}, \boldsymbol{u})$- and $(\boldsymbol{k}', \boldsymbol{u})$-planes, when effects connected with the electron drift show up particularly strongly. In that case we have when $u \cong v_s$

$$\frac{d\Sigma_{l\to l'}}{d^2\boldsymbol{\omega}'} = 4n_0\left(\frac{e^2}{m_e c^2}\right)^2 \left(\frac{m_e c^2}{3T}\right)^{3/2} \frac{c}{v_s}\frac{\sqrt{\varepsilon(\omega)}}{\varepsilon(\omega')} \frac{\cos^2 2\theta}{|f_1 f_2 \cos^2\theta|}, \tag{12.2.2.7}$$

where

$$f_1(\theta, \theta', \varphi) = 1 - \frac{u}{v_s}\left(1 - \frac{1}{2}(\pi - \theta - \theta')\tan\theta - \frac{1}{8}\varphi^2 \tan^2\theta\right),$$

$$f_2(\theta, \theta', \varphi) = 1 - \frac{u}{v_s}\left(1 - \frac{1}{8}\varphi^2\tan^2\theta\right); \tag{12.2.2.8}$$

we have assumed here that $\theta \neq \pi/2$.

Besides the singularity as $f_2 \to 0$, which is caused by the critical fluctuations, expression (12.2.2.7) also tends to infinity as $f_1 \to 0$. This is connected with the fact that we have neglected the damping of the sound oscillations. In order to take the damping into account we must replace the δ-function in eqn. (12.2.2.6) by the expression

$$\delta(\Delta\omega^2 - q^2 v_s^2) \to \frac{\gamma}{2\pi q v_s}\{[(\Delta\omega - q v_s)^2 + \gamma^2]^{-1} + [(\Delta\omega + q v_s)^2 + \gamma^2]^{-1}\}, \tag{12.2.2.9}$$

where γ is the damping rate of the oscillations,

$$\gamma = \sqrt{\frac{\pi m_e}{8 m_i}}\,|\Delta\omega - (\boldsymbol{q}\cdot\boldsymbol{u})|.$$

As a result of such a substitution we get the following expression for the scattering coefficient, which is valid as $f_1 \to 0$:

$$\frac{d\Sigma_{l\to l'}}{d^2\boldsymbol{\omega}'} = 4n_0\left(\frac{e^2}{m_e c^2}\right)^2 \left(\frac{m_e c^2}{3T}\right)^{3/2} \frac{c}{v_s}\frac{\sqrt{\varepsilon(\omega')}}{\varepsilon(\omega)} \frac{\cos^2 2\theta}{[\pi m_e/8m_i]^{1/4}|f_2|^{3/2}\sin\theta}. \tag{12.2.2.10}$$

We see that the scattering coefficient is proportional to the large factor $(m_i/m_e)^{1/4}$ and also contains the factor $|f_2|^{-3/2}$. Close to angles for which $f_1 = 0$ the quantity $d\Sigma/d^2\boldsymbol{\omega}'$ therefore tends to infinity as $f_2 \to 0$ much faster than when $f_1 \neq 0$.

Let us finally consider the transformation of Langmuir waves into transverse ones in a plasma through which a beam is passing. Of particular interest is then the case when the beam velocity is close to the critical velocity for which Langmuir oscillations with a wavevector equal to the change in wavevector in the scattering process begin to be unstable. Using (11.5.3.13) and (12.2.2.1) we easily obtain the following expression for the coefficient of the transformation of a Langmuir wave into a transverse one:

$$\left.\begin{aligned} d\Sigma_{\mathrm{l}\to\mathrm{t}} &= \frac{1}{4}n_0\left(\frac{e^2}{m_e c^2}\right)^2\left(\frac{m_e c^2}{3T}\right)^{1/2}\frac{\omega'^2}{\omega^2}\sqrt{\left[\frac{\varepsilon(\omega')}{\varepsilon(\omega)}\right]}\sin^2\vartheta\frac{A}{|\Delta\omega-(\boldsymbol{q}\cdot\boldsymbol{u})|}\delta(\Delta\omega^2-\omega_{pq}^2)\,d\omega'\,d^2\boldsymbol{\omega}' \\ A &= \frac{1}{64}\left(3-\frac{k^2}{q^2}\right)^2 q^2 r_D^2(\omega_{pe}^2+k^2c^2)\,\varepsilon(\omega'), \quad \omega_{pq}^2 = \omega_{pe}^2\left[1+3q^2r_D^2-\frac{n_0'}{n_0}\frac{\omega_{pe}^2}{q^2v_e'^2}\right], \end{aligned}\right\} \tag{12.2.2.11}$$

where $\Delta\omega = \sqrt{(\omega_{pe}^2+k'^2c^2)}-\omega_{pe}-\frac{3}{2}(k^2v_e^2/\omega_{pe})$. As $|(\boldsymbol{q}\cdot\boldsymbol{u})| \to \omega_{pq}$ the coefficient in front of the δ-function in this expression grows without bounds.

When determining the transformation and scattering coefficients $d\Sigma/d^2\boldsymbol{\omega}'$ we assumed that the plasma was uniform and we considered only the interaction between the waves and the fluctuations. In the case of a non-uniform plasma one must consider yet another possible kind of transformation and scattering of waves—the transformation and scattering by static inhomogeneities. To illustrate what is the characteristic order of magnitude of that kind of transformation coefficients we give here the expression for the coefficient for the transformation of Langmuir waves into transverse waves due to spatial inhomogeneities (see Zheleznyakov, 1970):

$$\frac{d\Sigma_{\mathrm{l}\to\mathrm{t}}}{d^2\boldsymbol{\omega}'} \sim k\frac{T_e}{m_e c^2}\frac{\omega}{\nu}\left|\frac{3}{2}\frac{c\,\mathrm{grad}\,\varepsilon}{\omega}\right|^{3/2},$$

where ν is the collision frequency. If we compare this expression with (12.2.2.1) we see that for many actual plasmas—such as, for instance, the solar corona—the contribution from transformations by inhomogeneities to the total transformation coefficient is small. The contribution from transformations by inhomogeneities is also small in the case of a turbulent plasma or a plasma which is nearly unstable, as in those cases the transformations by turbulent or critical fluctuations are very strong.

We note that the transformation by static inhomogeneities does not affect the differential coefficients $d\Sigma/d\omega' d^2\boldsymbol{\omega}'$ with $\omega' \neq \omega$ at all, as the wave frequency does not change in the interaction with static inhomogeneities.

It is also possible to encounter cases where the transformation by static inhomogeneities is much stronger than the transformation by thermal fluctuations. Such a situation occurs, in particular, in the so-called 100% transformation (Moiseev, 1967). As an example of a 100% transformation we note the transformation of an extra-ordinary wave incident from the outside on an inhomogeneous plasma layer situated in a non-uniform magnetic field, if $|\mathrm{grad}\,B^2| > m_e c^2\,|\mathrm{grad}\,n|$, where n is the plasma density.

We shall not consider any further effects connected with the interaction between waves and static inhomogeneities.

12.2.3. SPONTANEOUS EMISSION BY A NON-EQUILIBRIUM PLASMA

To conclude this section we shall consider a peculiar effect occurring in a non-equilibrium plasma—the *spontaneous emission* connected with the transformation of two fluctuating longitudinal waves into a transverse one (Akhiezer, Daneliya, and Tsintsadze, 1964). In the case of equilibrium the whole of the radiation from the plasma reduces to Rayleigh emission, as any additional effects connected with the interaction between the waves vanish due to detailed balancing. The intensity of such a spontaneous emission turns out to be anomalously large if the plasma state is nearly unstable.

Let us determine the amplitude of the secondary wave which is produced as the result of the interaction between two waves propagating through the plasma. We shall start from the complete set of equations describing the plasma: kinetic equations for all kinds of particles and the Maxwell equations. Assuming the amplitudes of the interacting waves to be small and expanding the particle distribution functions and the electrical and magnetic fields in power series in these amplitudes we get for the nth order terms ($n = 1, 2, \ldots$):

$$\left\{\frac{\partial}{\partial t}+(\boldsymbol{v}\cdot\nabla)\right\} F_\alpha^{(n)}+\frac{e_\alpha}{m_\alpha}\left(\left\{\boldsymbol{E}^{(n)}+\frac{1}{c}[\boldsymbol{v}\wedge\boldsymbol{B}^{(n)}]\right\}\cdot\frac{\partial}{\partial\boldsymbol{v}}\right)F_\alpha^{(n)}$$
$$=-\frac{e_\alpha}{m_\alpha}\sum_{n'=1}^{n-1}\left(\left\{\boldsymbol{E}^{(n')}+\frac{1}{c}[\boldsymbol{v}\wedge\boldsymbol{B}^{(n')}]\right\}\cdot\frac{\partial}{\partial\boldsymbol{v}}\right)F_\alpha^{(n-n')}, \tag{12.2.3.1}$$

$$\operatorname{curl}\boldsymbol{E}^{(n)}=-\frac{1}{c}\frac{\partial\boldsymbol{B}^{(n)}}{\partial t},\quad \operatorname{curl}\boldsymbol{B}^{(n)}=\frac{1}{c}\frac{\partial\boldsymbol{E}^{(n)}}{\partial t}+\frac{4\pi}{c}\sum_\alpha e_\alpha\int\boldsymbol{v}F_\alpha^{(n)}\,d^3\boldsymbol{v}. \tag{12.2.3.2}$$

The non-linear wave–wave interaction effect is clearly described by the right-hand sides of the kinetic eqns. (12.2.3.1).

Solving eqns. (12.2.3.1) and (12.2.3.2) and restricting ourselves to the case when both primary waves are longitudinal we get the following expression for the field of the secondary wave:

$$E_i^{(2)}(\boldsymbol{k}',\omega')=i\sum_j\Lambda_{ij}^{-1}(\boldsymbol{k}',\omega')\int C_j(\omega,\boldsymbol{k}:\omega'-\omega,\boldsymbol{k}'-\boldsymbol{k})\,\varphi^{(1)}(\boldsymbol{k},\omega)\,\varphi^{(1)}(\boldsymbol{k}'-\boldsymbol{k},\omega'-\omega)\,d^3\boldsymbol{k}\,d\omega, \tag{12.2.3.3}$$

where $\varphi^{(1)}(\boldsymbol{k},\omega)=i(\{\boldsymbol{k}/k^2\}\cdot\boldsymbol{E}^{(1)}(\boldsymbol{k},\omega))$ is the potential of the primary wave, and

$$\left.\begin{aligned}\Lambda_{ij}^{-1}(\boldsymbol{k},\omega)&=\frac{k_ik_j}{k^2}\,\varepsilon_l^{-1}(\boldsymbol{k},\omega)+\left(\delta_{ij}-\frac{k_ik_j}{k^2}\right)\left[\frac{k^2c^2}{\omega^2}-\varepsilon_t(\boldsymbol{k},\omega)\right]^{-1},\\ \boldsymbol{C}(\omega_1,\boldsymbol{k}_1,\omega_2,\boldsymbol{k}_2)&=\sum_\alpha\frac{4\pi e_\alpha^3}{m_\alpha^2(\omega_1+\omega_2)}\int\frac{\boldsymbol{v}}{\omega_1+\omega_2-(\{\boldsymbol{k}_1+\boldsymbol{k}_2\}\cdot\boldsymbol{v})}\left(\boldsymbol{k}_1\cdot\frac{\partial}{\partial\boldsymbol{v}_1}\right)\\ &\quad\times\frac{1}{\omega_2-(\boldsymbol{k}_2\cdot\boldsymbol{v})}\left(\boldsymbol{k}_2\cdot\frac{\partial}{\partial\boldsymbol{v}}\right)F_\alpha^{(0)}\,d^3\boldsymbol{v}.\end{aligned}\right\} \tag{12.2.3.4}$$

When due to the interaction of the two longitudinal waves a transverse wave is formed, the intensity of the radiation will be characterized by a change I in the energy of the secondary wave per unit time,

$$I=\frac{V}{(2\pi)^3}\int Q(\boldsymbol{k}')\,d^3\boldsymbol{k}', \tag{12.2.3.5}$$

where V is the volume of the system. If both colliding waves are fluctuation waves, we have from (12.2.3.3) and (12.2.3.4)

$$Q(\boldsymbol{k}')=\frac{1}{(4\pi)^5}\frac{\omega'^2}{k'^2}\int|(\boldsymbol{k}'\cdot\boldsymbol{C}(\omega,\boldsymbol{k};\omega'-\omega,\boldsymbol{k}'-\boldsymbol{k}))|^2\langle\varphi^2\rangle_{\boldsymbol{k}\omega}\langle\varphi^2\rangle_{\boldsymbol{k}'-\boldsymbol{k}\,\omega'-\omega}\,d^3\boldsymbol{k}\,d\omega, \tag{12.2.3.6}$$

where $\langle\varphi^2\rangle_{k\omega}$ is the spectral density of the correlation function of the potential. In deriving this formula we used the fact that if we neglect higher-order correlations we can write the fourth correlation function in the form

$$\langle\varphi_{k_1\omega_1}\varphi_{k_2\omega_2}\overset{*}{\varphi}_{k_3\omega_3}\overset{**}{\varphi}_{k_4\omega_4}\rangle = (2\pi)^3\langle\varphi^2\rangle_{k_1\omega_1}\langle\varphi^2\rangle_{k_2\omega_2}\{\delta(\boldsymbol{k}_1-\boldsymbol{k}_3)\,\delta(\boldsymbol{k}_2-\boldsymbol{k}_4)\,\delta(\omega_1-\omega_3)\,\delta(\omega_2-\omega_4) + \delta(\boldsymbol{k}_1-\boldsymbol{k}_4)\,\delta(\boldsymbol{k}_2-\boldsymbol{k}_3)\,\delta(\omega_1-\omega_4)\,\delta(\omega_2-\omega_3)\}, \tag{12.2.3.7}$$

Let us consider a two-temperature plasma in which the electrons move relative to the ions and let us determine in that case the increase in the energy of the transverse waves as a result of the scattering of fluctuation Langmuir oscillations by low-frequency sound fluctuations in the plasma. We note that if one of the colliding waves is a Langmuir wave, $\omega_1 = \omega_{pe}$, while the second one is a low-frequency one, $\omega_2 \ll \omega_{pe}$, the function $\boldsymbol{C}$ occurring in (12.2.3.6) becomes simple:

$$\boldsymbol{C}(\omega_1, \boldsymbol{k}_1; \omega_2, \boldsymbol{k}_2) \cong \frac{e}{T_e}\boldsymbol{k}_1. \tag{12.2.3.8}$$

Bearing in mind that the term in the expression for the correlation function of the potential which describes the Langmuir oscillations has the form

$$\langle\varphi^2\rangle_{q\omega} = 8\pi^2\frac{T\omega_{pe}}{q^2}\,\delta(\tilde{\omega}^2-\omega_{pe}^2-3q^2v_e^2), \tag{12.2.3.9}$$

where $\tilde{\omega} = \omega-(\boldsymbol{q}\cdot\boldsymbol{u})$ is the frequency in the rest frame of the electrons, and using eqns. (12.2.3.6), (12.2.3.8), and (11.5.3.17), we find

$$Q(\boldsymbol{k}') = \frac{1}{8\pi}\frac{e^2}{m_e}T\omega_{pe}\int\frac{[\boldsymbol{k}\wedge\boldsymbol{k}']^2}{k^2k'^2}\,\frac{q^2v_s^2}{|\Delta\omega-(\boldsymbol{q}\cdot\boldsymbol{u})|}\,\delta(\Delta\omega^2-q^2v_s^2)\,\delta\{[\omega-(\boldsymbol{k}\cdot\boldsymbol{u})]^2-\omega_{pe}^2\}\,d\omega\,d^3\boldsymbol{k}, \tag{12.2.3.10}$$

where $\Delta\omega = \sqrt{(\omega_{pe}+k^2c^2)}-\omega$ and $\boldsymbol{q} = \boldsymbol{k}'-\boldsymbol{k}$. One sees easily that the function $Q(\boldsymbol{k}')$ tends to infinity when $u = v_s$, if $\cos\theta' \to k'c^2/2\omega_{pe}u$. We can estimate the coefficient in front of the resonance denominator $\{(k'^2c^2/2\omega_{pe})-(\boldsymbol{k}'\cdot\boldsymbol{u})\}^{-1}$ by integrating (12.2.3.10) over k up to $k \sim 1/a_1$ where a_1 is a quantity of the order of a few Debye radii:

$$Q(\boldsymbol{k}') \cong \frac{e^2T}{m_e a_1^3}\,\frac{\sin^2\theta'}{(k'^2c^2/2\omega_{pe})-(\boldsymbol{k}'\cdot\boldsymbol{u})}\,. \tag{12.2.3.11}$$

We note that the radiation from the plasma caused by the transformation of a fluctuating Langmuir wave into a transverse wave is, according to formulae (12.2.3.10) and (12.2.3.11) anomalously large only in the long wavelength region where $k' \sim \omega_{pe}u/c^2$.

12.3. Incoherent Reflection of Electromagnetic Waves from a Plasma

12.3.1. REFLECTION COEFFICIENT

An electromagnetic wave falling upon a bounded plasma undergoes not only the normal reflection described by the Fresnel formulae, but also incoherent reflection where the frequency of the reflected wave is not equal to that of the incident wave and the angle of reflection is different from the angle of incidence (Akhiezer, 1964a). The incoherent reflection is caused by the interaction of the electromagnetic wave with the fluctuations in the plasma and can be used for an experimental determination of the spectral distribution of the fluctuations and to find in that way the behaviour of various plasma parameters.

If we want to study the reflection of electromagnetic waves from a plasma, which we shall assume to fill the half-space $z > 0$, we must find the electrical field $\boldsymbol{E}$ which satisfies the Maxwell equations and the following boundary conditions: the field $\boldsymbol{E}$ must be a superposition of the incident wave $\boldsymbol{E}_0 e^{i(\boldsymbol{k}\cdot\boldsymbol{r})-i\omega t}(k_z > 0)$ and reflected waves, as $z \to -\infty$; as $z \to +\infty$ the field must vanish.

We shall assume that the components of the wavevectors of the incident and of the reflected wave normal to the boundary satisfy the inequalities $k_z r_D, k_z' r_D \ll 1$, where r_D is the electron Debye radius. We can then neglect the structure of the boundary layer with a depth of the order of r_D and consider the boundary to be a discontinuity plane on which the usual boundary conditions of macroscopic electrodynamics are satisfied, that is, the quantities $\boldsymbol{E}_t$, curl $\boldsymbol{E}_t$, and $(\partial E_z/\partial t)+4\pi j_z$, with $\boldsymbol{E}_t$ the component of the vector $\boldsymbol{E}$ which is parallel to the boundary, must be continuous.

If the phase velocities of the incident and reflected waves are large compared to the thermal velocities of the particles in the plasma we can use the expression $\boldsymbol{j} = en\boldsymbol{v}$ for the current, where n is the electron density and $\boldsymbol{v}$ their hydrodynamic velocity which is connected with the fields $\boldsymbol{E}$ and $\boldsymbol{B}$ through the equation

$$\left[\frac{d}{dt}+\frac{1}{\tau}\right]\boldsymbol{v} = -\frac{e}{m_e}\left(\boldsymbol{E}+\frac{1}{c}[\boldsymbol{v}\wedge\boldsymbol{B}]\right), \tag{12.3.1.1}$$

where τ is the average time between collisions which we shall, whenever possible, assume to be infinite. Putting $n = n_0+\delta n$, where n_0 is the average value and δn the fluctuation in the density, we can look for the field $\boldsymbol{E}$ determined by the Maxwell equations and eqn. (12.3.1.1) in the form of a power series in the density fluctuation:

$$\boldsymbol{E} = \boldsymbol{E}^{\circ}+\boldsymbol{E}'+\ldots$$

Once we have determined the field $\boldsymbol{E}$ for $z < 0$, we can find the component of the Poynting vector which is normal to the boundary and hence find the reflected energy flux by averaging this component over the plasma fluctuations. Dividing the energy flux, averaged over the fluctuations, dS of the reflected waves, which have wavevectors in the range $\boldsymbol{k}'$ to $\boldsymbol{k}'+d\boldsymbol{k}'$, by the incident energy flux S_0 we get the differential reflection coefficient dR:

$$dR = \frac{dS}{S_0}. \tag{12.3.1.2}$$

If there were no fluctuations in the plasma, the field $\boldsymbol{E}'$ would vanish and one would only have the normal reflection of electromagnetic waves from the surface in which $\boldsymbol{k}'$ is uniquely connected with $\boldsymbol{k}$ through the relations $\boldsymbol{k}_t' = \boldsymbol{k}_t$, $k_z' = -k_z$. Therefore, dR contains a term proportional to $\delta(\boldsymbol{k}_t'-\boldsymbol{k}_t)\,\delta(k_z'+k_z)$. Integrating that term over $\boldsymbol{k}'$ we get the usual reflection coefficient R_0 given by the Fresnel formulae. In particular, for unpolarized incident radiation

R_0 has the form

$$R_0 = \left|\frac{k_z - g}{k_z + g}\right|^2 + \left|\frac{\varepsilon k_z - g}{\varepsilon k_z + g}\right|^2, \tag{12.3.1.3}$$

where ε is the dielectric constant of the plasma,

$$\varepsilon = 1 - \frac{\omega_{pe}^2}{\omega^2}\left(1 - \frac{i}{\omega\tau}\right),$$

and $g = \sqrt{(\varepsilon k^2 - k_t^2)}$, $\operatorname{Im} g > 0$.

However, due to fluctuations in the plasma $\boldsymbol{E}' \neq 0$ and dR contains, apart from the δ-functionlike term corresponding to the normal reflection of the radiation from the surface an additional term which one can show to have the following form in the case of unpolarized incident radiation:

$$dR = \frac{l}{4\pi c}\left(\frac{e^2}{m_e\omega}\right)^2 G(\theta, \theta', \varphi)\, \phi(\boldsymbol{\Delta k}, \Delta\omega; l)\, d^3\boldsymbol{k}'. \tag{12.3.1.4}$$

Here ϕ is the Fourier–Laplace transform of the electron density correlation function of the plasma,

$$\phi(\boldsymbol{\Delta k}, \Delta\omega; l) = \int_0^\infty \frac{dZ}{l} e^{-Z/l} \int d^2\boldsymbol{r}_t\, dt \int_{-2Z}^{2Z} dz\, e^{-i(\boldsymbol{\Delta k}\cdot\boldsymbol{r}) + i\Delta\omega t} \langle \delta n(\boldsymbol{r}_1, t_1)\, \delta n(\boldsymbol{r}_2, t_2)\rangle, \tag{12.3.1.5}$$

$\boldsymbol{r} = \boldsymbol{r}_2 - \boldsymbol{r}_1$, $Z = \frac{1}{2}(z_1 + z_2)$, $\boldsymbol{k}'$ and ω' are the wavevector and frequency of the reflected wave,

$$\boldsymbol{\Delta k}_t = \boldsymbol{k}_t' - \boldsymbol{k}_t, \quad \Delta k_z = -\operatorname{Re}(g + g'), \quad \Delta\omega = \omega' - \omega, \quad l^{-1} = 2\operatorname{Im}(g + g'),$$
$$g' = \sqrt{(\varepsilon(\omega')\, k'^2 - k_t'^2)},$$

and the brackets $\langle \ldots \rangle$ indicate averaging over the fluctuations.

The function G which depends on the angles of incidence θ and reflection θ', as well as on the angle φ between the vectors $\boldsymbol{k}_t$ and $\boldsymbol{k}_t'$ has the form

$$G(\theta, \theta', \varphi) = |b_1|^2 \frac{\cos^2\theta'}{\cos\theta}\left\{|\tilde{c}_1|^2 \cos^2\theta \cos^2\varphi + |\tilde{c}_2|^2 \sin^2\varphi + \left|\frac{\tilde{c}_3}{(\varepsilon' - \sin^2\theta')^{1/2}}\right|^2 \sin^2\theta \sin^2\theta' \right.$$
$$\left. - \operatorname{Re}\frac{\tilde{c}_1^*\tilde{c}_3}{(\varepsilon' - \sin^2\theta')^{1/2}} \sin 2\theta \sin\theta' \cos\varphi\right\} + \frac{|b_2|^2}{\cos\theta}\{|c_1|^2 \cos^2\theta \sin^2\varphi + |c_2|^2 \cos^2\varphi\}, \tag{12.3.1.6}$$

where

$$\left.\begin{aligned} &\varepsilon' = \varepsilon(\omega'); \quad b_1 = \frac{2g'}{g' - \varepsilon' k_z'}, \quad b_2 = \frac{2|k_z'|}{g' - k_z'}; \quad c_1 = \frac{2g}{g + \varepsilon k_z}, \quad c_2 = \frac{2k_z}{g + k_z}, \\ &c_3 = \frac{2k_z}{g + \varepsilon k_z}; \quad \tilde{c}_{1,2} = \left[1 + 2\frac{\Delta\omega}{\omega'}\frac{k_t'^2}{\Delta k^2}\right] c_{1,2}, \quad \tilde{c}_3 = \left[1 + 2\frac{\Delta\omega}{\omega'}\frac{g'^2}{\Delta k^2}\right] c_3. \end{aligned}\right\} \tag{12.3.1.7}$$

12.3.2. SPECTRAL DISTRIBUTION OF THE REFLECTED EMISSION

The differential reflection coefficient is determined by the electron density correlation function of the plasma. We must take the boundary into account when determining this correlation function. However, in the case of interest to us when $k_z r_D, k_z' r_D \ll 1$ when the long-wavelength fluctuations are important, we can neglect the influence of the boundary and use eqn. (11.2.6.3) for the spectral density of the electron density fluctuations in an unbounded plasma. The function ϕ which occurs in (12.3.1.4) then takes the form

$$\phi(\boldsymbol{\Delta k}, \Delta\omega; l) = \frac{1}{\pi} \int_{-\infty}^{+\infty} \frac{(2l)^{-1} \langle \delta n^2 \rangle_{q\Delta\omega}}{(2l)^{-1} + (q_z - \Delta k_z)^2} \, dq_z, \tag{12.3.2.1}$$

where $\boldsymbol{q}_t = \boldsymbol{\Delta k}_t$. If $l^{-1} \ll \Delta k_z$, the reflection coefficient is given by the formula

$$dR = \frac{1}{4\pi} \left(\frac{e^2}{m_e \omega} \right)^2 \frac{l}{c} G(\theta, \theta', \varphi) \langle \delta n^2 \rangle_{q\Delta\omega} \, d^3 \boldsymbol{k}', \tag{12.3.2.2}$$

where $\boldsymbol{q} = \boldsymbol{\Delta k}$. The spectrum of the reflected radiation—like the spectrum of the scattered radiation; see Section 12.1—consists in this case of a Doppler-broadened main line and sharp maxima at $\Delta\omega = \pm\omega_{pe}$. In the case of a strongly non-isothermal plasma there appear in the spectrum of the reflected radiation additional sharp maxima connected with the existence of non-isothermal sound in the plasma. The differential reflection coefficient is a smooth function of $\Delta\omega$, if the condition $l^{-1} \ll \Delta k_z$ is not satisfied.

We note that when eqn. (12.3.2.2) is valid, that is, when there are sharp maxima in the spectral distribution of the reflected radiation, there is no need whatever for the incident wave to be weakly damped over a wavelength when going into the plasma. In particular, if we have a wave which is normally incident upon the plasma with a frequency $\omega = \omega_{pe} - \xi$ ($\xi \ll \omega_{pe}$), which has a damping rate $\text{Im } g = \sqrt{(2\omega_{pe}\xi)}/c$, the reflection coefficient for the normally reflected wave has in a non-isothermal plasma sharp maxima at $\Delta\omega = \pm q v_s$, where v_s is the non-isothermal sound velocity, near which

$$dR = \frac{e^2}{32\pi m_e \sqrt{(2\omega_{pe}\xi)}} G(\theta, \theta', \varphi) \{\delta(\Delta\omega - q v_s) + \delta(\Delta\omega + q v_s)\} \, d^3 \boldsymbol{k}'. \tag{12.3.2.3}$$

If the conditions

$$\cos^2 \theta \gg |1 - \varepsilon|, \qquad \cos^2 \theta' \gg |1 - \varepsilon'|, \qquad |\Delta\omega| \ll \omega,$$

are satisfied, we have

$$G(\theta, \theta', \varphi) = \frac{1 + \cos^2 \vartheta}{\cos \theta}, \tag{12.3.2.4}$$

where ϑ is the angle between the vectors $\boldsymbol{k}$ and $\boldsymbol{k}'$, and the reflection coefficient differs merely by a normalization coefficient from the scattering coefficient in an unbounded plasma—for angles larger than $\pi/2$,

$$dR = \frac{l}{\cos \theta} d\Sigma. \tag{12.3.2.5}$$

If even only one of the above conditions is not satisfied, the reflection coefficient will, according to (12.3.1.4) and (12.3.1.6), differ appreciably from the scattering coefficient. The difference between these quantities is caused by the following two circumstances. Firstly, the transverse wave which appears in the plasma through scattering by fluctuations excites not only transverse but also longitudinal waves. At the boundary the longitudinal waves are transformed into transverse waves and they contribute considerably to the reflection coefficient when $\Delta\omega \sim \omega$. Secondly, the incident wave penetrating into the plasma and the scattered wave coming out of the plasma are refracted and it is necessary to take this into account if $\cos^2\theta \lesssim |1-\varepsilon|$, or $\cos^2\theta' \lesssim |1-\varepsilon'|$, that is, if $k^2 r_D^2 \cos^2\theta \lesssim v_e^2/c^2$ or $k'^2 r_D^2 \times \cos^2\theta' < v_e^2/c^2$, where v_e is the electron thermal velocity in the plasma. At frequencies $\omega \lesssim \omega_{pe}$ these conditions are satisfied for all angles of incidence and reflection, so that in that region of frequencies the reflection coefficient depends strongly on the angles θ and θ', and not only on the angle ϑ between the wavevectors of the incident and the reflected waves. In particular, for normal incidence ($\theta = 0$) and for normal reflection ($\theta' = 0$) we have

$$\left.\begin{aligned} dR &= \frac{1}{\pi}\left(\frac{e^2}{m_e\omega}\right)^2 \frac{l}{c}|1+\sqrt{\varepsilon}|^{-2}\left\{\left|b_1\frac{\tilde{c}_1}{c_1}\right|^2\cos^2\theta' + |b_2|^2\right\}\langle\delta n^2\rangle_{q\Delta\omega}\, d^3\mathbf{k}',\ \theta = 0; \\ dR &= \frac{1}{\pi}\left(\frac{e^2}{m_e\omega}\right)^2 \frac{l}{c}|1+\sqrt{\varepsilon}|^{-2}\frac{|c_1|^2\cos^2\theta + |c_2|^2}{\cos\theta}\langle\delta n^2\rangle_{q\Delta\omega}\, d^3\mathbf{k}', \quad \theta' = 0. \end{aligned}\right\} \qquad (12.3.2.6)$$

In the case of glancing incidence or reflection of waves it is necessary to take the boundary into account for all frequencies. In particular, if $\omega, \omega' \gg \omega_{pe}$, $\theta \approx \pi/2 - \omega_{pe}/\omega$ and $\pi/2 - \theta' \gg \omega_{pe}/\omega'$, we have

$$dR = \frac{1}{\pi}\left(\frac{e^2}{m_e\omega}\right)^2 \frac{l}{c}\,\frac{1+\cos^2\vartheta}{\cos\theta}\langle\delta n^2\rangle_{q\Delta\omega}\, d^3\mathbf{k}'. \qquad (12.3.2.7)$$

We see that in that case the reflection coefficient is four times larger than the coefficient dR calculated using eqn. (12.3.2.5) and neglecting the plasma boundary.

We have already indicated that eqns. (12.3.2.5) to (12.3.2.7) determine the reflection coefficients, averaged over the polarization, for the case of unpolarized incident radiation. Without going into a detailed discussion of the polarization effects when radiation is reflected from the boundary of a plasma, we shall merely mention the "total polarization" phenomenon which occurs in this case: a wave which is reflected at a well-defined angle is polarized in the plane at right angles to the plane of reflection. In particular, if the plane of reflection is the same as the plane of incidence ($\varphi = 0$), the angle of reflection at which complete polarization of the reflected wave occurs is given by the equation (we assume that $\Delta\omega \ll \omega$)

$$\theta' = \psi(\theta), \quad \sin^2\psi = \varepsilon' - \frac{\varepsilon'}{\varepsilon}\sin^2\theta. \qquad (12.3.2.8)$$

If the incident wave is polarized in the plane of incidence, a wave which is reflected at an angle $\theta' = \psi(\theta, \varphi)$, where

$$\sin^2\psi = \varepsilon'\,\frac{\varepsilon - \sin^2\theta}{\sin^2\theta\tan^2\varphi + \varepsilon}, \qquad (12.3.2.9)$$

will be totally polarized.

12.4. Scattering and Transformation of Waves in a Plasma in a Magnetic Field

12.4.1. FIELD OF THE SCATTERED WAVES; SCATTERING AND TRANSFORMATION CROSS-SECTIONS

Let us now consider the effect of a magnetic field on the scattering of waves in a plasma (Akhiezer, Akhiezer, and Sitenko, 1962; Farley, Dougherty, and Barron, 1961; Sitenko and Kirochkin, 1963, 1964, 1966). We showed in Chapter 5 that a magnetic field leads to a splitting of the plasma oscillation frequency and to the apperance of new kinds of eigenfrequencies in the low-frequency region (Alfvén and magneto-sound waves). There occur therefore in the scattered radiation in a plasma in a magnetic field additional maxima connected with the scattering by Alfvén and magneto-sound oscillations. The interaction of waves propagating through the plasma with fluctuation oscillations can also lead to a transformation of the waves. The intensity of Raman scattering and of wave transformation is determined by the amplitude of the fluctuations. Under non-equilibrium conditions these intensities can grow anomalously, if the plasma is close to the region of kinetic instability.

To describe the scattering and transformation processes in a plasma in a magnetic field we shall, as before, use the kinetic eqns. (12.1.1.7) and the Maxwell equations with the current (12.1.1.4); we shall take for $\boldsymbol{B}$ the total magnetic field in the plasma, including the constant external field $\boldsymbol{B}_0$.

The field of the incident wave is described by the equation

$$\text{curl curl}\, \boldsymbol{E} + \frac{\hat{\varepsilon}}{c^2}\frac{\partial^2 \boldsymbol{E}}{\partial t^2} = 0, \tag{12.4.1.1}$$

where $\hat{\varepsilon}$ is the dielectric permittivity tensor of the plasma in a magnetic field. Taking into account the non-linear terms in the kinetic eqn. (12.1.1.7) we find for the field of the scattered waves the equation

$$\text{curl curl}\, \boldsymbol{E}' + \frac{\hat{\varepsilon}}{c^2}\frac{\partial^2 \boldsymbol{E}}{\partial t^2} = -\frac{4\pi}{c^2}\frac{\partial \boldsymbol{I}}{\partial t}, \tag{12.4.1.2}$$

where $\boldsymbol{I}$ is the current caused by the field of the incident wave and by the fluctuations in the plasma. If the incident wave is a plane monochromatic wave,

$$\boldsymbol{E}^0(\boldsymbol{r}, t) = \boldsymbol{E}^0 e^{i(\boldsymbol{k}\cdot\boldsymbol{r}) - i\omega t},$$

we have for the Fourier components of the current $\boldsymbol{I}$ ($\boldsymbol{q} = \boldsymbol{k}' - \boldsymbol{k}$, $\Delta\omega = \omega' - \omega$)

$$\begin{aligned}
\boldsymbol{I}_{k'\omega'} = -\sum_\alpha \frac{e_\alpha^2}{m_\alpha \omega_{B\alpha}} \int d^3v \Bigg\{ \Bigg[\left(\left\{ \boldsymbol{E}^0 + \frac{1}{c}[\boldsymbol{v} \wedge \boldsymbol{B}^0] \right\} \cdot \frac{\partial}{\partial \boldsymbol{v}} \right) \delta f_\alpha(\boldsymbol{q}, \Delta\omega) \\
+ \left(\left\{ \delta \boldsymbol{E}_{q\Delta\omega} + \frac{1}{c}[\boldsymbol{v} \wedge \delta \boldsymbol{B}_{q\Delta\omega}] \right\} \cdot \frac{\partial f_\alpha^0}{\partial \boldsymbol{v}} \right) \Bigg] \exp\left[-\frac{i}{\omega_{B\alpha}} \int_0^\varphi \{(\boldsymbol{k}'\cdot\boldsymbol{v}) - \omega'\}\, d\varphi \right] \\
\times \int^\varphi \boldsymbol{v}\, d\varphi \exp\left[\frac{i}{\omega_{B\alpha}} \int_0^\varphi \{(\boldsymbol{k}'\cdot\boldsymbol{v}) - \omega'\}\, d\varphi \right] \Bigg\}.
\end{aligned} \tag{12.4.1.3}$$

Bearing in mind the large difference between the electron and ion masses and the presence of a fast oscillating factor in (12.4.1.3), we can restrict ourselves to taking only the electron terms into account in (12.4.1.3). Equation (12.4.1.2) describes all scattering and transformation processes for waves in a magneto-active plasma.

Let us determine the average increase I in the energy of the field of the scattered waves, per unit time. To do this we note that the total energy transferred to the plasma by the current $\boldsymbol{I}$ is equal to

$$P = -\frac{1}{2}\,\mathrm{Re}\int (\boldsymbol{I}^*(\boldsymbol{r}, t)\cdot\boldsymbol{E}(\boldsymbol{r}, t))\, d^3\boldsymbol{r}\, dt = -\frac{1}{2}\,\mathrm{Re}\int (\boldsymbol{I}^*_{k\omega}\cdot\boldsymbol{E}_{k\omega})\,\frac{d^3\boldsymbol{k}\, d\omega}{(2\pi)^4}. \tag{12.4.1.4}$$

The electrical field strength of the scattered waves is, according to (12.4.1.2), connected with the current, which is excited, through the relation

$$\boldsymbol{E}_{k\omega} = -\frac{4\pi i}{\omega}\,\frac{\hat{\lambda}\boldsymbol{I}_{k\omega}}{\Lambda(\boldsymbol{k}, \omega)}, \tag{12.4.1.5}$$

where $\sum_j \lambda_{ij}\Lambda_{jk} = \Lambda\delta_{ik}$, $\Lambda \equiv \mathrm{Det}(\Lambda)_{ij}$. We also remind ourselves that

$$\lambda_{ij}(\boldsymbol{k}, \omega) = e_i e_j^*\,\mathrm{Tr}\,\lambda(\boldsymbol{k}, \omega), \tag{12.4.1.6}$$

where $\boldsymbol{e}$ is the polarization vector of the $\boldsymbol{k}$, ω-oscillation.

We shall substitute (12.4.1.5) into (12.4.1.4) and perform a statistical average. Noting that

$$\langle I_i^*(\boldsymbol{k}, \omega)\, I_j(\boldsymbol{k}, \omega)\rangle = TV\langle I_i I_j\rangle_{k\omega}, \tag{12.4.1.7}$$

where V is the volume of the plasma and T the interaction time, we can write the average radiation intensity in the form

$$I \equiv \frac{\langle P\rangle}{T} = \frac{V}{8\pi^2}\int \langle|(\boldsymbol{e}\cdot\boldsymbol{I}^*)|^2\rangle_{k\omega}\,\frac{|\mathrm{Tr}\,\lambda|}{|\omega|}\,\delta\{\Lambda(\boldsymbol{k}, \omega)\}\, d^3\boldsymbol{k}\, d\omega. \tag{12.4.1.8}$$

Dividing the intensity of the radiation I by the energy flux density S_0 in the propagation direction of the incident wave and by the magnitude of the scattering volume V, we find the coefficient for the scattering or the transformation of the waves:

$$\Sigma = \frac{I}{VS_0}. \tag{12.4.1.9}$$

We note that we can also define the quantity Σ in a different way, taking for S_0 in (12.4.1.9) the total energy flux. The results will then only differ by a normalization factor.

If as a result of a scattering or a transformation process low-frequency waves are formed, the radiation intensity is equal to

$$I = \frac{V}{8\pi^2}\int \frac{1}{|\omega'|}\langle|(\boldsymbol{e}\cdot\boldsymbol{I}^*)|^2\rangle_{k'\omega'}\,\frac{1}{|\boldsymbol{e}|^2-(\boldsymbol{k}'\cdot\boldsymbol{e})|^2/k'^2}\,[\delta(n^2-n_+^2)+\delta(n^2-n_-^2)]+\delta(A)\Big\}\, d^3\boldsymbol{k}'\, d\omega', \tag{12.4.1.10}$$

where $A = \sum_{i,j} \varepsilon_{ij}k_i' k_j'/k'^2$. The first term in the braces determines the increase in the energy of the high-frequency electromagnetic waves in the plasma and the second term the growth in energy of the scattered Langmuir waves. It follows from the energy and momentum conservation laws,

$$\omega' = \omega + \Delta\omega, \qquad \boldsymbol{k}' = \boldsymbol{k} + \boldsymbol{q}, \tag{12.4.1.11}$$

that the high-frequency waves can be excited by both high-frequency and low-frequency incident waves. In the first case the fluctuations can be either low-frequency or high-frequency ones. If, however, the incident wave is a low-frequency one, it can be transformed into a high-frequency wave only when interacting with high-frequency fluctuations.

As the phase velocities in the case of high-frequency waves are much larger than the electron thermal velocity in the plasma, we can simplify the expression for the current $\boldsymbol{I}_{k',\,\omega'}$, expanding the integrand in (12.4.1.3) in a power series in $(\boldsymbol{k}'\cdot\boldsymbol{v})/\omega'$:

$$I_i(\boldsymbol{k}', \omega') = i\omega' \sum_{j,k} \varkappa'_{ij} \left\{ \delta_{jk} \frac{\delta n(\boldsymbol{q}, \Delta\omega)}{n_0} + \frac{1}{\omega} \sum_l \left[k_j\,\delta_{kl} - \delta_{jk} k_l - 4\pi \frac{\omega^2}{\omega_{pe}^2} \left(\varkappa_{jk} k'_l + \delta_{jl} \sum_m k'_m \varkappa_{mk} \right) \right] \right.$$
$$\left. \times\, \delta v_l(\boldsymbol{q}, \omega) - \frac{i}{e n_0} \sum_l \left[k_l \varkappa_{lk} \delta E_j(\boldsymbol{q}, \Delta\omega) + \frac{\omega}{c} \sum_m \varepsilon_{jlm} \varkappa_{lk}\, \delta B_m(\boldsymbol{q}, \Delta\omega) \right] \right\} E^0_k, \tag{12.4.1.12}$$

where δn and $\delta\boldsymbol{v}$ are the fluctuations in the electron density and the electron macroscopic velocity, and $\delta\boldsymbol{E}$ and $\delta\boldsymbol{B}$ the fluctuations in the electrical and magnetic fields in the plasma. This expression follows also directly from a hydrodynamical consideration.

We can use (12.4.1.12) to study the following processes of scattering and transformation of high-frequency waves in a magneto-active plasma: scattering of (ordinary and extraordinary) electromagnetic waves, transformation of electromagnetic waves into Langmuir ones, scattering of Langmuir waves, and transformation of Langmuir into electromagnetic waves (Sitenko and Kirochkin, 1966). We can also use (12.4.1.12) to study the transformation of low-frequency into high-frequency waves (Sitenko and Kirochkin, 1966).

12.4.2. SCATTERING AND TRANSFORMATION OF ELECTROMAGNETIC WAVES BY INCOHERENT FLUCTUATIONS

Scattering of electromagnetic waves with a small change in frequency ($\Delta\omega \ll \omega$) occurs in a plasma in a magnetic field, as in an unmagnetized plasma, mainly by electron density fluctuations. Neglecting scattering caused by fluctuations in the electron velocity and also by field fluctuations, we can write the differential scattering cross-section in the form (Akhiezer, Akhiezer, and Sitenko, 1962; Sitenko and Kirochkin, 1963)

$$d\Sigma_{\mathrm{E}\to\mathrm{E}} = \frac{1}{2\pi} \left(\frac{e^2}{m_e c^2} \right)^2 \frac{\omega^2 \omega'^2}{\omega_{pe}^4} \mathcal{N} \langle \delta n^2 \rangle_{q\Delta\omega}\, d\omega'\, d^2\omega', \tag{12.4.2.1}$$

where

$$\left.\begin{aligned} \mathcal{N} &= n'^3 |\boldsymbol{e}'^*(\hat{\varepsilon} - \hat{1})\,\boldsymbol{e}|^2 / n[|\boldsymbol{e}|^2 - \{|(\boldsymbol{k}\cdot\boldsymbol{e})|^2/k^2\}]\,(\boldsymbol{e}'^*\hat{\varepsilon}'\boldsymbol{e}'), \\ \boldsymbol{e} &= \left(1,\ i\frac{\varepsilon_2}{n^2 - \varepsilon_1},\ \frac{n^2 \sin\vartheta \cos\vartheta}{n^2 \sin^2\vartheta - \varepsilon_3} \right), \\ \boldsymbol{e}' &= \left(\cos\varphi - i\frac{\varepsilon'_2}{n'^2 - \varepsilon'_1} \sin\varphi,\ \ \sin\varphi + i\frac{\varepsilon'_2}{n'^2 - \varepsilon'_1} \cos\varphi,\ \ \frac{n'^2 \sin\vartheta' \cos\vartheta'}{n'^2 \sin^2\vartheta' - \varepsilon'_3} \right); \end{aligned}\right\} \tag{12.4.2.2}$$

$\boldsymbol{e}$ and $\boldsymbol{e}'$ are the polarization vectors of the incident and the scattered waves and n and n' their refractive indices. The factor $\mathcal{M}$ depends on the directions of propagation of the incident and scattered waves relative to the magnetic field (the angles ϑ and ϑ'), and also on the difference φ of the azimuthal angles of the wavevectors $\boldsymbol{k}$ and $\boldsymbol{k}'$.

Equation (12.4.2.1) becomes the corresponding expression for the differential scattering cross-section in an isotropic plasma when $\boldsymbol{B}_0 = 0$. Indeed, if $\boldsymbol{B}_0 = 0$, we have

$$\varepsilon_{ik} - \delta_{ik} = -\frac{\omega_{\text{pe}}^2}{\omega^2}\delta_{ik}, \quad (\boldsymbol{k}\cdot\boldsymbol{e}) = (\boldsymbol{k}'\cdot\boldsymbol{e}') = 0, \quad n = \sqrt{\varepsilon(\omega)}.$$

Therefore, $\mathcal{M} = (\omega_{\text{pe}}/\omega)^4 e_\perp^2$, where $\boldsymbol{e}_\perp$ is the component of $\boldsymbol{e}$ which is at right angles to $\boldsymbol{k}'$, when $\boldsymbol{B}_0 = 0$, and eqn. (12.4.2.1) becomes the same as formula (12.1.2.7), after averaging over the possible orientations of the vector $\boldsymbol{e}$.

The spectral density of the electron density fluctuations is in the general case of a non-isothermal magneto-active plasma given by the formula

$$e^2\langle\delta n^2\rangle_{\boldsymbol{q}\Delta\omega} = \frac{2}{\Delta\omega}\,\text{Im}\,\Big\{T_\text{e}\sum_{m,\,n}\Big(q_m - 4\pi\sum_{i,\,k} q_i\varkappa_{ik}^{\text{i}}\Lambda_{km}^{-1}\Big)^*\Big(q_n - 4\pi\sum_{j,\,l} q_j\varkappa_{jl}^{\text{i}}\Lambda_{ln}^{-1}\Big)\varkappa_{mn}^{\text{e}*}$$
$$+16\pi^2 T_\text{i}\sum_{i,\,j,\,k,\,l,\,m,\,n}(q_i\varkappa_{ik}^{\text{e}}\Lambda_{km}^{-1})^*(q_j\varkappa_{jl}^{\text{e}}\Lambda_{ln}^{-1})\,\varkappa_{mn}^{\text{i}*}\Big\}. \quad (12.4.2.3)$$

If $\Delta\omega^2 \ll q^2c^2$ we can greatly simplify the correlation function (12.4.2.3) and it becomes the same as in the case of an unmagnetized plasma:

$$e^2\langle\delta n^2\rangle_{\boldsymbol{q}\Delta\omega} = \frac{2q^2}{\Delta\omega\,|\,\varepsilon\,|^2}\{T_\text{e}\,|\,1+4\pi\varkappa^\text{i}\,|^2\,\text{Im}\,\varkappa^\text{e} + 16\pi^2 T_\text{i}\,|\,\varkappa^\text{e}\,|^2\,\text{Im}\,\varkappa^\text{i}\}, \quad (12.4.2.4)$$

where we must take for $\varkappa$ and ε the longitudinal components of the respective tensors.

If the change in wavevector in the scattering process, $\boldsymbol{q}$, is parallel to $\boldsymbol{B}_0$, the spectrum of the scattered radiation will have the same characteristic frequencies as in the case of an isotropic plasma. The intensity of the scattering depends strongly on the strength of the magnetic field. If the direction of $\boldsymbol{q}$ is not the same as that of $\boldsymbol{B}_0$, the magnetic field affects the spectrum of the scattered radiation.

In the case of a non-isothermal plasma the spectrum of the scattered radiation is for angles between $\boldsymbol{q}$ and $\boldsymbol{B}_0$ different from $\pi/2$ characterized by a sharp maximum at $\Delta\omega = 0$, as in the case when there is no magnetic field. This maximum is caused by the interaction of the electromagnetic wave with the incoherent electron density fluctuations in the plasma. (For small shifts in frequency we can neglect the interaction of the incident wave with the fluctuations in the electron velocity, the electrical field, and the magnetic field.) Although the scattering is by electron density fluctuations, the Doppler broadening of the main maximum is determined by the ion thermal velocity as the ions and electrons interact through the self-consistent field.

If $T_\text{e} = T_\text{i}$ and $\omega \gg \omega_\pm$, where $\omega_\pm$ are the Langmuir oscillation frequencies in a magnetic field, we can use eqn. (12.4.2.3) to find the total scattering coefficient for electromagnetic waves in the plasma:

$$d\Sigma = \frac{1}{2}n_0\left(\frac{e^2}{m_\text{e}c^2}\right)^2\frac{\omega^4}{\omega_{\text{pe}}^4}\mathcal{M}_{\omega'=\omega}\frac{1+q^2r_\text{D}^2}{2+q^2r_\text{D}^2}\,d^2\boldsymbol{\omega}. \quad (12.4.2.5)$$

If $q^2 r_D^2 \gg 1$ we can consider this formula to be a generalization of the well-known Rayleigh formula for the case of a magneto-active medium.

The maximum in the spectrum of the scattered radiation caused by the interaction with incoherent fluctuations is greatly lowered in the case of a non-isothermal plasma. In a strongly non-isothermal plasma ($T_e \gg T_i$) the height of the maximum is lower by a factor $\sqrt{(m_i/m_e)}$ than for the case of an isothermal plasma.

The interaction between the electromagnetic waves and the electron density fluctuations can also lead to a transformation of these waves into Langmuir waves. The differential cross-section for the transformation of an ordinary or an extra-ordinary wave into a Langmuir wave is, according to (12.4.1.10), for $\Delta\omega \ll \omega$ equal to

$$d\Sigma_{\mathrm{E}\to\mathrm{L}} = \frac{1}{2\pi}\left(\frac{e^2}{m_e c^2}\right)^2 \frac{\omega^2\omega'^2}{\omega_{\mathrm{pe}}^4} \frac{n_{\mathrm{L}}'^3\,|\boldsymbol{k}'(\hat{\varepsilon}-\hat{1})\,\boldsymbol{e}\,|^2}{n[\,|\boldsymbol{e}\,|^2-\{\,|(\boldsymbol{k}\cdot\boldsymbol{e})\,|^2/k^2\}](\boldsymbol{k}'\hat{\varepsilon}'\boldsymbol{k}')} \langle\delta n^2\rangle_{\boldsymbol{q}\Delta\omega}\, d\omega'\, d^2\boldsymbol{\omega}'. \quad (12.4.2.6)$$

As in the case of scattering, the interaction with incoherent fluctuations plays the main role in the case of transformation of electromagnetic waves with a small change in frequency. As $\boldsymbol{B}_0 \to 0$, eqn. (12.4.2.6) changes to expression (12.2.1.3). In eqn. (12.4.2.6) n_{L}' is the refractive index of the Langmuir waves.

The ratio of the transformation coefficient (12.4.2.6) to the scattering coefficient (12.4.2.1) is of the order of c^3/v_e^3. In the range of frequencies close to $\omega_+(\vartheta)$ and $\omega_-(\vartheta)$ the absorption connected with the transformation of electromagnetic waves into Langmuir ones is thus much more important than the scattering of the electromagnetic waves.

We note that we could have obtained eqn. (12.4.2.6) directly from formula (12.4.2.1) if if in the latter we had taken for the scattered wave a Langmuir wave, $n' = n_{\mathrm{L}}'$ and $\boldsymbol{e}' = \boldsymbol{k}'/k'$.

12.4.3. SCATTERING AND TRANSFORMATION OF ELECTROMAGNETIC WAVES BY RESONANCE FLUCTUATIONS

Apart from the main maximum at $\Delta\omega = 0$ there are also maxima in the spectrum of the scattered radiation which are connected with the scattering and transformation of electromagnetic waves by resonance (coherent) fluctuations in the plasma. We can use (11.5.2.8) to express the scattering current (12.4.1.12) solely in terms of the electrical field fluctuations. The current correlation function which occurs in the general formula (12.4.1.10) for the intensity of the emitted waves has then the form

$$\langle|(\boldsymbol{e}\cdot\boldsymbol{I}^*)\,|^2\rangle_{k'\omega'} = \frac{\omega'^2\Delta\omega^2}{e^2 n_0^2 c^2}\,|\,B\,|^2\,\langle|\,\delta E\,|^2\rangle_{\boldsymbol{q}\Delta\omega}\,|\,\boldsymbol{E}^\circ\,|^2, \quad (12.4.3.1)$$

where

$$B = \sum_{i,j} e_i'^*\varkappa_{ij}'\Big\{\sum_{k,l}\tilde{n}e_j\tilde{\varkappa}_k\tilde{\varkappa}_{kl}\tilde{e}_l + n\sum_{k,l}(\varkappa_j e_k - e_j\varkappa_k)\,\tilde{\varkappa}_{kl}\tilde{e}_l + \frac{\omega}{\Delta\omega}\sum_{k,l}[n\tilde{e}_j\varkappa_k\varkappa_{kl}e_l$$
$$+\tilde{n}(\tilde{\varkappa}_j\tilde{e}_k - \tilde{e}_j\tilde{\varkappa}_k)\,\varkappa_{kl}e_l] - 4\pi n'\frac{\omega\omega'}{\omega_{\mathrm{pe}}^2}\sum_{k,l,m}[\varkappa_{jk}e_k\varkappa_l'\tilde{\varkappa}_{lm}\tilde{e}_m + \tilde{\varkappa}_{jk}\tilde{e}_k\varkappa_l'\varkappa_{lm}e_m]\Big\}, \quad (12.4.3.2)$$

where $\boldsymbol{e}$, $\boldsymbol{\varkappa}$, and n are the polarization vector, the unit vector in the propagation direction, and the refractive index of the incident wave, while $\tilde{\boldsymbol{e}}$, $\tilde{\boldsymbol{\varkappa}}$, $\tilde{\boldsymbol{n}}$ and $\boldsymbol{e}'$, $\boldsymbol{\varkappa}'$, n' are the same quantities for the fluctuation waves and the scattered waves. We bear in mind that the frequencies and wavevectors of the incident, scattered, and fluctuation waves are interconnected through the relations

$$\omega' = \omega + \Delta\omega, \quad \boldsymbol{k}' = \boldsymbol{k} + \boldsymbol{q}, \tag{12.4.3.3}$$

which express the energy and momentum conservation laws. Knowing the spectral density of the resonance fluctuations and using the properties of the incident and the excited waves we can easily separate in (12.4.3.1) the terms corresponding to well-defined kinds of transitions.

Let us first of all consider the scattering and transformation of electromagnetic waves by high-frequency Langmuir fluctuations. We shall assume that the refractive index of the fluctuating Langmuir oscillations $\tilde{n}' \gg 1$ while the refractive indices of the incident wave, n, and of the scattered wave, n', are of the order of unity. It follows from the conservation laws (12.4.3.3) that $\omega' \cong \omega \gg \Delta\omega$. One sees easily that in that case the interaction between the incident wave and the electron density fluctuations—the first term in (12.4.3.2)—plays the main role. The cross-section for the scattering of electromagnetic waves will therefore be given by formula (12.4.2.1). The spectral density (12.4.2.4) has in the case $q^2 r_D^2 \ll 1$ δ-functionlike maxima at frequencies $\omega_+(\tilde{\vartheta})$ and $\omega_-(\tilde{\vartheta})$, where $\tilde{\vartheta}$ is the angle between $\boldsymbol{q}$ and $\boldsymbol{B}_0$, which is connected with the angles ϑ, ϑ', and φ through the relation

$$\tan^2\tilde{\vartheta} = \frac{k^2\sin^2\vartheta + k'^2\sin^2\vartheta' + 2kk'\sin\vartheta\sin\vartheta'\cos\varphi}{(k\cos\vartheta - k'\cos\vartheta')^2}. \tag{12.4.3.4}$$

The cross-section for the scattering of electromagnetic waves when the frequency shifts $\Delta\omega$ lies close to the frequencies of the Langmuir oscillations, $\omega_\pm(\tilde{\vartheta})$, has the form

$$d\Sigma_{\mathrm{E}+\tilde{\mathrm{L}}\to\mathrm{E}} = \frac{1}{2} n_0 \left(\frac{e^2}{m_e c^2}\right)^2 \frac{\omega^2\omega'^2}{\omega_{pe}^4} \mathfrak{N} q^2 r_D^2 \frac{|\Delta\omega^2 - \omega_B^2|}{\tilde{\omega}_+^2 - \tilde{\omega}_-^2}$$
$$\times\{\delta(\Delta\omega - \tilde{\omega}_+) + \delta(\Delta\omega + \tilde{\omega}_+) + \delta(\Delta\omega - \tilde{\omega}_-) + \delta(\Delta\omega + \tilde{\omega}_-)\}\, d\omega'\, d^2\boldsymbol{\omega}'. \tag{12.4.3.5}$$

The cross-section for the transformation of electromagnetic waves by Langmuir fluctuations into Langmuir waves is equal to

$$d\Sigma_{\mathrm{E}+\tilde{\mathrm{L}}\to\mathrm{L}} = \frac{1}{2} n_0 \left(\frac{e^2}{m_e c^2}\right)^2 \frac{\omega^2\omega'^2}{\omega_{pe}^4} q^2 r_D^2 \frac{n_L'^3 |\boldsymbol{k}'\hat{Q}^L \boldsymbol{e}|^2}{n[|\boldsymbol{e}|^2 - \{|(\boldsymbol{k}\cdot\boldsymbol{e})|^2/k^2\}][(\boldsymbol{k}'\hat{\varepsilon}'\boldsymbol{k}')} \frac{|\Delta\omega^2 - \omega_B^2|}{\tilde{\omega}_+^2 - \tilde{\omega}_-^2}$$
$$\times\{\delta(\Delta\omega - \tilde{\omega}_+) + \delta(\Delta\omega + \tilde{\omega}_+) + \delta(\Delta\omega - \tilde{\omega}_-) + \delta(\Delta\omega + \tilde{\omega}_-)\}\, d\omega'\, d^2\boldsymbol{\omega}', \tag{12.4.3.6}$$

where

$$Q_{ij}^L = \sum_k (\varepsilon_{ik}' - \delta_{ik}) \left\{\delta_{kj} + \frac{\omega\Delta\omega}{\omega_{pe}^2} \sum_{l,\,m} [(\varepsilon_{kj} - \delta_{kj})\, \varkappa_l'(\tilde{\varepsilon}_{lm} - \delta_{lm})\, \tilde{\varkappa}_m + (\tilde{\varepsilon}_{kl} - \delta_{kl})\, \tilde{\varkappa}_l \varkappa_m'(\varepsilon_{mj} - \delta_{mj})]\right\}. \tag{12.4.3.7}$$

In the case of a non-equilibrium plasma—for instance, a plasma through which a beam of charged particles is passing—we must take into account an extra factor $R\{1 - (u/\tilde{u})\cos\tilde{\vartheta}\}^{-1}$, which is caused by replacing the temperature T by the effective temperature (11.5.4.4), in the cross-sections (12.4.3.5) and (12.4.3.6).

The relative contribution from Raman scattering by Langmuir fluctuations which is given by (12.4.3.5) to the total scattering cross-section (12.4.2.5) is in an equilibrium plasma of the order $q^2 r_D^2$. Under non-equilibrium conditions the cross-section for the Raman scattering of electromagnetic waves, like the cross-section for the transformation of electromagnetic waves into Langmuir ones, can increase anomalously, provided the plasma is close to the threshold for kinetic instability.

Raman scattering of electromagnetic waves can also occur in a magneto-active plasma by low-frequency magneto-sound and Alfvén fluctuations. Using the general expression (12.4.1.10) for the intensity and eqns. (11.4.3.1), (11.4.3.3), and (11.4.3.4) for the spectral densities of the low-frequency fluctuations we can easily study different actual cases of scattering and transformation.

It turns out that the scattering and transformation of electromagnetic waves by the slow magneto-sound fluctuations are the most important processes in a non-isothermal plasma. In that case the electron density fluctuations play the main role in (12.4.1.12). The cross-sections for the scattering and transformation of electromagnetic waves by slow magneto-sound waves are determined by the equation

$$d\Sigma_{\mathrm{E}+\tilde{\mathrm{S}}\to\mathrm{E,\,L}} = \frac{1}{2} n_0 \left(\frac{e^2}{m_e c^2}\right)^2 \frac{\omega^4}{\omega_{pe}^4} \frac{n^3 \,|\, \boldsymbol{e}'^*(\hat{\varepsilon}'-1)\,\boldsymbol{e}\,|^2}{n[\,|\,\boldsymbol{e}\,|^2 - \{|\,(\boldsymbol{k}\cdot\boldsymbol{e})\,|^2/k^2\}](\boldsymbol{e}'^*\hat{\varepsilon}'\boldsymbol{e}')}$$
$$\times\{\delta(\Delta\omega - q v_s \cos\tilde{\vartheta}) + \delta(\Delta\omega + q v_s \cos\tilde{\vartheta})\}\, d\omega'\, d^2\boldsymbol{\omega}', \qquad (12.4.3.8)$$

where we must take for n and $\boldsymbol{e}$ the refractive index and polarization vector of the ordinary extra-ordinary, and Langmuir waves. The differential scattering cross-section (12.4.3.8) differs from the corresponding cross-section (12.1.3.4) in an isotropic plasma by the dispersion law for the fluctuation oscillations.

The ratio of the cross-section for the scattering of electromagnetic waves by slow magneto-sound fluctuations, integrated over the frequencies, in a strongly non-isothermal plasma to the total cross-section (12.4.2.5) for the scattering of electromagnetic waves by incoherent fluctuations in a non-isothermal plasma is of the order of unity. In a strongly non-isothermal plasma the main line in the spectrum of the scattered radiation is thus split into two lines, connected with the scattering by slow magneto-sound fluctuations.

In the case of scattering and transformation of electromagnetic waves by fast magneto-sound fluctuations we must take into account not only the density fluctuations, but also the magnetic field fluctuations. The corresponding cross-sections for the scattering and transformation of electromagnetic waves are equal to

$$d\Sigma_{\mathrm{E}+\tilde{\mathrm{F}}\to\mathrm{E,L}} = \frac{1}{2} n_0 \left(\frac{e^2}{m_e c^2}\right)^2 \frac{\omega^4}{\omega_{pe}^4} \frac{v_s^2}{v_A^2} \frac{n'^3 |\, \boldsymbol{e}'^* \tilde{Q}^{\mathrm{F}} \boldsymbol{e}\,|^2}{n[\,|\,\boldsymbol{e}\,|^2 - \{|\,(\boldsymbol{k}\cdot\boldsymbol{e})\,|^2/k^2\}]\,(\boldsymbol{e}'^*\hat{\varepsilon}'\boldsymbol{e}')}$$
$$\times\{\delta(\Delta\omega - q v_A) + \delta(\Delta\omega + q v_A)\}\, d\omega'\, d^2\boldsymbol{\omega}', \qquad (12.4.3.9)$$

where

$$Q_{ij}^{\mathrm{F}} = \sum_k (\varepsilon'_{ik} - \delta_{ik}) \left\{ -i \sin\tilde{\vartheta}\, \delta_{kj} + \frac{\omega\omega_B}{\omega_{pe}^2} \sum_l (\tilde{\varkappa}_k \tilde{e}_l - \tilde{\varkappa}_l \tilde{e}_k)(\varepsilon_{lj} - \delta_{lj}) \right\}. \qquad (12.4.3.10)$$

The ratio of the scattering cross-section (12.4.3.9), integrated over the frequency, to the cross-section (12.4.2.5) is of the order v_s^2/v_A^2.

In the case of scattering and transformation of electromagnetic waves by Alfvén fluctuations the magnetic field fluctuations play the main role as Alfvén oscillations are not accompanied by changes in density. The cross-sections for scattering and transformation of electromagnetic waves by Alfvén fluctuations are equal to

$$d\Sigma_{\mathrm{E}+\tilde{\mathrm{A}}\to\mathrm{E,L}} = \frac{1}{2} n_0 \left(\frac{e^2}{m_e c^2}\right)^2 \frac{\omega^6}{\omega_{\mathrm{pe}}^6} \frac{v_{\mathrm{e}}^2}{c^2 \cos^2 \tilde{\vartheta}} \frac{n'^3 \,|\, \boldsymbol{e}'^* \tilde{Q}^{\mathrm{A}} \boldsymbol{e} \,|^2}{n[\,|\,\boldsymbol{e}\,|^2 - \{|\,(\boldsymbol{k}\cdot\boldsymbol{e})\,|^2/k^2\}](\boldsymbol{e}'^* \hat{\varepsilon}' \boldsymbol{e}')}$$
$$\times\{\delta(\Delta\omega - q v_{\mathrm{A}} \cos\tilde{\vartheta}) + \delta(\Delta\omega + q v_{\mathrm{A}} \cos\tilde{\vartheta})\}\, d\omega'\, d^2\boldsymbol{\omega}', \qquad (12.4.3.11)$$

where

$$Q_{ij}^{\mathrm{A}} = \sum_{k,l} (\varepsilon'_{ik} - \delta_{ik})(\tilde{\varkappa}_k \tilde{e}_l - \tilde{\varkappa}_l \tilde{e}_k)(\varepsilon_{lj} - \delta_{lj}). \qquad (12.4.3.12)$$

Equation (12.4.3.11) is valid both for an isothermal and for a non-isothermal plasma.

The ratio of the cross-section for the scattering of electromagnetic waves by Alfvén fluctuations, integrated over the frequency, to the cross-section (12.4.2.5) is of the order of v_{e}^2/c^2.

The cross-sections for scattering and transformation of electromagnetic waves by low-frequency fluctuations can, as in the case of Langmuir fluctuations, grow strongly in a non-equilibrium plasma, if it is close to the region of kinetic instability.

12.4.4. SCATTERING AND TRANSFORMATION OF LANGMUIR WAVES

The general formulae (12.4.1.10) and (12.4.1.12) enable us also to study the scattering of Langmuir waves and the transformation of Langmuir waves into high-frequency electromagnetic waves (Sitenko and Kirochkin, 1966). Choosing a Langmuir wave for the incident wave and using for the incident energy flux density the expression

$$\left.\begin{aligned} S_0 &= \frac{c}{8\pi} n\xi \,|\, \boldsymbol{E}^{\mathrm{o}} \,|^2, \quad \xi = \omega^2 \left(\frac{dA_0}{d\omega^2}\right) \Big/ n \frac{d(\omega n)}{d\omega}, \quad n^2 = \frac{A_0}{\psi}, \\ A_0 &= 1 - \frac{\omega_{\mathrm{pe}}^2}{\omega^2 - \omega_B^2} \sin^2\vartheta - \frac{\omega_{\mathrm{pe}}^2}{\omega^2} \cos^2\vartheta, \\ \psi &= \frac{v_{\mathrm{e}}^2}{c^2} \frac{\omega_{\mathrm{pe}}^2}{\omega^2} \left\{ \frac{3\omega^4}{(\omega^2 - \omega_B^2)(\omega^2 - 4\omega_B^2)} \sin^4\vartheta + \omega^2 \frac{6\omega^4 - 3\omega^2\omega_B^2 + \omega_B^4}{(\omega^2 - \omega_B^2)^3} \right. \\ &\qquad \left. \times \sin^2\vartheta \cos^2\vartheta + \cos^4\vartheta \right\} \end{aligned}\right\} \qquad (12.4.4.1)$$

we can apply (12.4.1.10) to find easily concrete expressions for the cross-sections for various scattering and transformation processes.

In the case of scattering and transformation of Langmuir waves by incoherent fluctuations in an isothermal plasma, as in the case of scattering of low-frequency electromagnetic waves, the interaction between the incident wave and the electron density fluctuations plays the main role. The cross-sections for the scattering and transformation of Langmuir waves by incoherent fluctuations are equal to

$$d\Sigma_{\mathrm{L}\to\mathrm{L,E}} = \frac{1}{2\pi} \left(\frac{e^2}{m_e c^2}\right)^2 \frac{\omega^4}{\omega_{\mathrm{pe}}^4} \frac{n'^3 \,|\, \hat{\boldsymbol{e}}'^* (\hat{\varepsilon}' - \hat{1}) \boldsymbol{\varkappa} \,|^2}{n\xi (\boldsymbol{e}'^* \hat{\varepsilon}' \boldsymbol{e}')} \langle \delta n^2 \rangle_{q\Delta\omega}\, d\omega'\, d^2\boldsymbol{\omega}', \qquad (12.4.4.2)$$

where n and $\boldsymbol{e}$ are determined by eqns. (12.4.2.1) in the case of scattering, and by eqns. (12.4.2.2) in the case of transformation, while $\boldsymbol{\varkappa} = \boldsymbol{k}/k$. If $\boldsymbol{B}_0 = 0$, formula (12.4.4.2) becomes eqn. (12.2.2.2).

The spectral density of the incoherent fluctuations is characterized by a sharp maximum in the region of small changes in frequency, with a width determined by the ion thermal velocity. Equation (12.4.4.2) is applicable only in the region of the incoherent maximum. The ratio of the coefficient for the transformation of a Langmuir wave into a high-frequency electromagnetic wave to the coefficient for the scattering of Langmuir waves with small change in frequency is equal to v_e^3/c^3.

One can easily use eqns. (12.4.1.10) and (12.4.3.1) to study the scattering and transformation of Langmuir waves by resonance fluctuations.

Only the electron density and magnetic field fluctuations turn out to be important for the scattering and transformation of Langmuir waves by low-frequency fluctuation oscillations. The cross-sections for different kinds of scattering and transformation of Langmuir waves are similar to the corresponding cross-sections for the scattering and transformation of electromagnetic waves. We shall give the final expressions for the cross-sections for various kinds of scattering and transformation of Langmuir waves.

The cross-sections for scattering and transformation of Langmuir waves by slow magneto-sound fluctuations in a strongly non-isothermal plasma are equal to

$$d\Sigma_{L+\tilde{S}\to L,E} = \frac{1}{2} n_0 \left(\frac{e^2}{m_e c^2}\right)^2 \frac{\omega^4}{\omega_{pe}^4} \frac{n'^3 |e'^*(\hat{\varepsilon}'-\hat{1})\boldsymbol{\varkappa}|^2}{n\xi(e'^*\hat{\varepsilon}'e')} \{\delta(\Delta\omega - qv_s \cos\tilde{\vartheta}) + \delta(\Delta\omega + qv_s \cos\hat{\vartheta})\}\, d\omega'\, d^2\boldsymbol{\omega}'. \tag{12.4.4.3}$$

The cross-sections for the scattering and transformation of Langmuir waves by fast magneto-sound fluctuations are equal to

$$d\Sigma_{L+\tilde{F}\to L,E} = \frac{1}{2} n_0 \left(\frac{e^2}{m_e c^2}\right) \frac{\omega^4}{\omega_{pe}^4} \frac{v_s^2}{v_A^2} \frac{n'^3 |e'^*\tilde{Q}^F\boldsymbol{\varkappa}|^2}{n\xi(e'^*\hat{\varepsilon}'e')} \{\delta(\Delta\omega - qv_A) + \delta(\Delta\omega + qv_A)\}\, d\omega'\, d^2\boldsymbol{\omega}'. \tag{12.4.4.4}$$

The cross-sections for the scattering and transformation of Langmuir waves by Alfvén fluctuations are equal to

$$d\Sigma_{L+\tilde{A}\to L,E} = \frac{1}{2} n_0 \left(\frac{e^2}{m_e c^2}\right)^2 \frac{\omega^6}{\omega_{pe}^6} \frac{v_e^2}{c^2 \cos^2\vartheta} \frac{n'^3 |e'^*\tilde{Q}^A\boldsymbol{\varkappa}|^2}{n\xi(e'^*\hat{\varepsilon}'e')} \times\{\delta(\Delta\omega - qv_A \cos\tilde{\vartheta}) + \delta(\Delta\omega + qv_A \cos\tilde{\vartheta})\}\, d\omega'\, d^2\boldsymbol{\omega}'. \tag{12.4.4.5}$$

The quantities Q^F and Q^A are given by the same expressions as for the case of scattering of electromagnetic waves. We note that the ratio of the cross-section for the transformation of Langmuir waves to the scattering cross-section for all kinds of low-frequency fluctuations is of the order of v_e^2/c^2.

It turns out that it is important to take into account the electron density and electrical field fluctuations in the case of the interaction of a Langmuir wave with high-frequency fluctuations. The incident Langmuir wave can interact both with Langmuir fluctuations and with high-frequency electromagnetic fluctuations.

Scattering of Langmuir waves by Langmuir fluctuations is impossible because of the conservation laws (12.4.1.11). The transformation of a Langmuir wave by Langmuir fluctuations into a high-frequency ordinary or extra-ordinary electromagnetic wave is determined by the cross-section

$$d\Sigma_{\mathrm{L}+\tilde{\mathrm{L}}\to\mathrm{E}} = \frac{1}{2} n_0 \left(\frac{e^2}{m_e c^2}\right)^2 \frac{\omega'^4}{\omega_{pe}^4} q^2 r_D^2 \frac{n'^3 |\boldsymbol{e}'^* \hat{Q}^{\mathrm{L}} \boldsymbol{\varkappa}|^2}{n\xi(\boldsymbol{e}'^* \hat{\varepsilon}' \boldsymbol{e}')} \frac{|\Delta\omega^2 - \omega_B^2|}{\tilde{\omega}_+^2 - \tilde{\omega}_-^2}$$

$$\times \{\delta(\Delta\omega - \tilde{\omega}_+)\,\delta(\Delta\omega + \tilde{\omega}_+) + \delta(\Delta\omega - \tilde{\omega}_-) + \delta(\Delta\omega + \tilde{\omega}_-)\}\, d\omega'\, d^2\boldsymbol{\omega}',$$

$$Q_{ij}^{\mathrm{L}} = \sum_k (\varepsilon'_{ik} - \delta_{ik})(\delta_{kj} + \tilde{\varkappa}_k \varkappa_j). \qquad (12.4.4.6)$$

This cross-section is larger by a factor c/v_e than the cross-section (12.4.2.5). The frequencies of the waves formed as a result of such a process will lie close to the sum of the modified Langmuir frequencies $\omega_\pm(\vartheta) \pm \omega_\pm(\tilde{\vartheta})$.

The cross-sections for the scattering and transformation of Langmuir waves by electromagnetic fluctuations are equal to

$$d\Sigma_{\mathrm{L}+\tilde{\mathrm{E}}\to\mathrm{L,E}} = \frac{1}{2} n_0 \left(\frac{e^2}{m_e c^2}\right)^2 \frac{\omega'^4}{\omega_{pe}^4} k^2 r_D^2 \frac{n'^3 |\boldsymbol{e}' \hat{Q}^{\mathrm{E}} \tilde{\boldsymbol{e}}|^2}{n\xi(\boldsymbol{e}'^* \hat{\varepsilon}' \boldsymbol{e}')\{|\tilde{\boldsymbol{e}}|^2 - |(\tilde{\boldsymbol{\varkappa}} \cdot \tilde{\boldsymbol{e}})|^2\}}$$

$$\left\{\delta\left(\Delta\omega - \frac{qc}{\tilde{n}}\right) + \delta\left(\Delta\omega + \frac{qc}{\tilde{n}}\right)\right\} d\omega'\, d^2\boldsymbol{\omega}', \qquad (12.4.4.7)$$

where

$$Q_{ij}^{\mathrm{E}} = \begin{cases} \sum_k (\varepsilon'_{ik} - \delta_{ik}) \left\{\delta_{kj} + \dfrac{\omega\Delta\omega}{\omega_{pe}^2} \left[\sum_{l,m} (\varepsilon_{kl} - \delta_{kl}) \varkappa_l \varkappa'_m (\tilde{\varepsilon}_{mj} - \delta_{mj}) + (\tilde{\varepsilon}_{kj} - \delta_{kj}) \sum_{l,m} \varkappa_l (\varepsilon^0_{lm} - \delta_{lm}) \varkappa^0_m\right]\right\}, \\ \sum_k (\varepsilon'_{ik} - \delta_{ik}) \left\{\delta_{kj} - \dfrac{\Delta\omega}{\omega} \dfrac{\tilde{n}}{n} \sum_l (\tilde{\varepsilon}_{lj} - \delta_{lj}) \tilde{\varkappa}_l \varkappa_k\right\}. \end{cases}$$

The cross-sections for the scattering and transformation of Langmuir waves are in a non-equilibrium plasma characterized by the same anomalies in the region of critical fluctuations as the corresponding cross-sections for electromagnetic waves.

12.4.5. TRANSFORMATION OF LOW-FREQUENCY WAVES BY LANGMUIR FLUCTUATIONS

In concluding this section we shall now dwell upon the transformation, caused by Langmuir fluctuations, of low-frequency waves into high-frequency ones in a magneto-active plasma (Sitenko and Kirochkin, 1966). The most important terms in expression (12.4.1.12) for the current are in the case of an incident low-frequency wave those which are connected with the magnetic field and the electron density in the incident wave. Using eqn. (12.4.1.10) for the intensity of the emission of high-frequency waves, and dividing it by the energy flux density connected with the incident low-frequency wave, can easily find explicit expressions for the cross-sections for various kinds of wave transformation processes.

The cross-sections for transformation of a slow magneto-sound wave by Langmuir fluctuations into the excitation of an ordinary, extra-ordinary, or Langmuir wave are given by the formula

$$d\Sigma_{\mathrm{S}+\tilde{\mathrm{L}}\to\mathrm{E,L}} = \frac{1}{2} n_0 \left(\frac{e^2}{m_e c^2}\right)^2 \frac{\omega'^4}{\omega_{pe}^4} \frac{cn^3 |\boldsymbol{e}'^*(\hat{\varepsilon}' - 1)\tilde{\boldsymbol{\varkappa}}|^2}{v_s |\cos\vartheta| (\boldsymbol{e}'^* \hat{\varepsilon}' \boldsymbol{e}')} \frac{|\Delta\omega^2 - \omega_B^2|}{\tilde{\omega}_+^2 - \tilde{\omega}_-^2}$$

$$\times \{\delta(\Delta\omega - \tilde{\omega}_+) + \delta(\Delta\omega + \tilde{\omega}_+) + \delta(\Delta\omega - \tilde{\omega}_-) + \delta(\Delta\omega + \tilde{\omega}_-)\}\, d\omega'\, d^2\boldsymbol{\omega}'. \qquad (12.4.5.1)$$

The cross-sections for the transformation of a fast magneto-sound wave by Langmuir fluctuations into an ordinary, extra-ordinary, or Langmuir wave are equal to

$$d\Sigma_{F+\tilde{L}\to E,L} = \frac{1}{2}n_0\left(\frac{e^2}{m_e c^2}\right)^2 \frac{\omega\omega'^4}{\omega_B\omega_{pe}^4}\frac{v_e^2}{cv_A}\frac{n'^3|e'^*\hat{P}^L e|^2}{(e'^*\hat{\varepsilon}'e')}\frac{|\Delta\omega^2-\omega_B^2|}{\tilde{\omega}_+^2-\tilde{\omega}_-^2}$$
$$\times\{\delta(\Delta\omega-\tilde{\omega}_+)+\delta(\Delta\omega+\tilde{\omega}_+)+\delta(\Delta\omega-\tilde{\omega}_-)+\delta(\Delta\omega+\tilde{\omega}_-)\}\,d\omega'\,d^2\boldsymbol{\omega}', \quad (12.4.5.2)$$

$$P_{ij}^L = \sum_{k,l}(\varepsilon'_{ik}-\delta_{ik})\left\{\tilde{\varkappa}_k\varkappa_i\varkappa_{lj}^e+\frac{\Delta\omega}{\omega}(\varkappa_k\delta_{lj}-\varkappa_l\delta_{kj})\sum_m(\tilde{\varepsilon}_{lm}-\delta_{lm})\tilde{\varkappa}_m.\right\}.$$

Finally, the cross-sections for the transformation of an Alfvén wave by Langmuir fluctuations into an ordinary, extra-ordinary, or Langmuir wave are given by the formula

$$d\Sigma_{A+\tilde{L}\to E,L} = \frac{1}{2}n_0\left(\frac{e^2}{m_e c^2}\right)^2 \frac{\omega'^4\Delta\omega^2}{\omega_{pe}^6}\frac{v_e^2}{cv_A}\frac{n'^3|e'^*\hat{R}^L e|^2}{|\cos\vartheta|^3(e'^*\hat{\varepsilon}e')}\frac{|\Delta\omega^2-\omega_B^2|}{\tilde{\omega}_+^2-\tilde{\omega}_-^2}$$
$$\times\{\delta(\Delta\omega-\tilde{\omega}_+)+\delta(\Delta\omega+\tilde{\omega}_+)+\delta(\Delta\omega-\tilde{\omega}_-)+\delta(\Delta\omega+\tilde{\omega}_-)\}\,d\omega'\,d^2\boldsymbol{\omega}', \quad (12.4.5.3)$$

$$R_{ij}^L = \sum_{k,l,m}(\varepsilon'_{ik}-\delta_{ik})(\varkappa_k\delta_{lj}-\varkappa_l\delta_{kj})(\tilde{\varepsilon}_{lm}-\delta_{lm})\tilde{\varkappa}_m.$$

We note that of all these processes the transformation of a slow magneto-sound wave into a Langmuir wave has the largest cross-section. The ratio of the cross-section for that process, integrated over the frequency, to (12.4.2.5) is of the order c^4/v_e^4.

12.5. Scattering and Transformation of Waves in a Partially Ionized Plasma in an External Electrical Field

12.5.1. SCATTERING OF TRANSVERSE WAVES WHEN THERE IS NO EXTERNAL MAGNETIC FIELD

We now turn to a study of wave scattering and transformation processes in a partially ionized plasma in an external constant, uniform electrical field (Angeleĭko and Akhiezer, 1968). We shall first of all consider Raman scattering of electromagnetic waves by ion-sound oscillations in a partially ionized plasma in an external electrical field $\boldsymbol{E}^0$.

Substituting expression (11.7.1.1) into the general formula (12.1.2.7) for the differential cross-section for the scattering of electromagnetic waves and restricting ourselves to the case of long-wavelength incident radiation ($kr_D \ll 1$) we get

$$d\Sigma = 4\pi\zeta n_0\sqrt{Z}\left(\frac{e^2}{m_e c^2}\right)^2(1+\cos^2\vartheta)\left\{\frac{\delta(\Delta\omega-2kv_s\sin\frac{1}{2}\vartheta)}{R-\cos\chi}+\frac{\delta(\Delta\omega+2kv_s\sin\frac{1}{2}\vartheta)}{R+\cos\chi}\right\}$$
$$\times d\omega'\,d^2\boldsymbol{\omega}', \qquad \zeta = \Gamma(\tfrac{1}{4})\{32\pi\,2^{1/4}\,3^{1/2}\,\Gamma(\tfrac{3}{4})\}^{-1}\approx 1.4\times10^{-2}, \quad (12.5.1.1)$$

where χ is the angle between the vector $\boldsymbol{q} = \boldsymbol{k}-\boldsymbol{k}'$ and the direction of the external electrical field,

$$\cos\chi = \frac{\cos\theta-\cos\theta'}{2\sin\frac{1}{2}\vartheta}, \quad (12.5.1.2)$$

R is the quantity given by formula (7.2.1.11), ϑ the scattering angle, that is, the angle between $\boldsymbol{k}$ and $\boldsymbol{k}'$, while θ and θ' are the angles between $\boldsymbol{k}$ and $\boldsymbol{k}'$ and $\boldsymbol{E}_0$. We draw attention to the fact that the angle χ in this section is defined as the angle between the direction of the electron current and the wavevector of the fluctuations, in contrast to Chapter 7 or Section 11.7, where χ was the angle between the direction of the electron current and the group velocity of the wave. If $\Delta\omega > 0$, the new and the old definitions are the same; if $\Delta\omega < 0$, the sign of $\cos\chi$ is different for the old and the new definitions of χ.

We see that two narrow lines occur in the spectrum of the scattered radiation: ion-sound satellites with frequencies $\omega' = \omega - 2kv_s \sin\frac{1}{2}\vartheta$ (Stokes satellite) and $\omega' = \omega + 2kv_s \sin\frac{1}{2}\vartheta$ (anti-Stokes satellite).

Integrating eqn. (12.5.1.1) over the frequency ω' we find the scattering intensity per unit solid angle,

$$\frac{d\Sigma}{d^2\omega'} = 4\pi\zeta\, n_0 \sqrt{Z}\left(\frac{e^2}{m_e c^2}\right)^2 (1+\cos^2\vartheta)\left\{f\left(\chi, \frac{E_0}{E_c}\right) + f_+\left(\chi, \frac{E_0}{E_c}\right)\right\}, \tag{12.5.1.3}$$

where f and f_+ are functions, characterizing the intensities of the Stokes and the anti-Stokes satellites,

$$f(\chi, E_0/E_c) = f_+(\pi-\chi, E_0/E_c) = \frac{1}{R-\cos\chi}, \tag{12.5.1.4}$$

while E_c is the critical value of the electrical field, given by equation (7.2.1.12). We sketch in Fig. 12.5.1 f as function of χ for different values of the external electrical field, and in Fig. 12.5.2 f as function of E_0 for $\chi = 0$.

We see that the angular distribution of the scattered radiation depends strongly on the strength of the external electrical field. When $E_0 \ll E_c$ this distribution is almost isotropic—

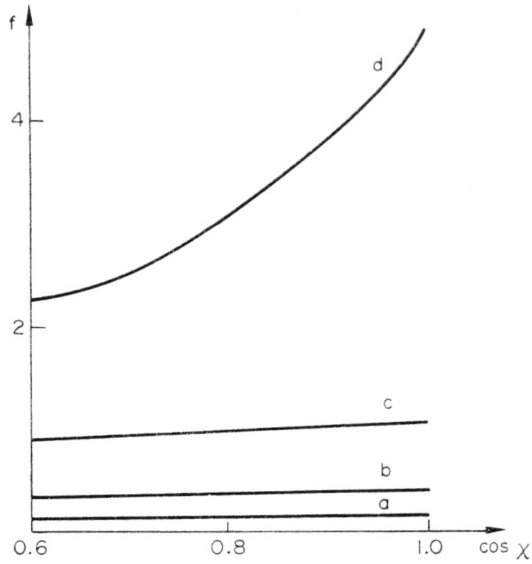

FIG. 12.5.1. Angular distribution of the scattered radiation. We plot the quantity $f(\chi, E_0/E_c)$ as function of $\cos\chi = (\cos\theta - \cos\theta')/2\sin\frac{1}{2}\vartheta$. The curves a, b, c, and d refer to the cases $E_0/E_c = 0.5, 0.7, 0.8$, and 0.9; $m_0 = m_i = 10^4\, m_e$; $Z = 1$.

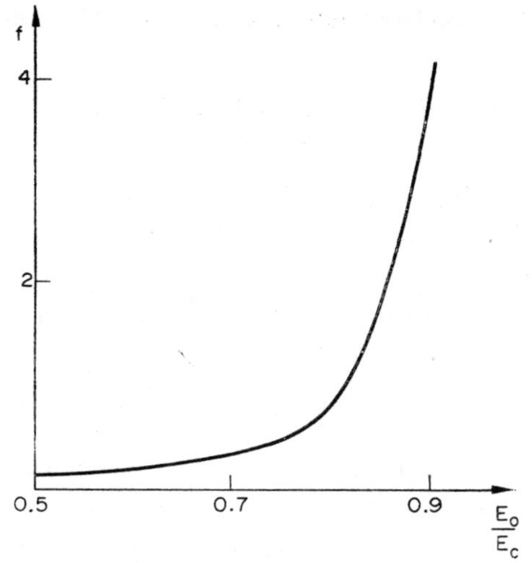

FIG. 12.5.2. The scattering intensity as function of the strength of the external electrical field. We plot $f(\chi, E_0/E_c)$ as function of the ratio E_0/E_c for $\chi = 0$; $m_0 = m_i = 10^4\, m_e$; $Z = 1$.

if we neglect the factor

$$1+\cos^2 \vartheta.$$

When E_0 increases, the intensity of waves scattered over angles $\theta'_{\pm}$, with

$$\cos \theta'_{\pm} = \cos \theta \pm 2 \sin \tfrac{1}{2} \vartheta, \tag{12.5.1.5}$$

ncreases steeply. As $E_0 \to E_c$ the scattering cross-section, per unit solid angle, becomes—in the framework of the linear theory—infinite for $\theta' = \theta'_{\pm}$.

The total cross-section for scattering of electromagnetic waves by ion-sound oscillations increases slowly when the external electrical field strength increases, and tends logarithmically to infinity as $E_0 \to E_c$. The steep increase in the intensity of the scattering of electromagnetic waves as $E_0 \to E_c$ is connected with the instability of ion sound when $E_0 \geq E_c$. This effect is of the same nature as the critical opalescence in a collisionless plasma which is at the boundary of the stability region, considered in Subsection 12.1.4.

We emphasize that all expressions obtained in the present and subsequent subsections for the cross-sections for the scattering of electromagnetic waves are valid in the stability region, when $E_0 < E_c$. If the plasma oscillations grow, one needs take into account non-linear effects which impose a limit on the growth of the fluctuations, and hence on the growth of the cross-sections for the scattering and transformation of waves.

12.5.2. SCATTERING OF TRANSVERSE WAVES WHEN THERE IS AN EXTERNAL MAGNETIC FIELD PRESENT

Let us now consider the Raman scattering of electromagnetic waves in a partially ionized plasma in external electrical and magnetic fields (Akhiezer and Angeleĭko, 1969a; Angeleĭko and Akhiezer, 1969). Assuming that the change in frequency $\Delta\omega$ is small compared to the frequency ω we get, from the general formula (12.4.2.1), for the differential cross-section for the scattering of transverse electromagnetic waves by density fluctuations, the following expression

$$d\Sigma = \frac{1}{2\pi}\left(\frac{e^2}{m_e c^2}\right)^2\left(\frac{\omega}{\omega_{pe}}\right)^4 \mathcal{N}^{\lambda\lambda'}\langle n_e^2\rangle_{q\Delta\omega}\, d\omega'\, d^2\boldsymbol{\omega}', \tag{12.5.2.1}$$

where the quantity $\mathcal{N}^{\lambda\lambda'}$ which depends on the wavevectors and polarizations of the incident and the scattered waves is given by formula (12.4.2.2); $\lambda = \pm$, where the upper sign corresponds to a fast and the lower sign to a slow transverse wave; see formula (7.1.3.6). We see that when there is a magnetic field present in the plasma there are, in principle, depending on the polarization of the incident and the scattered wave, four possible kinds of scattering processes for transverse waves: $t_+ \to t_+$, $t_- \to t_-$, $t_+ \to t_-$, and $t_- \to t_+$. For arbitrary ratios of the frequencies ω, ω_{pe}, and ω_{Be} such processes involving ion-sound or magneto-sound fluctuations may turn out to be forbidden, as the frequency of a sound wave with wavevector $\boldsymbol{q} = \boldsymbol{k} - \boldsymbol{k}'$ can turn out to be small compared with $\Delta\omega = \omega_t(\boldsymbol{k}) - \omega_t(\boldsymbol{k}')$. To fix the ideas we shall restrict ourselves to considering the case of high-frequency incident radiation, $\omega \gg \omega_{pe}, \omega_{Be}$; all four $t \to t$ processes then turn out to be possible.

We note that it follows from the form of the quantities $\mathcal{N}^{\lambda\lambda'}$ that for arbitrary scattering angles ϑ the differential cross-sections for all four $t \to t$ processes will be of the same order of magnitude; however, if the scattering angle is small ($\vartheta \ll 1$) the cross-sections for the processes $t_+ \to t_-$ and $t_- \to t_+$ are small—proportional to ϑ^4.

If the magnetic field is not too strong so that only the electron component of the plasma is strongly magnetized ($\omega_{Bi} \ll kv_s \ll \omega_{Be}$), the correlation function of the electron density fluctuations is in the sound region ($T_i/m_i \ll (\Delta\omega/k)^2 \ll T_e/m_e$) given by eqn. (11.7.2.2). Substituting this expression into eqn. (12.5.2.1) we get the following expression for the differential cross-section for the scattering of transverse electromagnetic waves by ion-sound oscillations in the case of long-wavelength incident radiation

$$d\Sigma = 8\pi\zeta n\sqrt{Z}\,\mathcal{N}^{\lambda\lambda'}\left(\frac{\omega}{\omega_{pe}}\right)^4\left(\frac{e^2}{m_e c^2}\right)^2\left\{\frac{\delta(\Delta\omega - 2kv_s\sin\frac{1}{2}\vartheta)}{\mu^\pm(R^\pm - \cos\chi)} + \frac{\delta(\Delta\omega + 2kv_s\sin\frac{1}{2}\vartheta)}{\mu^\mp(R^\mp + \cos\chi)}\right\} \times d\omega'\, d^2\boldsymbol{\omega}', \tag{12.5.2.2}$$

where χ is the angle between the vectors $\boldsymbol{q} = \boldsymbol{k} - \boldsymbol{k}'$ and $\boldsymbol{B}_0$, which is connected with the scattering angle ϑ and the angles θ, θ' between the vectors $\boldsymbol{k}$, $\boldsymbol{k}'$ and the direction of the external magnetic field through relation (12.5.1.2), while the quantities $R^\pm$ and $\mu^\pm$ are given by eqns. (7.2.2.7); the upper and lower signs refer, respectively, to the cases $\chi < \pi/2$ and $\chi > \pi/2$. We see that two ion-sound satellites with frequencies $\omega' = \omega \pm 2kv_s \sin\frac{1}{2}\vartheta$ occur in the spectrum of the scattered radiation.

Integrating expression (12.5.2.2) over the frequency we find the scattering intensity per unit solid angle,

$$\frac{d\Sigma}{d^2\boldsymbol{\omega}'} = 16\pi\zeta n_0\sqrt{Z}\,\mathcal{N}^{\lambda\lambda'}\left(\frac{\omega}{\omega_{pe}}\right)^4\left(\frac{e^2}{m_e c^2}\right)^2\{[\mu^\pm(R^\pm - \cos\chi)]^{-1} + [\mu^\mp(R^\mp + \cos\chi)]^{-1}\}. \tag{12.5.2.3}$$

In the case of a very strong magnetic field when not only the electron, but also the ion component of the plasma is strongly magnetized ($kv_s \ll \omega_{Bi}$) we get by substituting eqn.

(11.7.2.4) for the electron density correlator into formula (12.5.2.1)

$$d\Sigma = 8\pi\zeta n_0\sqrt{Z}\,\mathcal{N}^{\lambda\lambda'}\cos^{-4}\chi\left(\frac{\omega}{\omega_{pe}}\right)^4\left(\frac{e^2}{m_e c^2}\right)^2\left\{\frac{\delta(\Delta\omega - kv_s[\cos\theta - \cos\theta'])}{\xi - \operatorname{sgn}\cos\chi} + \frac{\delta(\Delta\omega + kv_s[\cos\theta - \cos\theta'])}{\xi + \operatorname{sgn}\cos\chi}\right\} d\omega'\, d^2\boldsymbol{\omega}', \tag{12.5.2.4}$$

where the quantity ξ is given by formula (7.2.2.10). In that case there occur in the spectrum of the scattered radiation two magneto-sound satellites with frequencies

$$\omega' = \omega \pm kv_s(\cos\theta - \cos\theta').$$

Integrating expression (11.5.2.4) over the frequency we find the intensity of the scattering per unit solid angle:

$$\frac{d\Sigma}{d^2\boldsymbol{\omega}'} = 16\pi\zeta n_0\sqrt{Z}\,\frac{\mathcal{N}^{\lambda\lambda'}}{\cos^4\chi}\left(\frac{\omega}{\omega_{pe}}\right)^4\left(\frac{e^2}{m_e c^2}\right)^2\frac{\xi}{\xi^2-1}\,. \tag{12.5.2.5}$$

We note that expressions (12.5.2.4) and (12.5.2.5)—like eqn. (11.7.2.4) for the correlation function of the electron density fluctuations—are valid provided the angle χ lies not too close to $\pi/2$.

Let us investigate how the intensity and the angular distribution of the scattered radiation changes with the strengths of the external fields $\boldsymbol{E}_0$ and $\boldsymbol{B}_0$. If the fields $\boldsymbol{E}_0$ and $\boldsymbol{B}_0$ are not too strong ($E_0 \ll E_c$, $\omega_{Bi} \ll kv_s$) the angular distribution of the scattered radiation is nearly isotropic—if we forget about the polarization-dependent factor $\mathcal{N}^{\lambda\lambda'}$.

In the case of a not very strong magnetic field ($\omega_{Bi} \ll k_0 v_s$) the angular distribution of the scattered radiation becomes more and more anisotropic when the electrical field strength increases, and the largest intensity corresponds to waves scattered over the angles $\theta'_\pm$ given by formula (12.5.1.5). As $E_0 \to E_c$ where $E_c(\boldsymbol{B}_0)$ is the critical value of the electrical field given by formula (7.2.2.8), the scattering cross-section per unit solid angle tends—in the framework of the linear theory—to infinity for $\theta' = \theta'_\pm$. The total cross-section for the scattering of electromagnetic waves by ion-sound oscillations increases with increasing E_0 and tends—logarithmically—to infinity as $E_0 \to E_c$.

In the case of a very strong magnetic field ($\omega_{Bi} \gg kv_s$) the angular distribution of the scattered radiation is strongly anisotropic for all external electrical field strengths: the quantity $d\Sigma/d^2\boldsymbol{\omega}'$ is, according to (12.5.2.5), in that case proportional to $\cos^{-4}\chi$. When E_0 increases, the differential scattering cross-section increases, tending to infinity as $E_0 \to E_c(\boldsymbol{B}_0)$ for all directions of propagation of the scattered waves at the same time.

The steep growth of the intensity of the scattering of electromagnetic waves as $E_0 \to E_c$ —the critical opalescence effect—is connected with the instability of the magneto-sound (or ion-sound) oscillations when $E_0 \geq E_c$.

We draw attention to the essential difference in the nature of the critical opalescence in the cases of strongly and weakly magnetized ions. In the latter case the critical opalescence must occur only for waves scattered at angles $\theta'_\pm$, while in the former case, this effect must be observed for scattering over any angle θ'. Because of this the total cross-section for the scattering of electromagnetic waves will when E_0 is close to $E_c(\boldsymbol{B}_0)$ be much larger in the case of strongly magnetized ions than in the case when they are weakly magnetized. Indeed, one easily sees, integrating expressions (12.5.2.3) and (12.5.2.5) over all directions that in the case of strongly magnetized ions the total scattering cross-sections is proportional to $(E_c - E_0)^{-1}$ as $E_0 \to E_c$, while in the case of weakly magnetized ions it is proportional to $\ln\{1 - [E_0/E_c(\boldsymbol{B}_0)]\}$.

12.5.3. CRITICAL OPALESCENCE WHEN LONGITUDINAL WAVES ARE SCATTERED OR TRANSFORMED

Apart from the Raman scattering processes for transverse electromagnetic waves (t → t) which were considered in the preceding subsections, there may occur in the plasma also other processes of scattering and transformation of electromagnetic oscillations when they interact with sound fluctuations. When there is no magnetic field, three such processes are possible: the scattering of high-frequency longitudinal waves (l → l), the transformation of transverse waves into longitudinal ones (t → l), and the inverse process of the transformation of longitudinal into transverse waves (l → t). We shall now turn to a study of these processes (Angeleĭko, 1968).

The intensity of the transformation (scattering) will as before be characterized by the transformation (scattering) coefficient

$$d\Sigma_{\lambda \to \lambda'} = \frac{dI^{(\lambda \to \lambda')}}{VS_0}, \tag{12.5.3.1}$$

where dI is the average increase in energy of the field of the transformed (scattered) wave per unit time and S_0 the energy flux density of the incident wave; the index λ characterizes the kind of oscillation: for a Langmuir wave $\lambda = \mathrm{l}$, and for a transverse wave $\lambda = \mathrm{t}$; V is the volume of the system.

We can easily express the coefficients for the transformation and scattering of high-frequency waves in terms of the correlation function of the electron density fluctuations. Using eqns. (12.2.1.3), (12.2.2.1), and (12.2.2.2) we get

$$d\Sigma_{\mathrm{t}\to\mathrm{l}} = \frac{\omega_{\mathrm{pe}}^3}{c^2 k n_0} \sin^2\vartheta \langle n_{\mathrm{e}}^2\rangle_{\boldsymbol{q}\Delta\omega} \frac{d^3\boldsymbol{k}'}{(4\pi)^3}, \tag{12.5.3.2}$$

$$d\Sigma_{\mathrm{l}\to\mathrm{t}} = \frac{2\omega_{\mathrm{pe}}}{3r_{\mathrm{D}}^2 k n_0^2} \sin^2\vartheta \langle n_{\mathrm{e}}^2\rangle_{\boldsymbol{q}\Delta\omega} \frac{d^3\boldsymbol{k}'}{(4\pi)^3}, \tag{12.5.3.3}$$

$$d\Sigma_{\mathrm{l}\to\mathrm{l}} = \frac{2\omega_{\mathrm{pe}}}{3r_{\mathrm{D}}^2 k n_0^2} \cos^2\vartheta \langle n_{\mathrm{e}}^2\rangle_{\boldsymbol{q}\Delta\omega} \frac{d^3\boldsymbol{k}'}{(4\pi)^3}, \tag{12.5.3.4}$$

where $\boldsymbol{q} = \boldsymbol{k} - \boldsymbol{k}'$, $\Delta\omega = \omega - \omega'$; $\boldsymbol{k}$, ω and $\boldsymbol{k}'$, ω' are the wavevectors and frequencies of the incident and the scattered (transformed) waves and ϑ is the angle between the vectors $\boldsymbol{k}$ and $\boldsymbol{k}'$.

Let us first of all consider the transformation of a transverse wave into a longitudinal one (t → l) by ion-sound oscillations. Substituting expression (11.7.1.2) for the correlation function of the electron density fluctuations into eqn. (12.5.3.2) and assuming that $kr_{\mathrm{D}} \ll 1$, we find

$$d\Sigma_{\mathrm{t}\to\mathrm{l}} = \zeta \frac{e^2 r_{\mathrm{D}}^2 \omega_{\mathrm{pe}} \sqrt{Z}}{kc^2} \sin^2\vartheta \left\{ \frac{\delta(\Delta\omega - qv_{\mathrm{s}})}{R - \cos\chi} + \frac{\delta(\Delta\omega + qv_{\mathrm{s}})}{R + \cos\chi} \right\} d^3k', \tag{12.5.3.5}$$

where χ is the angle between the vectors $\boldsymbol{q}$ and $\boldsymbol{E}_0$.

We remind ourselves that we have shown in Chapter 7 that if the plasma is situated in an external electrical field Langmuir oscillations may occur with frequencies $\omega_l(\boldsymbol{k})$, given by formula (7.1.3.3), and with frequencies $-\omega_l(-\boldsymbol{k})$. There are thus two $\mathrm{t} \to \mathrm{l}$ processes possible: a process ($\mathrm{t} \to \mathrm{l}_1$) in which the transformed wave has the frequency $\omega_l(\boldsymbol{k}')$, and a process ($\mathrm{t} \to \mathrm{l}_2$) in which the transformed wave has the frequency $-\omega_l(-\boldsymbol{k}')$.

From (12.5.3.5) we get for the coefficient for the transformation $\mathrm{t} \to \mathrm{l}_1$

$$d\Sigma_{\mathrm{t}\to\mathrm{l}_1} = \frac{Ze^2}{16\pi m_e c^2}\sqrt{\frac{m_0}{m_i}\frac{\omega^2{}_{pe}k'}{c^2k^3\alpha}}\sin^2\vartheta\left\{\frac{\delta(k'-k_1')}{R+\cos\theta'}+\frac{\delta(k'-k_2')}{R-\cos\theta'}\right\}d^3\boldsymbol{k}', \quad (12.5.3.6)$$

where θ' is the angle between the vectors $\boldsymbol{k}'$ and $\boldsymbol{E}_0$,

$$k_{1,2}' = \frac{c^2k^2\alpha}{2\omega_{pe}v_s(\cos\theta' \pm\alpha)}, \qquad \alpha \equiv \frac{v_s}{u} = \left[\frac{3\Gamma^3(\frac{3}{4})}{\pi\Gamma(\frac{1}{4})}Z\frac{m_0}{m_i}\right]^{-1/2} \approx 0.7\sqrt{\frac{Zm_0}{m_i}}, \quad (12.5.3.7)$$

where we have assumed that in the transformation of a transverse into a longitudinal wave $k' \gg k$.

If $\alpha > 1$, there appears, according to these formulae, one line with wavenumber k_1' in the spectrum of the transformed waves; however, if $\alpha < 1$, there appears a second line with wavenumber k_2'. The angular distribution of the transformed radiation is then characterized by a strong anisotropy. As $E_0 \to E_c$, in the case when $\alpha > 1$ the intensity of waves with wavenumber k_1' for which $\theta' = \pi$ (and in the case when $\alpha < 1$ the intensity of waves with wavenumber k_2' for which $\theta' = 0$) tends—in the framework of the linear theory—to infinity. Such a steep increase in the transformation coefficient—critical opalescence when the waves are transformed—is connected with the existence of critical fluctuations which occur in the plasma near the boundary of the stability region.

Integrating eqn. (12.5.3.6) over k' we find the coefficient for the transformation $\mathrm{t} \to \mathrm{l}_1$ per unit solid angle. Assuming that $\alpha > 1$ to fix the ideas, we get

$$\frac{d\Sigma_{\mathrm{t}\to\mathrm{l}_1}}{d^2\boldsymbol{\omega}'} = \frac{Ze^2}{16\pi m_e c^4}\sqrt{\frac{m_0}{m_i}}\frac{\omega_{pe}^3\sin^2\vartheta}{k^3\alpha}\frac{k_1'^3}{R+\cos\theta'}. \quad (12.5.3.8)$$

We can determine the total transformation coefficient $U^{(\mathrm{t}\to\mathrm{l}_1)}$, which characterizes the total intensity of the transformed waves, by integrating over $\boldsymbol{k}'$. For a plasma near the boundary of the stability region ($E_c - E_0 \ll E_c$) we then get

$$\Sigma_{\mathrm{t}\to\mathrm{l}_1} = \frac{Ze^2k_0'^3\omega_{pe}^2}{8m_ec^4k^3\alpha}\sqrt{\frac{m_0}{m_i}}\sin^2\theta\ln(R-1), \quad (12.5.3.9)$$

where $k_0' = c^2k^2\alpha(2\omega_{pe}v_s)^{-1}\,|\,1-\alpha\,|^{-1}$ while θ is the angle between the vectors $\boldsymbol{k}$ and $\boldsymbol{E}_0$.

We can similarly study the second possible process for the transformation of transverse waves into longitudinal ones—the $\mathrm{t} \to \mathrm{l}_2$ process. One sees easily that one can obtain all equations describing that process from the corresponding eqns. (12.5.3.6) to (12.5.3.9) by changing $\boldsymbol{E}_0$ to $-\boldsymbol{E}_0$, or, by making the substitutions $\theta \to \pi-\theta$, $\theta' \to \pi-\theta'$, $\chi \to \pi-\chi$; we shall therefore not give these equations.

Let us now consider the transformation of longitudinal into transverse waves. Assuming that in that case $k' \ll k$ and using (12.5.3.3), we get for the coefficient for the transformation

$l_1 \to t$ the expression

$$d\Sigma_{l_1 \to t} = \frac{Ze^2}{48\pi\alpha m_e c^2 r_D^2 kk'} \sqrt{\frac{m_0}{m_i}} \sin^2 \vartheta \left\{ \frac{\delta(k'-k_3')}{R-\cos\theta} + \frac{\delta(k'-k_4')}{R+\cos\theta} \right\} d^3\boldsymbol{k}', \quad (12.5.3.10)$$

where, as before, θ is the angle between the vectors $\boldsymbol{k}$ and $\boldsymbol{E}_0$ and

$$k_{3,4}' = c^{-1} \sqrt{\{2ku\omega_{pe}(\cos\theta \mp \alpha)\}}. \quad (12.5.3.11)$$

We shall also give the expression for the coefficient for the transformation $l_1 \to t$ per unit solid angle and for the total transformation coefficient:

$$\frac{d\Sigma_{l_1 \to t}}{d^2\boldsymbol{\omega}'} = \frac{Ze^2}{48\pi\alpha m_e c^2 r_D^2 k} \sqrt{\frac{m_0}{m_i}} \omega_{pe} \sin^2 \theta' \left\{ \frac{k_3'}{R-\cos\theta} + \frac{k_4'}{R+\cos\theta} \right\}, \quad (12.5.3.12)$$

$$\Sigma_{l_1 \to t} = \frac{Ze^2}{18\alpha m_e c^2 r_D^2 k} \sqrt{\frac{m_0}{m_i}} \left\{ \frac{k_3'}{R-\cos\theta} + \frac{k_4'}{R+\cos\theta} \right\}, \quad (12.5.3.13)$$

where θ' is the angle between the vectors $\boldsymbol{k}'$ and $\boldsymbol{u}$; if $k_{3,4}'^2 < 0$, the corresponding terms in these expressions must be omitted.

We see that as $E_0 \to E_c$ the coefficients for the transformation $l_1 \to t$ tend—in the framework of the linear theory—to infinity for waves with $\theta = 0$ (if $\alpha < 1$) or for waves with $\theta = \pi$ (if $\alpha > 1$), while in both cases $k' = k_0' \equiv c^{-1}\sqrt{(2ku\omega_{pe}|1-\alpha|)}$. Without giving the expressions for the coefficients for the transformation $l_2 \to t$, we note that one can obtain them from the corresponding coefficients for the transformation $l_1 \to t$ by the substitutions $\theta \to \pi-\theta$, $\theta' \to \pi-\theta'$, $\chi \to \pi-\chi$.

Let us now turn to a study of the scattering of longitudinal waves. Substituting expression (11.7.1.2) into eqn. (12.5.3.4) and integrating over the absolute magnitude of the vector $\boldsymbol{k}'$, we get

$$\frac{d\Sigma_{l_1 \to l_1}}{d^2\boldsymbol{\omega}'} = \frac{d\Sigma_{l_2 \to l_2}}{d^2\boldsymbol{\omega}'} = \frac{Ze^2\alpha}{72\pi m_i v_s^2 r_D^2} \sqrt{\frac{2m_i}{m_0}} \cos^2 \vartheta \frac{R}{R^2-\cos^2\chi}, \quad (12.5.3.14)$$

where χ is the angle between the vectors $\boldsymbol{q}$ and $\boldsymbol{E}_0$; we restrict ourselves to the case of waves with a not too long wavelength, $\sqrt{(m_e/m_i)} \ll kr_D \ll 1$. We note that then the processes $l_1 \rightleftarrows l_2$ are impossible, if they involve ion-sound fluctuations.

We see that the angular distribution of the scattered Langmuir waves depends strongly on the external electrical field strength. When $E_0 \ll E_c$, this distribution is almost isotropic —apart from a factor $\cos^2 \vartheta$. When E_0 increases the intensity of the waves scattered over angles $\theta_\pm'$ given by formula (12.5.1.5) increases steeply. As $E_0 \to E_c$ the scattering cross-section per unit solid angle tends—in the framework of the linear theory—to infinity for $\theta' = \theta_\pm'$. Critical opalescence can therefore occur also in the case of scattering of longitudinal waves.

The total cross-section for the scattering of longitudinal waves by ion-sound oscillations increases slowly with increasing external electrical field strength and tends, logarithmically, to infinity as $E_0 \to E_c$.

We have considered the transformation and scattering of external Langmuir waves, that is, Langmuir waves excited by external sources. Of course, even if there are no external sources, there are always Langmuir waves in the plasma with amplitudes determined by the level of the plasma fluctuations: fluctuation Langmuir waves. The existence of fluctuation Langmuir waves leads to the peculiar effect of spontaneous emission by a non-equilibrium plasma caused by the transformation of these waves into transverse electromagnetic waves by ion-sound fluctuations. We shall now study this effect.

The increase per unit time of the energy of transverse waves with wavevectors in the range $\boldsymbol{k}'$, $\boldsymbol{k}'+d\boldsymbol{k}'$ is clearly connected with the transformation coefficient $d\sum_{l\to t}$ through the relation

$$dP = Q\frac{Vd^3\boldsymbol{k}'}{(2\pi)^3}, \quad Q = \int \frac{3\omega_{pe}r_D^2}{8\pi}\langle E^2\rangle^L_{k,\,\omega}\frac{d\sum_{l\to t}}{d^3\boldsymbol{k}'}\frac{d^3\boldsymbol{k}\,d\omega}{2\pi}, \tag{12.5.3.15}$$

where $\langle E^2\rangle^L$ is the term in the expression for the correlation function of the electrical field fluctuations which describes the fluctuation Langmuir waves. We use now eqns. (11.7.1.3) and (12.5.3.4) and the fact that

$$\langle E^2\rangle_{k\omega} = (4\pi)^2 k^{-2}\langle \varrho^2\rangle_{k\omega}.$$

Noting that for the $l \to t$ process we have $k' \ll k$ and integrating over ω, we get

$$Q = \frac{Ze^2}{m_e c^2}\frac{16\zeta\omega_{pe}}{k'^2}\int d^3\boldsymbol{k}\,k\sin^2\vartheta\left\{T_L(k,\cos\theta)\left[\frac{\delta(k-k_1)}{R-\cos\theta}+\frac{\delta(k-k_2)}{R+\cos\theta}\right]\right.$$
$$\left.+T_L(k,-\cos\theta)\left[\frac{\delta(k-\bar{k}_1)}{R+\cos\theta}+\frac{\delta(k-\bar{k}_2)}{R-\cos\theta}\right]\right\}, \tag{12.5.3.16}$$

where T_L is the effective temperature of the Langmuir waves,

$$k_{1,2} = \frac{c^2k'^2}{2\omega_{pe}u(\cos\theta\mp\alpha)},$$

while the quantities $\bar{k}_{1,2}$ are obtained from $k_{1,2}$ through the substitution $\theta\to\pi-\theta$; θ is again the angle between the vectors $\boldsymbol{k}$ and $\boldsymbol{E}_0$.

If $E_c-E_0 \ll E_c$, eqn. (12.5.3.16) becomes

$$Q = 6\pi\frac{Ze^2}{m_e c^2}\zeta T_L(k_0, 1)\sin^2\theta'\ln(R-1), \tag{12.5.3.17}$$

where θ' is the angle between the vectors $\boldsymbol{k}'$ and $\boldsymbol{E}_0$ and $k_0 = c^2k'^2(2\omega_{pe}u)^{-1}|1-\alpha|^{-1}$. We see that as $E_0 \to E_c$ the intensity of the spontaneous emission increases steeply, tending to infinity—in the framework of the linear theory.

We can similarly consider the processes of transformation and scattering of longitudinal waves in a partially ionized plasma which is not only in an external electrical field, but also in an external magnetic field (Angeleĭko and Akhiezer, 1969). Without giving the corresponding expressions we note that in that case also critical opalescence occurs: the scattering and transformation coefficients grow steeply when the external electrical field approaches its critical value $E_c(\boldsymbol{B}_0)$. It is important that in the case of a very strong magnetic field ($qv_s \ll \omega_{Bi}$) the critical opalescence effect is much stronger than in the case of a weak field. In fact, as $E_0 \to E_c$, the total cross-sections for the $1 \to 1$ and $t \to 1$ processes tend to infinity as $(E_c - E_0)^{-1}$, and not as $\ln|1-(E_0/E_c)|$ as in the case when there is no magnetic field or the magnetic field is weak; however, the cross-section for the $1 \to t$ process tends to infinity for all values $\theta < \pi/2$ (for $\alpha < 1$) or $\theta > \pi/2$ (for $\alpha > 1$), and not solely for $\theta = \pi$ or $\theta = 0$, when there is no or a weak magnetic field; θ is again the angle between the average directed electron velocity $\boldsymbol{u}$ and the wavevector of the incident wave.

12.6. Scattering and Transformation of Waves in a Turbulent Plasma

12.6.1. TRANSFORMATION OF LONGITUDINAL INTO TRANSVERSE WAVES

We have seen that the intensity of the processes for scattering and transformation of electromagnetic waves in a plasma is determined by the level of fluctuations in it. In a turbulent plasma, in which the energy of the random waves appreciably exceeds the thermal level, the intensity of the scattered (or transformed) radiation must thus turn out to be much larger than in the case of a quiescent plasma.

The detailed features of the interaction of the waves with the plasma, in particular, the spectral and angular distributions of the scattered (or transformed) radiation are determined by the nature of the spectral and angular distributions of the fluctuations in the plasma. This can enable us, in principle, to reconstruct the turbulence spectrum once we know the cross-sections for the transformation and scattering of waves in a turbulent plasma.

Turning to a study of the processes of scattering and transformation of electromagnetic waves in a turbulent plasma we shall, first of all, consider the transformation of a Langmuir wave into a transverse one as the result of interacting with turbulent sound waves. According to eqn. (12.5.3.3) thc coefficient for the transformation is given by the expression

$$\frac{d\sum_{1\to t}}{d^3\boldsymbol{k}'} = \frac{\pi e^2\omega_{pe}}{6kr_D^2T_e^2}\sin^2\vartheta\left\{I(-\boldsymbol{k})\,\delta\left(\frac{c^2k'^2}{2\omega_{pe}}-\frac{3}{2}\omega_{pe}k^2r_D^2-(\boldsymbol{k}\cdot\boldsymbol{u})-kv_s\right)\right.$$
$$\left.+I(\boldsymbol{k})\,\delta\left(\frac{c^2k'^2}{2\omega_{pe}}-\frac{3}{2}\omega_{pe}k^2r_D^2-(\boldsymbol{k}\cdot\boldsymbol{u})+kv_s\right)\right\}; \qquad (12.6.1.1)$$

to fix the ideas, we have assumed in deriving the formula that the wavevector of the Langmuir wave is not too small, $k^2r_D^2 \gg T_e^2(m_em_ic^4)^{-1}$.

The function $I(\boldsymbol{q})$ which characterizes the level and the distribution of the random sound waves is connected with the correlation function of the electrical potential fluctuations in the sound region ($qv_i \ll \omega \ll qv_e$, $qr_D \ll 1$) through the relation

$$\langle\varphi^2\rangle_{q\omega} = (2\pi)^4\{I(\boldsymbol{q})\,\delta(\omega-qv_s)+I(-\boldsymbol{q})\,\delta(\omega+qv_s)\}. \qquad (12.6.1.2)$$

In the case of a plasma with a directed electron motion the function $I(\boldsymbol{q})$ depends strongly on the angle χ between the wavevector $\boldsymbol{q}$ and the electron current velocity $\boldsymbol{u}$. In the stability region ($\cos\chi < v_s/u$) this function is determined by the well-known formula from the linear fluctuation theory (see Section 11.7)

$$I(\boldsymbol{q}) = \frac{r_D^2 T_e}{(2\pi)^2}\left(1-\frac{u}{v_s}\cos\chi\right)^{-1}. \tag{12.6.1.3}$$

Near the boundary of the stability region, as $\cos\chi \to v_s/u$, the quantity I grows steeply.

Of most interest is the transformation of Langmuir waves for which the angle θ between the vectors $\boldsymbol{k}$ and $\boldsymbol{u}$ satisfies the condition

$$\cos^2\theta \geqslant \left(\frac{v_s}{u}\right)^2. \tag{12.6.1.4}$$

This inequality is the condition for the sound wave involved in the transformation to be turbulent.

The transverse waves which occur as a result of the transformation have, according to eqn. (12.6.1.1), wavevectors k'_+ or k'_-, where

$$k'^2_\pm = 3k^2\frac{T_e}{m_e c^2}\left\{1+\frac{2}{3}\sqrt{\frac{m_e}{m_i}}\frac{1}{kr_D}\left(\pm 1-\frac{u}{v_s}\cos\theta\right)\right\}. \tag{12.6.1.5}$$

Integrating eqn. (12.6.1.1) over $\boldsymbol{k}'$ we get the following expression:

$$\Sigma_{l\to t} = \left(\frac{2\pi}{3}\right)^2\frac{e^2\omega_{pe}^2}{r_D^2 k T_e^2 c^2}\{k'_+ I(-\boldsymbol{k})+k'_- I(\boldsymbol{k})\}. \tag{12.6.1.6}$$

The differential transformation coefficient is connected with the function $\Sigma_{l\to t}$ through the relation

$$\frac{d\Sigma_{l\to t}}{d^2\boldsymbol{\omega}'} = \frac{3}{8\pi}\Sigma_{l\to t}\sin^2\vartheta. \tag{12.6.1.7}$$

Let us investigate how the emitted power changes with the angle θ between the vectors $\boldsymbol{k}$ and $\boldsymbol{u}$. If condition (12.6.1.4) is not satisfied, we get by substituting expression (12.6.1.3) into eqn. (12.6.1.6) the following formula:

$$\Sigma_{l\to t} = \frac{e^2\omega_{pe}^2}{9kT_e c^2}\left\{k'_+\left(1+\frac{u}{v_s}\cos\theta\right)^{-1}+k'_-\left(1-\frac{u}{v_s}\cos\theta\right)^{-1}\right\}. \tag{12.6.1.8}$$

If $kr_D < \sqrt{(2m_e/3m_i)}\,\{1-(u/v_s)\cos\theta\}$, we have $k'^2_- < 0$; in that case we must omit the last term in the braces in eqn. (12.6.1.8).

When the vector $\boldsymbol{k}$ approaches the surface of the critical cone defined by eqn. (12.6.1.4), the quantity $\sum_{l\to t}$ increases steeply. If the vector $\boldsymbol{k}$ lies inside the cone (12.6.1.4), eqn. (12.6.1.6) becomes

$$\sum_{l\to t} = \left(\frac{2\pi}{3}\right)^2 \frac{e^2\omega_p^2}{r_D^2 k T_e^2 c^2}{}_{e}\, k'_{\pm} I(\mp\boldsymbol{k}), \tag{12.6.1.9}$$

where the upper (lower) sign corresponds to the case $\cos\theta < 0$ ($\cos\theta > 0$). In deriving this formula we neglected the contribution from the non-turbulent sound waves. Of course, if $kr_D < \sqrt{(2m_e/3m_i)}\{-1+(u/v_s)|\cos\theta|\}$, we have $k'^2_{\mp} < 0$, and expression (12.6.1.9) therefore vanishes when $\cos\theta < 0$; the function $\sum_{l\to t}$ is in that case determined by the second term in eqn. (12.6.1.8) and is of the same order as in the non-turbulent region.

We see that the quantity $\sum_{l\to t}(\boldsymbol{k})$—apart from the factor $k'_{\pm}$—is proportional to $I(\mp\boldsymbol{k})$; we can therefore use that quantity directly to reconstruct the turbulence spectrum. Especially convenient for the reconstruction of the turbulence spectrum is a study of the transverse radiation distribution which is the result of the transformation of Langmuir oscillations with not too long wavelengths, $k^2r_D^2 \gg (m_e/m_i)\{1-(v/_s u)\}^2$. In that case the function I has the same angular dependence as the function P, differing from the latter merely by a factor proportional to the wavelength,

$$I(\boldsymbol{k}) = \left(\frac{3}{2\pi}\right)^2 \frac{r_D^2 T_e^2 c^2}{e^2\,\omega_{pe}^2} \sqrt{\left(\frac{m_e c^2}{3T_e}\right)} \sum_{l\to t}(\mp\boldsymbol{k}), \tag{12.6.1.10}$$

where the upper (lower) sign refers to the case $\cos\theta < 0$ ($\cos\theta > 0$).

Let us consider in a little more detail the transformation of longitudinal into transverse waves in the case when the main mechanism leading to the establishing of the ion-sound turbulence is the non-linear damping of ion sound (Akhiezer, 1965e). In that case the function $I(\boldsymbol{k})$ is determined by the formulae of Section 10.3. In order to estimate the order of magnitude of the power of the transverse radiation which is the result of the interaction of the Langmuir waves with the turbulent ion-sound oscillations we substitute expression (10.3.2.7) into eqn. (12.6.1.6):

$$\sum_{l\to t} \sim \frac{\omega_{pe}^2 k'_{\pm}}{c^2 r_D^2 k^4} f\left[\frac{m_e T_e}{m_i T_i \omega_{pe}\tau_e}\right]^{1/2}, \tag{12.6.1.11}$$

where f is a function which changes with k more slowly than a power.

We see that if the wavevector of the incident longitudinal waves lies within the critical cone (12.6.1.4), the power of the transverse radiation exceeds the analogous quantity for the case of a non-turbulent plasma by a factor of the order R, where

$$R \sim \frac{T_e}{e^2k^3r_D^2} f\left[\frac{m_e T_e}{m_i T_i \omega_{pe}\tau_e}\right]^{1/2}. \tag{12.6.1.12}$$

The emitted power increases here as $k_{\pm}/k^3$ the wavenumber of the incident wave decreases

The transformation of longitudinal into transverse waves can be of interest also as one possible mechanism for the escape of energy from a plasma. Indeed, there are always random Langmuir waves in a plasma; their amplitude is determined by the electron temperature.

Through interacting with the turbulent sound waves the random Langmuir waves can be transformed into transverse waves which after that may leave the plasma. The power P of such a spontaneous emission can easily be determined by putting $|E_0|^2_k \sim T_e$ into eqn. (12.6.1.1) and integrating over $\boldsymbol{k}$,

$$P \sim \frac{V}{r_D^3}\omega_{pe}T_e\left(\frac{T_e}{m_e c^2}\right)^{3/2}\left[\frac{m_e T_e}{m_i T_i \omega_{pe}\tau_e}\right]^{1/2}. \tag{12.6.1.13}$$

This quantity is larger than the power carried away by the transverse waves when there is no turbulence by a factor R_0, where

$$R_0 = \frac{r_D T_e}{e^2}\left[\frac{m_e T_e}{m_i T_i \omega_{pe}\tau_e}\right]^{1/2}. \tag{12.6.1.14}$$

12.6.2. SCATTERING OF ELECTROMAGNETIC WAVES BY TURBULENT ION-SOUND OSCILLATIONS

Let us now consider the scattering of transverse electromagnetic waves by turbulent ion-sound oscillation. The scattering coefficient $d\Sigma$ which is the ratio of the intensity of the scattered wave to the energy flux density of the incident wave can be expressed in terms of the correlation function of the electron density fluctuations through eqn. (12.1.2.7). In the sound region—medium-range frequencies and long wavelengths, $qv_i \ll \omega \ll qv_e$, $qr_D \ll 1$—this correlation function is connected with the function $I(\boldsymbol{q})$ through the relation

$$e^2\langle n_e^2\rangle_{\boldsymbol{q}\omega} = \pi^2 r_D^{-4}\{I(q)\,\delta(\omega-qv_s)+I(-q)\,\delta(\omega+qv_s)\}. \tag{12.6.2.1}$$

Substituting this expression into eqn. (12.1.2.7) and bearing in mind that the relative change in frequency is small in a scattering process, $\Delta\omega \ll \omega$, we get

$$d\Sigma = \pi^2\left(\frac{e^2}{m_e c^2}\right)^2 n_0\frac{1+\cos^2\vartheta}{r_D^2 T_e}\left\{I(\boldsymbol{q})\,\delta\left(\Delta\omega-2kv_s\sin\frac{1}{2}\vartheta\right)\right.$$
$$\left.+I(-\boldsymbol{q})\,\delta\left(\Delta\omega+2kv_s\sin\frac{1}{2}\vartheta\right)\right\}d^2\boldsymbol{\omega}'\,d\omega', \tag{12.6.2.2}$$

where n_0 is the unperturbed electron density, ϑ the scattering angle, that is, the angle between the vectors $\boldsymbol{k}$ and $\boldsymbol{k}'$, $\boldsymbol{q} = \boldsymbol{k}'-\boldsymbol{k}$, and $\Delta\omega = \omega'-\omega$. The length of the vector $\boldsymbol{q}$ and the angle χ between this vector and the direction of the electron velocity $\boldsymbol{u}$ are determined by the expressions

$$q = 2k\sin\frac{1}{2}\vartheta, \quad \cos\chi = \frac{\cos\theta'-\cos\theta}{2\sin\frac{1}{2}\vartheta}, \tag{12.6.2.3}$$

where $\theta\,(\theta')$ is the angle between $\boldsymbol{k}\,(\boldsymbol{k}')$ and $\boldsymbol{u}$.

The quantity $d\Sigma$ depends differently on the wavevectors of the incident and the scattered waves and is of a completely different order of magnitude depending on whether or not the following inequality is satisfied:

$$|(\boldsymbol{k}\cdot\boldsymbol{u})-(\boldsymbol{k}'\cdot\boldsymbol{u})| \gg |\boldsymbol{k}-\boldsymbol{k}'|\,v_s,$$

which is the condition that the sound wave taking part in the scattering be turbulent. Introducing the angle φ between the $(\boldsymbol{k}, \boldsymbol{u})$- and $(\boldsymbol{k}', \boldsymbol{u})$-planes, it is convenient to write this inequality in the form

$$4\sin^2\frac{1}{2}\varphi \leqslant \frac{1}{\sin\theta\sin\theta'}\left\{\left(\frac{u^2}{v_s^2}-1\right)(\cos\theta-\cos\theta')^2-(\sin\theta-\sin\theta')^2\right\}. \quad (12.6.2.4)$$

Integrating expression (12.6.2.2) over ω' we get the scattering coefficient per unit solid angle,

$$\frac{d\Sigma}{d^2\boldsymbol{\omega}'} = \pi^2\left(\frac{e^2}{m_e c^2}\right)^2 n_0 \frac{1+\cos^2\vartheta}{r_D^2 T_e}\{I(\boldsymbol{q})+I(-\boldsymbol{q})\}. \quad (12.6.2.5)$$

Let us investigate how the quantity $d\Sigma/d^2\boldsymbol{\omega}'$ varies with the direction of the vector $\boldsymbol{k}'$. If the angle φ is sufficiently large so that condition (12.6.2.4) is not satisfied, we get by substituting expression (12.6.1.3) into eqn. (12.6.2.5) the well-known linear-theory formula for the scattering coefficient (compare eqn. (12.1.4.2))

$$\frac{d\Sigma}{d^2\boldsymbol{\omega}'} = \frac{1}{2}n_0\left(\frac{e^2}{m_e c^2}\right)^2(1+\cos^2\vartheta)\left\{1-\frac{u^2(\cos\theta-\cos\theta')^2}{(2v_s\sin\frac{1}{2}\vartheta)^2}\right\}^{-1}. \quad (12.6.2.6)$$

As the vector $\boldsymbol{k}'$ approaches the surface of the critical cone, defined by eqn. (12.6.2.4), the quantity $d\Sigma/d^2\boldsymbol{\omega}'$ grows steeply. If the vector $\boldsymbol{k}'$ lies inside the critical cone (12.6.2.4), eqn. (12.6.2.5) becomes

$$\frac{d\Sigma}{d^2\boldsymbol{\omega}'} = \pi^2\left(\frac{e^2}{m_e c^2}\right)^2 n_0 \frac{1+\cos^2\vartheta}{r_D^2 T_e} I(\pm\boldsymbol{q}), \quad (12.6.2.7)$$

where the upper (lower) sign corresponds to the case $\theta' < \theta$ $(\theta' > \theta)$ while the quantities $\boldsymbol{q}$ and χ are given by eqns. (12.6.2.3).

Equations (12.6.2.4) and (12.6.2.7) can be considerably simplified for the case of a turbulent plasma which is just critical, $1-(v_s/u) \ll 1$. For a strong interaction between an electromagnetic wave and the turbulent fluctuations in such a plasma it is necessary that the angle φ be small and that the angle θ' lie close to $\pi-\theta$. Condition (12.6.2.4) then becomes

$$(\theta+\theta'-\pi)^2+\varphi^2\tan^2\theta \leqslant 8\left(1-\frac{v_s}{u}\right), \quad (12.6.2.8)$$

and the scattering coefficient per unit solid angle is given by eqn. (12.6.2.7) in which we must substitute

$$\cos^2\vartheta = \cos^2 2\theta, \quad q = 2k\,|\cos\theta|, \quad \cos\chi = (-\operatorname{sgn}\cos\theta)\{1-\tfrac{1}{8}[\theta+\theta'-\pi)^2+\varphi^2\tan^2\theta]\}. \quad (12.6.2.9)$$

In principle, we can use eqns. (12.6.2.7) and (12.6.2.3) to find the function $I(\boldsymbol{q})$, once we know the function $d\Sigma/d^2\boldsymbol{\omega}'$ which characterizes the angular distribution of the scattered radiation. In particular, if $\varphi = 0$ and $\theta > \theta'$, we have

$$I(\boldsymbol{q}) = \frac{r_D^2 T_e}{\pi^2 n_0}\left(\frac{e^2}{m_e c^2}\right)^{-2} \frac{d\Sigma/d^2\boldsymbol{\omega}'}{1+\cos^2(\theta-\theta')}, \tag{12.6.2.10}$$

where we must substitute for k and θ' the functions $k(\boldsymbol{q})$ and $\theta'(\boldsymbol{q})$ determined by the relations

$$\sin \tfrac{1}{2}(\theta+\theta') = \cos \chi, \quad \sin \frac{1}{2}|\theta-\theta'| = \frac{q}{2k}.$$

The frequency-dependence of the quantity $d\Sigma/d^2\boldsymbol{\omega}'$ (for fixed θ) thus directly determines the q-dependence of I; the angular dependence of the function I can easily be reconstructed from the angular dependence of the function $d\Sigma/d^2\boldsymbol{\omega}'$.

We note that the change in frequency in the scattering process is uniquely determined by the frequency of the incident wave and the scattering angle,

$$|\Delta\omega| = 2\omega \frac{v_s}{c} \sin \frac{1}{2}\vartheta.$$

Hence, by measuring the magnitude of $\Delta\omega$ we can easily find the ion-sound velocity and from this the electron temperature.

Let us consider in somewhat more detail the scattering of transverse electromagnetic waves in the case when the main mechanism establishing stationary ion-sound turbulence is the non-linear ion-sound damping (Akhiezer, 1965e). In order to estimate the order of magnitude of the cross-section for the scattering of electromagnetic waves by turbulent ion-sound oscillations we substitute expression (10.3.2.7) into eqn. (12.6.2.7):

$$\frac{d\Sigma}{d^2\boldsymbol{\omega}} = n_0^2\left(\frac{e^2}{m_e c^2}\right)^2 f \frac{1}{k^3}\left[\frac{m_e T_e}{m_i T_i \omega_{pe}\tau_e}\right]^{1/2}. \tag{12.6.2.11}$$

It follows from this formula that if the wavevector $\boldsymbol{k}'$ of the scattered wave lies inside the critical cone (12.6.2.4), the scattering cross-section exceeds the analogous quantity for the case of a non-turbulent plasma by a factor of the order R where R is given by formula (12.6.1.12). Comparing eqns. (12.6.2.6) and (12.6.2.11) we see that the energy flux connected with the scattered waves is almost totally concentrated inside the critical cone.

It follows from eqn. (12.6.2.11) that if the vector $\boldsymbol{k}'$ lies inside the critical cone, the quantity $d\Sigma/d^2\boldsymbol{\omega}'$ will be proportional to ω^{-3} where ω is the frequency of the incident wave; the total cross-section for the scattering of electromagnetic waves in a turbulent plasma is also proportional to ω^{-3}. We note that if there is no turbulence—and also in a turbulent plasma if the vector $\boldsymbol{k}'$ lies outside the critical cone—the cross-section for the scattering of transverse waves by ion-sound oscillations is independent of the frequency of the incident wave.

12.6.3. SCATTERING OF ELECTROMAGNETIC WAVES BY TURBULENT HIGH-FREQUENCY OSCILLATIONS

Let us now consider the scattering of transverse electromagnetic waves by turbulent Langmuir oscillations. As in Chapter 10 we shall characterize the intensity of the turbulent Langmuir oscillations by a function $I(\boldsymbol{q})$ which is connected with the correlator of the electrostatic potential through the relation

$$\langle \varphi^2 \rangle_{\boldsymbol{q}\omega} = (2\pi)^4 \{ I(\boldsymbol{q})\, \delta(\omega - \omega_l(\boldsymbol{q})) + I(-\boldsymbol{q})\, \delta(\omega + \omega_l(\boldsymbol{q})) \}, \qquad (12.6.3.1)$$

where $\omega_l(\boldsymbol{q})$ is the frequency of the Langmuir wave.

Using the general formula (12.1.2.7) we can easily obtain the cross-section for the scattering of electromagnetic waves by turbulent Langmuir oscillations, bearing in mind that the correlation function of the electron density fluctuations is in the high-frequency region connected with the correlation function of the potential fluctuations through the relation

$$e^2 \langle n_e^2 \rangle_{\boldsymbol{q}\omega} = (4\pi)^{-2}\, q^4 \langle \varphi^2 \rangle_{\boldsymbol{q}\omega}.$$

If the turbulence in the plasma is caused by an electron current, the Langmuir waves for which $\cos\chi > 0$, where χ is the angle between the vector $\boldsymbol{q}$ and the direction of the electron current, will be turbulent. Using this we get for the cross-section per unit solid angle for scattering of electromagnetic waves by turbulent Langmuir oscillations the following expression:

$$\frac{d\Sigma}{d^2\boldsymbol{\omega}'} = \frac{4\pi e^2 k^4}{(m_e c^2)^2} (1 + \cos^2\vartheta) \sin^4 \frac{1}{2}\vartheta\, I(\pm \boldsymbol{q}), \qquad (12.6.3.2)$$

where the upper (lower) sign refers to the case $\theta > \theta'$ ($\theta < \theta'$); $\theta(\theta')$ is again the angle between the vector $\boldsymbol{k}$ ($\boldsymbol{k}'$) and the direction of the electron current (to fix the ideas we have assumed that $\omega \gg \omega_{pe}$). Equation (12.6.3.2) together with relations (12.6.2.3) for q and χ make it, in principle, possible to find the function $I(\boldsymbol{q})$ once we know the function $d\Sigma/d^2\boldsymbol{\omega}'$ characterizing the angular distribution of the scattered radiation.

Let us now consider the specific nature of the scattering of light by growing fluctuations in a plasma through which a beam of charged particles is passing with a velocity larger than the critical one (Akhiezer, 1963). We showed in Section 11.6 that in the correlation functions for such a system the times t, t' occur not only in the combination $t - t' = \Delta t$, but also separately; the correlators then contain terms growing both with increasing $\bar{t} \equiv \frac{1}{2}(t + t')$ and with increasing $|\Delta t|$. It is important that the correlators oscillate fast with changing Δt—with a frequency of the order $(\boldsymbol{q} \cdot \boldsymbol{u})$, where $\boldsymbol{q}$ is the wavevector of the fluctuations and $\boldsymbol{u}$ the beam velocity—while slowly growing—with a growth rate $\gamma \sim \omega_{pe}(n'/n_0)^{1/2}$ in the non-resonance case and a growth rate $\gamma \sim \omega_{pe}(n'/n_0)^{1/3}$ in the resonance case, where n' is the beam density—with increasing Δt and $\bar{t}$. Therefore, if all the frequencies ω, ω', and $\Delta\omega$ are large compared with the growth rate γ we can use again formula (12.1.2.7) for the scattering cross-section, neglecting the non-oscillating Δt-dependence of the correlator—that is, putting $\gamma\Delta t \to 0$—and assuming the Fourier component of the correlator with respect to Δt to be a slowly varying function of $\bar{t}$.

We thus arrive at the conclusion that when electromagnetic waves are scattered in a beam-plasma system, the spectrum of the scattered radiation contains not only the Doppler broadened main line and the Langmuir satellites, but also an additional line, connected with the fact that in such a plasma oscillations can propagate with a frequency close to $(\boldsymbol{q}\cdot\boldsymbol{u})$. This line has the greatest intensity when $|(\boldsymbol{q}\cdot\boldsymbol{u})| \approx \omega_{pe}$; it is then superimposed upon the Langmuir line.

In contrast to the scattering of light in a plasma without beams, the scattering of light in the plasma-beam system is strongly anisotropic. The scattering will be particularly strong when the direction of the vector $\boldsymbol{k}'$ satisfies the condition $(\boldsymbol{k}'\cdot\boldsymbol{u}) = (\boldsymbol{k}\cdot\boldsymbol{u})\pm\omega_{pe}$. Introducing the angle θ (θ') between the vectors $\boldsymbol{k}$ ($\boldsymbol{k}'$) and $\boldsymbol{u}$ and assuming, to fix the ideas, that $\omega \gg \omega_{pe}$, we can write this relation conveniently in the form

$$\cos\theta' = \cos\theta \pm \frac{c\omega_{pe}}{u\omega}. \tag{12.6.3.3}$$

If condition (12.6.3.3) is satisfied, the differential scattering cross-section has the form

$$d\Sigma = \{1+h(\bar{t})\}\, d\Sigma^0, \tag{12.6.3.4}$$

where $d\Sigma^0$ is the cross-section for the scattering of electromagnetic waves by Langmuir waves in a plasma without beams, which is given by formula (12.1.3.6), while

$$h(t) = \tfrac{1}{9}\{2\cosh(2\gamma t)+4\cosh(\gamma t)+3\}, \quad \gamma = \frac{\sqrt{3}}{2}\,\omega_{pe}\left(\frac{n'}{2n_0}\right)^{1/3}. \tag{12.6.3.5}$$

The term $h(\bar{t})$ is connected with the scattering by the specific oscillations in the plasma-beam system which are absent in the case of a plasma without beams. If $\bar{t} = 0$, $h = 1$; with increasing $\bar{t}$, $h(\bar{t})$ increases. The increase in the scattering cross-section occurs until the amplitudes of the growing fluctuations reach saturation, determined by non-linear effects.

12.7. Echoes in a Plasma

12.7.1. UNDAMPED OSCILLATIONS OF THE DISTRIBUTION FUNCTION AND ECHO EFFECTS IN A PLASMA

We have studied in the preceding sections processes for the scattering and transformation of waves in a plasma, which are caused by non-linear wave interactions. We shall now consider one more effect, caused by the non-linear wave interactions, the so-called *plasma echo*.

Let us first of all remind ourselves that even when there are no collisions the oscillations of macroscopic quantities in the plasma are exponentially damped with time, while deviations of the distribution functions from their equilibrium values can undergo undamped oscillations. The existence of such undamped oscillations of the distribution functions is caused by the fact that if there are no binary collisions it is impossible to establish equilibrium by the action of only the self-consistent field, since the self-consistent field does not change the entropy. The presence of undamped oscillations in the distribution functions also leads to the possibility for the occurrence of echo effects (Gould, O'Neil, and Malmberg, 1967; O'Neil and Gould, 1968).

It is easy to explain qualitatively the mechanism to produce an echo in a plasma. Let there be excited oscillations in the electrical field of the form $\exp(ik_1x)$ at time $t = 0$. These oscillations will be damped exponentially with time (Landau damping) but they will lead to the appearance of undamped oscillations in the electron distribution function in the plasma of the form

$$f_1^{(1)} \sim f_1(v)\exp(ik_1x - ik_1vt). \tag{12.7.1.1}$$

After a sufficiently long time the oscillations in the distribution function will not be accompanied by oscillations in the macroscopic quantities as the integration of the deviation (12.7.1.1) of the distribution function over the velocities will give zero because of the fast-changing nature of the factor $\exp(-ik_1vt)$. Therefore, after a time t, much larger than γ^{-1}, where γ is the Landau damping rate of the corresponding oscillation, there will be practically no electrical field connected with the initial perturbation in the plasma, although a memory of the initial perturbation will be stored in the form of the undamped oscillations of the distribution function.

If we now at time $\tau(\tau \gg \gamma^{-1})$ again excite oscillations of the electrical field in the form $\exp(ik_2x)$, they will lead to the appearance of undamped oscillations of the distribution function of the form

$$f_2^{(1)} \sim f_2(v)\exp\{ik_2x - ik_2v(t-\tau)\}. \tag{12.7.1.2}$$

If we neglect non-linear effects, these oscillations in the distribution function will exist independent of the earlier excited oscillations (12.7.1.1) and when $t-\tau \gg \gamma^{-1}$ the macroscopic manifestations of these oscillations will be absent. However, if we take the non-linear wave–wave interactions into account, there occurs the possibility of the appearance of echo effects. Indeed, the effect of non-linearities in the kinetic equation means that the second perturbation leads not only to the appearance of the oscillations (12.7.1.2) of the distribution function, but also to the appearance of oscillations at the sum-frequencies:

$$f^{(2)} \sim f_1(v)f_2(v)\exp\{i(k_1+k_2)x - i(k_1+k_2)vt + ik_2v\tau\}. \tag{12.7.1.3}$$

If $k_1 < 0$ and $t = k_2\tau/(k_2 - |k_1|)$, the exponent in (12.7.1.3) will be independent of v, and the integral of (12.7.1.3) over the velocities will be non-zero. This means that at the time $t = k_2\tau/(k_2 - |k_1|)$ there arise again macroscopic oscillations—electrical field oscillations. These oscillations are echo oscillations.

In order that echo oscillations can occur it is clearly necessary that the period $t-\tau$ is much longer than the inverse of the Landau damping rate, γ^{-1}:

$$\frac{k_2}{k_2 - |k_1|}\tau - \tau \gg \gamma^{-1}. \tag{12.7.1.4}$$

This condition is satisfied, if $|k_1|/(k_2 - |k_1|)$ is of the order of unity.

We see that the plasma echo is an essentially non-linear effect. The example considered here corresponded to taking second-order non-linearity into account. However, in principle, higher-order echo effects are also possible in a plasma.

Apart from temporal echo effects one can also have spatial echo effects. If electrical field oscillations with a given frequency ω_1 are discontinuously excited at a well-defined spot in the plasma, while at a different place at a distance l from the original point, where $l \gg v_{\text{ph}}/\gamma$ with v_{ph} the phase velocity of the oscillations considered, oscillations are excited with a different frequency ω_2 ($> \omega_1$), at a distance $\omega_2 l/(\omega_2-\omega_1)$ from the first source echo oscillations will occur with frequency $\omega_2-\omega_1$. Higher-order spatial echo effects are also possible: in fact, at distances

$$l_{pq} = \frac{p\omega_2}{\omega_{pq}}\, l \tag{12.7.1.5}$$

from the first source echo oscillations with a sum-frequency

$$\omega_{pq} = p\omega_2 - q\omega_1 > 0, \tag{12.7.1.6}$$

where p and q are integers, will occur.

12.7.2. LONGITUDINAL FIELD ECHO OSCILLATIONS

Let us now turn to a quantitative consideration of temporal echo effects in a plasma (O'Neil and Gould, 1968; Sitenko, Nguen Van Chong, and Pavlenko, 1970a, b). Limiting our considerations to longitudinal oscillations only we shall start from the non-linear kinetic equations for the electron and ion distribution functions and the equation to determine the self-consistent electrical field,

$$\begin{gathered}\frac{\partial F_\alpha}{\partial t} + (\boldsymbol{v}\cdot\nabla)F_\alpha + \frac{e_\alpha}{m_\alpha}\left(\boldsymbol{E}\cdot\frac{\partial F_\alpha}{\partial \boldsymbol{v}}\right) = 0,\\ \operatorname{div}\boldsymbol{E} = 4\pi\left\{\sum_\alpha e_\alpha \int F_\alpha\, d^3v + \varrho^0\right\},\end{gathered} \tag{12.7.2.1}$$

where ϱ^0 is the given external charge density, and $\alpha = $ e, i. We shall assume that the unperturbed state of the plasma is uniform and in equilibrium.

We choose the external charge density ϱ^0 in the form

$$\varrho^0(\boldsymbol{r}, t) = \varrho_1 \exp\{i(\boldsymbol{k}_1\cdot\boldsymbol{r})\}\,\delta\{\omega_0(t-0)\} + \varrho_2 \exp\{i(\boldsymbol{k}_2\cdot\boldsymbol{r})\}\,\delta\{(\omega_0(t-\tau)\}, \tag{12.7.2.2}$$

that is, we shall assume that external perturbations occur in the plasma at times $t = 0$ and $t = \tau$ ($\tau \gg \gamma^{-1}$) and that the spatial behaviour of the perturbations are in the form of plane waves; ω_0 is an arbitrary quantity with the dimensions of a frequency.

Performing a spatial Fourier and a temporal Laplace transformation in the set (12.7.2.1) we get

$$\left.\begin{aligned}&[p + i(\boldsymbol{k}\cdot\boldsymbol{v})]\,F_{\boldsymbol{k}p} + \frac{e}{m}\int\frac{dp'}{2\pi i}\int\frac{d^3\boldsymbol{k}'}{(2\pi)^3}\left(\boldsymbol{E}_{\boldsymbol{k}-\boldsymbol{k}'\;p-p'}\cdot\frac{\partial F_{\boldsymbol{k}'p'}}{\partial \boldsymbol{v}}\right) = 0,\\ &i(\boldsymbol{k}\cdot\boldsymbol{E}_{\boldsymbol{k}p}) = 4\pi\left\{\sum e\int F_{\boldsymbol{k}p}\, d^3v + \varrho^0_{\boldsymbol{k}p}\right\},\end{aligned}\right\} \tag{12.7.2.3}$$

where we have dropped the index α and where

$$\varrho^0_{kp} = \frac{(2\pi)^3}{\omega_0}\{\varrho_1\delta(\boldsymbol{k}-\boldsymbol{k}_1)+\varrho_2 e^{-p\tau}\,\delta(\boldsymbol{k}-\boldsymbol{k}_2)\}. \tag{12.7.2.4}$$

Assuming the external perturbation to be small, we shall solve the set (12.7.2.3) by the method of successive approximations, that is, we shall look for the distribution functions F and the field $\boldsymbol{E}$ in the form

$$F = f_0+f^{(1)}+f^{(2)}+\ldots, \qquad \boldsymbol{E} = \boldsymbol{E}^{(1)}+\boldsymbol{E}^{(2)}+\ldots,$$

where the f_0 are the unperturbed functions for which we shall take Maxwell distributions, while $f^{(1)}$ and $f^{(2)}$ (respectively, $\boldsymbol{E}^{(1)}$ and $\boldsymbol{E}^{(2)}$) are corrections which are linear and quadratic in the external perturbation (12.7.2.4).

In the linear approximation we find (we now put α = e)

$$E^{(1)}_{kp} = -4\pi i\,\frac{\boldsymbol{k}\varrho^0_{kp}}{k^2\varepsilon(\boldsymbol{k}, ip)}, \quad f^{(1)}_{kp} = -\frac{e}{m_e}\left(\boldsymbol{E}^{(1)}_{kp}\cdot\frac{\partial f_0/\partial \boldsymbol{v}}{p+i(\boldsymbol{k}\cdot\boldsymbol{v})}\right), \tag{12.7.2.5}$$

where $\varepsilon(\boldsymbol{k}, ip)$ is the longitudinal dielectric permittivity of the plasma. By using the inverse Laplace transform we easily find the asymptotic temporal behaviour of the quantities (12.7.2.5). The perturbation (12.7.2.2) leads in the plasma to damped electrical field oscillations $\boldsymbol{E}^{(1)}_k(t)$ with frequencies equal to the eigen frequencies of the plasma and there also occur oscillations of the distribution functions $f^{(1)}_k$ with frequencies $\omega = (\boldsymbol{k}\cdot\boldsymbol{v})$.

Taking into account the terms in the kinetic equations which are non-linear in the external perturbation we find the field amplitude in the second approximation

$$\boldsymbol{E}^{(2)}_{kp} = -\frac{(4\pi)^3 e^3\boldsymbol{k}}{m_e^2k^2\varepsilon(\boldsymbol{k}, ip)}\int d^3\boldsymbol{v}\int\limits_{\sigma'-i\infty}^{\sigma'+i\infty}\frac{dp'}{2\pi i}$$
$$\times\int\frac{d^3\boldsymbol{k}'}{(2\pi)^3}\,\frac{(\boldsymbol{k}\cdot\{\boldsymbol{k}-\boldsymbol{k}'\})\,\varrho_{k-k'\;p-p'}\,\varrho_{k'p'}(\boldsymbol{k}'\cdot\{\partial f_0/\partial\boldsymbol{v}\})}{k'^2(\boldsymbol{k}\cdot\boldsymbol{k}')^2\,[p+i(\boldsymbol{k}\cdot\boldsymbol{v})]^2\,[p'+i(\boldsymbol{k}'\cdot\boldsymbol{v})]\,\varepsilon(\boldsymbol{k}', ip')\,\varepsilon(\boldsymbol{k}-\boldsymbol{k}', ip-ip')}\,; \tag{12.7.2.6}$$

we shall here and henceforth assume the plasma to be a one-component plasma, for the sake of simplicity. As the plasma echo is caused by the interference between the first and the second perturbation, we need retain in the product of the components of the external charge density only the cross terms, that is,

$$\varrho_{k-k'\;p-p'}\,\varrho_{k'p'} \to \frac{(2\pi)^6}{\omega_0^2}\,\varrho_1\varrho_2\delta(\boldsymbol{k}-\boldsymbol{k}_1-\boldsymbol{k}_2)\{\delta(\boldsymbol{k}'-\boldsymbol{k}_2)\,e^{-p'\tau}+\delta(\boldsymbol{k}'-\boldsymbol{k}_1)\,e^{-(p-p')\tau}\}. \tag{12.7.2.7}$$

From (12.7.2.7) it follows that the wavevector of the quadratic signal is the sum of the wavevectors of the successive perturbations,

$$\boldsymbol{k} = \boldsymbol{k}_1+\boldsymbol{k}_2. \tag{12.7.2.8}$$

Performing the inverse Laplace transform and using (12.7.2.6) and (12.7.2.7) we find

$$E_k^{(2)}(t) = -\frac{8(2\pi)^6 e^3 \varrho_1\varrho_2}{m_e^2\omega_0^2 k^2 k_1^2 k_2^2}\, \boldsymbol{k}\delta(\boldsymbol{k}-\boldsymbol{k}_1-\boldsymbol{k}_2)\int d^3\boldsymbol{v}\int\limits_{\sigma-i\infty}^{\sigma+i\infty}\frac{dp}{2\pi i}\,\frac{e^{pt}}{[p+i(\boldsymbol{k}\cdot\boldsymbol{v})]^2\,\varepsilon(k, ip)}$$

$$\times\int\limits_{\sigma'-i\infty}^{\sigma'+i\infty}\frac{dp'}{2\pi i}\left\{\frac{(\boldsymbol{k}\cdot\boldsymbol{k}_1)(\boldsymbol{k}_2\cdot\{\partial f_0/\partial\boldsymbol{v}\})\,e^{-p'\tau}}{[p'+i(\boldsymbol{k}_2\cdot\boldsymbol{v})]\,\varepsilon(\boldsymbol{k}_1, ip-ip')\varepsilon(\boldsymbol{k}_2, ip')}+\frac{(\boldsymbol{k}\cdot\boldsymbol{k}_2)(\boldsymbol{k}_1\cdot\{\partial f_0/\partial\boldsymbol{v}\})\,e^{-(p-p')\tau}}{[p'+i(\boldsymbol{k}_1\cdot\boldsymbol{v})]\,\varepsilon(\boldsymbol{k}_1, ip')\,\varepsilon(\boldsymbol{k}_2, ip-ip')}\right\},\tag{12.7.2.9}$$

where $\sigma > \sigma' > 0$.

We can use Cauchy's theorem to integrate in (12.7.2.9) over p' and p, completing the integration contour by an infinitely large semicircle, taken to the right or to the left of the straight lines $\operatorname{Re} p' = \sigma'$ or $\operatorname{Re} p = \sigma$, depending on the sign of the quantities in front of p' or p in the exponent. We must take the semicircle for the integral over p' of the first term in the braces in (12.7.2.9) to the right of the line $\operatorname{Re} p' = \sigma'$, and the integral vanishes. The semicircle for the integral of the second term in the braces in (12.7.2.9) must be chosen to the left of the line $\operatorname{Re} p' = \sigma'$. If the quantity τ and the interval between the time the echo appears and the second perturbation is large compared to γ^{-1}, we need take into account in the integral over p' only the contribution from the pole at $p' = -i(\boldsymbol{k}_1\cdot\boldsymbol{v})$. The contributions from the poles where the dielectric permittivity vanishes can be neglected by virtue of the condition $\gamma\tau \gg 1$. Integrating similarly over p and only taking into account the second-order pole at $p = -i(\boldsymbol{k}\cdot\boldsymbol{v})$, we find

$$E_k^{(2)}(t) = -\frac{8(2\pi)^6 e^3\varrho_1\varrho_2}{m_e^2\omega_0^2 k_1^2 k_2^2 k^2}(\boldsymbol{k}_2\cdot\boldsymbol{k})\,\boldsymbol{k}\delta(\boldsymbol{k}-\boldsymbol{k}_1-\boldsymbol{k}_2)\int\frac{\left(\boldsymbol{k}_1\cdot\dfrac{\partial f_0}{\partial\boldsymbol{v}}\right)e^{-i(\boldsymbol{k}_1\cdot\boldsymbol{v})\tau}}{\varepsilon(\boldsymbol{k}_1, (\boldsymbol{k}_1\cdot\boldsymbol{v}_1))}$$

$$\times\frac{d}{dp}\left[\frac{e^{p(t-\tau)}}{\varepsilon(\boldsymbol{k}, ip)\,\varepsilon(\boldsymbol{k}_2, ip-(\boldsymbol{k}_1\cdot\boldsymbol{v}))}\right]_{p=-i(\boldsymbol{k}\cdot\boldsymbol{v})}d^3\boldsymbol{v} = -\frac{8(2\pi)^6 e^3\varrho_1\varrho_2}{m_e^2\omega_0^2 k_1^2 k_2^2 k^2}(\boldsymbol{k}_2\cdot\boldsymbol{k})\,\boldsymbol{k}(t-\tau)$$

$$\times\delta(\boldsymbol{k}-\boldsymbol{k}_1-\boldsymbol{k}_2)\int\frac{\left(\boldsymbol{k}_1\cdot\dfrac{\partial f_0}{\partial\boldsymbol{v}}\right)e^{-i(\boldsymbol{k}\cdot\boldsymbol{v})t+i(\boldsymbol{k}_2\cdot\boldsymbol{v})\tau}}{\varepsilon(\boldsymbol{k}_1, (\boldsymbol{k}_1\cdot\boldsymbol{v}))\,\varepsilon(\boldsymbol{k}_2, (\boldsymbol{k}_2\cdot\boldsymbol{v}))\,\varepsilon((\boldsymbol{k}, (\boldsymbol{k}\cdot\boldsymbol{v}))}\,d^3\boldsymbol{v}.\tag{12.7.2.10}$$

We can write the exponential in the integral in this equation in the form $e^{-i(\boldsymbol{k}\cdot\boldsymbol{v})(t-\tau')}$ where

$$\tau' = \frac{(\boldsymbol{k}_2\cdot\boldsymbol{v})}{(\boldsymbol{k}\cdot\boldsymbol{v})}\,\tau.\tag{12.7.2.11}$$

At time $t = \tau'$ this exponential becomes unity and the field strength $E_k^{(2)}(t)$ attains its largest absolute magnitude. When t, however, is not equal to τ', the expression for the field $\boldsymbol{E}^{(2)}$ vanishes after integrating over the velocities because of the fast oscillations of the exponential factor.

The vectors $\boldsymbol{k}_1$, $\boldsymbol{k}_2$, and $\boldsymbol{k}$ are connected through eqn. (12.7.2.8). It is clear that in order that an echo can occur it is necessary that the inequality $\tau' > \tau$ is satisfied. Therefore, if the wavevector $\boldsymbol{k}_2$ is parallel to $\boldsymbol{k}_1$, so that, from (12.7.2.11), $\tau' < \tau$, the echo can not occur. One can show by explicit calculations that an echo can occur only when the vector $\boldsymbol{k}_2$ is antiparallel to the vector $\boldsymbol{k}_1$ and the absolute magnitude of $\boldsymbol{k}_2$ is larger than that of $\boldsymbol{k}_1$: in that case $\tau' > \tau$. One can show that when there is an echo, the angle between the vectors $\boldsymbol{k}_1$ and $\boldsymbol{k}_2$ may differ from π only by a small amount of the order $\sqrt{(\omega_{pe}/\tau)}(kv_e)^{-1}$.

12.7.3. ECHO WHEN THE WAVEVECTORS OF THE PERTURBATIONS ARE ANTIPARALLEL

We shall assume that the vector $\boldsymbol{k}_2$ is antiparallel to the vector $\boldsymbol{k}_1$ and that $k_2 > k_1$. In that case $k = k_2 - k_1$ and

$$\tau' = \frac{k_2}{k_2 - k_1}\tau. \tag{12.7.3.1}$$

To simplify the calculations we shall assume that $k_1 = \frac{1}{2}k_2$, so that $k = k_1$. Integrating in (12.7.2.10) over the velocity components at right angles to $\boldsymbol{k}$ we find for the field $\boldsymbol{E}^{(2)}(\boldsymbol{r}, t)$ $(\boldsymbol{k} = \boldsymbol{k}_1 + \boldsymbol{k}_2)$:

$$\boldsymbol{E}^{(2)}(\boldsymbol{r}, t) = \frac{4(2\pi)^3 e^3 \varrho_1 \varrho_2}{m_e^2 \omega_0^2 k^3} (t-\tau)\, \boldsymbol{k} e^{i(\boldsymbol{k}\cdot\boldsymbol{r})} \int\limits_{-\infty}^{+\infty} \frac{(\partial f_0/\partial v)\, e^{-ikv(t-\tau')}}{\varepsilon(\boldsymbol{k}_1, -(\boldsymbol{k}_1\cdot\boldsymbol{v}))\, \varepsilon(\boldsymbol{k}_2, (\boldsymbol{k}_2\cdot\boldsymbol{v}))\, \varepsilon(\boldsymbol{k}, (\boldsymbol{k}\cdot\boldsymbol{v}))}\, dv. \tag{12.7.3.2}$$

The factors $\varepsilon^{-1}(\boldsymbol{k}_1, -(\boldsymbol{k}_1\cdot\boldsymbol{v}))$, $\varepsilon^{-1}(\boldsymbol{k}_2, (\boldsymbol{k}_2\cdot\boldsymbol{v}))$, and $\varepsilon^{-1}(\boldsymbol{k}, (\boldsymbol{k}\cdot\boldsymbol{v}))$ in eqn. (12.7.2.3) have a simple physical meaning. They describe the effect of the dielectric properties of the plasma on the external perturbation and on the echo field. If $kr_D \ll 1$, the factor $\varepsilon^{-1}(\boldsymbol{k}, (\boldsymbol{k}\cdot\boldsymbol{v}))$ has a steep maximum for velocities v close to the phase velocity of the corresponding waves in the plasma.

We can perform the remaining integration in (12.7.3.2) over the velocity component along $\boldsymbol{k}$ by going into the complex v-plane and completing the integration contour by a semicircle of infinitely large radius, lying either in the upper or in the lower half-plane. If $t < \tau'$, we complete the integration contour with a semicircle in the upper half-plane (the contribution from the integral along this semicircle vanishes) and we find that only the pole—which lies inside the contour—corresponding to the condition $\varepsilon(\boldsymbol{k}_1, -(\boldsymbol{k}_1\cdot\boldsymbol{v})) = 0$ contributes to the result:

$$\boldsymbol{E}^{(2)}(\boldsymbol{r}, t) = \frac{64\pi^3 e^3 \varrho_1 \varrho_2 \boldsymbol{k}}{m_e^2 \omega_0^2 k_1 k_2 k} (t-\tau)\, e^{i(\boldsymbol{k}\cdot\boldsymbol{r})}$$

$$\times \sum_{\pm} \frac{\left.\dfrac{\partial f_0}{\partial v}\right|_{v=\pm\omega_{pe}/k_1} \exp\left\{\dfrac{k}{k_1}(\pm i\omega_{pe} - \gamma_1)(\tau' - t)\right\}}{\left.\dfrac{\partial}{\partial v}\varepsilon(\boldsymbol{k}_1, -k_1 v)\right|_{v=\pm\omega_{pe}/k_1} \varepsilon\{\boldsymbol{k}_2, (k_2/k_1)(\pm\omega_{pe} - i\gamma_1)\}\, \varepsilon\{\boldsymbol{k}, (k/k_1)(\pm\omega_{pe} - i\gamma_1)\}}. \tag{12.7.3.3}$$

If $t > \tau'$, we must complete the integration contour by a semi-circle in the lower v-half-plane. The poles corresponding to the conditions $\varepsilon(\boldsymbol{k}_2, (\boldsymbol{k}_2\cdot\boldsymbol{v})) = 0$ and $\varepsilon(\boldsymbol{k}, (\boldsymbol{k}\cdot\boldsymbol{v})) = 0$ then contribute to the result:

$$E^{(2)}(\mathbf{r}, t) = \frac{64\pi^3 e^3 \varrho_1 \varrho_2 \mathbf{k}}{m_e^2 \omega_0^2 k_1 k_2 k} (t-\tau)\, e^{i(\mathbf{k}\cdot\mathbf{r})}$$

$$\times \left\{ \sum_{\pm} \frac{\left.\frac{\partial f_0}{\partial v}\right|_{v=\pm\omega_{pe}/k_2} \exp\left[\frac{k}{k_2}(\pm i\omega_{pe}-\gamma_2)(t-\tau')\right]}{\varepsilon\left(\mathbf{k}_1, \frac{k_1}{k_2}(\pm\omega_{pe}-i\gamma_2)\right) \left.\frac{\partial}{\partial v}\varepsilon(\mathbf{k}_2, k_2 v)\right|_{v=\pm\omega_{pe}/k_2} \varepsilon\left(\mathbf{k}, \frac{k}{k_2}(\pm\omega_{pe}-i\gamma_2)\right)} \right.$$

$$\left. + \sum_{\pm} \frac{\left.\frac{\partial f_0}{\partial v}\right|_{v=\pm\omega_{pe}/k} \exp\left[(\pm i\omega_{pe}-\gamma)(t-\tau')\right]}{\varepsilon\left(\mathbf{k}_1, \frac{k_1}{k}(\pm\omega_{pe}-i\gamma)\right) \varepsilon\left(\mathbf{k}_2, \frac{k_2}{k}(\pm\omega_{pe}-i\gamma)\right) \left.\frac{\partial}{\partial v}\varepsilon(\mathbf{k}, kv)\right|_{v=\pm\omega_{pe}/k}} \right\}. \quad (12.7.3.4)$$

We must note the asymmetric nature of the echo signal as function of time. The growth of the echo oscillations is given by the exponent $\exp\{-\gamma_1(\tau'-t)\}$, while its damping is given by the terms with the exponents $\exp\{-\gamma_2(t-\tau')\}$ and $\exp\{-\gamma(t-\tau')\}$, where γ_1, γ_2, and γ are the Landau damping coefficients for waves with wavevectors $\mathbf{k}_1$, $\mathbf{k}_2$, and $\mathbf{k}$.

Let us determine the form of the echo signal for the case when all three oscillations correspond to Langmuir frequencies and $kr_D \ll 1$. As $\gamma \sim \exp(-\frac{1}{2}k^{-2}r_D^{-2})$ and $k_2 = 2k$, we have $\gamma_2 \gg \gamma$. We can thus neglect in (12.7.3.4) the contribution from the first term. Using also expression (4.3.4.2) for $\varepsilon(\mathbf{k}, \omega)$ we can obtain the following expression for the amplitude of the echo oscillations:

$$E^{(2)}(\mathbf{r}, t) = -i\frac{32\pi^2 e\omega_{pe}}{m_e\omega_0^2 k^2} \mathbf{k}\tau\varrho_1\varrho_2 \sin\varphi\, e^{i(\mathbf{k}\cdot\mathbf{r})-\gamma|t-\tau'|} \cos[\omega_{pe}(t-\tau')+\varphi], \quad (12.7.3.5)$$

where the phase φ is determined by the relation

$$\tan\varphi = \frac{2k}{k-k_1}\frac{\gamma}{\omega_{pe}}.$$

We can similarly easily find the form of the echo signal in a non-isothermal plasma in the case of sound oscillations:

$$E^{(2)}(\mathbf{r}, t) = 8\pi^2 i\frac{\omega_{pe}^2 e\varrho_1\varrho_2\tau k^6 r_D^6}{m_i\omega_0^2\gamma_s} \mathbf{k} \exp[i(\mathbf{k}\cdot\mathbf{r})-\gamma_s|t-\tau'|] \cos[kv_s(t-\tau')]. \quad (12.7.3.6)$$

We note that in this case the shape of the signal is symmetric in time.

Concluding this section we shall give expressions for the amplitudes of the echo oscillations when echo sound oscillations are produced as the result of the superposition of

Langmuir oscillations:

$$E^{(2)}(r, t) = -\frac{32\pi^2 i}{3}\frac{e\varrho_1\varrho_2}{m_e\omega_{pe}^2 k^2}\omega_{pe}\tau k e^{i(k\cdot r)}\begin{cases}\sin\varphi\, e^{\gamma_1(t-\tau')}\cos[\omega_{pe})(t-\tau')+\varphi], & t<\tau';\\ -\left\{\frac{2}{3}\frac{\gamma_2}{\omega_{pe}}\exp\left[-\frac{k}{k_2}\gamma_2(t-\tau')\right]\cos\left[\frac{k}{k_2}\omega_{pe}(t-\tau')\right]\right. \\ \left.+\frac{3}{4}k^4 r_D^4\frac{k^2 v_s^2}{\gamma_s\omega_{pe}}\exp[-\gamma_s(t-\tau')]\cos[kv_s(t-\tau')],\right. & t>\tau';\end{cases} \tag{12.7.3.7}$$

and when echo Langmuir oscillations are formed as the result of the superposition of sound oscillations on Langmuir oscillations:

$$E^{(2)}(r, t) = -\frac{32\pi^2 i}{3}\frac{e\varrho_1\varrho_2}{m_e\omega_{pe}^2 k^2}\omega_{pe}\tau k e^{i(k\cdot r)}\begin{cases}-\frac{3}{4}k^4 r_D^4\frac{k^2 v_s^2}{\omega_{pe}\gamma_s}\exp[\gamma_s(t-\tau')]\cos[kv_s(t-\tau')], & t<\tau';\\ \sin\varphi\, e^{-\gamma(t-\tau')}\cos[\omega_{pe}(t-\tau')+\varphi], & t>\tau'.\end{cases} \tag{12.7.3.8}$$

The echo oscillations (12.7.3.7) and (12.7.3.8) are symmetric in time.

We note that taking binary collisions into account leads to a damping of the oscillations of the distribution functions and hence to a weakening of the echo effects in a plasma.

CHAPTER 13

Scattering of Charged Particles in a Plasma

13.1. The Passage of Charged Particles Through an Unmagnetized Plasma

13.1.1. THE FIELD OF A CHARGE IN A PLASMA

We mentioned already in Section 1.1 that the field of a charged test particle which is introduced into the plasma is screened due to the polarization effect. If the particle moves, there is apart from the screening of its field yet another effect: the deceleration of the particle caused by the excitation of waves in the plasma by the moving particle.†

We now turn to a study of these effects. We shall first of all determine the field produced by a charged particle moving in the plasma. To do this we must use the Maxwell equations and take into account the extra charge density ϱ_0 and current density $\boldsymbol{j}_0$ connected with the moving particle. Assuming the particle velocity $\boldsymbol{v}$ to be sufficiently large—it must be much larger than the thermal velocities of the particles in the plasma—we can write ϱ_0 and $\boldsymbol{j}_0$ in the form

$$\varrho_0 = Ze\,\delta(\boldsymbol{r}-\boldsymbol{v}t), \quad \boldsymbol{j}_0 = Ze\boldsymbol{v}\,\delta(\boldsymbol{r}-\boldsymbol{v}t), \tag{13.1.1.1}$$

where Ze is the charge of the particle.

Expanding the fields and currents in Fourier integrals and eliminating the magnetic induction from the equations we obtain the following equations for the electrical field strength:

$$\sum_j \left\{ n^2\left(\frac{k_i k_j}{k^2}-\delta_{ij}\right)+\varepsilon_{ij}(\boldsymbol{k},\omega)\right\} E_j(\boldsymbol{k},\omega) = \frac{4\pi}{i\omega} j_{0i}(\boldsymbol{k},\omega), \tag{13.1.1.2}$$

where $\boldsymbol{j}_0(\boldsymbol{k},\omega)$ is the Fourier component of the current density connected with the moving charge,

$$\boldsymbol{j}_0(\boldsymbol{k},\omega) = \frac{1}{(2\pi)^3} Ze\boldsymbol{v}\,\delta\{(\omega-(\boldsymbol{k}\cdot\boldsymbol{v})\}. \tag{13.1.1.3}$$

† Bohr (1948), Fermi (1940), and Tamm and Frank (1937) developed the general theory of the passage of charged particles through a medium. Several authors (Vlasov, 1950; Akhiezer and Sitenko, 1952; Bohm and Pines, 1951; Pines and Bohm, 1952; Lindhard, 1954; Neufeld and Ritchie, 1955; Sitenko and Stepanov, 1958; Larkin, 1960) have studied the energy loss of particles moving through an unmagnetized plasma.

In the case of an isotropic plasma the dielectric permittivity tensor is given by eqn. (4.3.4.2) and the solution of eqn. (13.1.1.2) has the form

$$E_i(\boldsymbol{k}, \omega) = \frac{4\pi}{i\omega}\sum_j \left\{\frac{k_i k_j}{k^2}\frac{1}{\varepsilon_l(\boldsymbol{k}, \omega)} + \left[\delta_{ij} - \frac{k_i k_j}{k^2}\right]\frac{1}{\varepsilon_t(\boldsymbol{k}, \omega) - n^2}\right\} j_{0j}(\boldsymbol{k}, \omega). \quad (13.1.1.4)$$

Using the inverse Fourier transformation we get from this the following expression for the electrical field of a point charge moving with a constant velocity through the plasma ($\omega = (\boldsymbol{k}\cdot\boldsymbol{v})$):

$$\boldsymbol{E}(\boldsymbol{r}, t) = \frac{Ze}{2\pi^2 i}\int \left\{\frac{\boldsymbol{k}}{k^2\varepsilon_l(\boldsymbol{k}, \omega)} + \frac{\boldsymbol{v} - \boldsymbol{k}[(\boldsymbol{k}\cdot\boldsymbol{v})/k^2]}{\omega[\varepsilon_t(\boldsymbol{k}, \omega) - (k^2c^2/\omega^2)]}\right\} e^{i(\boldsymbol{k}\cdot\boldsymbol{r}) - i\omega t}\, d^3\boldsymbol{k}, \quad (13.1.1.5)$$

The first term in (13.1.1.5) describes the potential (longitudinal) part of the field of a moving charge and the second term corresponds to the rotational (transverse) part of the field.

If the particle is at rest ($\boldsymbol{v} = 0$) there is no rotational part in (13.1.1.5) and the field is completely longitudinal. If we introduce for that case the scalar potential of the field,

$$\boldsymbol{E} = -\text{grad}\, \varphi,$$

we find from (13.1.1.5)

$$\varphi(\boldsymbol{r}) = \frac{Ze}{2\pi^2}\int \frac{e^{i(\boldsymbol{k}\cdot\boldsymbol{r})}}{k^2\varepsilon_l(\boldsymbol{k}, 0)}\, d^3\boldsymbol{k}. \quad (13.1.1.6)$$

Substituting in (13.1.1.6) the static value of the longitudinal plasma dielectric permittivity,

$$\varepsilon_l(\boldsymbol{k}, 0) = 1 + (kr_D)^{-2},$$

and integrating, we get finally the following expression for the potential of the field of a charge at rest in the plasma:

$$\varphi(\boldsymbol{r}) = \frac{Ze}{r}\exp(-r/r_D), \quad (13.1.1.7)$$

where we have assumed the charge to be at the origin. We obtained this result already earlier (eqn. (1.1.1.3)).

13.1.2. POLARIZATION ENERGY LOSSES WHEN A CHARGED PARTICLE MOVES THROUGH THE PLASMA

When a charge particle moves through a plasma it loses part of its energy due to its interaction with neighbouring particles (in what follows we shall follow the paper by Akhiezer and Sitenko, 1952). We can clearly consider the energy losses as the work done by the braking force acting on the particle due to the electromagnetic field produced by the particle itself. The change in the particle energy per unit time is equal to

$$\frac{d\mathcal{E}}{dt} = Ze(\boldsymbol{v}\cdot\boldsymbol{E})\bigg|_{\boldsymbol{r} = \boldsymbol{v}t}, \quad (13.1.2.1)$$

where $\boldsymbol{E}$ is the field produced by the particle at its own position.

If the particle moves with a high velocity its energy loss will be relatively small and the particle velocity will practically remain constant. Under those conditions the field $\boldsymbol{E}$ is given by eqn. (13.1.1.5) and the expression for $d\mathcal{E}/dt$ becomes ($\omega = (\boldsymbol{k}\cdot\boldsymbol{v})$)

$$\frac{d\mathcal{E}}{dt} = \frac{Z^2e^2}{2\pi^2 i}\int\left\{\frac{\omega}{k^2\varepsilon_l(\boldsymbol{k},\omega)}+\frac{v^2-(\omega^2/k^2)}{\omega[\varepsilon_t(\boldsymbol{k},\omega)-(k^2c^2/\omega^2)]}\right\}d^3\boldsymbol{k}. \qquad (13.1.2.2)$$

We must here integrate over k up to some maximum value k_0 for which we can still employ safely the macroscopic arguments for the interaction between the particle and the plasma which we have used here.

Using the fact that the real parts of ε_l and ε_t are even functions of the frequency and the imaginary parts odd functions, we can rewrite eqn. (13.1.2.2) in the form ($\omega = (\boldsymbol{k}\cdot\boldsymbol{v})$)

$$\frac{d\mathcal{E}}{dt} = \frac{Z^2e^2}{2\pi^2}\,\mathrm{Im}\int\left\{\frac{\omega}{k^2\varepsilon_l(\boldsymbol{k},\omega)}+\frac{v^2-(\omega^2/k^2)}{\omega[(\varepsilon_t\boldsymbol{k},\omega)-(k^2c^2/\omega^2)]}\right\}d^3\boldsymbol{k}. \qquad (13.1.2.3)$$

The first term in (13.1.2.3) corresponds to the interaction between the particle and the longitudinal field and determines the polarization energy losses of the particle. The second term in (13.1.2.3) takes into account the interaction of the particle with the transverse field and corresponds in the general case to the energy losses connected with the emission of transverse waves (Cherenkov emission). As the phase velocities of the transverse waves in an isotropic plasma exceed the velocity of light the condition for Cherenkov emission is not satisfied and in that case the second term in (13.1.2.3) vanishes.

We note that both the polarization energy losses (if we neglect the particle recoil) and the losses connected with Cherenkov emission, are independent of the mass of the moving particle. In particular, these losses will be non-vanishing even for a particle which is infinitely heavy.

Let us evaluate the polarization energy losses for a charged particle moving through an unmagnetized equilibrium plasma. Noting that $d^3\boldsymbol{k} = 2\pi\varkappa d\varkappa\, d\omega/v$, where $\varkappa$ is the component of $\boldsymbol{k}$ at right angles to the velocity $\boldsymbol{v}$ we can write the expression for the polarization losses in the form ($\varphi(z) = 2z\exp(-z^2)\int_0^z \exp(x^2)\,dx$; $z = \omega/kv_e\sqrt{2}$)

$$\frac{d\mathcal{E}}{dt} = \frac{Z^2e^2}{\pi v}\,\mathrm{Im}\int_0^{\varkappa_0}\varkappa\,d\varkappa\int_{-\infty}^{+\infty}\frac{\omega\,d\omega}{\varkappa^2+(\omega^2/v^2)+r_D^2[1-\varphi(z)+i\sqrt{\pi}\,z\exp(-z^2)]} \qquad (13.1.2.4)$$

Expression (13.1.2.4) diverges as $\varkappa_0 \to \infty$. This is connected with the fact that if we describe the plasma by macroscopic dielectric permittivities we take in fact only the long-range interactions into account and neglect the close collisions of the particle with the electrons in the plasma. One may neglect these collisions if the collision parameter b is much larger than v/ω_{pe}; however, if $b < v/\omega_{pe}$ the main role is played by collisions between the charged particle and the separate electrons in the plasma so that the macroscopic description of the interaction between the particle and the plasma loses its meaning. Small values of b correspond to large values of $\varkappa_0$ so that, as we have noted already, the integration over $\varkappa$ in (13.1.2.4) must be taken up to some maximum value $\varkappa_0$.

If the particle velocity is much larger than the thermal velocity v_e of the electrons in the plasma, the important z-values in equation (13.1.2.4) are those for which $z \gg 1$. Using the asymptotic expansion for the function $\varphi(z)$ we can write the energy losses caused by far collisions in the form

$$\frac{d\mathcal{E}}{dt} = \frac{Z^2e^2\omega_{pe}^2}{\pi v} \operatorname{Im} \int_0^{\varkappa_0} \varkappa \, d\varkappa \int_{-\infty}^{+\infty} \frac{\omega \, d\omega}{[\varkappa^2+(\omega^2/v^2)]\,[\omega^2-\omega_{pe}^2+2i\gamma\omega]}, \tag{13.1.2.5}$$

where γ is the damping rate (4.2.2.7) of the plasma waves.

The integral over ω along the real axis is equal to the residue of the only pole $\omega = i\varkappa v$ which lies in the upper half-plane; we have thus

$$\frac{d\mathcal{E}}{dt} = -\frac{Z^2e^2\omega_{pe}^2}{v} \int_0^{\varkappa_0} \frac{\varkappa \, d\varkappa}{\varkappa^2+(\omega_{pe}/v)^2}. \tag{13.1.2.5a}$$

Assuming that $\varkappa_0 v \gg \omega_{pe}$ we get finally the following expression for the energy losses caused by far collisions:

$$\frac{d\mathcal{E}}{dt} = -\frac{4\pi n_0 Z^2 e^4}{m_e v} \ln \frac{\varkappa_0 v}{\omega_{pe}}. \tag{13.1.2.6}$$

To obtain the total energy losses of the particle we must add to this expression the additional energy losses of the particles in close collisions. Generally speaking, we must describe close collisions quantum-mechanically, and one can use a classical description only in the case of a very short de Broglie wavelength. More precisely, in order to be able to use a classical description it is necessary that the particle de Broglie wavelength,

$$\lambda\!\!\!^{-} = \frac{\hbar}{(Mm_e/[M+m_e])\,v},$$

where M is the particle mass and m_e the electron mass, is much shorter than the effective impact parameter b between the moving particle and the plasma particles (electrons). In classical mechanics the quantity b is given by

$$b \sim \frac{Z^2e^2}{(Mm_e/[M+m_e])\,v^2}.$$

The criterion for the applicability of the classical discussion,

$$\lambda\!\!\!^{-} \ll b,$$

can thus be written as

$$v \ll \frac{Ze^2}{\hbar}. \tag{13.1.2.7}$$

We have also assumed that the velocity of the particle is large compared to the average thermal velocity of the plasma particles,

$$v \gg v_e.$$

If $v > Ze^2/\hbar$ we must use a quantum-mechanical treatment of the close collisions which we shall give later on, but now we shall consider the case when condition (13.1.2.7) is satisfied, that is, when the classical discussion is valid.

First of all, we connect the upper limit $\varkappa_0$ of the integral which determines the energy losses of the particle in distant collisions with the corresponding value b_0 of the collision parameter. To do this we shall consider the energy loss in distant collisions as the energy flux of the electromagnetic field which passes through a cylindrical surface of radius b_0 around the trajectory of the charge:

$$\frac{d\mathcal{E}}{dt} = \frac{c}{4\pi}\int ([\boldsymbol{E} \wedge \boldsymbol{B}]\cdot d^2\boldsymbol{S}),$$

where $d^2\boldsymbol{S}$ is a surface area element of that cylindrical surface. Using the axial symmetry of the field we can rewrite this expression in the form

$$\frac{d\mathcal{E}}{dt} = -\frac{1}{2} b_0 c \int_{-\infty}^{\infty} E_z B \, dz, \tag{13.1.2.8}$$

where we have taken the z-axis along the motion of the particle. If $v \gg v_e$ the spatial dispersion is unimportant. There is in that case no Cherenkov emission and the energy losses (13.1.2.8) are therefore solely connected with the longitudinal part of the electrical field. Using (13.1.1.5) we can write the components of the longitudinal electrical and the magnetic field in the form

$$\left.\begin{aligned} E_z^l &= -\frac{iZe}{\pi v^2}\int_{-\infty}^{+\infty} \frac{\omega}{\varepsilon(\omega)} K_0\left(\frac{b|\omega|}{v}\right) \exp\left[i\omega\left(\frac{z}{v}-t\right)\right] d\omega, \\ B &= \frac{Ze}{\pi c}\int_{-\infty}^{+\infty} kK_1(kb) \exp\left[i\omega\left(\frac{z}{v}-t\right)\right] d\omega, \quad k = \frac{|\omega|}{c}\sqrt{\left[1-\frac{v^2}{c}\varepsilon(\omega)\right]}. \end{aligned}\right\} \tag{13.1.2.9}$$

Substituting these expressions into (13.1.2.7) and noting that if we neglect the spatial dispersion $\varepsilon(\omega) = 1-(\omega_{pe}/\omega)^2$, we find after integration

$$\frac{d\mathcal{E}}{dt} = -\frac{Z^2e^2\omega_{pe}^3 b_0}{vc} K_0\left(\frac{b_0\omega_{pe}}{v}\right) K_1\left(\frac{b_0\omega_{pe}}{v}\right). \tag{13.1.2.10}$$

Assuming that $\omega_{pe} b_0/v \ll 1$ and using the asymptotic expressions for the Macdonald functions for small values of the argument,

$$K_0(x) \approx \ln\frac{2}{\gamma x}, \quad K_1(x) \approx \frac{1}{x},$$

where $\gamma \approx 1.78$, we get

$$\frac{d\mathcal{E}}{dt} = -\frac{4\pi n_0 Z^2 e^4}{m_e v} \ln\left(\frac{2v}{\gamma b_0 \omega_{pe}}\right). \tag{13.1.2.11}$$

Comparing this expression with expression (13.1.2.6) we see that

$$\varkappa_0 = \frac{2}{\gamma}\frac{1}{b_0} \approx \frac{1.123}{b_0}. \tag{13.1.2.12}$$

Let us now determine the energy losses of a particle caused by close collisions in the classical case, that is, when condition (13.1.2.7) is satisfied. If n_0 is the electron density in the plasma and b the collision parameter, the average number of collisions during a time interval dt, for which b lies between b and $b+db$, is equal to $2\pi n_0 vb\, db\, dt$. The energy lost by the particle in collision with the electrons in the plasma is equal to

$$\Delta\mathcal{E} = -\frac{2Z^2e^4}{m_e v^2}\frac{1}{b^2+\xi^2}, \quad \xi = \frac{Ze^2(M+m_e)}{Mm_e v^2}, \tag{13.1.2.13}$$

where Ze is the charge and M the mass of the particle. Multiplying this expression by $2\pi n_0 vb\, db\, dt$ we find the energy losses caused by close collisions between the particle and the electrons in the plasma,

$$\frac{d\mathcal{E}}{dt} = -\frac{4\pi n_0 Z^2 e^4}{m_e v}\int_0^{b_0}\frac{b\, db}{b^2+\xi^2}, \tag{13.1.2.14}$$

where the integration over b is from zero to some value b_0 for which the electrons can still be considered to be free. Assuming that $b_0 \gg \xi$ we find

$$\frac{d\mathcal{E}}{dt} = -\frac{4\pi n_0 Z^2 e^4}{m_e v}\ln\left[\frac{Mm_e v^2}{Ze^2(M+m_e)}b_0\right]. \tag{13.1.2.15}$$

Adding expressions (13.1.2.11) and (13.1.2.15) we find the total energy losses of a moving particle per unit time:

$$\frac{d\mathcal{E}}{dt} = -\frac{4\pi n_0 Z^2 e^4}{m_e v}\ln\left[\frac{2}{\gamma}\frac{Mm_e v^3}{Ze^2(M+m_e)\,\omega_{pe}}\right]. \tag{13.1.2.16}$$

This formula is valid in the classical case, that is, when the inequality $v \ll Ze^2/\hbar$ is satisfied; moreover, the particle velocity must be much higher than the thermal velocity of the electrons in the plasma, $v \gg v_e$.

In the quantal case, when $v \approx Ze^2/\hbar$, the upper limit in eqn. (13.1.2.5a) must be connected with the maximum transfer of momentum from the moving particle to an electron in the medium. For sufficient large transfers one can assume the electrons to be free, and then the maximum momentum transfer is, according to (13.1.2.17), equal to

$$\hbar\varkappa_0 = \frac{2Ze^2}{b_0 v}. \tag{13.1.2.17}$$

Substituting the value $\varkappa_0 = 2Ze^2/\hbar b_0 v$, thus obtained, into expression (13.1.2.6) and adding the result to expression (13.1.2.15) we find the total energy losses of the particle for the quantal case, $v \gtrsim Ze^2/\hbar$:

$$\frac{d\mathcal{E}}{dt} = -\frac{4\pi n_0 Z^2 e^4}{m_e v} \ln \frac{2Mm_e v^2}{(M+m_e)\,\hbar\omega_{pe}}. \tag{13.1.2.18}$$

We note that we could have obtained this formula directly from the general formula (13.1.2.3) by using the quantum-mechanical expression for the plasma dielectric permittivity ε_l and taking the particle recoil into account.

Equations (13.1.2.16) and (13.1.2.18) determine the total energy losses of a charged particle only when its mass is much larger than the electron mass. If a light particle (an electron) moves through the plasma, the interaction of the particle with the fluctuation field in the plasma, which we neglected in deriving (13.1.2.16) and (13.1.2.18), plays an essential role.

13.1.3. CHANGE IN THE ENERGY OF A MOVING CHARGE, CAUSED BY THE FLUCTUATIONS OF THE FIELD IN THE PLASMA

We have already mentioned that eqns. (13.1.2.16) and (13.1.2.18) did not take into account the fluctuations of the electromagnetic field in the plasma; moreover, when deriving these formulae we neglected the change in the velocity of the particle, that is, we neglected the particle recoil. We shall now consider these effects, starting with the classical case, when $v \ll Ze^2/\hbar$ (Sitenko and Tszyan' Yu-Taĭ, 1963; Sitenko, 1964; Kalman and Ron, 1961).

We shall start from the following expression for the average change in the energy of the moving particle per unit time:

$$\frac{d\mathcal{E}}{dt} = Ze\langle(\boldsymbol{v}(t)\cdot\boldsymbol{E}(\boldsymbol{r}(t), t))\rangle, \tag{13.1.3.1}$$

where $\boldsymbol{E}(\boldsymbol{r}(t), t)$ is the electrical field at the spot where the particle is situated ($\boldsymbol{r}(t)$ is the radius vector determining the position of the particle at time t), $\boldsymbol{v}(t)$ the particle velocity, while the brackets $\langle\ldots\rangle$ indicate a statistical average.

In the classical case, when we can use the concept of a trajectory, the equation of motion of the particle has the form

$$\dot{\boldsymbol{v}}(t) = \frac{Ze}{M}\boldsymbol{E}(\boldsymbol{r}(t), t). \tag{13.1.3.2}$$

Formally integrating this equation we find

$$\left.\begin{aligned} \boldsymbol{v}(t) &= \boldsymbol{v}_0 + \frac{Ze}{M}\int_{t_0}^{t} dt'\,\boldsymbol{E}(\boldsymbol{r}(t'), t'), \\ \boldsymbol{r}(t) &= \boldsymbol{r}_0 + \boldsymbol{v}_0(t-t_0) + \frac{Ze}{M}\int_{t_0}^{t} dt' \int_{t_0}^{t'} dt''\,\boldsymbol{E}(\boldsymbol{r}(t''), t''), \end{aligned}\right\} \tag{13.1.3.3}$$

where $\boldsymbol{r}_0$ and $\boldsymbol{v}_0$ are the radius vector and the velocity of the particle at t_0.

We shall choose a time interval Δt sufficiently long compared with the period of the random fluctuations of the electrical field in the plasma, but small compared to the time interval after which the motion of the particle is appreciably changed. As during this time interval Δt the particle trajectory differs little from a straight line, we can write for the particle velocity and the field acting upon the particle at time $t = t_0+\Delta t$ approximately

$$\left.\begin{aligned} \boldsymbol{v}(t) &\cong \boldsymbol{v}_0+\frac{Ze}{M}\int_{t_0}^{t} dt'\,\boldsymbol{E}(\boldsymbol{r}_0(t'), t'), \\ \boldsymbol{E}(\boldsymbol{r}(t), t) &\cong \boldsymbol{E}(\boldsymbol{r}_0(t), t)+\frac{Ze}{M}\int_{t_0}^{t} dt'\int_{t_0}^{t'} dt''\sum_j E_j(\boldsymbol{r}_0(t''), t'')\frac{\partial}{\partial x_{0j}}\boldsymbol{E}(\boldsymbol{r}_0(t), t), \end{aligned}\right\} \quad (13.1.3.4)$$

where $\boldsymbol{r}_0(t)$ is the radius vector of the particle along its uniform motion along a straight line,

$$\boldsymbol{r}_0(t) = \boldsymbol{r}_0+\boldsymbol{v}_0(t-t_0). \quad (13.1.3.5)$$

Substituting (13.1.3.4) into (13.1.3.1) we get the following expression for the average change in the energy of the moving particle per unit time, up to terms in e^2:

$$\frac{d\mathcal{E}}{dt} = Ze\langle(\boldsymbol{v}_0\cdot\boldsymbol{E}(\boldsymbol{r}_0(t), t))\rangle+\frac{Z^2e^2}{M}\left\langle\int_{t_0}^{t} dt'\int_{t_0}^{t'} dt''\sum_j E_j(\boldsymbol{r}_0(t''), t'')\frac{\partial}{\partial x_{0j}}(\boldsymbol{v}_0\cdot\boldsymbol{E}(\boldsymbol{r}_0(t), t))\right\rangle$$
$$+\frac{Z^2e^2}{M}\left\langle\int_{t_0}^{t} dt'(\boldsymbol{E}(\boldsymbol{r}_0(t'), t')\cdot\boldsymbol{E}(\boldsymbol{r}_0(t), t))\right\rangle. \quad (13.1.3.6)$$

It is clear that the average value of the fluctuating part of the field vanishes so that the quantity $\langle\boldsymbol{E}(\boldsymbol{r}, t)\rangle$ is the same as the field strength produced by the particle itself in the plasma. The first term in (13.1.3.6) is thus the polarization energy loss of the moving particle which we found in the preceding subsection and which is given by formula (13.1.2.3). We note that the first term in (13.1.3.6), like the second and third terms, is proportional to the square of the charge of the moving particle, as the average value of the field strength is proportional to Ze.

The second and third terms in (13.1.3.6) determine the change in the energy of the moving particle which is connected with the fluctuations in the electrical field in the plasma and with the change in the particle velocity under the influence of that field. Let us first of all consider the second term which determines the dynamic friction of the particle caused by the presence of space–time correlations between the fluctuations in the electrical field in the plasma. The presence of such correlations leads to extra energy losses of the moving particle —apart from the polarization losses.

Changing the order of integration over t' and t'' in the second term in (13.1.3.6), integrating over t' and introducing a new variable $\xi = t-t''$, we get

$$\left.\frac{d\mathcal{E}}{dt}\right|_{\mathrm{II}} = \frac{Z^2e^2}{M}\int_{0}^{\Delta t}\xi\,d\xi\sum_{ij}\left\langle E_j(\boldsymbol{r}_0(t-\xi), t-\xi)\frac{\partial}{\partial x_{0j}}v_{0i}E_i(\boldsymbol{r}_0(t), t)\right\rangle. \quad (13.1.3.7)$$

As the correlation functions for the fluctuations of the field are exponentially small for large Δt we can put the upper limit in this integral equal to infinity ($\Delta t \to \infty$). Taking the space–time Fourier transforms of the field components $E_i(\mathbf{r}, t)$, we get the following expression for the loss of directed motion energy of the particle ($\omega = (\mathbf{k}\cdot\mathbf{v}_0)$):

$$\left.\frac{d\mathcal{E}}{dt}\right|_{\mathrm{II}} = \frac{Z^2e^2}{16\pi^3 M}\int \omega \frac{\partial}{\partial\omega}\langle E_1^2\rangle_{\mathbf{k}\omega}\, d^3\mathbf{k}, \tag{13.1.3.8}$$

where E_1 is the longitudinal electrical field in the plasma.

The third term in (13.1.3.6) determines the average change in the particle energy which is connected with the correlations between the fluctuation change in the velocity of the particle itself and the fluctuation electrical field in the plasma. The presence of such correlations leads on average to an increase in the energy of the moving particle. Fourier transforming, one can easily show that ($\omega = (\mathbf{k}\cdot\mathbf{v}_0)$)

$$\left.\frac{d\mathcal{E}}{dt}\right|_{\mathrm{III}} = \frac{Z^2e^2}{16\pi^3 M}\int \langle E_1^2\rangle_{\mathbf{k}\omega}\, d^3\mathbf{k}. \tag{13.1.3.9}$$

Adding expressions (13.1.3.8) and (13.1.3.9) we get the change in the energy of the moving particle which are connected with the fluctuations in the field and take the particle recoil into account (Sitenko and Tszyan' Yu-Taĭ, 1963) ($\omega = (\mathbf{k}\cdot\mathbf{v}_0)$)

$$\left(\frac{d\mathcal{E}}{dt}\right)_{\mathrm{fl}} = \frac{Z^2e^2}{16\pi^3 M}\int \frac{\partial}{\partial\omega}[\omega\langle E_1^2\rangle_{\mathbf{k}\omega}]\, d^3\mathbf{k}. \tag{13.1.3.10}$$

To obtain the total energy losses of the particle we must add to this expression the polarization losses given by formula (13.1.2.3).

In the case of an isotropic plasma the correlation function of the longitudinal field $\langle E_1^2\rangle_{\mathbf{k}\omega}$ depends only on the absolute magnitude of the wavevector $\mathbf{k}$. Integrating for that case over the angles in (13.1.3.10) we can write the formula for the fluctuation energy loss in the following form:

$$\left(\frac{d\mathcal{E}}{dt}\right)_{\mathrm{fl}} = \frac{Z^2e^2}{4\pi^2 M v_0^3}\int_0^{k_0v_0} \omega^2\langle E_1^2\rangle_{\omega/v_0,\,\omega}\, d\omega, \tag{13.1.3.11}$$

where k_0 is the maximum value of the wavevector which determines the region where the macroscopic discussion is applicable.

As the spectral density $\langle E_1^2\rangle_{\mathbf{k}\omega}$ of the field fluctuations is positive, the fluctuation change in the energy of a particle moving in an isotropic plasma is, according to (13.1.3.11), also positive, that is, the particle energy increases in that case. When a charged particle moves through an anisotropic plasma, that is, through a plasma in a magnetic field or through a plasma through which a beam of charged particles is moving, the interaction between the particle and the fluctuation field can lead to a decrease in the particle energy.

In contrast to the polarization energy losses which are independent of the mass of the moving particle, the fluctuation energy changes are inversely proportional to the mass of the moving particle. When a heavy particle with a mass much larger than the electron mass moves through the plasma, the fluctuation effects are therefore unimportant. However, if a light particle (an electron) moves through the plasma, the fluctuation energy change will be of the same order of magnitude as the polarization losses.

In concluding this subsection, let us evaluate the polarization losses and the fluctuation energy change of a moving particle in the case of an equilibrium plasma. Using eqns. (11.2.2.6) and (4.3.4.3) for the spectral density of the field fluctuations and for the longitudinal dielectric permittivity and integrating in (13.1.3.9) and (13.1.3.10) over the absolute magnitude k of the wavevector from zero to a maximum value k_0—for which the macroscopic discussion is still valid—we get the following general formulae:

$$\left.\begin{aligned}\left(\frac{d\mathcal{E}}{dt}\right)_{\text{pol}} &= -\frac{16\sqrt{\pi}\, n_0 Z^2 e^4}{m_e v_0}\int_0^{\zeta} z^2\, dz\, e^{-z^2} L(z, k_0 r_D),\\ \left(\frac{d\mathcal{E}}{dt}\right)_{\text{fl}} &= \frac{8\sqrt{\pi}\, n_0 Z^2 e^4}{M v_0}\,\zeta\, e^{-\zeta^2} L(\zeta, k_0 r_D),\end{aligned}\right\} \tag{13.1.3.12}$$

where $\zeta = v_0/v_e\sqrt{2}$ and

$$L(z, k_0 r_D) = \ln k_0 r_D - \frac{1}{4}\ln\{[1-\varphi(z)]^2 + \pi z^2 e^{-2z^2}\}$$
$$-\frac{1}{2\sqrt{\pi}}\frac{[1-\varphi(z)]\, e^{z^2}}{z}\left\{\frac{\pi}{2} - \arctan\frac{1-\varphi(z)}{z\sqrt{\pi}}\, e^{z^2}\right\}. \tag{13.1.3.13}$$

We can take for k_0 the reciprocal of the minimum impact parameter for distant collisions, $k_0 \cong M m_e v_0^2/Ze^2(M+m_e)$.

At sufficiently high temperatures and low electron densities $k_0 r_D \gg 1$. If in (13.1.3.12) we restrict ourselves to the main term ($\sim \ln k_0 r_D$) we find ($\zeta = v_0/v_e\sqrt{2}$)

$$\left.\begin{aligned}\left(\frac{d\mathcal{E}}{dt}\right)_{\text{pol}} &= \frac{-2\sqrt{2}\,\pi n_0 Z^2 e^4}{m_e v_e}\,\frac{\Phi(\zeta)-\zeta\Phi'(\zeta)}{\zeta}\ln(k_0 r_D),\\ \left(\frac{d\mathcal{E}}{dt}\right)_{\text{fl}} &= \frac{4\sqrt{(2\pi)}\, n_0 Z^2 e^4}{M v_0}\, e^{-\zeta^2}\ln(k_0 r_D),\end{aligned}\right\} \tag{13.1.3.14}$$

where $\Phi(\zeta)$ is the error function,

$$\Phi(\zeta) = \frac{2}{\sqrt{\pi}}\int_0^{\zeta} dz\, e^{-z^2}.$$

We see that the ratio $(d\mathcal{E}/dt)_{\text{fl}}/(d\mathcal{E}/dt)_{\text{pol}}$ is in an equilibrium plasma of the order of m_e/M.

In the limiting cases of low and high particle velocities the following approximate expressions are valid:

$$\left.\begin{aligned}\left(\frac{d\mathcal{E}}{dt}\right)_{\text{pol}} &\cong -\frac{4\sqrt{(2\pi)}\, n_0 Z^2 e^4}{3 m_e v_e}\,\frac{v_0^2}{v_e^2}\ln k_0 r_D, \quad v_0 \ll v_e,\\ \left(\frac{d\mathcal{E}}{dt}\right)_{\text{pol}} &\cong -\frac{4\pi n_0 Z^2 e^4}{m_e v_0}\ln k_0 r_D, \quad v_0 \gg v_e.\end{aligned}\right\} \tag{13.1.3.15}$$

We show in Fig. 13.1.1 the velocity-dependence of the polarization losses, the fluctuation energy changes, and the total energy changes of an electron per unit time. We see that in the case of a particle moving in an equilibrium plasma with a velocity much larger than the average thermal electron velocity ($v_0 \gg v_e$) we can neglect the effect of the fluctuation field.

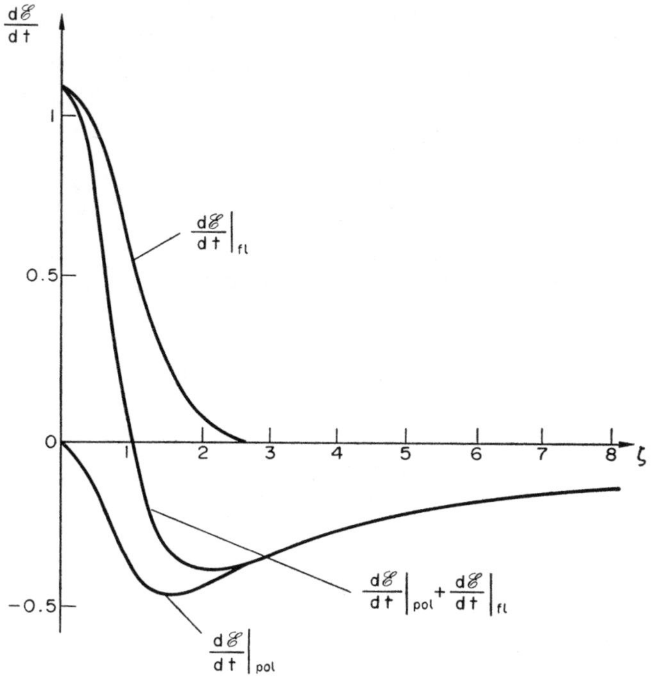

FIG. 13.1.1. Polarization energy losses $(d\mathcal{E}/dt)_{pol}$, fluctuation energy change $(d\mathcal{E}/dt)_{fl}$, and total change of energy per unit time of an electron as function of its velocity; $d\mathcal{E}/dt$ in units $(4\pi n_0 Z^2 e^4/m_e v_e \sqrt{2}) \ln k_0 r_D$; $\zeta = v_0/v_e\sqrt{2}$.

The total change in the particle energy reduces then to the polarization losses in the directed energy. If, however, the directed velocity of the charged particle is comparable to the thermal electron velocity, the fluctuation interaction begins to play an important role. In particular, if the particle velocity is less than the thermal electron velocity, the energy of the particle moving through the plasma increases due to the fluctuation interaction.

We note that eqns. (13.1.3.14) are good approximations if the argument of the logarithm, $k_0 r_D$, is large. This condition is satisfied if the interaction energy of the particles in the plasma is small compared to their kinetic energy.

13.1.4. SCATTERING PROBABILITY AND ENERGY LOSSES OF A PARTICLE

We determined in the preceding subsections the energy losses of a particle moving through a plasma, assuming that the condition for classical behaviour, $v \ll Ze^2/\hbar$, is satisfied. We shall now study the problem of the passage of a particle through a plasma without making this assumption. To do this we determine the probability for the scattering of the particle considered caused by the interaction between the particle and the electrical field in the plasma.

We shall assume that the particle is non-relativistic. The energy of its interaction with the field in the plasma then has the form

$$V(\boldsymbol{r}, t) = Ze\,\varphi(\boldsymbol{r}, t),$$

where $\varphi(\boldsymbol{r}, t)$ is the scalar potential of the electrical field in the plasma, and the probability that the particle changes from a state with momentum $\boldsymbol{p}$ to one with momentum $\boldsymbol{p}'$ is given by the well-known quantum-mechanical formula,

$$dW_{\boldsymbol{p}\to\boldsymbol{p}'} = \frac{1}{\hbar^2}\left|\int V_{\boldsymbol{p}'\boldsymbol{p}}(t)\exp\left[\frac{i}{\hbar}(\varepsilon_{\boldsymbol{p}'}-\varepsilon_{\boldsymbol{p}})\,t\right]dt\right|^2\frac{Vd^3\boldsymbol{p}'}{(2\pi\hbar)^3}, \tag{13.1.4.1}$$

where $\varepsilon_p = p^2/2M$ is the energy of a particle with momentum $\boldsymbol{p}$, $V_{\boldsymbol{p}'\boldsymbol{p}}$ the matrix element of the interaction energy,

$$V_{\boldsymbol{p}'\boldsymbol{p}}(t) = Ze\int\varphi(\boldsymbol{r}, t)\exp\left[\frac{i}{\hbar}(\{\boldsymbol{p}-\boldsymbol{p}'\}\cdot\boldsymbol{r})\right]\frac{d^3\boldsymbol{r}}{V}, \tag{13.1.4.2}$$

and V is the normalization volume. As the potential φ is subject to fluctuations, we must average this expression over the fluctuations. Denoting the averaged probability by $\langle dW_{\boldsymbol{p}\to\boldsymbol{p}'}\rangle$ we have

$$\langle dW_{\boldsymbol{p}\to\boldsymbol{p}'}\rangle = \frac{Z^2e^2}{\hbar^2}\iint\langle\varphi(\boldsymbol{r}, t)\,\varphi(\boldsymbol{r}', t')\rangle\exp\left\{\frac{i}{\hbar}([\boldsymbol{p}-\boldsymbol{p}']\cdot[\boldsymbol{r}-\boldsymbol{r}'])+\frac{i}{\hbar}(\varepsilon_{\boldsymbol{p}'}-\varepsilon_{\boldsymbol{p}})(t'-t)\right\}$$
$$\times\, d^3\boldsymbol{r}\,dt\,d^3\boldsymbol{r}'\,dt'\,\frac{d^3\boldsymbol{p}'}{\sqrt{(2\pi\hbar)^3}}\,. \tag{13.1.4.3}$$

After averaging, the product of the potentials will depend only on the relative coordinate and the difference in the times. Therefore, changing from $\boldsymbol{r}, t, \boldsymbol{r}', t'$ to new integration variables $\boldsymbol{r}'-\boldsymbol{r}$, $\frac{1}{2}(\boldsymbol{r}'+\boldsymbol{r})$, $t'-t$, $\frac{1}{2}(t'+t)$, we get for $\langle dW_{\boldsymbol{p}\to\boldsymbol{p}'}\rangle$ a quantity which is proportional to the volume V and the observation time $\frac{1}{2}(t'+t)$. We can thus speak of the averaged scattering probability per unit time,

$$dw_{\boldsymbol{p}\to\boldsymbol{p}'} = \frac{Z^2e^2}{\hbar^2}\langle\varphi^2\rangle_{\boldsymbol{q}\omega}\frac{d^3\boldsymbol{p}'}{(2\pi\hbar)^3}, \tag{13.1.4.4}$$

where $\langle\varphi^2\rangle_{\boldsymbol{q}\omega}$ is the spectral density of the fluctuations of the potential of the electrical field and

$$\boldsymbol{q} = \frac{1}{\hbar}(\boldsymbol{p}'-\boldsymbol{p}), \qquad \omega = \frac{1}{\hbar}(\varepsilon_{\boldsymbol{p}'}-\varepsilon_{\boldsymbol{p}}).$$

We can express the potential fluctuations in terms of the charge density fluctuations

$$\langle\varphi^2\rangle_{\boldsymbol{q}\omega} = \left(\frac{4\pi}{q^2}\right)^2\langle\varrho^2\rangle_{\boldsymbol{q}\omega}, \tag{13.1.4.5}$$

whence

$$dw_{\boldsymbol{p}\to\boldsymbol{p}'} = \left(\frac{4\pi Ze}{\hbar q^2}\right)^2\langle\varrho^2\rangle_{\boldsymbol{q}\omega}\frac{d^3\boldsymbol{p}'}{(2\pi\hbar)^3}\,. \tag{13.1.4.6}$$

The probability for the scattering of a particle passing through the plasma is thus completely determined by the spectral density of the field (or charge density) fluctuations in the plasma, or, more precisely, by the value of that spectral density for $\boldsymbol{q} = (\boldsymbol{p}'-\boldsymbol{p})/\hbar$ and $\omega = (\varepsilon_{p'}-\varepsilon_p)/\hbar$.

Multiplying the scattering probability by $\hbar\omega$ and integrating over $\boldsymbol{p}'$ we find the change in the energy of the moving particle per unit time:

$$\frac{d\mathcal{E}}{dt} = \frac{2Z^2e^2}{\pi\hbar^4}\int \frac{\omega}{q^4}\langle\varrho^2\rangle_{q\omega}\, d^3\boldsymbol{p}'. \tag{13.1.4.7}$$

This general formula determines the energy loss of a particle which is caused both by the polarization of the plasma and by the fluctuations in it. It refers equally well to a plasma as to any other medium. In deriving it we assumed merely that the interaction between the moving particle and the plasma is weak so that we could use the first-order approximation of perturbation theory.

If we use this formula for actual evaluations of $d\mathcal{E}/dt$ we need know the spectral density $\langle\varrho^2\rangle_{q\omega}$. This quantity has, in general, a different form in quantum and classical regions. In what follows we shall be interested solely in the classical region, that is, the classical limit of formula (13.1.4.7). To find it we turn to eqn. (11.1.2.20) which determines the spectral density of fluctuations for equilibrium systems for which we have not yet taken the limit $\hbar \to 0$. This formula has as factor

$$\bar{N}_\omega = \begin{cases} \bar{N}_{|\omega|}, & \omega > 0, \\ -(\bar{N}_{|\omega|}+1), & \omega < 0, \end{cases}$$

where $\bar{N}_{|\omega|}$ is the Planck distribution function,

$$\bar{N}_{|\omega|} = \frac{1}{e^{\hbar|\omega|/T}-1}, \tag{13.1.4.8}$$

and T the temperature of the medium. The presence of this factor has a simple, but profound physical meaning. In fact, we saw in Section 11.2 that the spectral density of the fluctuations in the field and in other quantities have steep maxima—which in the transparency region are δ-functionlike—for $\omega = \pm\omega_r(\boldsymbol{q})$, where $\omega_r(\boldsymbol{q})$ is the frequency of the rth eigen oscillations with wavevector $\boldsymbol{q}$. Bearing in mind that $\hbar\omega = \varepsilon_{p'}-\varepsilon_p$ and $\hbar\boldsymbol{q} = \boldsymbol{p}'-\boldsymbol{p}$, one can say that a particle passing through a plasma emits and absorbs plasmons, that is, quasi-particles with energy $\hbar\omega_r(\boldsymbol{q})$ and momentum $\hbar\boldsymbol{q}$ while the probability for emission is proportional to $\bar{N}_{\omega_r(\boldsymbol{q})}+1$, and the probability for absorption to $\bar{N}_{\omega_r(\boldsymbol{q})}$, where $\bar{N}_{\omega_r(\boldsymbol{q})}$ is the average number of plasmons with frequency $\omega_r(\boldsymbol{q})$ at temperature T. The fact that the emission probability is proportional to $\bar{N}_{|\omega|}+1$, while the absorption probability is proportional to $\bar{N}_{|\omega|}$ is connected with the nature of the plasmon statistics which are bosons, like photons, and obey Bose–Einstein statistics.

As

$$\lim_{\hbar\to 0} \hbar\omega\bar{N}_\omega = T, \tag{13.1.4.9}$$

we have in the classical formulae for the spectral densities of the fluctuations in Section 11.2 the temperature T instead of the quantity $\hbar\omega\bar{N}_\omega$; but it is clear from our discussion that we can directly find the spectral densities of the fluctuations in the quasi-classical case using the classical formulae for these spectral densities if

$$\hbar\omega_r(q) \ll |\,\varepsilon_{p'}-\varepsilon_p\,|, \quad \hbar q \ll |\,\boldsymbol{p}'-\boldsymbol{p}\,|;$$

To do this we must replace in the classical formulae T by $\hbar\omega\bar{N}_\omega$, that is,

$$\langle\varrho^2\rangle_{q\omega} = \frac{\hbar\omega\bar{N}_\omega}{T}\langle\varrho^2\rangle^{\text{cl}}_{q\omega}, \tag{13.1.4.10}$$

where $\langle\varrho^2\rangle^{\text{cl}}_{q\omega}$ is the spectral density of the charge density fluctuations in the classical case; it satisfies the condition

$$\langle\varrho^2\rangle^{\text{cl}}_{q\omega} = \langle\varrho^2\rangle^{\text{cl}}_{-q-\omega}. \tag{13.1.4.11}$$

In the transparency region, that is, when $\omega \cong \omega_{\text{p}}(\boldsymbol{q})$ where ω_{p} is the frequency of a weakly damped eigen oscillation with wavevector $\boldsymbol{q}$, plasmons can not be in thermal equilibrium with the plasma particles. The frequency distribution of the plasmons may thus differ from the Planck distribution and is described, in general, by an arbitrary function $N_{\omega(\boldsymbol{q})}$. Instead of (13.1.4.10) we then have the relation

$$\langle\varrho^2\rangle_{q\omega} = \frac{\hbar\omega N_{\omega(\boldsymbol{q})}}{T^*(\boldsymbol{q},\omega)}\langle\varrho^2\rangle^{\text{cl}}_{q\omega}, \tag{13.1.4.12}$$

where T^* is a function of $\boldsymbol{q}$ and ω—in general depending also on $\hbar$—which satisfies the condition

$$\lim_{\hbar\to 0}\frac{\hbar\omega N_{\omega(\boldsymbol{q})}}{T^*(\boldsymbol{q},\omega)} = 1. \tag{13.1.4.13}$$

From this condition it follows that

$$\lim_{\hbar\to 0} T^*(\boldsymbol{q},\omega) = \tilde{T}(\boldsymbol{q},\omega), \tag{13.1.4.14}$$

where $\tilde{T}$ is the effective plasmon temperature, determined by formula (11.5.3.6′). Indeed, it follows from (11.5.3.9) that the energy density of the fluctuation plasma oscillations with wavevector $\boldsymbol{q}$ is equal to

$$\frac{1}{4\pi}\langle E^2\rangle_{\boldsymbol{q}} = \frac{1}{2}\int_{-\infty}^{+\infty} d\omega\tilde{T}(\boldsymbol{q},\omega)\,\{\delta(\omega-\omega_{\text{p}}(\boldsymbol{q}))+\delta(\omega+\omega_{\text{p}}(\boldsymbol{q}))\}. \tag{13.1.4.15}$$

However, the same energy density can, as $\hbar \to 0$, be written in the form

$$\frac{1}{4\pi}\langle E^2\rangle_{\boldsymbol{q}} = \int_0^\infty \hbar\omega\,N_\omega\,d\omega\;\delta(\omega-\omega_{\text{p}}(\boldsymbol{q})) \cong \frac{1}{2}\int_{-\infty}^{+\infty}\hbar\omega N_\omega\,d\omega\,\{\delta(\omega-\omega_{\text{p}}(\boldsymbol{q}))+\delta(\omega+\omega_{\text{p}}(\boldsymbol{q}))\}, \tag{13.1.4.16}$$

whence we obtain, using (13.1.4.13), the relation (13.1.4.14).

Substituting (13.1.4.10) into (13.1.4.6) and using formula (11.2.2.5), we find the probability for the scattering of a particle, passing through an equilibrium plasma, in the quasi-classical case:

$$dw_{p\to p'} = \frac{8\pi Z^2e^2}{\hbar q^2} N_\omega \frac{\text{Im } \varepsilon_l(\boldsymbol{q},\omega)|^2}{|\varepsilon_l(\boldsymbol{q},\omega)|^2} \frac{d^3\boldsymbol{p}'}{(2\pi\hbar)^3}, \tag{13.1.4.17}$$

where $\varepsilon_l(\boldsymbol{q}, \omega)$ is the longitudinal dielectric permittivity of the plasma, given by formula (4.3.4.3).

From this expression we get for the probability for the scattering of a fast particle ($v \gg v_e$, where v is the particle velocity and v_e the average electron thermal velocity in the plasma) for which the change in its energy is small compared to the energy itself ($\omega_p = \sqrt{(\omega_{pe}^2 + 3q^2v_e^2)}$) the following expression:

$$dw_{p\to p'} = \frac{4\pi^2Z^2e^2}{\hbar q^2} \{|\omega|(N_{|\omega|}+1)\,\delta(\omega+\omega_p)+\omega N_\omega\,\delta(\omega-\omega_p)\} \frac{d^3\boldsymbol{p}'}{(2\pi\hbar)^3}. \tag{13.1.4.18}$$

As we mentioned earlier, it can be seen that the scattering of a fast particle involving a small relative change in energy proceeds mainly with the emission and absorption of longitudinal electron oscillations by the particle; the first term in (13.1.4.18) correspond here to emission and the second to absorption of the oscillations.

Equation (13.1.4.18) was obtained under the assumption that the plasma was in a state of thermodynamic equilibrium. However, we can use it in the quasi-classical case also for a plasma with a non-equilibrium distribution of the plasma waves, provided we replace the Planck function $\bar{N}_\omega$ by the non-equilibrium distribution function N_ω of the plasma waves.

We can similarly find the probability for the scattering of a particle by low-frequency oscillations in a two-temperature plasma:

$$dw_{p\to p'} = \frac{4\pi^2Z^2e^2r_D^2}{\hbar(1+q^2r_D^2)} \{|\omega|(N_{|\omega|}+1)\,\delta(\omega+\omega_s)+\omega N_\omega\,\delta(\omega-\omega_s)\} \frac{d^3\boldsymbol{p}'}{(2\pi\hbar)^3}, \tag{13.1.4.19}$$

where $\omega_s = qv_s\{1+q^2r_D^2\}^{-1/2}$ with v_s the non-isothermal sound velocity and N_ω the distribution function of the low-frequency oscillations.

We have already mentioned that eqn. (13.1.4.12) is valid in the quasi-classical case when the energy and momentum transfers are small, $\hbar|\omega| \ll |\varepsilon_{p'}-\varepsilon_p|$, $\hbar q \ll |\boldsymbol{p}'-\boldsymbol{p}|$. For large momentum transfers—of the order of the particle momentum—it is not valid. However, in that case we can at once write down a simple expression for $\langle\varrho^2\rangle_{q\omega}$ using the fact that if $\hbar q \sim m_e v_e$, where v_e is the electron thermal velocity in the plasma, effects connected with the influence of the self-consistent field cannot play a role, that is, each particle in the plasma can only be correlated with itself when $\hbar q \sim m_e v_e$. This means that we can use eqn. (11.3.2.5) for the spectral density of the charge density fluctuations when $\hbar q \sim m_e v_e$ and take into account in them only the recoil effect, that is, replacing the δ-function $\delta\{\omega-(\boldsymbol{q}\cdot\boldsymbol{v})\}$ which occurs there by $\delta\{\omega-(\boldsymbol{q}\cdot\boldsymbol{v})-(\hbar q^2/2m_e)\}$:

$$\langle\varrho^2\rangle_{q\omega} = 2\pi e^2 \int f_0(v)\,\delta\{\omega-(\boldsymbol{q}\cdot\boldsymbol{v})-(\hbar q^2/2m_e)\}\,d^3\boldsymbol{v}. \tag{13.1.4.20}$$

We now turn again to the general formula (13.1.4.7) for the energy losses. If the velocity of the moving particle is much larger than the average electron thermal velocity of the plasma the main contribution to the expression for the energy losses comes from two regions for the momentum transfer $\hbar q$: the region $\hbar q \sim m_e v_e$ (large momentum transfers, that is, close collisions) and the region $q \ll r_D^{-1}$, where r_D is the Debye screening radius (small momentum transfers, that is, distant collisions). To evaluate the energy loss in these regions we choose a value q_0 in the range

$$r_D^{-1} < q_0 < \frac{m_e v_e}{\hbar}$$

and use for $q < q_0$ the quasi-classical expression (13.1.4.12) for $\langle \varrho^2 \rangle_{q\omega}$, and for $q > q_0$ expression (13.1.4.20) which refers to a perfect gas of non-interacting plasma particles. The intermediate region $q \sim q_0$ does not contribute significantly to the expression for the energy loses, and therefore we have

$$\frac{d\mathcal{E}}{dt} \cong \left(\frac{d\mathcal{E}}{dt}\right)_{q<q_0} + \left(\frac{d\mathcal{E}}{dt}\right)_{q>q_0}. \tag{13.1.4.21}$$

We can find at once an expression for the quantity $(d\mathcal{E}/dt)_{q>q_0}$:

$$\left(\frac{d\mathcal{E}}{dt}\right)_{q>q_0} = -\frac{Z^2 e^2 \omega_{pe}^2}{v} \ln \frac{2M m_e v}{\hbar q_0 (M+m_e)}, \tag{13.1.4.22}$$

and we therefore turn to the evaluation of $(d\mathcal{E}/dt)_{q<q_0}$. Substituting (13.1.4.12) into (13.1.4.7) and noting that

$$\delta\left(\omega - \frac{\varepsilon_{p'} - \varepsilon_p}{\hbar}\right) d\omega \to 1,$$

we can write $(d\mathcal{E}/dt)_{q<q_0}$ in the form

$$\left(\frac{d\mathcal{E}}{dt}\right)_{q<q_0} = \frac{2Z^2 e^2}{\pi} \iint \frac{\omega^2 N_\omega}{q^4 T^*(\boldsymbol{q}, \omega)} \langle \varrho^2 \rangle_{q\omega}^{\text{cl}} \, \delta\left(\omega - (\boldsymbol{q}\cdot\boldsymbol{v}) - \frac{\hbar q^2}{2M}\right) d\omega\, d^3\boldsymbol{q}, \tag{13.1.4.23}$$

or, using (13.1.4.11) and the relation $T^*(-\boldsymbol{q}, -\omega) = T^*(\boldsymbol{q}, \omega)$, in the form

$$\left(\frac{d\mathcal{E}}{dt}\right)_{q<q_0} = \frac{Z^2 e^2}{\pi} \iint \frac{\omega^2}{q^4 T^*(\boldsymbol{q}, \omega)} \langle \varrho^2 \rangle_{q\omega}^{\text{cl}} \left\{ N_\omega \delta\left(\omega - (\boldsymbol{q}\cdot\boldsymbol{v}) - \frac{\hbar q^2}{2M}\right) + N_{-\omega}\, \delta\left(\omega - (\boldsymbol{q}\cdot\boldsymbol{v}) + \frac{\hbar q^2}{2M}\right)\right\} d\omega\, d^3\boldsymbol{q}. \tag{13.1.4.24}$$

We are interested in the classical limit, $\hbar \to 0$, of this formula. To find it, we expand the δ-function in a power series in $\hbar$,

$$\delta\left\{\omega - (\boldsymbol{q}\cdot\boldsymbol{v}) - \frac{\hbar q^2}{2M}\right\} \to \delta\{\omega - (\boldsymbol{q}\cdot\boldsymbol{v})\} - \frac{\hbar q^2}{2M} \delta'\{\omega - (\boldsymbol{q}\cdot\boldsymbol{v})\},$$

and write $(d\mathcal{E}/dt)_{q<q_0}$ as a sum,

$$\left(\frac{d\mathcal{E}}{dt}\right)_{q<q_0} = \left(\frac{d\mathcal{E}}{dt}\right)_{\text{pol}} + \left(\frac{d\mathcal{E}}{dt}\right)_{\text{fl}}, \tag{13.1.4.25}$$

where

$$\left(\frac{d\mathcal{E}}{dt}\right)_{\text{pol}} = \frac{Z^2e^2}{\pi}\iint \frac{\omega^2}{q^4T^*(\boldsymbol{q},\omega)}\langle\varrho^2\rangle^{\text{cl}}_{\boldsymbol{q}\omega}(N_\omega+N_{-\omega})\,\delta\{\omega-(\boldsymbol{q}\cdot\boldsymbol{v})\}\,d\omega\,d^3\boldsymbol{q},$$

$$\left(\frac{d\mathcal{E}}{dt}\right)_{\text{fl}} = -\frac{Z^2e^2}{\pi}\frac{\hbar}{M}\iint \frac{\omega^2N_\omega}{q^4T^*(\boldsymbol{q},\omega)}\langle\varrho^2\rangle^{\text{cl}}_{\boldsymbol{q}\omega}\delta'\{\omega-(\boldsymbol{q}\cdot\boldsymbol{v})\}\,d\omega\,d^3\boldsymbol{q}.$$

Noting that

$$N_\omega+N_{-\omega} = -1,$$

and

$$\lim_{\hbar\to 0} = \omega N_\omega = \frac{T^*(\boldsymbol{q},\omega)}{\hbar},$$

we get for $(d\mathcal{E}/fdt)_{\text{pol}}$ and $(d\mathcal{E}/dt)_{\text{fl}}$ the expressions

$$\left(\frac{d\mathcal{E}}{dt}\right)_{\text{pol}} = -\frac{Z^2e^2}{\pi}\int \frac{(\boldsymbol{q}\cdot\boldsymbol{v})^2}{q^4\tilde{T}(\boldsymbol{q},(\boldsymbol{q}\cdot\boldsymbol{v}))}\langle\varrho^2\rangle^{\text{cl}}_{\boldsymbol{q}(\boldsymbol{q}\cdot\boldsymbol{v})}\,d^3\boldsymbol{q}, \tag{13.1.4.26}$$

$$\left(\frac{d\mathcal{E}}{dt}\right)_{\text{fl}} = \frac{Z^2e^2}{\pi M}\int \frac{d^3\boldsymbol{q}}{q^2}\frac{\partial}{\partial\omega}(\omega\langle\varrho^2\rangle^{\text{cl}}_{\boldsymbol{q}\omega}), \qquad \omega = (\boldsymbol{q}\cdot\boldsymbol{v}), \tag{13.1.4.27}$$

which do not contain $\hbar$. The next terms in the expansion of the δ-function in powers of $\hbar$ would lead to terms in $(d\mathcal{E}/dt)_{q<q_0}$ which vanish as $\hbar\to 0$.

The quantity $(d\mathcal{E}/dt)_{\text{pol}}$ gives the energy losses of the particle caused by the polarization of the medium—the energy losses due to Cherenkov emission also enter into $(d\mathcal{E}/dt)_{\text{pol}}$—while the quantity $(d\mathcal{E}/dt)_{\text{fl}}$ is the change in the particle energy caused by the interaction between the particle and the field fluctuations.

To verify the correctness of the first statement we assume that the plasma is in thermal equilibrium, in which case

$$\langle\varrho^2\rangle^{\text{cl}}_{\boldsymbol{q}\omega} = \frac{q^2}{2\pi}\frac{T}{\omega}\,\text{Im}\left\{-\frac{1}{\varepsilon_l(\boldsymbol{q},\omega)}\right\}, \tag{13.1.4.28}$$

and eqn. (13.1.4.26) becomes

$$\left(\frac{d\mathcal{E}}{dt}\right)_{\text{pol}} = \frac{Z^2e^2}{2\pi^2}\,\text{Im}\int \frac{(\boldsymbol{q}\cdot\boldsymbol{v})}{q^2\varepsilon_l(\boldsymbol{q},(\boldsymbol{q}\cdot\boldsymbol{v}))}\,d^3\boldsymbol{q}, \tag{13.1.4.29}$$

which is identical with (13.1.2.3).

It is clear that the same formula will also be valid for a non-equilibrium plasma if we neglect the fluctuations in the ion charge density; in that case eqn. (13.1.4.28) remains valid, but in it we need replace T by $\tilde{T}$.

We note that the polarization energy losses due to the excitation of plasma waves are independent of the number of waves N_ω. This is connected with the fact that the energy losses are determined by the spontaneous emission of waves by the particle, while the contributions to the energy losses from the induced emission and absorption—which are proportional to the number of waves N_ω—cancel one another.

To verify the validity of the second statement we use the relation

$$\langle \varrho^2 \rangle^{\text{cl}}_{q\omega} = \frac{q^2}{16\pi^2} \langle E_1^2 \rangle_{q\omega},$$

substitute it into (13.1.4.27), and we get the formula ($\omega = (\boldsymbol{q}\cdot\boldsymbol{v})$)

$$\left(\frac{d\mathcal{E}}{dt}\right)_{\text{fl}} = \frac{Z^2e^2}{16\pi^3 M} \int \frac{\partial}{\partial\omega} [\omega \langle E_1^2 \rangle_{q\omega}]\, d^3\boldsymbol{q}, \tag{13.1.4.30}$$

which is identical with (13.1.3.10).

If we neglect the fluctuations in the ion charge density, we have ($\omega = (\boldsymbol{q}\cdot\boldsymbol{v})$)

$$\left(\frac{d\mathcal{E}}{dt}\right)_{\text{fl}} = \frac{Z^2e^2}{2\pi M} \int \frac{\partial}{\partial\omega} \left[\frac{\omega}{|\omega|} \tilde{T}(\boldsymbol{q}, \omega)\, \delta\{\varepsilon_1(\boldsymbol{q}, \omega)\}\right] d^3\boldsymbol{q}. \tag{13.1.4.31}$$

Comparison of this formula with eqn. (13.1.4.29) for $(d\mathcal{E}/dt)_{\text{pol}}$ shows that

$$\frac{(d\mathcal{E}/dt)_{\text{fl}}}{(d\mathcal{E}/dt)_{\text{pol}}} \sim \frac{m_e}{M} \frac{\tilde{T}}{T}, \tag{13.1.4.32}$$

that is, the fluctuation losses are important only in the case of hot plasmons, when

$$\tilde{T} \gg T.$$

Using formula (13.1.4.29) and putting $\varepsilon_l = 1-(\omega_{\text{pe}}/\omega)^2$. we easily find the polarization energy losses for $q < q_0$:

$$\left(\frac{d\mathcal{E}}{dt}\right)_{q<q_0} = -\frac{Z^2e^2\omega_{\text{pe}}^2}{v} \ln \frac{q_0 v}{\omega_{\text{pe}}}. \tag{13.1.4.33}$$

Adding this expression to (13.1.4.22) we find the total polarization energy losses for a particle per unit time:

$$\frac{d\mathcal{E}}{dt} = -\frac{Z^2e^2\omega_{\text{pe}}^2}{v} \ln \frac{2Mm_e v^2}{\hbar\omega_{\text{pe}}(M+m_e)}. \tag{13.1.4.34}$$

This formula, which we have already met with, has the structure of the general Bohr formula for polarization losses. We draw attention to the fact that the parameter q_0 does not occur in it. This is connected with the fact that, as we noted earlier, momentum transfers of the order of $\hbar q_0$ do not contribute significantly to the energy losses.

We see that the polarization energy losses of a fast particle strongly depend on the plasma density, but are independent of its temperature. We can therefore use formula (13.1.4.34) whatever the particle distribution function in the plasma, provided the average plasma particle velocity is much lower than the velocity of the moving particle. The total energy losses are also independent of the number N_ω of the plasma waves, provided the average energy $\tilde{T}$ of the plasma waves is not too large, $\tilde{T} \ll TM/m_e$.

If the moving particle is an electron we must in our discussion of close collisions take into account that it is identical with the electrons in the plasma. This leads to an additional factor $e/2\sqrt{2}$ (where e is the base of the natural logarithms) in the argument of the logarithm in formula (13.1.4.33). The polarization losses of a fast electron passing through a plasma are thus given by the formula

$$\frac{d\mathcal{E}}{dt} = -\frac{e^2\omega_{\text{pe}}^2}{v}\left\{\ln\frac{m_e v^2}{2\sqrt{2}\,\hbar\omega_{\text{pe}}}+1\right\}. \tag{13.1.4.35}$$

We have already noted that eqn. (13.1.4.4) for the scattering probability is valid in the first Born approximation which is well known to be applicable in the case of a fast particle ($e^2/\hbar v \ll 1$) for all $\boldsymbol{q}$, and in the case of a slow particle only for small momentum transfers. Equation (13.1.4.33) for the energy losses in close collisions is therefore valid only if $e^2/\hbar v \ll 1$.

We note that formula (13.1.4.34)—which we obtained, neglecting terms $\propto v^2/c^2$—also determines the order of magnitude of the energy losses in the relativistic region, $v \sim c$. Taking relativistic effects into account only leads to a change in the argument of the logarithm, but leaves the coefficient $Z^2e^2\omega_{\text{pe}}^2/v$ unaltered.

13.2. Dynamic Friction and Diffusion Coefficients in a Plasma

13.2.1. THE FOKKER–PLANCK EQUATION FOR TEST PARTICLES

In the previous section we considered the motion through the plasma of a foreign particle (which we shall call a test particle) and we determined its energy losses. These losses which are caused by the interaction between the test particle (which we assume to be non-relativistic) and the electrical field in the plasma can be interpreted, as we indicated in Section 13.1, as the work done by the friction forces which occur when the particle moves through the plasma. Indeed, let us turn to the approximate equation of motion (13.1.3.2) of the test particle ($\boldsymbol{r}_0(t) = \boldsymbol{r}_0+\boldsymbol{v}(t-t_0)$):

$$\dot{v}_i(t) = \frac{Ze}{M}E_i(\boldsymbol{r}_0(t), t)+\frac{Z^2e^2}{M^2}\int\limits_{t_0}^{t}dt'\int\limits_{t_0}^{t'}dt''\sum_j E_j(\boldsymbol{r}_0(t''), t'')\frac{\partial}{\partial x_{0j}}E_i(\boldsymbol{r}_0(t), t), \tag{13.2.1.1}$$

where t lies in the range t_0, $t_0+\Delta t$ with Δt a length of time, which is large compared to the period of the field fluctuations but small compared to the characteristic times over which the velocity of the test particle changes significantly, and let us determine the friction force D_i per unit particle mass:

$$D_i = \frac{\langle \Delta v_i\rangle}{\Delta t}, \tag{13.2.1.2}$$

where Δv_i is the change in the ith component of the velocity of the test particle during the time Δt and the brackets $\langle\ldots\rangle$ indicate, as usual, averaging over the fluctuations. Using (13.2.1.1) we get in the case of an isotropic plasma (Sitenko and Tszyan' Yu-Taĭ, 1963)

$$D_i = \frac{v_i}{v}D, \quad D = -\frac{1}{Mv}\frac{d\mathcal{E}}{dt}, \tag{13.2.1.3}$$

where $d\mathcal{E}/dt$ is the loss in the energy of the directed motion of the test particle ($\omega = (\boldsymbol{k}\cdot\boldsymbol{v})$),

$$\frac{d\mathcal{E}}{dt} = \frac{Z^2e^2}{2\pi^2}\,\mathrm{Im}\int \frac{\omega}{k^2\varepsilon_l(\boldsymbol{k},\omega)}\,d^3\boldsymbol{k} + \frac{Z^2e^2}{16\pi^3 M}\int \omega\,\frac{\partial}{\partial\omega}\langle E_l^2\rangle_{\boldsymbol{k}\omega}\,d^3\boldsymbol{k}. \tag{13.2.1.4}$$

The quantity D characterizing the average change in the absolute magnitude of the velocity of the test particle per unit time may be called the dynamic friction coefficient.

The first term in the formula for $d\mathcal{E}/dt$ determines the dynamic friction of the test particle caused by its interaction with the electrical field occurring in the plasma through the motion of the particle itself. This interaction leads to the usual polarization energy losses of the particle and we can therefore call the dynamic friction connected with the polarization energy losses polarization friction.

The second term in $d\mathcal{E}/dt$ determines the additional dynamic friction of the particle connected with the presence of space–time correlations between the fluctuation electrical fields in the plasma. We have seen that the presence of such correlations leads to extra losses in the energy of the directed motion of the particle. We note that the dynamic friction of the particle is connected only with the fluctuations in the longitudinal electrical field in the plasma.

If in the plasma not just one, but a set of test particles, characterized by a distribution function $f(\boldsymbol{v}, t)$, is moving, we can from a knowledge of the dynamic friction coefficient of the test particles determine the change in their distribution function caused by the polarization and fluctuation losses:

$$\left(\frac{\partial f}{\partial t}\right)_{\mathrm{pol+fl}} = -\sum_i \frac{\partial}{\partial v_i}(D_i f). \tag{13.2.1.5}$$

This change proceeds very slowly as the energy of the interaction between the test particles and the field in the plasma is small compared to their kinetic energy.

The quantity $(\partial f/\partial t)_{\mathrm{pol+fl}}$ does not, however, determine the total change in the distribution function of the test particles as apart from the change in the distribution function caused by the polarization and fluctuation energy losses of the particle we must still take into account the change in the distribution function caused by the diffusion of the test particles in their velocity space. The diffusion of test particles is a slow process caused, like the fluctuation energy losses, by the field fluctuations in the plasma. It can be characterized by a diffusion tensor,

$$D_{ij} = \frac{\langle \Delta v_i\, \Delta v_j\rangle}{\Delta t}, \tag{13.2.1.6}$$

which is analogous to the diffusion tensor in coordinate space

$$D_{ij}^{(x)} = \frac{\langle \Delta x_i\, \Delta x_j\rangle}{\Delta t}.$$

By using the equation of motion for the test particles we can easily show that

$$D_{ij} = \frac{Z^2e^2}{M^2}\int_0^{\Delta t} d\xi \langle E_i(\boldsymbol{r}_0(t), t)\, E_j(\boldsymbol{r}_0(t-\xi), t-\xi)\rangle, \tag{13.2.1.7}$$

and as Δt by definition is large compared to the period of the field fluctuations we can here let Δt tend to infinity. Also Fourier transforming we get finally (Sitenko and Tszyan' Yu-Taĭ, 1963) ($\omega = (\boldsymbol{k}\cdot\boldsymbol{v})$)

$$D_{ij} = \frac{Z^2e^2}{8\pi^3M^2}\int \frac{k_ik_j}{k^2}\langle E_1^2\rangle_{\boldsymbol{k}\omega}\,d^3\boldsymbol{k}. \tag{13.2.1.8}$$

In an isotropic plasma we can write the diffusion tensor in the form ($\omega = (\boldsymbol{k}\cdot\boldsymbol{v})$)

$$\left.\begin{aligned} D_{ij} &= \frac{v_iv_j}{v^2}D_{||}+\frac{1}{2}\left(\delta_{ij}-\frac{v_iv_j}{v^2}\right)D_{\perp},\\ D_{||} &= \frac{Z^2e^2}{8\pi^3M^2v^2}\int\frac{\omega^2}{k^2}\langle E_1^2\rangle_{\boldsymbol{k}\omega}\,d^3\boldsymbol{k},\\ D_{\perp} &= \frac{Z^2e^2}{8\pi^3M^2}\int\left(1-\frac{\omega^2}{k^2v^2}\right)\langle E_1^2\rangle_{\boldsymbol{k}\omega}\,d^3\boldsymbol{k}. \end{aligned}\right\} \tag{13.2.1.9}$$

The longitudinal diffusion coefficient $D_{||}$ characterizes the average change in the square of the velocity component along the direction of motion of the particle, while the transverse diffusion coefficient $D_{\perp}$ characterizes the average change in the square of the velocity component in the perpendicular direction.

A comparison of these formulae with eqn. (13.1.3.10) for $(d\mathcal{E}/dt)_{\text{fl}}$ shows that, as to order of magnitude,

$$D_{||}, D_{\perp} \sim \frac{1}{M}\left(\frac{d\mathcal{E}}{dt}\right)_{\text{fl}}. \tag{13.2.1.10}$$

Knowing the diffusion coefficients D_{ij} we can determine the change in the distribution function of the test particles caused by their diffusion in velocity space:

$$\left(\frac{\partial f}{\partial t}\right)_{\text{diff}} = \frac{1}{2}\sum_{i,j}\frac{\partial^2}{\partial v_i\,\partial v_j}(D_{ij}f). \tag{13.2.1.11}$$

Adding this expression to expression (13.2.1.5) for $(\partial f/\partial t)_{\text{pol+fl}}$ we find the total change in the distribution function of the test particles per unit time:

$$\frac{\partial f}{\partial t} = -\sum_i\frac{\partial}{\partial v_i}(D_if)+\frac{1}{2}\sum_{i,j}\frac{\partial^2}{\partial v_i\,\partial v_j}(D_{ij}f). \tag{13.2.1.12}$$

This equation is the Fokker–Planck equation for test particles and describes the slow change of the distribution function in velocity space.

13.2.2. DYNAMIC FRICTION AND DIFFUSION COEFFICIENTS IN AN ELECTRON PLASMA

Let us first of all consider the dynamic friction and diffusion in an equilibrium plasma, neglecting the motion of the ions (Sitenko and Tszyan' Yu-Taĭ, 1963). The dynamic friction coefficient in a plasma is determined by the general formula (13.2.1.3). Using for the longitudinal dielectric permittivity and the spectral density of the field fluctuations eqns. (4.3.4.3) and (11.2.2.6) we get after integrating (13.2.1.3)

$$D \equiv D_{\text{pol}}+D_{\text{fl}} = -\frac{2}{\sqrt{\pi}}\frac{Z^2e^2\omega_{\text{pe}}^2}{Mv_{\text{e}}^2}\left(1+\frac{m_{\text{e}}}{M}\right)\zeta^{-2}\int_0^{\zeta} dz\, z^2e^{-z^2}L(z, k_0r_{\text{D}}), \quad (13.2.2.1)$$

where $\zeta = v/v_{\text{e}}\sqrt{2}$ and the function $L(z, k_0r_{\text{D}})$ is given by eqn. (13.1.3.13).

The polarization friction coefficient D_{pol} corresponds to the 1 in the bracket in (13.2.2.1) and the fluctuation friction coefficient D_{fl} to the term m_{e}/M. For a heavy test particle ($M \gg m_{\text{e}}$) the polarization friction D_{pol} is much larger than the fluctuation friction D_{fl}. If the test particle is an electron, $D_{\text{pol}} = D_{\text{fl}}$.

In the case of high temperatures and small electron densities ($k_0r_{\text{D}} \gg 1$) which is of most interest to us we can restrict ourselves in (13.1.3.13) to merely the main term, $L(z, k_0r_{\text{D}}) \cong \ln k_0r_{\text{D}}$. As a result we get from (13.2.2.1) the well-known Chandrasekhar formula

$$D = -\frac{Z^2e^2\omega_{\text{pe}}^2}{Mv_{\text{e}}^2}\left(1+\frac{m_{\text{e}}}{M}\right)G(\zeta)\ln k_0r_{\text{D}}, \quad (13.2.2.2)$$

where the function $G(\zeta)$ can be expressed in terms of the error function $\Phi(\zeta)$,

$$G(\zeta) = \frac{\Phi(\zeta)-\zeta\Phi'(\zeta)}{2\zeta^2}. \quad (13.2.2.3)$$

In the limiting cases of low and high velocities of the test particle we have

$$\left.\begin{aligned} D &= -\frac{2}{3\sqrt{\pi}}\frac{Z^2e^2\omega_{\text{pe}}^2}{Mv_{\text{e}}^2}\left(1+\frac{m_{\text{e}}}{M}\right)\zeta\ln(k_0r_{\text{D}}), \quad \zeta \ll 1, \\ D &= -\frac{1}{2}\frac{Z^2e^2\omega_{\text{pe}}^2}{Mv_{\text{e}}^2}\left(1+\frac{m_{\text{e}}}{M}\right)\frac{1}{\zeta^2}\ln(k_0r_{\text{D}}), \quad \zeta \gg 1. \end{aligned}\right\} \quad (13.2.2.4)$$

For slow particles the dynamic friction in the plasma increases linearly with the particle velocity, while for fast particles the friction decreases as the inverse square of the velocity.

The maximum of the dynamic friction is determined by the condition

$$\frac{d^2}{d\zeta^2}\left(\frac{\Phi(\zeta)}{\zeta}\right) = 0, \quad (13.2.2.5)$$

whence we find $\zeta \cong 0.97$. The friction will therefore be largest when the velocity of the test particle is almost the same as the average electron thermal velocity in the plasma.

We can in the same way find the longitudinal and transverse diffusion coefficients:

$$\left.\begin{aligned} D_{\parallel} &= \frac{4}{\sqrt{\pi}}\frac{Z^2e^2\omega_{\text{pe}}^2}{M^2v}\zeta^{-2}\int_0^{\zeta} dz\, z^2e^{-z^2}L(z, k_0r_{\text{D}}), \\ D_{\perp} &= \frac{4}{\sqrt{\pi}}\frac{Z^2e^2\omega_{\text{pe}}^2}{M^2v}\zeta^{-2}\int_0^{\zeta} dz(\zeta^2-z^2)\, e^{-z^2}L(z, k_0r_{\text{D}}). \end{aligned}\right\} \quad (13.2.2.6)$$

A comparison of eqns. (13.2.2.6) and (13.2.2.1) shows that the diffusion coefficients $D_{\parallel}$ and $D_{\perp}$ contain in the denominator an extra factor M as compared to the dynamic friction coefficient D. For heavy test particles ($M \gg m_e$) the slowing down caused by the dynamic friction is therefore much more important than the velocity dispersion connected with diffusion.

Retaining in (13.2.2.6) the main terms we get the Chandrasekhar formulae

$$D_{\parallel} = \frac{2Z^2e^2\omega_{pe}^2 m_e}{M^2 v} G(\zeta) \ln (k_0 r_D), \quad D_{\perp} = \frac{2Z^2e^2\omega_{pe}^2 m_e}{M^2 v} [\Phi(\zeta) - G(\zeta)] \ln (k_0 r_D). \tag{13.2.2.7}$$

If the velocity of the test particle is low ($v \ll v_e$) the difference between the parallel and perpendicular displacements—relative to the direction of the motion—disappears and

$$D_{\perp} = 2D_{\parallel}, \quad \zeta \ll 1. \tag{13.2.2.8}$$

In the case of high particle velocities ($v \gg v_e$) we have the relation

$$D_{\perp} = 2\zeta^2 D_{\parallel}, \quad \zeta \gg 1. \tag{13.2.2.9}$$

If the velocity of the test particles is larger than the average thermal velocity of the electrons in the plasma, the diffusion in velocity space will therefore by mainly lateral, that is, at right angles to the initial particle velocity.

13.2.3. FRICTION AND DIFFUSION COEFFICIENTS IN A TWO-TEMPERATURE PLASMA

The general formulae obtained in Subsection 13.2.1 enable us also to study the friction and diffusion in a non-equilibrium plasma provided we know the spectral density of the field fluctuations. In particular, we can use formulae (13.2.1.3) and (13.2.1.9) to find the dynamic friction and diffusion coefficients in a two-temperature electron–ion plasma (Sitenko and Tszyan' Yu-Taĭ, 1963). The dielectric permittivity and the spectral density of the fluctuations in the longitudinal electrical field are given by expressions (4.3.4.3) and (11.4.1.14) for a two-temperature plasma. Using (12.2.1.3) we get the following formulae for the dynamic friction coefficients D_{pol} and D_{fl} in a two-temperature plasma:

$$\left.\begin{aligned} D_{pol} &= -\frac{2}{\sqrt{\pi}} \frac{Z^2e^2\omega_{pe}^2}{Mv_e^2} \zeta^{-2} \int_0^{\zeta} dz\, z^2[e^{-z^2} + p\mu e^{-\mu^2 z^2}]\, L(z, p, \mu, k_0 r_D), \\ D_{fl} &= -\frac{2}{\sqrt{\pi}} \frac{Z^2e^2\omega_{pe}^2}{Mv_e^2} \frac{m_e}{M} \zeta^{-2} \int_0^{\zeta} dz\, z^2\, [e^{-z^2} + \mu^3 e^{-\mu^2 z^2}]\, L(z, p, \mu, k_0 r_D), \end{aligned}\right\} \tag{13.2.3.1}$$

where $p = T_e/T_i$, $\mu = \sqrt{(m_i/m_e)}$, and

$$L(z, p, \mu, k_0 r_D) = \ln(k_0 r_D) - \frac{1}{4} \ln \{[1 - \varphi(z) + p\{1 - \varphi(\mu z)\}]^2 + \pi z^2 (e^{-z^2} + p\mu e^{-\mu^2 z^2})^2\}$$
$$- \frac{1}{2\sqrt{\pi}} \frac{1 - \varphi(z) + p\{1 - \varphi(\mu z)\}}{z[\exp(-z)^2 + p\mu \exp(-\mu^2 z^2)]} \left\{\frac{\pi}{2} - \arctan \frac{1 - \varphi(z) + p\{1 - \varphi(\mu z)\}}{\sqrt{\pi} z[\exp(-z^2) + p\mu \exp(-\mu^2 z^2)]}\right. . \tag{13.2.3.2}$$

Restricting ourselves to the main terms for $k_0 r_D \gg 1$ we get for the dynamic friction coefficients the following approximate formulae

$$\left.\begin{aligned} D_{\text{pol}} &= -\frac{Z^2e^2\omega_{\text{pe}}^2}{Mv_{\text{e}}^2}\{G(\zeta)+pG(\mu\zeta)\}\ln(k_0r_D), \\ D_{\text{fl}} &= -\frac{Z^2e^2\omega_{\text{pe}}^2}{Mv_{\text{e}}^2}\frac{m_{\text{e}}}{M}\{G(\zeta)+\mu^2G(\mu\zeta)\}\ln(k_0r_D). \end{aligned}\right\} \quad (13.2.3.3)$$

Taking the motion of the ions into account leads to an appreciable increase in the role of the fluctuation friction. In fact, if the test particle is an electron, the fluctuation friction can be considerably larger than the polarization friction.

We note that for small velocities of the test particles both the polarization friction D_{pol} and the fluctuation friction D_{fl} are basically determined by the interaction between the particle and the ions in the plasma. We can similarly easily find the diffusion coefficients $D_{\|}$ and $D_{\perp}$ for a two-temperature plasma:

$$\left.\begin{aligned} D_{\|} &= \frac{4}{\sqrt{\pi}}\frac{Z^2e^2\omega_{\text{pe}}^2m_{\text{e}}}{M^2v}\zeta^{-2}\int_0^{\zeta} dz\, z^2[e^{-z^2}+\mu e^{-\mu^2z^2}]\, L(z, p, \mu, k_0r_D), \\ D_{\perp} &= \frac{4}{\sqrt{\pi}}\frac{Z^2e^2\omega_{\text{pe}}^2m_{\text{e}}}{M^2v}\zeta^{-2}\int_0^{\zeta} dz(\zeta^2-z^2)\,[e^{-z^2}+\mu e^{-\mu^2z^2}]\, L(z, p, \mu, k_0r_D). \end{aligned}\right\} \quad (13.2.3.4)$$

In the limiting case $k_0 r_D \gg 1$ we have

$$\left.\begin{aligned} D_{\|} &= \frac{2Z^2e^2\omega_{\text{pe}}^2m_{\text{e}}}{M^2v}[G(\zeta)+G(\mu\zeta)]\ln(k_0r_D), \\ D_{\perp} &= \frac{2Z^2e^2\omega_{\text{pe}}^2m_{\text{e}}}{M^2v}[\Phi(\zeta)-G(\zeta)+\Phi(\mu\zeta)-G(\mu\zeta)]\ln(k_0r_D). \end{aligned}\right\} \quad (13.2.3.5)$$

It follows from (13.2.3.5) that the transverse diffusion depends strongly on the ion motion. If $v \gg v_e$, the transverse diffusion is approximately equally caused by the interaction of the test particle with the electrons and with the ions in the plasma. If $v_i \ll v \ll v_e$, the transverse diffusion is mainly determined by the interaction with the ions.

Using the expressions which we have just obtained for the friction and diffusion coefficient we can estimate the time τ it takes to equalize the electron and ion temperatures T_e and T_i:

$$\frac{dT_{\text{e}}}{dt} = -\frac{T_{\text{e}}-T_{\text{i}}}{\tau}. \quad (13.2.3.6)$$

To do this we turn to the formula for the change in energy of the test particle per unit time

$$\frac{d\mathcal{E}}{dt} = \frac{1}{2}M(2vD+D_{\|}+D_{\perp}). \quad (13.2.3.7)$$

If the test particle is an electron, we have

$$\frac{d\mathcal{E}_{\text{e}}}{dt} = \frac{e^2\omega_{\text{pe}}^2}{v}\left\{-\Phi(\zeta)+2\zeta\Phi'(\zeta)-\frac{p}{\mu^2}\Phi(\mu\zeta)+\left(1+\frac{p}{\mu^2}\right)\mu\zeta\Phi'(\mu\zeta)\right\}. \quad (13.2.3.8)$$

The first and second terms here determine the change in the electron energy caused by the interaction with the other electrons and the third and fourth terms the change in energy caused by the interaction with the ions in the plasma.

Let us now assume that the test particles are characterized by a Maxwell distribution with temperature T_e. Averaging eqn. (13.2.3.8) over the Maxwell distribution we obtain clearly the change in the electron temperature per unit time:

$$\left\langle \frac{d\mathcal{E}_e}{dt} \right\rangle = \frac{3}{2}\frac{dT_e}{dt}. \tag{13.2.3.9}$$

One checks easily that the sum of the first two terms in (13.2.3.8) vanishes after averaging as the electrons in the plasma are in a state of thermal equilibrium. However, the change in energy caused by the electron–ion interaction is equal to

$$\left\langle \frac{d\mathcal{E}_e}{dt} \right\rangle = -\frac{4\sqrt{(2\pi)}\,n_0 e^4}{m_e m_i}\,\frac{T_e - T_i}{[(T_e/m_i)+(T_e/m_i)]^{3/2}}\ln(k_0 r_D). \tag{13.2.3.10}$$

Using (13.2.3.6) and (13.2.3.9) we get as a result the expression for the relaxation time of a non-isothermal plasma, encountered earlier:

$$\tau = \frac{3}{8\sqrt{(2\pi)}}\,\frac{m_e m_i}{n_0 e^4 \ln(k_0 r_D)}\left(\frac{T_e}{m_e}+\frac{T_i}{m_i}\right)^{3/2}. \tag{13.2.3.11}$$

13.3. Passage of Charged Particles Through an Equilibrium Plasma in a Magnetic Field

13.3.1. SCATTERING PROBABILITY IN A MAGNETO-ACTIVE PLASMA

Let us now turn to a discussion of the interaction of a charged test particle with a plasma in a constant and uniform magnetic field (Sitenko and Stepanov, 1958; Akhiezer, 1956; Akhiezer and Faĭnberg, 1962; Akhiezer, 1961; Kitsenko, 1963). The magnetic field leads, first of all, to a complication of the nature of the motion of the moving particle and, secondly, to changes in the correlation functions of the fluctuations in the plasma.

If the direction of motion of the particle before and after the scattering is close to the direction of the magnetic field, the magnetic field does not affect the motion of the particle. For this it is necessary that the following conditions are satisfied:

$$q_\perp v \sin\alpha \ll \omega_{BZ}, \qquad \frac{\hbar q_\perp^2}{M} \ll \omega_{BZ},$$

where $\hbar q_\perp$ is the component of the momentum transfer at right angles to the direction of the magnetic field $\boldsymbol{B}_0$; $\omega_{BZ} = ZeB_0/Mc$ and α is the angle between the direction of motion of the particle and direction of the magnetic field.

The influence of the magnetic field on the motion of the test particle can also be neglected when the twisting of the particle orbit along a characteristic path is small. For this it is necessary that one of the following conditions holds:

$$q_\perp v \sin\alpha \gg \omega_{BZ}, \qquad \frac{\hbar q_\perp^2}{M} \gg \omega_{BZ}.$$

Let us first of all consider the scattering of a particle in these cases. The probability for the scattering of a particle is then given by the general formula (13.1.4.6) in which we must take for $\langle \varrho^2 \rangle_{q\omega}$ the spectral density of the charge density fluctuations in a plasma in a magnetic field.

We showed in Section 11.2 that the function $\langle \varrho^2 \rangle_{q\omega}$ has sharp maxima at the frequencies of the plasma eigen oscillations; the corresponding terms in the expression for dw can be interpreted as the probabilities for the scattering of the particle with the excitation (or absorption) of a given kind of oscillation. In particular, the probability for the scattering of a particle with the excitation or absorption of longitudinal electron oscillations has the form

$$dw_{p \to p'} = \frac{4\pi^2 Z^2 e^2}{\hbar q^2} \frac{(\omega^2-\omega_{Be}^2)^2}{\omega^4 \sin^2\vartheta + (\omega^2-\omega_{Be}^2)^2 \cos^2\vartheta}$$

$$\times \{|\omega| (N_{|\omega|}+1) [\delta(\omega+\omega_+)+\delta(\omega+\omega_-)] + \omega N_\omega [\delta(\omega-\omega_+)+\delta(\omega-\omega_-)]\} \frac{d^3p'}{(2\pi\hbar)^3}, \tag{13.3.1.1}$$

where the frequencies $\omega_\pm$ are given by eqn. (5.1.2.6) and ϑ is the angle between the vector $\boldsymbol{q}$ and the direction of the magnetic field.

13.3.2. POLARIZATION ENERGY LOSSES OF A PARTICLE, CAUSED BY THE INTERACTION WITH THE LONGITUDINAL FIELD

Multiplying (13.3.1.1) by $\hbar\omega$ and integrating over $\boldsymbol{q}$ we find the energy transferred to the particle by plasma oscillations with wavevectors with absolute magnitude less than a certain value q_0 (Akhiezer, 1961):

$$\left(\frac{d\mathcal{E}}{dt}\right)_{q<q_0} = -\frac{Z^2 e^2 \omega_{pe}^2}{v} \left\{ \ln \frac{q_0 v}{\omega_{pe}} - f\left(\alpha, \frac{\omega_{pe}}{|\omega_{Be}|}\right)\right\}, \tag{13.3.2.1}$$

where $\omega_{Be} = -eB_0/m_e c$ and

$$\left.\begin{aligned} f(\alpha, u) &= \frac{1}{2\pi u^2}\left\{\int_0^{z_1} dz\, g(z) \ln z - \int_{z_2}^{z_3} dz\, g(z) \ln z\right\} - \ln u, \\ g(z) &= z(1-z)\{z(z-z_1)(z-z_2)(z_3-z)\}^{-1/2}, \\ z_{1,2} &= \tfrac{1}{2}(1+u^2) \mp \tfrac{1}{2}\sqrt{\{(1+u^2)^2 - 4u^2 \sin^2\alpha\}}, \quad z_3 = 1+u^2. \end{aligned}\right\} \tag{13.3.2.2}$$

Let us now find the total polarization energy losses of a fast particle moving through a plasma in a magnetic field. When considering the close collisions (large momentum transfers) we can neglect the difference between the motion of the particle and a rectilinear motion and use formula (13.1.4.22) for the energy loss per unit time. If the angle α between the particle velocity and the magnetic field satisfies one of the inequalities

$$\sin\alpha \gg \frac{\omega_{BZ}}{\max\{\omega_{pe}, |\omega_{Be}|\}}; \qquad \sin\alpha \ll \frac{\omega_{BZ}}{\max\{\omega_{pe}, |\omega_{Be}|\}}, \tag{13.3.2.3}$$

we can neglect the curvature of the particle also in considering the distant collisions. As in the case where there is no magnetic field, the scattering of a particle occurs mainly through the emission and absorption of plasma waves so that we can use formula (13.3.2.1) to determine the energy losses.

Adding the contributions from the close and the distant collisions we get finally the following expressions for the polarization energy losses of a fast particle:

$$\left.\begin{aligned}\left(\frac{d\mathcal{E}}{dt}\right)_{\text{pol}} &= -\frac{Z^2e^2\omega_{\text{pe}}^2}{v}\left\{\ln\frac{2Mm_ev^2}{\hbar\omega_{\text{pe}}(M+m_e)} - f\left(\alpha, \frac{\omega_{\text{pe}}}{|\omega_{Be}|}\right)\right\}, &\quad \frac{e^2}{\hbar v} \ll 1,\\ \left(\frac{d\mathcal{E}}{dt}\right)_{\text{pol}} &= -\frac{Z^2e^2\omega_{\text{pe}}^2}{v}\left\{\ln\left[\frac{2}{\gamma}\frac{Mm_ev^3}{Ze^2\omega_{\text{pe}}(M+m_e)} - f\left(\alpha, \frac{\omega_{\text{pe}}}{|\omega_{Be}|}\right)\right]\right\}, &\quad \frac{e^2}{\hbar v} \gg 1,\end{aligned}\right\} \tag{13.3.2.4}$$

where the function f is given by formula (13.3.2.2). The first terms in these expressions are the total energy losses of a fast particle in an unmagnetized plasma (compare (13.1.2.16) and (13.1.2.18)) while the second terms describe the influence of the magnetic field; they depend on the direction of motion of the particle relative to the direction of the magnetic field and vanish when $\omega_{Be} = 0$.

We see that the total polarization energy losses of a fast particle in a magnetic field are proportional to the plasma density and depend relatively weakly (logarithmically) on the magnetic field strength. As in the case of an unmagnetized plasma, the losses are independent of the temperature—provided the fluctuations in the plasma are not too different from the equilibrium ones.

In the case of a strong magnetic field ($|\omega_{Be}| \gg \omega_{\text{pe}}$) expressions (13.3.2.4) simplify considerably: the function f occurring in them then has the form (the angle α is assumed to be different from 0 or π)

$$f\left(\alpha, \frac{\omega_{\text{pe}}}{|\omega_{Be}|}\right) = \frac{1}{4}\sin^2\alpha\left(1+\ln\left[\frac{1}{4}\frac{\omega_{\text{pe}}^2}{\omega_{Be}^2}\sin^2\alpha\right]\right) - \ln\frac{\omega_{\text{pe}}}{|\omega_{Be}|}. \tag{13.3.2.5}$$

We have already mentioned that expressions (13.3.2.4) are valid provided the angle α satisfies one of the conditions (13.3.2.3). This requirement is satisfied, in particular, for particles of any mass, if they move along the magnetic field. We then get from (13.3.2.4) (to fix the ideas we assume that $e^2/\hbar v \ll 1$)

$$\left(\frac{d\mathcal{E}}{dt}\right)_{\text{pol}} = -\frac{Z^2e^2\omega_{\text{pe}}^2}{v}\ln\frac{2Mm_ev^2}{\hbar\sqrt{[\omega_{\text{pe}}^2+\omega_{Be}^2]}(M+m_e)}, \tag{13.3.2.6}$$

We can use eqns. (13.3.2.4) for practically any angle α in the case of a heavy particle. In particular, if such a particle moves at right angles to the field direction the expression for the energy losses has the form (we assume that $e^2/\hbar v \ll 1$)

$$\left(\frac{d\mathcal{E}}{dt}\right)_{\text{pol}} = -\frac{Z^2e^2\omega_{\text{pe}}^2}{v}\left\{\ln\frac{4m_ev^2}{\hbar\sqrt{[2\omega_{\text{pe}}|\omega_{Be}|]}} - \frac{1}{4}\right\}. \tag{13.3.2.7}$$

Equations (13.3.2.4) to (13.3.2.7) determine the energy losses of a non-relativistic particle. If a relativistic particle moves through the plasma, its energy losses are again proportional to $Z^2e\omega_{\text{pe}}^2/v$ and weakly dependent on the magnetic field.

Let us look at the expression for the quantity which characterizes the effect of the magnetic field on the energy losses of the particle (Kitsenko, 1963):

$$F(\omega_{Be}, \beta) = \left(\frac{d\mathcal{E}}{dt}\right)_{\text{pol}} - \left(\frac{d\mathcal{E}}{dt}\right)_{\text{pol}, \, \boldsymbol{B}_0 = 0}. \tag{13.3.2.8}$$

If the condition $(\omega_{Be}/\omega_{pe})^2 \leqslant 4\beta^2/(1-\beta^2)$, with $\beta = v/c$, is satisfied, we have

$$F(\omega_{Be}, \beta) = \frac{Z^2 e^2 \omega_{Be}^2}{4v}(1-\beta^2)(2-\beta^2). \tag{13.3.2.9}$$

If

$$\frac{4\beta^2}{1-\beta^2} \leqslant \frac{\omega_{Be}^2}{\omega_{pe}^2} \leqslant \frac{4\beta^2}{(1-\beta^2)^2},$$

we have

$$F(\omega_{Be}, \beta) = \frac{Z^2 e^2 \omega_{pe}^2}{v}\left\{\frac{1}{2}\ln\frac{1+x}{1-x} + \frac{(1-\beta^2)(2-\beta^2)\,\omega_{Be}^2}{4\omega_{pe}^2} - \frac{1-\beta^2}{4\beta^2}\frac{\omega_{Be}^2}{\omega_{pe}^2}x\right\}, \tag{13.3.2.10}$$

where $x = \sqrt{\{1-[4\beta^2/(1-\beta^2)]\,(\omega_{pe}^2/\omega_{Be}^2)\}}$.

Finally, if $(\omega_{Be}/\omega_{pe})^2 \geqslant 4\beta^2/(1-\beta^2)^2$, we have

$$F(\omega_{Be}, \beta) = \frac{Z^2 e^2 \omega_{pe}^2}{v}\left\{\frac{1}{2}\ln\frac{(1+x)(1+\beta^2-y)}{(1-x)(1+\beta^2+y)} + \frac{1-\beta^2}{4\beta^2}\frac{\omega_{Be}^2}{\omega_{pe}^2}[\beta^2(2-\beta^2)-x+(1-\beta^2)\,y]\right\}, \tag{13.3.2.11}$$

where $y = \sqrt{\{(1-\beta^2)^2 - 4\beta^2(\omega_{pe}^2/\omega_{Be}^2)\}}$.

Putting $\beta \ll 1$ in eqn. (13.3.2.11) and using (13.1.2.18), we can obtain eqn. (13.3.2.6) for the energy losses of a non-relativistic particle moving along the magnetic field.

13.3.3. TAKING THE CURVATURE INTO ACCOUNT

Let us now briefly consider the case when the inequalities (13.3.2.3) are not satisfied so that it is necessary to take into account that the motion of the particle is made more complicated by the magnetic field (Akhiezer, 1961; Sitenko and Radzievskiĭ, 1966). It is well known that the motion of a particle in a magnetic field is determined by its momentum p_z along the field direction, the quantum number ν characterizing the motion in the plane perpendicular to $\boldsymbol{B}_0$, and the coordinate of the centre of the Larmor circle. The energy of a particle in a magnetic field has the form $\mathcal{E}_{\nu p_z} = (\nu+\frac{1}{2})\hbar\omega_{BZ} + p_z^2/2M$.

As in the case when there is no field, the probability for the scattering of the particle can be connected with the spectral density of the charge density fluctuations. We shall not give here the derivation, but merely present the expression for the probability that the particle per unit time makes a transition from a state with quantum numbers ν, p_z to a state with quantum numbers ν', p_z' (averaged over the initial and summed over the final values of the coordinate of the centre of the Larmor circle of the particle):

$$dw_{\nu p_z \to \nu' p_z'} = w_{\nu p_z \to \nu' p_z'}\frac{dp_z'}{2\pi\hbar}, \quad w_{\nu p_z \to \nu' p_z'} = \frac{Z^2 e^2}{\hbar^2}\int \frac{d^2\boldsymbol{q}_\perp}{(2\pi)^2}\Lambda_{\nu\nu'}\left(\frac{\hbar q_\perp^2}{2M\omega_{BZ}}\right)\langle\varphi^2\rangle_{q\omega}, \tag{13.3.3.1}$$

where $\hbar\omega$ are $\hbar q_z$ are the changes in the energy and longitudinal momentum of the particle,

$$\hbar\omega = (\nu'-\nu)\,\hbar\omega_{BZ} + \frac{p_z'^2 - p_z^2}{2M}, \quad \hbar q_z = p_z' - p_z,$$

and

$$\Lambda_{\nu\nu'}(a) = \int_0^\infty dx\, J_0(2\sqrt{\{ax\}})\, L_\nu(x)\, L_{\nu'}(x)\, e^{-x},$$

with the L_ν Laguerre polynomials and J_0 a Bessel function.

Multiplying (13.3.3.1) by $\hbar\omega$, summing over ν', and integrating over p_z', we find the total energy losses of the particle per unit time:

$$\frac{d\mathcal{E}}{dt} = \frac{Z^2e^2}{8\pi^3\hbar} \sum_{\nu'} \int \omega \Lambda_{\nu\nu'}\left(\frac{\hbar q_\perp^2}{2M\omega_{BZ}}\right) \langle\varphi^2\rangle_{q\omega}\, \delta\left[\omega - (\nu'-\nu)\,\omega_{BZ} - \frac{p_z'^2 - p_z'}{2M\hbar}\right] d\omega\, d^3\boldsymbol{q}. \quad (13.3.3.2)$$

This formula takes quantum effects consistently into account, provided we use for $\langle\varphi^2\rangle_{q\omega}$ the quantum-mechanical expression for the spectral density of the potential fluctuations. We shall, however, restrict our discussion to the classical case.

Using (11.1.2.20) one can easily show that up to terms linear in $\hbar$ we have

$$\langle\varphi^2\rangle_{q\omega} = \langle\varphi^2\rangle_{q\omega}^{\rm cl} + \frac{4\pi\hbar}{\sum_{i,j} q_i \varepsilon_{ij}(\boldsymbol{q}, \omega)\, q_j}. \quad (13.3.3.3)$$

Proceeding as in Subsection 13.1.4 and expanding $\Lambda_{\nu\nu'}$, and the δ-function in powers of $\hbar$, and using the symmetry properties of the classical correlation function, $\langle\varphi^2\rangle_{q\omega}^{\rm cl} = \langle\varphi^2\rangle_{-q-\omega}^{\rm cl}$, we get the following formula for the change in the energy of a charged particle moving along the magnetic field (Sitenko and Radzievskiĭ, 1966):

$$\frac{d\mathcal{E}}{dt} = \frac{Z^2e^2}{2\pi^2}\,{\rm Im} \int \frac{\omega}{\sum_{i,j} q_i\varepsilon_{ij}(\boldsymbol{q}, \omega)\, q_j}\, \delta[\omega - (\boldsymbol{q}\cdot\boldsymbol{v})]\, d\omega\, d^3\boldsymbol{q} - \frac{Z^2e^2}{16\pi^3 M} \int \omega\langle\varphi^2\rangle_{q\omega}^{\rm cl}$$
$$\times \left\{\frac{q_\perp^2}{2\omega_{BZ}}\, [\delta\{\omega - (\boldsymbol{q}\cdot\boldsymbol{v}) + \omega_{BZ}\} - \delta\{\omega - (\boldsymbol{q}\cdot\boldsymbol{v}) - \omega_{BZ}\} + q_z^2\, \delta'[\omega - (\boldsymbol{q}\cdot\boldsymbol{v})]\right\} d\omega\, d^3\boldsymbol{q}, \quad (13.3.3.4)$$

where $\boldsymbol{v}$ is the particle velocity and $\varepsilon_{ij}(\boldsymbol{q}, \omega)$ the dielectric permittivity tensor of the plasma.

The first term in formula (13.3.3.4) determines the polarization energy losses of a charged particle moving through a magneto-active plasma, and the second term determines the change in energy connected with the fluctuations in the longitudinal electrical field in the plasma. This formula is valid for an equilibrium plasma and also for a non-equilibrium plasma, provided we neglect the ion motion.

13.3.4. CHERENKOV EMISSION OF A CHARGE MOVING IN A PLASMA IN A MAGNETIC FIELD

In concluding this section we shall now consider the problem of the Cherenkov emission when a charged particle moves through a magneto-active plasma (Sitenko and Kaganov, 1955; Sitenko and Kolomenskiĭ, 1956; Kolomenskiĭ, 1956; Ginzburg (1940) studied the problem of energy losses of a charged particle in an anisotropic medium). We noted already in Section 13.1 that the phase velocities of the transverse waves in a plasma are larger than the velocity of light when there is no magnetic field present; Cherenkov emission in an unmagnetized plasma is therefore impossible. When an external magnetic field is present the refractive index of the plasma is larger than unity in a well defined frequency region, that is, the phase velocities of the corresponding waves are less than the velocity of light and there is therefore the possibility for the emission of electromagnetic waves (Cherenkov emission) when a charged particle moves through the plasma.

Let us consider the Cherenkov emission in that case in some detail (Sitenko and Kolomenskiĭ, 1956). We shall assume that the particle velocity is much higher than the electron thermal velocity in the plasma. We can then neglect the thermal motion of the electrons and consider the plasma to be an optically active anisotropic medium characterized by a dielectric permittivity tensor:

$$\varepsilon_{ij} = \begin{bmatrix} \varepsilon_1 & -i\varepsilon_2 & 0 \\ i\varepsilon_2 & \varepsilon_1 & 0 \\ 0 & 0 & \varepsilon_3 \end{bmatrix}, \tag{13.3.4.1}$$

where

$$\varepsilon_1 = 1 - \frac{\omega_{pe}^2}{\omega^2 - \omega_{Be}^2}, \quad \varepsilon_2 = \frac{\omega_{Be}}{\omega}\,\frac{\omega_{pe}^2}{\omega^2-\omega_{Be}^2}, \quad \varepsilon_3 = 1 - \frac{\omega_{pe}^2}{\omega^2}, \tag{13.3.4.2}$$

$\omega_{Be} = -eB_0/m_e c$, $\boldsymbol{B}_0$ is the external field, and the z-axis is taken along the field.

The Fourier component of the field strength produced by the moving particle is given by the equation

$$E_i(\boldsymbol{k}, \omega) = -i\frac{Ze}{2\pi^2\omega}\sum_j \Lambda_{ij}^{-1} v_j\, \delta\{\omega - (\boldsymbol{k}\cdot\boldsymbol{v})\}. \tag{13.3.4.3}$$

To evaluate the energy losses of a moving charge we use again the relation

$$\frac{d\mathcal{E}}{dt} = Ze\left(\boldsymbol{v}\cdot\boldsymbol{E}\Big|_{r=vt}\right), \tag{13.3.4.4}$$

where the field is evaluated at the position of the charge. Using eqn. (13.3.4.3) for the field we get the following formula for the energy losses of the particle per unit time, caused by distant collisions:

$$\frac{d\mathcal{E}}{dt} = -i\frac{Z^2e^2}{2\pi^2}\iint_{k<k_0} \frac{1}{\omega}\sum_j \Lambda_{ij}^{-1}(\boldsymbol{k}, \omega)\, v_i v_j\, \delta\{\omega-(\boldsymbol{k}\cdot\boldsymbol{v})\}\, d\omega\, d^3\boldsymbol{k}. \tag{13.3.4.5}$$

The integration over $\boldsymbol{k}$ is here up to some maximum value k_0 of its absolute magnitude which is of the order of b_0^{-1}, where b_0 is the minimum value of the collision parameter.

We now use the explicit form (11.2.7.1) of the tensor Λ_{ij} and the definition (5.1.1.9) for the refractive index of the ordinary and the extra-ordinary waves. We then get the following formula for $d\mathcal{E}/dt$ if the particle moves along the magnetic field:

$$\frac{d\mathcal{E}}{dt} = -i\frac{Z^2e^2v}{2\pi^2c^3}\int_{-\infty}^{+\infty} d\omega \int d^2\omega \int_0^{n_0} n^2\, dn\, \omega\, \frac{n^4\cos^2\vartheta - n^2\varepsilon_1(1+\cos^2\vartheta)+\varepsilon_1^2-\varepsilon_2^2}{[\varepsilon_1\sin^2\vartheta+\varepsilon_3\cos^2\vartheta]\,[n^2-n_+^2(\vartheta)]\,[n^2-n_-^2(\vartheta)]}$$

$$\times\,\delta(n\beta\cos\vartheta - 1), \tag{13.3.4.6}$$

where we have used instead of k a new integration variable $n = kc/\omega$. Integrating in eqn. (13.3.4.6) over the angles can be done immediately, by using the δ-function and the integration over n must then be from β^{-1} to $n_0 = k_0 c/\omega$. As a result we get

$$\frac{d\mathcal{E}}{dt} = -\frac{Z^2e^2}{\pi v}\operatorname{Re} i\int_0^\infty \frac{(1-\varepsilon_1\beta^2)(n_+^2-\varepsilon_1)-\beta^2\varepsilon_2^2}{\varepsilon_1(n_+^2-n_-^2)}\ln\frac{(n_0^2-n_+^2)\,\beta^2}{1-n_+^2\beta^2}\,\omega\,d\omega$$

$$-\frac{Z^2e^2}{\pi v}\operatorname{Re} i\int_0^\infty \frac{(1-\varepsilon_1\beta^2)(n_-^2-\varepsilon_1)-\beta^2\varepsilon_2^2}{\varepsilon_1(n_-^2-n_+^2)}\ln\frac{(n_0^2-n_-^2)\,\beta^2}{1-n_-^2\beta^2}\,\omega\,d\omega, \quad (13.3.4.7)$$

where

$$n_\pm^2 = \{(\varepsilon_1^2-\varepsilon_2^2+\varepsilon_1\varepsilon_3)\,\beta^2+\varepsilon_1-\varepsilon_3\pm[(\varepsilon_1^2-\varepsilon_2^2-\varepsilon_1\varepsilon_3)^2\,\beta^4-2\varepsilon_1(\varepsilon_1^2-\varepsilon_2^2+\varepsilon_3^2)\,\beta^2+2\varepsilon_3(2\varepsilon_1^2+\varepsilon_2^2)\,\beta^2$$
$$+(\varepsilon_1-\varepsilon_3)^2]^{1/2}\}/2\varepsilon_1\,\beta^2 \quad (13.3.4.8)$$

are the values of the refractive index in the directions of the maximum emission which are determined by the equations

$$\cos^2\vartheta_\pm = \beta^{-2}n_\mp^{-2}(\vartheta_\pm).$$

It is clear that those frequency regions in which the arguments of the logarithms take on negative values and moreover the regions in which the integrands have poles will contribute to the expression (13.3.4.7) for the energy losses, as only for those regions will the real parts needed in (13.3.4.7) be non-vanishing. We thus find

$$\frac{d\mathcal{E}}{dt} = \frac{Z^2e^2}{v}\int \frac{(1-\varepsilon_1\beta^2)\,(n_+^2-\varepsilon_1)-\beta^2\varepsilon_2^2}{\varepsilon_1(n_+^2-n_-^2)}\,\omega\,d\omega+\frac{Z^2e^2}{v}\int \frac{(1-\varepsilon_1\beta^2)\,(n_-^2-\varepsilon_1)-\beta^2\varepsilon_2^2}{\varepsilon_1(n_-^2-n_+^2)}\,\omega\,d\omega$$
$$-\frac{Z^2e^2}{v}\sum_r \frac{\omega_r}{|\,d\varepsilon_1/d\omega\,|_r}\ln\left\{1+\left(\frac{\varepsilon_1}{\varepsilon_3}\right)_r\frac{k_0^2v^2}{\omega_r^2}\right\}, \quad (13.3.4.9)$$

where the integration in the first two terms is over the frequency regions determined by the inequalities

$$n_0^2 > n_+^2 > \beta^{-2}, \quad n_0^2 > n_-^2 > \beta^{-2}, \quad (13.3.4.10)$$

while in the third term the summation is over all frequencies ω_r for which ε_1, ε_2, and ε vanish at the same time. Writing down the explicit form of n_+^2 and n_-^2, we get finally (Sitenko and Kolomenskiĭ, 1956)

$$\frac{d\mathcal{E}}{dt} = -\frac{Z^2e^2}{2v}\int \omega\,d\omega\left(\beta^2-\frac{1}{\varepsilon_1}\right)$$
$$\times\left\{1\pm\frac{\varepsilon_1(\varepsilon_1^2-\varepsilon_2^2-\varepsilon_1\varepsilon_3)\,\beta^4-(2\varepsilon_1^2-2\varepsilon_1\varepsilon_3+\varepsilon_2^2)\,\beta^2+\varepsilon_1-\varepsilon_3}{(1-\varepsilon_1\beta^2)[(\varepsilon_1-\varepsilon_2-\varepsilon_1\varepsilon_3)^2\,\beta^4-2\varepsilon_1(\varepsilon_1^2-\varepsilon_2^2+\varepsilon_3^2)\,\beta^2+2\varepsilon_3(2\varepsilon_1^2+\varepsilon_2^2)+\beta^2(\varepsilon_1-\varepsilon_3)^2]^{1/2}}\right\}$$
$$-\frac{Z^2e^2}{2v}\sum_r\frac{\omega_r}{|\,d\varepsilon_1/d\omega\,|_r}\ln\left\{1+\left(\frac{\varepsilon_1}{\varepsilon_3}\right)_r\frac{k_0^2v^2}{\omega_r^2}\right\}, \quad (13.3.4.11)$$

where the integration region is determined by the inequalities (13.3.4.10). The first term here is the energy loss of the particle through Cherenkov emission and the second represents strictly polarization losses. To see this let us elucidate the nature of the energy losses and let us for this evaluate the energy flux through a cylindrical surface around the trajectory of the charge. Let us first of all determine the field occurring in an optically active anisotropic medium when a point charge moves through it. Using the inverse Fourier transform, we get from (13.3.4.3) and the relations

$$\int_0^{2\pi} e^{i\varkappa r\cos\vartheta}\,d\vartheta = 2\pi\, J_0(\varkappa r), \qquad \int_0^{\infty} \frac{\varkappa I_0(\varkappa r)}{\varkappa^2+k^2}\,d\varkappa = K_0(kr), \qquad \operatorname{Re} k > 0,$$

the following expressions for the components of the electrical field strength in a cylindrical coordinate system:

$$\left.\begin{aligned}
E_z(\mathbf{r}, t) &= -i\frac{Ze}{\pi v^2}\int_{-\infty}^{+\infty} \frac{\omega}{\varepsilon_1}\left\{\frac{(1-\varepsilon_1\beta^2)(n_+^2-\varepsilon_1)-\beta^2\varepsilon_2^2}{n_+^2-n_-^2}K_0(k_+r)\right.\\
&\quad\left.+\frac{(1-\varepsilon_1\beta^2)(n_-^2-\varepsilon_1)-\beta^2\varepsilon_2^2}{n_-^2-n_+^2}K_0(k_-r)\right\}\exp\left[i\omega\left(\frac{z}{v}-t\right)\right]d\omega,\\
E_r(\mathbf{r}, t) &= \frac{Ze}{\pi v}\int_{-\infty}^{+\infty}\frac{1}{\varepsilon_1}\left\{\frac{n_+^2-\varepsilon_1}{n_+^2-n_-^2}k_+K_1(k_+r)+\frac{n_-^2-\varepsilon_1}{n_-^2-n_+^2}k_-K_1(k_-r)\right\}\exp\left[i\omega\left(\frac{z}{v}-t\right)\right]d\omega,\\
E_\varphi(\mathbf{r}, t) &= i\frac{Ze}{\pi v}\int_{-\infty}^{+\infty}\frac{\varepsilon_2}{\varepsilon_1(n_+^2-n_-^2)}\{k_+K_1(k_+r)-k_-K_1(k_-r)\}\exp\left[i\omega\left(\frac{z}{v}-t\right)\right]d\omega,
\end{aligned}\right\} \quad (13.3.4.12)$$

where $k_\pm^2 = (\omega/v)^2(1-\beta^2 n_\pm^2)$, while the $n_\pm^2$ are given by eqn. (13.3.4.8).

Similarly we get for the magnetic field strength the formulae

$$\left.\begin{aligned}
B_z(\mathbf{r}, t) &= -\frac{Ze}{\pi\beta v^2}\int_{-\infty}^{+\infty}\frac{\omega\varepsilon_2}{\varepsilon_1}\left\{\frac{1-\beta^2 n_+^2}{n_+^2-n_-^2}K_0(k_+r)+\frac{1-\beta^2 n_-^2}{n_-^2-n_+^2}K_0(k_-r)\right\}\exp\left[i\omega\left(\frac{z}{v}-t\right)\right]d\omega,\\
B_r(\mathbf{r}, t) &= -i\frac{Ze}{\pi\beta v}\int_{-\infty}^{+\infty}\frac{\varepsilon_2}{\varepsilon_1(n_+^2-n_-^2)}\{k_+K_1(k_+r)-k_-K_1(k_-r)\}\exp\left[i\omega\left(\frac{z}{v}-t\right)\right]d\omega,\\
B_\varphi(\mathbf{r}, t) &= \frac{Ze}{\pi c}\int_{-\infty}^{+\infty}\frac{1}{\varepsilon_1}\left\{\frac{\varepsilon_1 n_+^2-\varepsilon_1^2+\varepsilon_2^2}{n_+^2-n_-^2}k_+K_1(k_+r)+\frac{\varepsilon_1 n_-^2-\varepsilon_1^2+\varepsilon_2^2}{n_-^2-n_+^2}k_-K_1(k_-r)\right\}\\
&\quad\times\exp\left[i\omega\left(\frac{z}{v}-t\right)\right]d\omega.
\end{aligned}\right\} \quad (13.3.4.13)$$

The optical activity of the plasma leads to the occurrence of the electrical field strength component E_φ and the magnetic field strength components B_z and B_r, which are not present when a charge moves through a non-active medium.

Using the Poynting vector we can now determine the amount of energy emitted by the charge per unit path length:

$$\frac{d\mathcal{E}}{dz} = \frac{c}{4\pi} \int_{-\infty}^{+\infty} (E_z B_\varphi - E_\varphi B_z)\, 2\pi r\, dt. \tag{13.3.4.14}$$

Substituting (13.3.4.12) and (13.3.4.13) into (13.3.4.14) we get after a few simple transformations

$$\frac{d\mathcal{E}}{dt} = -\frac{Z^2 e^2}{\pi v} 2r \operatorname{Re} i \int_0^\infty \left\{ \frac{(1-\varepsilon_1\beta^2)(n_+^2-\varepsilon_1)-\beta^2\varepsilon_2^2}{\varepsilon_1(n_+^2-n_-^2)} K_0(k_+ r)\, k_+^* K_1(k_+^* r) \right.$$
$$\left. + \frac{(1-\varepsilon_1\beta^2)(n_-^2-\varepsilon_1)-\beta^2\varepsilon_2^2}{\varepsilon_1(n_-^2-n_+^2)} K_0(k_- r)\, k_-^* K_1(k_-^* r) \right\} \omega\, d\omega. \tag{13.3.4.15}$$

If ε_1, ε_2, and ε_3 have no common zeroes, we get a contribution to the real part in this expression only from the frequencies for which k_1 or k_2 is imaginary. Noting that when k is imaginary we have

$$k^* K_1(k^* r)\, K_0(kr) - k K_1(kr)\, K_0(k^* r) = \tfrac{1}{2}\pi i,$$

we get for the loss through Cherenkov emission the expression

$$\frac{d\mathcal{E}}{dt} = \frac{Z^2 e^2}{v} \int_{\beta^2 n_+^2 > 1} \frac{(1-\varepsilon_1\beta^2)(n_+^2-\varepsilon_1)-\beta^2\varepsilon_2^2}{\varepsilon_1(n_+^2-n_-^2)}\, \omega\, d\omega$$
$$+ \frac{Z^2 e^2}{v} \int_{\beta^2 n_-^2 > 1} \frac{(1-\varepsilon_1\beta^2)(n_-^2-\varepsilon_1)-\beta^2\varepsilon_2^2}{\varepsilon_1(n_-^2-n_+^2)}\, \omega\, d\omega, \tag{13.3.4.16}$$

which is the same as the first term in (13.3.4.9).

We note that the second term in (13.3.4.16) diverges logarithmically. This is connected with the fact that in determining the fields $E_z(\boldsymbol{r}, t)$ and $E_\varphi(\boldsymbol{r}, t)$ we integrated over k from 0 to ∞, while macroscopic electrodynamics is inapplicable as $k \to \infty$. It is clear that we must restrict the frequencies in (13.3.4.16) to the region $\beta^2 n_0^2 > \beta^2 n_-^2 > 1$.

It is well known that the total polarization losses (including those from close collisions) are independent of the parameter $k_0 \sim b_0^{-1}$ in an isotropic medium. This parameter which occurs under the logarithm sign in the expression for the losses in close collisions and also in the expression for the polarization losses corresponding to distant collisions and taking into account the interaction of the moving particle with the longitudinal field falls out of the final expression for the total losses. The losses through close collisions are in an anisotropic medium determined by the same formula as in an isotropic medium and the parameter k_0 occurs in that expression under the logarithm sign. Apart from there, however, it also occurs (and also under the logarithm sign) in the expression for the losses through the emission of extra-ordinary waves—which are longitudinal as $n^2 \to \infty$. The total losses, including close collisions, in an anisotropic medium, as in an isotropic medium, are then independent of the parameter k_0.

Using (13.3.4.2) we can write the expression for the energy losses of a charged particle, due to distant collisions, in the form (Sitenko and Stepanov, 1958):

$$\frac{d\mathcal{E}}{dt} = -\frac{Z^2e^2}{2v}\int \left| \frac{(1-\beta^2)(\omega^2-\omega_{Be}^2)+\beta^2\omega_{pe}^2}{\omega_{pe}^2+\omega_{Be}^2-\omega^2} \right.$$

$$\times \left\{1 \mp \frac{\eta\omega_{Be}[(1-\beta^2)^2(\omega^2-\omega_{Be}^2)+\beta^2(3-\beta^2)\,\omega_{pe}^2]}{[(1-\beta^2)(\omega^2-\omega_{Be}^2)+\beta^2\omega_{pe}^2][(1-\beta^2)^2\,\omega_{Be}^2+4\beta^2(\omega^2-\omega_{pe}^2)]^{1/2}}\right\}\left|\,\omega\, d\omega, \right. \quad (13.3.4.17)$$

where $\eta = (\omega^2-\omega_{Be}^2)/|\,\omega^2-\omega_{Be}^2\,|$ and $\beta = v/c$. The integration in (13.3.4.17) is over the frequency region defined by the inequalities

$$\frac{k_0^2v^2}{\omega^2} \geqslant n_\pm^2\beta^2 \geqslant 1. \quad (13.3.4.18)$$

We see that when there is an external magnetic field present the energy losses due to distant collisions are Cherenkov emission losses.

Only in a few limiting cases is it possible to integrate (13.3.4.17) over the frequencies. If the particle moves with a non-relativistic velocity we have

$$\frac{d\mathcal{E}}{dt} = -\frac{Z^2e^2\omega_{pe}^2}{2v}\ln\frac{k_0^2v^2}{\omega_{pe}^2+\omega_{Be}^2}, \qquad \beta \ll 1, \quad \beta \ll \frac{\omega_{pe}}{|\,\omega_{Be}\,|}. \quad (13.3.4.19)$$

If $1-\beta^2 \gg \omega_{pe}/|\,\omega_{Be}\,|$ (strong magnetic field), we have

$$\frac{d\mathcal{E}}{dt} = -\frac{Z^2e^2\omega_{pe}^2}{2v}\left\{\ln\frac{k_0^2v^2}{(1-\beta^2)\,\omega_{Be}^2}-\beta^2\right\}. \quad (13.3.4.20)$$

Finally, in the ultrarelativistic case we have

$$\frac{d\mathcal{E}}{dt} = -\frac{Z^2e^2\omega_{pe}^2}{2v}\ln\frac{k_0^2v^2}{\omega_{pe}^2}, \qquad 1-\beta^2 \ll 1, \quad 1-\beta^2 \ll \frac{\omega_{pe}}{|\,\omega_{Be}\,|}.$$

13.4. The Interaction of Charged Particles with a Non-equilibrium Plasma

13.4.1. SCATTERING OF CHARGED PARTICLES BY CRITICAL FLUCTUATIONS

In the preceding sections we have mainly discussed the interaction of a test particle with a plasma which is in equilibrium. We shall now turn to a study of the interaction between charged particles and a non-equilibrium plasma and we shall show that if the plasma is in a state which is nearly unstable the energy losses of the particles through the excitation of collective oscillations can be anomalously large.

The probability for the transition of the particle from a state with momentum $\boldsymbol{p}$ to a state with momentum $\boldsymbol{p}'$ is connected through eqn. (13.1.4.6) with the spectral density of the charge density fluctuations. If the plasma consists of cold ions and hot electrons, a particle passing through the plasma can interact strongly with the sound oscillations. Substituting expression (11.5.3.20), which determines the amplitude of the random sound waves, into formula (13.1.4.6) we find the cross-section per electron in the plasma for the scattering of the particle involving the excitation and absorption of non-isothermal sound oscillations

$$d\sigma \equiv \frac{d\Sigma}{n_0} = \left[\frac{2\pi ZeT_e}{\omega_{pe}}\right]^2 \frac{v_s[(\Delta p)^2+p^2\vartheta^2]^{1/2}}{\hbar m_e v\,|v\Delta p-(\{\boldsymbol{p}'-\boldsymbol{p}\}\cdot\boldsymbol{u})|}\{\delta(v\Delta p-v_s\sqrt{\{(\Delta p)^2+p^2\vartheta^2\}})$$
$$+\delta(v\Delta p+v_s\sqrt{\{(\Delta p)^2+p^2\vartheta^2\}})\}\frac{d^3\boldsymbol{p}'}{n_0(2\pi\hbar)^3}, \tag{13.4.1.1}$$

where $\Delta \boldsymbol{p} = \boldsymbol{p}'-\boldsymbol{p}$, $\boldsymbol{v} = \boldsymbol{p}/M$, and ϑ the scattering angle, that is the angle between the vectors $\boldsymbol{p}$ and $\boldsymbol{p}'$. Equation (13.4.1.1) is valid when the scattering angle and the energy transfer are sufficiently small, $pr_D\vartheta \ll \hbar$, $r_D\Delta p \ll \hbar$ (under those conditions weakly damped sound oscillations are excited).

Integrating in eqn. (13.4.1.1) over the modulus of the vector $\boldsymbol{p}'$ we get the scattering cross-section per unit solid angle:

$$\frac{d\sigma}{d^2\boldsymbol{\omega}'} = 4\left(\frac{Ze^2}{\hbar v_e}\right)^2\left(\frac{T_e}{\hbar\omega_{pe}}\right)^2\frac{M^2}{m_e^2}r_D^2\frac{(\theta'-\theta)^2+\varphi^2\sin^2\theta}{(\theta'-\theta)^2[1-(u/v_s)^2\sin^2\theta]+\varphi^2\sin^2\theta}, \tag{13.4.1.2}$$

where θ and θ' are the angles between the vectors $\boldsymbol{p}$, $\boldsymbol{p}'$ and the vector $\boldsymbol{u}$, φ is the angle between the $(\boldsymbol{p}, \boldsymbol{u})$- and $(\boldsymbol{p}', \boldsymbol{u})$-planes, while $\theta' \approx \theta$ and $\varphi \ll 1$ (to fix the ideas we assume that $v \gg v_s$). One sees easily that if $u/v_s \approx 1$ and if the particle moves almost at right angles to $\boldsymbol{u}$, the quantity $d\sigma/d^2\boldsymbol{\omega}'$ can be anomalously large.

We note that the divergence of $d\sigma/d^2\boldsymbol{\omega}'$ has a meaning only in the linear theory, in the framework of which the spectral density for the charge density fluctuations was determined. In fact, non-linear effects must lead to the saturation of the critical fluctuations and as a result the scattering cross-section must turn out to be finite. Similarly, non-linear effects impose limitations on the anomalous growth of the intensity of the Cherenkov emission, the energy loss of the particles, and the coefficients for the scattering and transformation of waves. However, we shall stick here to the linear theory.

Using eqns. (13.1.4.6) and (11.5.3.20) we can determine the energy loss by the particle per unit time through exciting sound oscillations:

$$dP = \left(\frac{eZT_e}{\omega_{pe}}\right)^2\frac{(qv_s)^2}{m_e}\left\{\frac{\delta\{(\boldsymbol{q}\cdot\boldsymbol{v})+qv_s+(\hbar q^2/2M)\}}{qv_s+(\boldsymbol{q}\cdot\boldsymbol{u})}-\frac{\delta\{(\boldsymbol{q}\cdot\boldsymbol{v})-qv_s+(\hbar q^2/2M)\}}{qv_s-(\boldsymbol{q}\cdot\boldsymbol{u})}\right\}\frac{d^3\boldsymbol{q}}{2\pi\hbar}. \tag{13.4.1.3}$$

The first term in this equation describes the stimulated emission and the second term the absorption of oscillations by the particle—as we are interested in plasma states close to instability when the number of sound oscillations in it is very large we can neglect the spontaneous emission.

Integrating expression (13.4.1.3) over the angles we find the intensity of the Cherenkov emission per unit frequency range. In the simplest case when $v \parallel \boldsymbol{u}$, we have

$$\frac{dP}{d\omega} = \left[\frac{ZeT_e\omega}{\omega_{pe}v_s^2}\right]^2\frac{u\omega}{m_eM(v-u)^2}. \tag{13.4.1.4}$$

This formula is valid if $u < v_s$—the sound oscillations are then still stable so that we can use eqn. (11.5.3.20) for the spectral density of their fluctuations; we have also, to fix the ideas, in deriving (13.4.1.4) assumed that $v-v_s \gg \hbar/Mr_D$.

Integrating expression (13.4.1.4) over the frequency we find the energy losses of the particle through exciting sound oscillations:

$$P = \left(\frac{ZeT_e^{\omega}}{2\omega_{pe}a_1^2}\right)^2 \frac{u}{m_e M(v-u)^2}, \tag{13.4.1.5}$$

where a_1 is a length of the order of a few Debye radii. If the particle velocity v, the sound velocity v_s, and the directed electron velocity u are close to one another the quantity P is very large and may exceed the energy losses of the particle caused by binary collisions.

In the general case when the angle θ_0 between the direction of the particle motion and the direction of $\boldsymbol{u}$ is different from zero, the intensity of the Cherenkov emission may be anomalously large, if $v \cos\theta_0 \approx u \approx v_s$. In that case we have

$$\frac{dP}{d\omega} = \left(\frac{ZeT_e^{\omega}}{\omega_{pe}v_s^2}\right)^2 \frac{\omega v_s}{m_e M(v\cos\theta_0-u)^2}. \tag{13.4.1.6}$$

We get hence for the energy losses of the particle through the excitation of sound oscillations

$$P = \left(\frac{ZeT_e}{2\omega_{pe}a_1^2}\right)^2 \frac{v_s}{m_e M(v\cos\theta_0-u)^2}. \tag{13.4.1.7}$$

If

$$(v\cos\theta_0-u)^2 < \frac{m_e uv}{4M}\left(\frac{r_D}{a_1}\right)^4,$$

we have $P > P_0$, where $P_0 = (Ze\omega_{pe})^2/v$ is a quantity of the order of the energy losses of the particle due to binary collisions (neglecting the Coulomb logarithm).

Let us also give the expression for the angular distribution of the Cherenkov emission:

$$\frac{dP}{d^2\omega'} = \frac{\omega_{pe}T_e}{6\pi}\left(\frac{r_D}{a_1}\right)^3 \frac{Z^2e^2}{\hbar v_e}\frac{v_s}{v}\left\{\frac{\delta[\cos\chi+(v_s/v)]}{1+(u/v_s)\cos\theta} - \frac{\delta[\cos\chi-(v_s/v)]}{1-(u/v_s)\cos\theta}\right\}, \tag{13.4.1.8}$$

where θ is the angle between the vectors $\boldsymbol{q}$ and $\boldsymbol{u}$ and χ the angle between the vectors $\boldsymbol{q}$ and $\boldsymbol{v}$. We can use eqn. (13.4.1.8) when $|\cos\theta| < v_s/u$ as in that case the sound oscillations considered do not grow. One sees easily that as $|\cos\theta| \to v_s/u$ the coefficient in front of one of the δ-functions in formula (13.4.1.8) increases without bound.

Let us now consider the problem of the scattering of charged particles in a plasma through which a compensated beam of charged particles is passing. We shall assume that the beam velocity u is much higher than the electron thermal velocity in the plasma and that the beam temperature T' is not too low so that the damping rate of the Langmuir plasma oscillations is basically determined through their interaction with the electrons in the beam. Under those conditions the spectral density of the charge density fluctuations has, in the high-frequency region ($\omega \gg qv_e$), the form

$$\langle \varrho^2\rangle_{q\omega} = \frac{T'q^2\omega_{pe}^2}{2|\omega-(\boldsymbol{q}\cdot\boldsymbol{u})|}\,\delta(\omega^2-\omega_{pe}^2). \tag{13.4.1.9}$$

It is well known that the presence of a hot beam always leads to the growth of Langmuir oscillations with a sufficiently short wavelength for which $|(\boldsymbol{q}\cdot\boldsymbol{u})| > \omega_{pe}$. Neglecting the non-linear effects of the interaction between the fluctuations, we can use eqn. (13.4.1.9) in the wavevector region in which the plasma oscillations do not yet grow. Substituting (13.4.1.9) into formula (13.1.4.6) we find the cross-section per plasma electron for the scattering of a fast particle ($v \gg v_e$) involving the excitation and absorption of Langmuir oscillations

$$d\sigma = \frac{Z^2e^2}{2\pi\hbar v n_0}\frac{T'\omega_{pe}}{(\Delta p)^2+p^2\vartheta^2}\left\{\frac{\delta(v\Delta p-\hbar\omega_{pe})}{\hbar\omega_{pe}-([\boldsymbol{p}'-\boldsymbol{p}]\cdot\boldsymbol{u})}+\frac{\delta(v\Delta p+\hbar\omega_{pe})}{\hbar\omega_{pe}+([\boldsymbol{p}'-\boldsymbol{p}]\cdot\boldsymbol{u})}\right\}d^3\boldsymbol{p}'. \quad (13.4.1.10)$$

This formula is valid under the same restrictions on Δp and ϑ as eqn. (13.4.1.1).

Integrating expression (13.4.1.10) over the modulus of the vector $\boldsymbol{p}'$ we get the cross-section for the scattering of the particle per unit solid angle

$$\frac{d\sigma}{d^2\boldsymbol{\omega}'} = \left(\frac{Ze}{\hbar v\vartheta}\right)^2\frac{T'}{2\pi n_0}\left\{\left|1-\frac{u}{v}\cos\theta'-\frac{pu}{\hbar\omega_{pe}}(\cos\theta'-\cos\theta)\right|^{-1}\right.$$
$$\left.+\left|1-\frac{u}{v}\cos\theta'+\frac{pu}{\hbar\omega_{pe}}(\cos\theta'-\cos\theta)\right|^{-1}\right\}, \quad (13.4.1.11)$$

where, to fix the ideas we have assumed that $\vartheta \gg \hbar\omega_{pe}/pv$. One sees easily that for some well-defined values of θ', close to θ, the expression for $d\sigma/d^2\boldsymbol{\omega}'$ becomes anomalously large.

Using eqns. (13.1.4.6) and (13.4.1.9) we can determine the energy lost by the particle per unit time through the excitation of Langmuir oscillations

$$dP = \frac{Z^2e^2\omega_{pe}^2T'}{2\pi\hbar q^2}\left\{\frac{\delta[(\boldsymbol{q}\cdot\boldsymbol{v})+\omega_{pe}+(\hbar q^2/2M)]}{\omega_{pe}+(\boldsymbol{q}\cdot\boldsymbol{u})}-\frac{\delta[(\boldsymbol{q}\cdot\boldsymbol{v})-\omega_{pe}+(\hbar q^2/2M)]}{\omega_{pe}-(\boldsymbol{q}\cdot\boldsymbol{u})}\right\}d^3\boldsymbol{q}. \quad (13.4.1.12)$$

This formula takes correctly into account the interaction of the particle with the non-growing oscillations, that is, with the oscillations with wavevectors satisfying the condition $|(\boldsymbol{q}\cdot\boldsymbol{u})| < \omega_{pe}$.

Integrating expression (13.4.1.12) over the angles we find the intensity of the Cherenkov emission per unit wavenumber range. In the simplest case when $\boldsymbol{v} \parallel \boldsymbol{u}$, we have

$$\frac{dP}{dq} = \frac{Z^2e^2T'qu}{M(v-u)^2}, \quad (13.4.1.13)$$

where, to fix the ideas, we have assumed that $v-u \gtrsim \hbar u/M\omega_{pe}r_D^2$. Integrating expression (13.4.1.13) over q we find the energy losses of the particle through the excitation of Langmuir oscillations

$$P = \frac{Z^2e^2T'u}{2a_1^2M(v-u)^2}. \quad (13.4.1.14)$$

If the particle velocity v and the beam velocity u lie close to one another, the quantity P will be very large and can exceed both the energy losses due to binary collisions, and the energy losses through the excitation of Langmuir oscillations in the absence of a beam. For that the following condition must hold:

$$(v-u)^2 < u^2 \frac{m_e T'}{M T_e} \left(\frac{r_D}{2a_1} \right)^2 .$$

If the angle between $\boldsymbol{v}$ and $\boldsymbol{u}$ is different from zero, apart from the stable oscillations, also the turbulent Langmuir oscillations, that is, the oscillations which in the framework of the linear theory are unstable, will contribute to the change in the particle energy.

Let us also give the angular distribution of the Langmuir waves emitted by the particle

$$\frac{dP}{d^2\boldsymbol{\omega}'} = \frac{Z^2 e^2 \omega_{pe} T'}{2\pi\hbar(u \cos\theta - v\cos\chi)}, \tag{13.4.1.15}$$

where θ is the angle between $\boldsymbol{q}$ and $\boldsymbol{u}$ and χ the angle between $\boldsymbol{q}$ and $\boldsymbol{v}$. We can use this formula if $v \cos^2\chi > u \cos\theta \cos\chi$, as then the Langmuir oscillations considered do not grow. One sees easily that if $|\cos\chi - (u/v)\cos\theta| \ll 1$, the quantity $dP/d^2\boldsymbol{\omega}'$ is anomalously large.

In concluding this subsection let us consider the interaction of the particle with the short-wavelength oscillations of a two-temperature plasma. If $r_D^{-1} \ll q \ll r_{Di}^{-1}$ ($r_{Di} = (T_i/T_e)^{1/2} r_D$ is the ion Debye radius) the spectral density of the charge density fluctuations has, in the range of medium-range frequencies, the form

$$\langle \varrho^2 \rangle_{q\omega} = \frac{T_e q^2 \omega_{pi}^2}{2\,|\omega - (\boldsymbol{q}\cdot\boldsymbol{u})|}\, \delta(\omega^2 - \omega_{pi}^2), \tag{13.4.1.16}$$

where $\omega_{pi} = (m_e/m_i)^{1/2}\omega_{pe}$ is the ion plasma frequency. Substituting this expression into (13.1.4.6) we easily find the cross-section for the scattering of the particle and can then determine its energy losses through the excitation of ion Langmuir oscillations. We can obtain all equations which then occur by replacing in eqns. (13.4.1.10) to (13.4.1.15) ω_{pe} by ω_{pi}, r_D by r_{Di}, and T' by T_e; we shall therefore not give these expressions. We note merely that the energy losses are particularly large if the particle velocity is, both as regards magnitude and as regards direction, close to the directed electron velocity. Under those conditions the energy losses of the particle through the excitation of ion Langmuir oscillations is equal to

$$P = \frac{Z^2 e^2 T_e^2 u}{2a_1^2 M T_i (v-u)^2}, \tag{13.4.1.17}$$

where, as before, a_1 is a length of the order of a few electron Debye radii.

13.4.2. INTERACTION OF CHARGED PARTICLES WITH A TURBULENT PLASMA

Let us now turn to the study of the interaction of charged particles with a turbulent plasma. We shall determine the change per unit time in energy and momentum of a particle caused by the emission and absorption of turbulent oscillations and we shall show that the detailed properties of the turbulence spectrum do not affect the way these quantities depend on the direction and magnitude of the particle velocity; the level of the turbulent fluctuations mainly determines (Akhiezer, 1964b, 1965a) the general coefficient in the expressions for the changes in energy and momentum.

The probability for the transition of the particle from a state with momentum $\boldsymbol{p}$ to a state with momentum $\boldsymbol{p}'$ is connected with the spectral density of the charge density fluctuations by the general relation (13.1.4.6). If the plasma consists of cold ions and hot electrons which are moving relative to the ions we can in the long-wavelength region ($qr_D \ll 1$) and the "medium-range frequency region" ($qv_i \ll \omega \ll qv_e$) write the spectral density of the charge density fluctuations in the form

$$\langle \varrho^2 \rangle_{q\omega} = \tfrac{1}{4} q^2 (qr_D)^2 \{T(\boldsymbol{q})\, \delta(\omega - qv_s) + T(-\boldsymbol{q})\, \delta(\omega + qv_s)\}, \tag{13.4.2.1}$$

where $T(\boldsymbol{q})$ is the effective temperature of the sound oscillations and r_D the electron Debye radius. The function $T(\boldsymbol{q})$ characterizing the level and distribution of the random sound oscillations is connected with the function $I(\boldsymbol{q})$ introduced in Chapter 10 by the obvious relation

$$I(\boldsymbol{q}) = \frac{r_D^2}{(2\pi)^2} T(\boldsymbol{q}). \tag{13.4.2.2}$$

Bearing in mind that the effective temperature of the oscillations is a function of the two scalars q and $\eta = (\boldsymbol{q} \cdot \boldsymbol{u})$, where $\boldsymbol{u}$ is the directed electron velocity, we shall denote the effective temperature by $T(q, (\boldsymbol{q} \cdot \boldsymbol{u}))$.

Using eqns. (13.4.2.1) and (13.4.2.2) we can determine the energy lost by the particle per unit time through exciting sound oscillations with wavevectors in the range $\boldsymbol{q}, \boldsymbol{q} + d\boldsymbol{q}$:

$$dP = \frac{Z^2 e^2 r_D^2 q v_s}{2\pi\hbar} \{T(q, (\boldsymbol{q}\cdot\boldsymbol{u}))\, \delta[(\boldsymbol{q}\cdot\boldsymbol{v}) - qv_s + (\hbar q^2/2M)]$$
$$- T(q, (\boldsymbol{q}\cdot\boldsymbol{u}))\, \delta(\boldsymbol{q}\cdot\boldsymbol{v}) + qv_s + (\hbar q^2/2M)]\}\, d^3\boldsymbol{q}, \tag{13.4.2.3}$$

where M, v, and Ze are the mass, velocity, and charge of the particle. The first term in this expression describes the absorption and the second the stimulated emission of the oscillations by the particle—for the states of the plasma which are of interest to us, where the number of sound waves is very large, we can neglect in eqn. (13.4.2.3) the additional term corresponding to the spontaneous emission, which is independent of the number of waves.

Integrating expression (13.4.2.3) over the angles we find the intensity of the Cherenkov emission per unit frequency range:

$$\frac{dP}{d\omega} = \frac{Z^2 e^2 r_D^2 u q^4}{\pi M v^2} D(\boldsymbol{q}) \tag{13.4.2.4}$$

where

$$D(\boldsymbol{q}) = \int_0^\pi d\varphi \left\{ \cos\theta - \frac{\sin\theta \cos\varphi}{\sqrt{[(v/v_s)^2 - 1]}} \right\} \frac{\partial T(q, \eta)}{\partial \eta}, \tag{13.4.2.5}$$

θ is the angle between the vectors $\boldsymbol{v}$ and $\boldsymbol{u}$ and the quantity η is connected with the integration variable φ through the relation

$$\eta = qu \left\{ \frac{v_s}{v} \cos\theta + \sqrt{\left[1 - \frac{v_s^2}{v^2}\right]} \sin\theta \cos\varphi \right\}. \tag{13.4.2.6}$$

We see that the energy losses of the particle are determined by the function $\partial T(q, \eta)/\partial\eta$ ($\eta = (\boldsymbol{q}\cdot\boldsymbol{u})$); to evaluate the energy losses we must therefore know how the effective temperature depends on the quantity $(\boldsymbol{q}\cdot\boldsymbol{u})$. For small values of $(\boldsymbol{q}\cdot\boldsymbol{u})$ when the oscillations with wavevector $\boldsymbol{q}$ are damped ($(\boldsymbol{q}\cdot\boldsymbol{u}) < qv_s$), we found in subsection 11.5.3 that the effective temperature is given by the formula

$$T(q,(\boldsymbol{q}\cdot\boldsymbol{u})) = \frac{T_e}{1-[(\boldsymbol{q}\cdot\boldsymbol{u})/qv_s]}. \tag{13.4.2.7}$$

As $(\boldsymbol{q}\cdot\boldsymbol{u}) \to qv_s$ the effective temperature increases steeply. When $(\boldsymbol{q}\cdot\boldsymbol{u}) > qv_s$ the quantity $T(q, (\boldsymbol{q}\cdot\boldsymbol{u}))$ is, in general, for given q a smooth function of $(\boldsymbol{q}\cdot\boldsymbol{u})$.

It follows from this nature of the $(\boldsymbol{q}\cdot\boldsymbol{u})$-dependence of the effective temperature of the oscillations that the derivative $\partial T/\partial(\boldsymbol{q}\cdot\boldsymbol{u})$ has a steep maximum for some value of $(\boldsymbol{q}\cdot\boldsymbol{u})$ close to qv_s: $(\boldsymbol{q}\cdot\boldsymbol{u}) = \eta_0 \approx qv_s$. The existence of such a maximum enables us to evaluate the function $D(\boldsymbol{q})$ without knowing the detailed way in which T depends on $(\boldsymbol{q}\cdot\boldsymbol{u})$ for $(\boldsymbol{q}\cdot\boldsymbol{u}) > qv_s$.

Indeed, let us expand the function $(\partial T/\partial\eta)^{-1}$ in a power series in $\eta-\eta_0$. Noting that the function $(\partial T/\partial\eta)^{-1}$ has a minimum for $\eta = \eta_0$ we find

$$\left[\frac{\partial T(q, \eta)}{\partial\eta}\right]^{-1} = \frac{qv_s}{T_e}\left\{\xi^2(q)+\lambda^2(q)\left[1-\frac{\eta}{\eta_0}\right]^2\right\}, \tag{13.4.2.8}$$

where

$$\xi^2(q) = T_e\left[qv_s\frac{\partial T}{\partial\eta}\right]^{-1}_{\eta=\eta_0}, \qquad \lambda^2(q) = \frac{T_e\eta_0^2}{2qv_s}\frac{\partial^2}{\partial\eta^2}\left[\frac{\partial T}{\partial\eta}\right]^{-1}_{\eta=\eta_0},$$

We note that the function $\xi^2(q)$ is proportional to the small parameter which is the ratio of the quantity T_e^2 to the function $T^2(q, (\boldsymbol{q}\cdot\boldsymbol{u}))$ for $(\boldsymbol{q}\cdot\boldsymbol{u}) > qv_s$; on the other hand, the function $\lambda^2(q)$ is of the order of unity.

Substituting this expansion into eqn. (13.4.2.5) and remembering that $\eta_0 \approx qv_s$ we see that the main contribution to the integral which determines $D(\boldsymbol{q})$ comes from angles φ close to φ_0 where

$$\cos\varphi_0 = \left[\frac{v}{u}-\cos\theta\right](\sin\theta)^{-1}\left[\frac{v^2}{v_s^2}-1\right]^{-1/2}. \tag{13.4.2.9}$$

If $\cos^2\varphi_0 < 1$ (and $\sin\varphi_0$ is not too small) we get for D:

$$D(\boldsymbol{q}) = \frac{\pi T_e}{\lambda\xi qu}\left[1-\frac{v_s^2}{v^2}\right]^{-3/2}\frac{\cos\theta-(v_s^2/uv)}{\sin\theta\sin\varphi_0}. \tag{13.4.2.10}$$

We see that the quantity $|D|$ is proportional to the large parameter ξ^{-1} and increases strongly when $\sin\varphi_0$ decreases.

For very small values of $|\sin\varphi_0|$—when $|\sin^2\varphi_0| \ll \xi(v_s/u)\{1-(v_s/v)^2\}^{-1/2}\{\sin\theta\}^{-1}$; $\cos^2\varphi_0$ can then be both less than and larger than unity—we find by using eqns. (13.4.2.5) and (13.4.2.8) that

$$D(\boldsymbol{q}) = \pm\frac{\pi T_e v^{3/2}(u^2-v_s^2)^{1/2}}{2q(\lambda v_s\sin\theta_\pm)^{1/2}(\xi u)^{3/2}(v^2-v_s^2)^{3/4}}. \tag{13.4.2.11}$$

The plus sign corresponds here to the case $\theta \approx \theta_+$ and the minus sign to the case $\theta \approx \theta_-$ where $\theta_\pm$ are the two values of the angle θ for which $\sin \varphi_0$ vanishes:

$$\cos \theta_\pm = \frac{v_s^2 \pm \sqrt{\{(u^2 - v_s^2)(v^2 - v_s^2)\}}}{uv}. \tag{13.4.2.12}$$

We see that for very small values of $|\sin \varphi_0|$ the quantity D is proportional to $\xi^{-3/2}$.

As $v \to u$, $\sin \theta_+ \to 0$; eqn. (13.4.2.11) ceases in that case to be valid. If $\sin \theta_+ \ll \xi v_s (v^2 - v_s^2)^{-1/2}$, we have instead of formula (13.4.2.11)

$$D(\boldsymbol{q}) = \frac{\pi T_e \cos \theta}{\xi^2 q v_s}. \tag{13.4.2.13}$$

In that case D is particularly large and proportional to ξ^{-2}.

We note that in deriving eqns. (13.4.2.10) and (13.4.2.11) we assumed that the particle velocity v lies not too close to the sound velocity v_s. We can verify by using eqns. (13.4.2.5) and (13.4.2.7) that if $1 - (v_s/v)^2 \ll (\xi v_s/u)^2 (\sin \theta)^{-1}$ the function D is given by eqn. (13.4.2.13) provided $\cos \theta \approx v_s/u$, and decreases strongly when $|\cos \theta - (v_s/u)|$ increases.

The case when $\cos \varphi_0 > 1$—and the difference $\cos \varphi_0 - 1$ not too small—corresponds to such values of the particle velocity that it cannot interact with the turbulent sound waves. It is in that case impossible to use for the determination of $D(\boldsymbol{q})$ the expansion (13.4.2.8); one can, however, simply evaluate $D(\boldsymbol{q})$ directly by substituting expression (13.4.2.7) for the effective temperature into formula (13.4.2.5):

$$D(\boldsymbol{q}) = \frac{\pi T_e v_s^2}{q u^3} \left[1 - \frac{v_s^2}{v^2}\right]^{-3/2} \frac{\cos \theta - (u/v)}{\sin^3 \theta \, |\sin \varphi_0|^3}. \tag{13.4.2.14}$$

In that case D does not contain the large parameter ξ^{-1}; none the less, if $|\sin \varphi_0| \ll 1$, this quantity will be very large—although much smaller than when $\cos^2 \varphi_0 \lessdot 1$. This is connected with the strong interaction of the particle with sound waves for which $qv_s - (\boldsymbol{q} \cdot \boldsymbol{u}) \ll qv_s$ and for which the effective temperature is large according to eqn. (13.4.2.7).

The case $\cos \varphi_0 < -1$ corresponds to such a motion of the particle that it interacts strongly with turbulent sound waves with $(\boldsymbol{q} \cdot \boldsymbol{u}) > qv_s$. As we noted already, the quantity $\partial T/\partial \eta$ is small in the region $(\boldsymbol{q} \cdot \boldsymbol{u}) > qv_s$; the function $D(\boldsymbol{q})$ is therefore also small when $\cos \varphi_0 < -1$.

We note that eqns. (13.4.2.10) to (13.4.2.14)—and therefore also the expressions for the changes in the energy and momentum of the particle—remain true, even if the assumption that the quantity $\partial T/\partial \eta$ is small in the region $\eta > \eta_0$ is not satisfied. Indeed, the contribution from the quantity $\partial T/\partial \eta$ with $\eta > \eta_0$ to the expression for D can only slightly change the function D for $\theta_+ < \theta < \theta_-$, without changing D for $\theta \approx \theta_\pm$ and, hence, not changing the nature of the function D for $\theta_+ \leq \theta \leq \theta_-$. In particular, the function D is positive for $\theta = \theta_+$ and negative for $\theta = \theta_-$ and must therefore vanish for some value of the angle $\theta = \theta_0$, where $\theta_+ < \theta_0 \lessdot \theta_-$—here θ_0 will be slightly different from $\arccos (v_s^2/uv)$.

According to (13.4.2.4) the function D determines the spectral distribution of the energy emitted by the particle. Substituting D into formula (13.4.2.4) and integrating over the frequency we get an expression for the change in energy P of the particle per unit time caused through its interaction with sound waves. If the angle θ between the directions of $\boldsymbol{v}$ and $\boldsymbol{u}$ lies in the range $\theta_+ < \theta < \theta_-$, where the "critical" angles $\theta_\pm$ are given by formula (13.4.2.12), P has the form

$$P = \frac{Z^2e^2\omega_{\text{pe}}^2 m_e}{Mv_s} Af(\boldsymbol{v}), \tag{13.4.2.15}$$

where A is a large quantity which is proportional to the ratio of the effective temperature of the turbulent sound waves to the electron temperature of the plasma,

$$A = r_D \int (qr_D)^3 (\lambda\xi)^{-1}\, dq, \tag{13.4.2.16}$$

while the function $f(\boldsymbol{v})$ is given by the formula

$$f(\boldsymbol{v}) = \frac{(\boldsymbol{u}\cdot\boldsymbol{v}) - v_s^2}{v^2 - v_s^2}\left\{\left(\frac{u^2}{v_s^2} - 1\right)\left(\frac{v^2}{v_s^2} - 1\right) - \left(\frac{(\boldsymbol{u}\cdot\boldsymbol{v})}{v_s^2} - 1\right)^2\right\}^{-1/2}. \tag{13.4.2.17}$$

Equations (13.4.2.15) to (13.4.7.17) determine explicitly how the change in the particle energy depends on the magnitude and the direction of its velocity. One sees easily that if $\theta_+ < \theta < \theta_0$, where

$$\cos\theta_0 = \frac{v_s^2}{uv},$$

the particle loses energy; if $\theta_0 < \theta < \theta_-$ the interaction with the sound oscillations leads to an increase in the particle energy. For $\theta = \theta_0$ the change in the particle energy equals zero. As $\theta \to \theta_\pm$ the quantity $|P|$ increases steeply. The way the particle energy losses depend on the angle between the particle velocity and the direction of the electron current is sketched in Fig. 13.4.1.

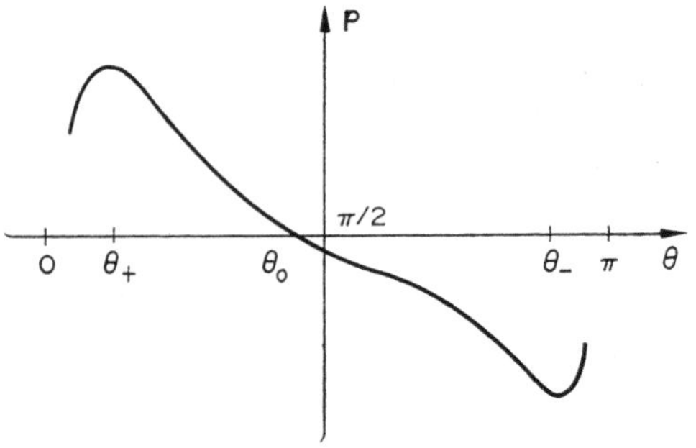

FIG. 13.4.1. Sketch of the way the particle energy losses P depend on the angle θ between the particle velocity and the direction of the electron current.

For angles θ very close to $\theta_\pm$ formula (13.4.2.15) ceases to be valid. If

$$|\theta - \theta_\pm| \ll \xi\frac{v_s}{u}\left[1 - \frac{v_s^2}{v^2}\right]^{1/2},$$

where ξ is a quantity of the order of the ratio of the electron temperature to the effective temperature of the waves in the turbulent region, we get by substituting expression (13.4.2.11) into formula (13.4.2.4) and integrating over the frequencies

$$P = \frac{Z^2 e^2 \omega_{pe}^2 m_e}{M v_s} A_1 f_\pm, \qquad (13.4.2.18)$$

where A_1 is a large quantity, proportional to $\xi^{-3/2}$,

$$A_1 = \tfrac{1}{2} r_D \int (q r_D)^3 \xi^{-3/2} \lambda^{-1/2} dq, \quad f_\pm = \pm v_s^{3/2} (u^2 - v_s^2)^{1/2} (uv \sin\theta_\pm)^{-1/2} (v^2 - v_s^2)^{-3/4}, \qquad (13.4.2.19)$$

where the $\pm$-signs correspond to the cases $\theta \approx \theta_\pm$. We see that when $\theta \approx \theta_\pm$ the change in the particle energy per unit time is particularly large.

For very small values of θ_+ (corresponding to $u \approx v$) eqn. (13.4.2.18) ceases to be valid. If $\theta_+ \ll \xi v_s (v^2 - v_s^2)^{-1/2}$ we have instead of formula (13.4.2.18)

$$P = \frac{Z^2 e^2 \omega_{pe}^2 m_e}{M v_s} A_2; \quad A_2 = r_D \int (q r_D)^3 \xi^{-2}\, dq. \qquad (13.4.2.20)$$

We must emphasize that the large change in the particle energy per unit time—proportional to ξ^{-1}, $\xi^{-3/2}$, or ξ^{-2}—occurs for angles between $\boldsymbol{v}$ and $\boldsymbol{u}$ which lie in the range $\theta_+ \leq \theta \leq \theta_-$ In order that such a range of angles occurs, it is necessary that the inequalities $v > v_s$ and $u > v_s$ are satisfied. The first of these inequalities is the condition for a strong interaction of the particle with the sound oscillations (Cherenkov condition), while the second inequality guarantees the existence of turbulent sound oscillations.

As $u \to v_s$ we have $\theta_+ \to \theta_- \to \arccos(v_s/v)$. Expressions (13.4.2.15) and (13.4.2.18) for the function P tend to zero for $(\boldsymbol{v} \cdot \boldsymbol{u}) = v_s^2$ and $u = v_s$: the particle energy losses are in that case proportional to less than the first power in $1/\xi$. Indeed, substituting expression (13.4.2.8) into formulae (13.4.2.4) and (13.4.2.5) we get

$$P = \frac{Z^2 e^2 \omega_{pe}^2 m_e}{M v_s} A_0 \left(\frac{v^2}{v_s^2} - 1\right)^{-3/2}; \quad A_0 = \tfrac{1}{2} r_D \int (q r_D)^3 \xi^{-1/2} \lambda^{-3/2}\, dq. \qquad (13.4.2.21)$$

In deriving eqns. (13.4.2.15) to (13.4.2.21) we assumed that the particle velocity v does not lie too close to the sound velocity v_s: $\{1 - (v_s/v)^2\} \gg \xi^n$; in the general case described by formula (13.4.2.15), $n = 1$; in the case when $\theta \approx \theta_\pm$, $n = \frac{2}{3}$. If $\{1 - (v_s/v)^2\} \gg \xi^n$, the particle energy losses for $\theta = \arccos(v_s/u)$ are determined by formula (13.4.2.20) and proportional to ξ^{-2}; for other angles the quantity P does not contain the large parameter $1/\xi$.

If $\theta > \theta_-$ and also if $\theta < \theta_+$ and $u < v$, we can obtain the change in the particle energy by substituting expression (13.4.2.7) into formula (13.4.2.5). Integrating over the frequency up to some maximum frequency v_s/a_1 where a_1 is a length of the order of a few Debye radii, we get

$$P = \frac{Z^2 e^2 \omega_{pe}^2 m_e}{M v_s} \alpha g(\boldsymbol{v}), \qquad (13.4.2.22)$$

where $\alpha = \frac{1}{4}(r_D/a_1)^4$ and

$$g(\boldsymbol{v}) = \frac{(\boldsymbol{u} \cdot \boldsymbol{v}) - u^2}{v_s^2} \left\{ \left(\frac{(\boldsymbol{u} \cdot \boldsymbol{v})}{v_s^2} - 1 \right)^2 - \left(\frac{u^2}{v_s^2} - 1 \right) \left(\frac{v^2}{v_s^2} - 1 \right) \right\}^{-3/2}. \qquad (13.4.2.23)$$

We note that this formula takes into account the stimulated emission and the absorption of sound oscillations by the particle and neglects other processes—such as close collisions, spontaneous emission of sound waves, or the emission and absorption of other kinds of plasma oscillations—leading to a change in the particle energy. Expression (13.4.2.22) therefore determines the total particle energy losses only for angles which lie sufficiently close to $\theta_{\pm}$ when the relative contribution from other processes to the energy change is small.

If $\theta < \theta_+$ and $u > v$ the change in the particle energy per unit time will, in general, be smaller than when $\theta_+ < \theta < \theta_-$. If we want to establish the v-dependence of P in that case, we need know in detail how the function $T(q, (\boldsymbol{q}\cdot\boldsymbol{u}))$ behaves for $(\boldsymbol{q}\cdot\boldsymbol{u}) > qv_s$.

We see that the change in the particle energy is particularly large when the angle θ between $\boldsymbol{v}$ and $\boldsymbol{u}$ lies close to one of the critical angles $\theta_{\pm}$. The quantity P is then, according to formula (13.4.2.18) which was obtained neglecting the sound dispersion, proportional to $\xi^{-3/2}$. Using (13.4.2.11) one can show that that formula must be employed when $A_3/A_1 \gg r$, where

$$A_3 = r_D \int (qr_D)^2(\lambda\xi)^{-1}\, dq; \quad r = |v-u|\,(uv\cos\theta_{\pm})^{-1/2}\{(v/v_s)^2-1\}^{-1/4}, \tag{13.4.2.24}$$

in which case we can neglect the smearing out of the critical angles due to the sound dispersion. If $A_3/A_1 \ll r$, we must replace eqn. (13.4.2.18) for P by the following expression:

$$P = \pm\frac{Z^2e^2\omega_{pe}^2 m_e}{M\,|\boldsymbol{v}-\boldsymbol{u}|}\left[\frac{u^2-v_s^2}{v^2-v_s^2}\right]^{1/2} A_3, \tag{13.4.2.25}$$

where the upper (lower) sign corresponds to the case $\theta \to \theta_+$ ($\theta \to \theta_-$).

If $\boldsymbol{v} \to \boldsymbol{u}$, the presence of sound dispersion does not impose any limitation on the growth of P near θ_+. We must in that case, independent of the ratio of A_3 to A_1, use formula (13.4.2.20).

Let us now consider the problem of the change in momentum of the charged particle caused by the emission and absorption of sound waves. Using formulae (13.1.4.6) and (13.4.2.1.) we can determine the momentum transferred per unit time by the particle to sound oscillations with wavevectors in the range $\boldsymbol{q}$, $\boldsymbol{q}+\boldsymbol{dq}$:

$$\boldsymbol{dQ} = -\frac{Z^2e^2r_D^2}{2\pi\hbar}\,\boldsymbol{q}\left\{T(q, (\boldsymbol{q}\cdot\boldsymbol{u}))\,\delta\left[(\boldsymbol{q}\cdot\boldsymbol{v})-qv_s+\frac{\hbar q^2}{2M}\right]\right.$$
$$\left.+T((q, -(\boldsymbol{q}\cdot\boldsymbol{u}))\,\delta\left[(\boldsymbol{q}\cdot\boldsymbol{v})+qv_s+\frac{\hbar q^2}{2M}\right]\right\} d^3\boldsymbol{q}.$$

Integrating this expression over $\boldsymbol{q}$ we get

$$\boldsymbol{Q} = \frac{Z^2e^2r_D^2uv_s}{\pi Mv^2}\int q^4\, dq\{\boldsymbol{I}D(q)+\boldsymbol{i}_u\, d_u(q)+\boldsymbol{i}_v\, d_v(q)\}, \tag{13.4.2.26}$$

where

$$\boldsymbol{I} = \frac{\boldsymbol{u}[v^2-(\boldsymbol{u}\cdot\boldsymbol{v})]+\boldsymbol{v}[u^2-(\boldsymbol{u}\cdot\boldsymbol{v})]}{(uv\sin\theta)^2}, \quad \boldsymbol{i}_u = \frac{\boldsymbol{u}v^2-\boldsymbol{v}(\boldsymbol{u}\cdot\boldsymbol{v})}{(uv\sin\theta)^2}, \quad \boldsymbol{i}_v = \frac{\boldsymbol{v}u^2-\boldsymbol{u}(\boldsymbol{u}\cdot\boldsymbol{v})}{(uv\sin\theta)^2}. \tag{13.4.2.27}$$

The function D is given by formula (13.4.2.5) and

$$\left.\begin{aligned} d_u(\boldsymbol{q}) &= \frac{1}{qv_s}\int_0^{\pi} d\varphi \left[\cos\theta - \frac{\sin\theta\cos\varphi}{\sqrt{\{(v/v_s)^2-1\}}}\right]\left[T+(\eta-qv_s)\frac{\partial T}{\partial\eta}\right], \\ d_v(\boldsymbol{q}) &= \frac{v}{qv_s u}\int_0^{\pi} d\varphi\, T(q,\eta), \end{aligned}\right\} \quad (13.4.2.28)$$

while the quantity η is connected with the integration variable φ through relation (13.4.2.6)

To elucidate the nature of the function D, it is sufficient, as we showed earlier, to know the behaviour of the function $T(q, (\boldsymbol{q}\cdot\boldsymbol{u}))$ near $(\boldsymbol{q}\cdot\boldsymbol{u}) \approx qv_s$; for the calculation of the quantities d_u and d_v, however, it is necessary to know the form of the function T also for $(\boldsymbol{q}\cdot\boldsymbol{u}) \neq qv_s$. One can, however, simply check that if $|\theta-\theta_{\pm}| \ll \sin\theta_{\pm}$, where the critical angles $\theta_{\pm}$ are given by eqn. (13.4.1.12), $|\boldsymbol{i}_u d_u| \ll |\boldsymbol{I}D|$ and $|\boldsymbol{i}_v d_v| \ll |\boldsymbol{I}D|$. Using formula (13.4.2.4) we can thus connect the change in particle momentum $\boldsymbol{Q}$ for $\theta \approx \theta_{\pm}$ with the change P in its energy:

$$\boldsymbol{Q} = \boldsymbol{I}P. \quad (13.4.2.29)$$

This equation, together with eqns. (13.4.2.15) to (13.4.2.25) for P, enables us to find how $\boldsymbol{Q}$ depends on the magnitude and direction of the particle velocity $\boldsymbol{v}$.

According to formula (13.4.2.29) the momentum components in the directions $\pm\boldsymbol{I}$ show the fastest change when the particle is moving through the plasma,

$$\frac{(\boldsymbol{Q}\cdot\boldsymbol{I})}{|\boldsymbol{I}|} = \frac{|\boldsymbol{u}-\boldsymbol{v}|}{uv\sin\theta}P,$$

while the momentum component in the direction $\boldsymbol{v}-\boldsymbol{u}$ is not changed when the particle moves through the plasma, $(\boldsymbol{Q}\cdot[\boldsymbol{v}-\boldsymbol{u}]) = 0$.

We note that the case $|\theta-\theta_{\pm}| \ll \sin\theta_{\pm}$, to which eqn. (13.4.2.29) refers, is that of the greatest interest when we study the interaction of a particle with a non-equilibrium plasma, as in that case the change in the particle momentum (and in its energy) per unit time is particularly large.

Let us also give an expression for the function $\boldsymbol{Q}$ for the case when the particle velocity lies close to the beam velocity both in magnitude and in direction,

$$\frac{|\boldsymbol{u}-\boldsymbol{v}|\,v_s}{u^2} < \theta \ll \frac{\xi}{\sqrt{[(v/v_s)^2-1]}}.$$

In that case $\boldsymbol{Q} = P\boldsymbol{v}/v^2$, where P is given by formula (13.4.2.20). The change in the particle momentum per unit time is in that case proportional to ξ^{-2} and thus particularly large.

It is well known that in a plasma consisting of hot electrons and cold ions one can excite not only sound oscillations with a linear dispersion law but also short-wavelength ion oscillations. The spectral density of the charge density fluctuations has, in the region of "medium-range" frequencies ($qv_i \ll \omega \ll qv_e$) and if we take into account the short-wavelength oscillations, the form ($q^2r_D^2 \ll T_e/T_i$)

$$\langle \varrho^2 \rangle_{q\omega} = \frac{1}{4}\frac{q^2(qr_D)^2}{1+q^2r_D^2}\{T((q, (\boldsymbol{q}\cdot\boldsymbol{u}))\,\delta(\omega-\omega_q)+T(q, -(\boldsymbol{q}\cdot\boldsymbol{u}))\,\delta(\omega+\omega_q)\}, \quad (13.4.2.30)$$

where ω_q is the frequency

$$\omega_q = \frac{qv_s}{\sqrt{(1+q^2r_D^2)}},$$

while $T(q, (\boldsymbol{q}\cdot\boldsymbol{u}))$ is the effective temperature of the oscillations which for $(\boldsymbol{q}\cdot\boldsymbol{u}) < \omega_q$ is given by the formula

$$T(q, (\boldsymbol{q}\cdot\boldsymbol{u})) = \frac{T_e}{1-[(\boldsymbol{q}\cdot\boldsymbol{u})/\omega_q]}.$$

As $(\boldsymbol{q}\cdot\boldsymbol{u}) \to \omega_q$ the quantity T increases steeply. In the region $(\boldsymbol{q}\cdot\boldsymbol{u}) > \omega_q$ the effective temperature changes smoothly with changing $(\boldsymbol{q}\cdot\boldsymbol{u})$.

Using formulae (13.1.4.6) and (13.4.2.30) we get the following expression for the change in particle energy per unit time caused by the excitation and absorption of ion oscillations with wavevectors in the range $\boldsymbol{q}$, $\boldsymbol{q}+d\boldsymbol{q}$:

$$\frac{dP}{dq} = \frac{Z^2e^2ur_D^2q^3\omega_q}{\pi Mv^2(1+q^2r_D^2)}\left\{D(\boldsymbol{q})\,\theta\left[1-\frac{\omega_q}{qv}\right]-\frac{\pi}{u}T(q, qu\cos\theta)\,\delta\left(q-\frac{\omega_q}{v}\right)\right\}, \quad (13.4.2.31)$$

where the function D is given by formula (13.4.2.5) with the substitution $v_s \to \omega_q/q$ while $\theta[x] = \frac{1}{2}(1+\operatorname{sgn} x)$.

Substituting expression (13.4.2.10) for the function D into (13.4.2.3) we see easily that for all values of the wavevector—except those values for which $\omega_q = qv$—the behaviour of the quantity dP/dq is independent of the detailed behaviour of the function T for $(\boldsymbol{q}\cdot\boldsymbol{u}) > \omega_q$. (In the case when $\omega_q = qv$ and $v > u\cos\theta$, when we can neglect the second term in eqn. (13.4.2.31), it is also not necessary to know the exact way T depends on $(\boldsymbol{q}\cdot\boldsymbol{u})$ for $(\boldsymbol{q}\cdot\boldsymbol{u}) > \omega_q$ in order to study the nature of the function dP/dq.)

Integrating expression (13.4.2.31) over q we can determine the change per unit time in the particle energy P caused through the interaction between the particle and the plasma oscillations. In the case of oscillations with a non-linear dispersion law we need, in general, to know the explicit form of the functions $\xi(q)$ and $\lambda(q)$ to find the explicit way P depends on the particle velocity $\boldsymbol{v}$. None the less, one can reach a number of conclusions about P which are independent of the explicit form of the functions ξ and λ.

First of all, the change in the particle energy will be large—proportional to at least the first power of $1/\bar{\xi}$—if the following condition holds:

$$\sin^2\theta \gg \frac{T_i\,|\boldsymbol{v}-\boldsymbol{u}|^2}{m_i v^2 u^2},$$

which is the condition that there exist oscillations with wavevectors satisfying the relations $(\boldsymbol{q}\cdot\boldsymbol{v}) = \omega_q$, $(\boldsymbol{q}\cdot\boldsymbol{u}) > \omega_q$.

Furthermore, if the angle between the direction of motion of the particle and the direction of the electron current is small, there is only a strong interaction between the particle and the oscillations if $v \approx u$.

If the condition

$$\frac{v_s\,|\boldsymbol{v}-\boldsymbol{u}|}{u^2} < \theta \ll \xi\frac{v_s}{u}$$

is satisfied, the particle energy losses will be particularly large and proportional to ξ^{-2}:

$$P = \frac{Z^2e^2\omega_{pe}^2m_e}{Mv}B_2, \quad B_2 = r_D\int\frac{(qr_D)^3}{1+q^2r_D^2}\xi^{-2}\,dq. \tag{13.4.2.32}$$

Finally, we can simply find the sign of P if $v \gg u$. The function P is positive, that is, the particle energy decreases, if $\theta < \pi/2$, and is negative, that is, the particle energy increases, if $\theta > \pi/2$.

If $v \gg v_s$ and $u \gg v_s$ the way the change in the energy of the particle depends on the magnitude and direction of its velocity can be obtained in explicit form. If $\sin\theta$ is not too small ($\theta \gg v_s\,|u-v|/uv, \pi-\theta \gg v_s(u+v)/uv$) we can use formulae (13.4.2.31) and (13.4.2.10) to obtain

$$P = \frac{Z^2e^2\omega_{pe}^2m_ev_s}{Mv^2}B\cot\theta, \tag{13.4.2.33}$$

where B is a large quantity, proportional to ξ^{-1},

$$B = r_D\int(qr_D)^3(1+q^2r_D^2)^{-3/2}(\lambda\xi)^{-1}\,dq.$$

If $\sin\theta \ll \mathrm{Min}\{\sqrt{(\xi v_s/u)}, \xi u/v_s\}$, we have from formula (13.4.2.11)

$$P = \pm\frac{Z^2e^2\omega_{pe}^2um_ev^{3/2}}{M\sqrt{[u\mp v]}}B_1, \quad B_1 = \frac{1}{2}r_D\int\frac{(qr_D)^3}{1+q^2r_D^2}\xi^{-3/2}\lambda^{-1/2}\,dq, \tag{13.4.2.34}$$

where the upper (lower) sign refers to the case of angles θ which are small (close to π). We see that the change in the energy of a particle moving in the direction of the electron current (or in the opposite direction) is proportional to $\xi^{-3/2}$ and thus particularly large. If the particle velocity is close to the current velocity, not only in magnitude, but also in direction, the energy losses—which in that case are given by eqn. (13.4.2.32)—are proportional to ξ^{-2}.

Let us now consider the case of a plasma through which passes a hot electron beam with a velocity u exceeding the thermal velocity of the electrons in the plasma. The spectral density of the charge fluctuations in such a plasma has in the high-frequency region ($\omega \gg qv_e$) and long wavelengths ($qr_D \ll 1$) the form

$$\langle\varrho^2\rangle_{q\omega} = \tfrac{1}{4}q^2\{T(q, (\boldsymbol{q}\cdot\boldsymbol{u}))\,\delta(\omega-\omega_{pe})+T(q, -(\boldsymbol{q}\cdot\boldsymbol{u}))\,\delta(\omega+\omega_{pe})\}, \tag{13.4.2.35}$$

where T is the effective temperature of the electron Langmuir oscillations. If $(\boldsymbol{q}\cdot\boldsymbol{u}) < \omega_{pe}$ the quantity T is determined by the temperature T' of the beam electrons,

$$T(q, (\boldsymbol{q}\cdot\boldsymbol{u})) = \frac{T'}{1-[(\boldsymbol{q}\cdot\boldsymbol{u})/\omega_{pe}]}\,.$$

If $(\boldsymbol{q}\cdot\boldsymbol{u}) > \omega_{pe}$ the effective temperature is determined by the level of the turbulent fluctuations and is much larger than T'.

Using formulae (13.1.4.6) and (13.4.2.10) we find the energy given by the particle per unit time to electron Langmuir oscillations with wavevectors in the range $\boldsymbol{q}$, $\boldsymbol{q}+d\boldsymbol{q}$:

$$\frac{dP}{dq} = \frac{Z^2e^2uq\omega_{pe}}{\pi Mv^2}\left\{D(\boldsymbol{q})\,\theta\left[1-\frac{\omega_{pe}}{qv}\right]-\frac{\pi}{u}T(q, qu\cos\theta)\,\delta\left(q-\frac{\omega_{pe}}{v}\right)\right\}, \quad (13.4.2.36)$$

where the function D is given by formula (13.4.2.5) with the substitution $v_s \to \omega_{pe}/q$.

One sees easily that all conclusions about the interaction of a particle with ion oscillations can be generalized to the case of electron oscillations through the substitution

$$T_e \to T', \quad \omega_{\boldsymbol{q}} \to \omega_{pe}, \quad T_i \to T_e, \quad m_i \to m_e.$$

In particular, the energy losses are particularly large—proportional to ξ^{-2}—if the condition $|1-(v/u)| < \theta \ll \xi$ is satisfied. In that case

$$P = \frac{Z^2e^2\omega_{pe}^2m_e}{Mv}\tilde{B}_2, \quad \tilde{B}_2 = r_D^2\int \xi^{-2}\,q\,dq. \quad (13.4.2.37)$$

13.4.3. INTERACTION OF CHARGED PARTICLES WITH A TURBULENT PLASMA IN A MAGNETIC FIELD

Let us now consider the interaction of charged particles with a turbulent plasma in a constant and uniform magnetic field. We shall assume the plasma to consist of cold ions at rest and hot electrons which move along the magnetic field with a velocity above the critical velocity for which the instability of the plasma oscillations sets in. We shall study how the change in the particle energy per unit time depends on the magnitude and direction of the particle velocity and we shall show that the details of the turbulence spectrum do not affect this dependence strongly, as in the case when there is no magnetic field (Akhiezer, 1965d).

The probability for the transition per unit time of a particle from a state with quantum numbers ν and p_z to a state with quantum numbers ν' and p_z' is given by eqn. (13.3.3.1). Expressing the spectral density of the potential fluctuations in terms of the spectral density of the charge density fluctuations we can rewrite eqn (13.3.3.1) in the form

$$w_{\nu p_z \to \nu' p_z'} = \left(\frac{4\pi Ze}{\hbar}\right)^2\int\frac{\langle\varrho^2\rangle_{\boldsymbol{q}\omega}}{q^4}A_{\nu\nu'}\left(\frac{\hbar q_\perp^2}{2M\omega_{Be}}\right)\frac{d^2\boldsymbol{q}_\perp}{(2\pi)^2}. \quad (13.4.3.1)$$

Multiplying expression (13.4.3.1) by $\hbar\omega$, summing over ν and integrating over p_z' we can determine the particle energy losses per unit time:

$$P = -\frac{(4\pi Ze)^2}{\hbar}\sum_{\nu'}\int\frac{\omega}{q^4}\langle\varrho^2\rangle_{\boldsymbol{q}\omega}\,\delta\left[\omega-\omega_{Be}(\nu'-\nu)-q_zv_z-\frac{\hbar q_z^2}{2M}\right]A_{\nu\nu'}\left(\frac{\hbar q_\perp^2}{2M\omega_{Be}}\right)d\omega\,\frac{d^2\boldsymbol{q}_\perp}{(2\pi)^3}, \quad (13.4.3.2)$$

where v_z is the component of the particle velocity along the magnetic field.

We can simplify eqn. (13.4.3.2) considerably in the case where we can neglect the "twisting" of the particle trajectory by the magnetic field. Assuming that $(v_A/c)\sin\theta \ll Mv/m_iv_s$, where v is the particle velocity, θ the angle between the direction of motion of the particle and the magnetic field, $v_A = B_0/4\pi n_0 m_i$ the Alfvén velocity, $v_s = \sqrt{(T_e/m_i)}$ the sound velocity, T_e and T_i the electron and ion temperatures, and m_e and m_i the electron and ion masses, and using the asymptotic relation for the function $A_{\nu\nu'}$, we get

$$P = \frac{-Z^2e^2}{\hbar}\int\left(\frac{4\pi}{q^2}\right)^2 \omega\langle\varrho^2\rangle_{q\omega}\,\delta\left(\omega-(\boldsymbol{q}\cdot\boldsymbol{v})-\frac{\hbar q^2}{2M}\right)d\omega\,\frac{d^3\boldsymbol{q}}{(2\pi)^3}. \tag{13.4.3.3}$$

Let us first of all consider the case of not very strong magnetic fields ($v_A \ll c$). If the plasma consists of cold ions at rest and hot electrons which move along the magnetic field with a velocity $\boldsymbol{u}$, the spectral density of the charge density fluctuations can in the sound region,

$$qv_i \ll \omega \ll qv_e, \quad qr_D \ll 1,$$

be written in the form ($v_A/c \ll qr_D \ll 1$)

$$\langle\varrho^2\rangle_{q\omega} = \tfrac{1}{4}q^2(qr_D)^2\left\{T(q, (\boldsymbol{q}\cdot\boldsymbol{u}))\,\delta(\omega-qv_s)+T(q, -(\boldsymbol{q}\cdot\boldsymbol{u}))\,\delta(\omega+qv_s)\right\}, \tag{13.4.3.4}$$

where T is the effective temperature of the sound waves and r_D the electron Debye radius. It is important that in eqn. (13.4.3.4) only the effective temperature of the waves depends on the magnetic field; when $v_A \ll c$, the argument of the δ-functions is independent of the magnetic field.

Substituting expression (13.4.3.4) into (13.4.3.3) and bearing in mind that in the turbulence region, $(\boldsymbol{q}\cdot\boldsymbol{u}) > qv_s$, we have the inequality $T \gg Mv^2$, we find

$$P = \frac{Z^2e^2r_D^2uv_s}{\pi Mv^2}\int q^4\,dq\int_0^{\pi} d\varphi\left\{\cos\theta-\frac{\sin\theta\cos\varphi}{\sqrt{[(v/v_s)^2-1]}}\right\}\frac{\partial T(q,\eta)}{\partial\eta}, \tag{13.4.3.5}$$

where

$$\eta \equiv (\boldsymbol{q}\cdot\boldsymbol{u}) = qu\left\{\frac{v_s}{v}\cos\theta+\sqrt{\left[1-\frac{v_s^2}{v^2}\right]}\sin\theta\cos\varphi\right\}.$$

Equation (13.4.3.5) is formally the same as the expression for the change in the energy of a particle passing through a turbulent plasma when there is no magnetic field. The way the change in the particle energy depends on the magnitude and direction of its velocity is thus of the same nature in a magnetic field as when there is no such field, and we see that

1. There exist critical angles between the directions of the particle motion and of the magnetic field

$$\theta_\pm = \arccos\frac{v_s^2\pm[(v^2-v_s^2)(u^2-v_s^2)]^{1/2}}{uv}, \tag{13.4.3.6}$$

close to which the change in the particle energy is particularly large,

$$P = \pm\frac{Z^2e^2(T^*)^{3/2}\,v_s^{1/2}\alpha_1\sqrt{(u^2-v_s^2)}}{r_D^2M(uvT_e\sin\theta_\pm)^{1/2}(v^2-v_s^2)^{3/4}}, \tag{13.4.3.7}$$

where T^* is a quantity of the order of the effective temperature of the waves in the turbulence region ($T^* \gg T_e$) while α_1 is a quantity of the order unity (the $\pm$-signs correspond to $\theta \approx \theta_\pm$).

2. In the region $\theta_+ < \theta < \theta_-$ the quantity P is given by the formula

$$P = \frac{Z^2e^2T^*v_s}{r_D^2Muv}\alpha\frac{(\boldsymbol{u}\cdot\boldsymbol{v})-v_s^2}{v^2-v_s^2}(\cos\theta_+-\cos\theta)^{-1/2}(\cos\theta-\cos\theta_-)^{-1/2}, \quad (13.4.3.8)$$

where α is a quantity of the order unity. This formula is exact if the effective temperature of the turbulent waves changes little when the angle ϑ between $\boldsymbol{q}$ and $\boldsymbol{B}_0$ varies. If T changes strongly with changing ϑ eqn. (13.4.3.8) will strictly be satisfied only for $|\theta-\theta_\pm| \ll 1$, while in the region $|\theta-\theta_\pm| \approx 1$ it is a good interpolation formula. We note that for $\theta < \theta_0$, where θ_0 is a critical angle ($\theta_0 \sim \arccos(v_s^2/uv)$) the particle energy decreases, while it increases for $\theta > \theta_0$.

3. The particle energy losses are particularly large when $\boldsymbol{v} \approx \boldsymbol{u}$, and also when $v \approx v_s$ and $(\boldsymbol{u}\cdot\boldsymbol{v}) \sim v_s^2$. In both cases P is proportional to the square of the effective temperature,

$$P = \frac{(ZeT^*)^2}{r_D^2MvT_e}\alpha_2, \quad (13.4.3.9)$$

where α_2 is a quantity of order unity.

4. If $\theta > \theta_-$ and also if $\theta < \theta_+$ and $\cos\theta < v/u$, the expression for the energy losses does not contain the large parameter T^*. If $|\theta-\theta_\pm| \ll 1$, the quantity P is, nevertheless, proportional to $|\theta-\theta_\pm|^{-3/2}$, and hence large.

5. As $u \to v_s$, we see from formula (13.4.3.6) that $\theta_\pm \to \theta_c = \arccos(v_s/v)$. When $\theta \approx \theta_c$ the energy losses are proportional to $(T^*)^{1/2}$,

$$P = \frac{Z^2e^2v_s^2\sqrt{(T^*T_e)}}{r_D^2M(v^2-v_s^2)^{3/2}}\alpha_0, \quad (13.4.3.10)$$

where $\alpha_0 \sim 1$.

We emphasize that eqns. (13.4.3.6) to (13.4.3.10) are formally the same as the corresponding equations in subsection 13.4.2; the presence of even a strong magnetic field affects only the magnitude of the effective temperature T^*, but not the way P depends on $\boldsymbol{v}$—provided the condition $v_A \ll c$ is satisfied.

We note that we must use eqn. (13.4.3.7) when we can neglect the smearing-out of the angles $\theta_\pm$ due to the sound dispersion, that is, when the condition

$$\int T(qr_D)^2\,dq \gg rT_e^{-1/2}\int T^{3/2}(qr_D)^3\,dq,$$

with r given by eqn. (13.4.2.24), is satisfied. In the opposite case one must use for P formula (13.4.2.25).

Let us now consider the case of a very strong magnetic field ($v_A \gg c$). In that case the spectral density of the charge density fluctuations has in the "sound" region sharp maxima corresponding to the possibility of the propagation of slow magneto-sound waves

$$\langle\varrho^2\rangle_{q\omega} = \tfrac{1}{4}q^2(qr_D)^2\{T(q,(\boldsymbol{q}\cdot\boldsymbol{u}))\,\delta(\omega-qv_s|\cos\vartheta|)+T(q,-(\boldsymbol{q}\cdot\boldsymbol{u}))\,\delta(\omega+qv_s|\cos\vartheta|)\}, \quad (13.4.3.11)$$

where ϑ is the angle between $\boldsymbol{q}$ and $\boldsymbol{B}_0$. It is essential that when $u > v_s$ all waves moving in the same direction as the electron current are turbulent and therefore characterized by a very high effective temperature (if $u < v_s$ all magneto-sound waves are stable).

Substituting expression (13.4.3.11) into formula (13.4.3.3) we get

$$P = -\frac{Z^2e^2r_D^2v_s}{\pi M v^2}\int q^3\,dq\int_0^{\pi} d\varphi\left|\frac{v_s}{v}-\cos\theta\right|\left\{\left(\frac{v_s}{v}-\cos\theta\right)^2+\sin^2\theta\cos^2\varphi\right\}^{-1}$$

$$\times\left\{\frac{\partial}{\partial y}\frac{yT(q,\,quy)}{(v_s/v)-\cos\theta+y(1-y^2)^{-1/2}\sin\theta\cos\varphi}\right\}_{y=y_0}, \qquad (13.4.3.12)$$

where

$$y_0 = \frac{\sin\theta\,|\cos\varphi|}{\{[(v_s/v)-\cos\theta]^2+\sin^2\theta\cos^2\varphi\}^{1/2}}.$$

Assuming that the effective temperature in the turbulence region is weakly dependent on the angle between $\boldsymbol{q}$ and $\boldsymbol{B}_0$, we can write expression (13.4.3.12) in the form

$$P = \frac{Z^2e^2v_s}{r_D^2M}T^*\frac{v\cos\theta-v_s}{[v^2-2vv_s\cos\theta+v_s^2]^{3/2}}, \qquad (13.4.3.13)$$

where

$$T^* = \tfrac{1}{2}r_D\int(qr_D)^3\,T\,dq.$$

From eqn. (13.4.3.13) it follows that a particle which moves at an angle θ to the magnetic field which is less than $\theta_c = \arccos(v_s/v)$ loses energy; if $\theta > \theta_c$ the particle energy increases.

We note that expression (13.4.3.13) tends to infinity if simultaneously $v \to v_s$ and $\theta \to 0$. If we want to determine P in that case, we must take the sound dispersion into account. As a result we get

$$P = \frac{Z^2e^2}{r_D^2Mv_s}T_1^*,$$

where

$$T_1^* \approx r_D\int T\,qr_D\,dq.$$

Bearing in mind that for ion sound $qr_D \ll 1$, we see that there is in this case a considerable increase in the particle energy losses.

REFERENCES†

ABRIKOSOV, A. A. and KHALATNIKOV, I. M. (1958) *Soviet Phys. JETP* **7**, 135.
ADAMSON, T. (1960) *Phys. Fluids* **3**, 706.
ADLAM, J. and ALLEN, J. (1958) *Phil. Mag.* **3**, 448.
AHIEZER, A. I. (1956) *Nuovo Cim. Suppl.* **3**, 591.
AKHIEZER, A. I. AKHIEZER, I. A. and POLOVIN, R. V. (1965) *High-frequency Plasma Properties*, Kiev, p. 133.
AKHIEZER, A. I., AKHIEZER, I. A., POLOVIN, R. V., SITENKO, A. G. and STEPANOV, K. N. (1967) *Collective Oscillations in a Plasma*, Pergamon Press, Oxford.
AKHIEZER, A. I., AKHIEZER, I. A. and SITENKO, A. G. (1962) *Soviet Phys. JETP* **14**, 462.
AKHIEZER, A. I., ALEKSIN, V. F., BAR'YAKHTAR, V. G. and PELETMINSKIĬ, S. V. (1962) *Soviet Phys. JETP* **15**, 386.
AKHIEZER, A. I., ALEKSIN, V. F. and KHODUSOV, V. D. (1971) *Nucl. Fusion* **11**, 403.
AKHIEZER, A. I. and FAĬNBERG, YA. B. (1949) *Dokl. Akad. Nauk SSSR* **69**, 555.
AKHIEZER, A. I. and FAĬNBERG, YA. B. (1951a) *Zh. Eksp. Teor. Fiz.* **21**, 1262.
AKHIEZER, A. I. and FAĬNBERG, YA. B. (1951b) *Usp. Fiz. Nauk* **44**, 321.
AKHIEZER, A. I. and FAĬNBERG, YA. B. (1962) *Theory of Linear Accelerators*, Gosatomizdat, Moscow, p. 320.
AKHIEZER, A. I. and LYUBARSKIĬ, G. YA. (1951) *Dokl. Akad. Nauk SSSR* **80**, 193.
AKHIEZER, A. I. and LYUBARSKIĬ, G. YA. (1955) *Proc. Phys. Faculty Khar'kov Univ.* **6**, 13.
AKHIEZER, A. I., LYUBARSKIĬ, G. YA. and POLOVIN, R. V. (1959) *Soviet Phys. JETP* **8**, 507.
AKHIEZER, A. I., LYUBARSKIĬ, G. YA. and POLOVIN, R. V. (1960) *Soviet Phys. Tech. Phys.* **4**, 849.
AKHIEZER, A. I., LYUBARSKIĬ, G. YA. and POLOVIN, R. V. (1961) *Soviet Phys. JETP* **13**, 673.
AKHIEZER, A. I., LYUBARSKIĬ, G. YA. and POLOVIN, R. V. (1963) *Plasma Physics and Controlled Thermonuclear Fusion*, Kiev, **3**, 151.
AKHIEZER, A. I. and PARGAMANIK, L. E. (1948) *Proc. Khar'kov Univ.* **2**, 75.
AKHIEZER, A. I. and POLOVIN, R. V. (1955) *Dokl. Akad. Nauk SSSR* **102**, 919.
AKHIEZER, A. I. and POLOVIN, R. V. (1956) *Soviet Phys. JETP* **3**, 696.
AKHIEZER, A. I., PROKHODA, I. G. and SITENKO, A. G. (1958) *Soviet Phys. JETP* **6**, 576.
AKHIEZER, A. I. and SITENKO, A. G. (1952) *Zh. Eksp. Teor. Fiz.* **23**, 161.
AKHIEZER, A. I. and SITENKO, A. G. (1959) *Soviet Phys. JETP* **8**, 82.
AKHIEZER, I. A. (1961) *Soviet Phys. JETP* **13**, 667.
AKHIEZER, I. A. (1962) *Soviet Phys. JETP* **15**, 406.
AKHIEZER, I. A. (1963) *Plasma Physics and Controlled Thermonuclear Fusion*, Kiev, **2**, 28.
AKHIEZER, I. A. (1964a) *Soviet Phys. Tech. Phys.* **8**, 699.
AKHIEZER, I. A. (1964b) *Phys. Lett.* **12**, 201.
AKHIEZER, I. A. (1965a) *Soviet Phys. JETP* **20**, 445.
AKHIEZER, I. A. (1965b) *Soviet Phys. JETP* **20**, 637.
AKHIEZER, I. A. (1965c) *Soviet Phys. JETP* **20**, 1519.
AKHIEZER, I. A. (1965d) *Ukr. Fiz. Zh.* **10**, 581.
AKHIEZER, I. A. (1965e) *Soviet Phys. JETP* **21**, 774.
AKHIEZER, I. A. and ANGELEĬKO, V. V. (1969a) *Soviet Phys. JETP* **28**, 1216.
AKHIEZER, I. A. and ANGELEĬKO, V. V. (1969b) *Ukr. Phys. J.* **13**, 1445.
AKHIEZER, I. A. and BOLOTIN, YU. L. (1963) *Nucl. Fusion* **3**, 271.
AKHIEZER, I. A. and BOLOTIN, YU. L. (1964) *Soviet Phys. JETP* **19**, 902.
AKHIEZER, I. A. and BOROVIK, A. E. (1967) *Soviet Phys. JETP* **24**, 823.

† As far as possible the English translations of Russian papers and the originals of other papers are given (Translator). This list is common to Volumes 1 and 2 and thus contains papers not referred to in Volume 2.

AKHIEZER, I. A. and BOROVIK, A. E. (1968) *Ukr. Phys. J.* **13,** 6.
AKHIEZER, I. A., DANELIYA, I. A. and TSINTSADZE, N. L. (1964) *Soviet Phys. JETP* **19,** 208.
AKHIEZER, I. A. and POLOVIN, R. V. (1959) *Soviet Phys. JETP* **9,** 1316.
AKHIEZER, I. A. and POLOVIN, R. V. (1960) *Soviet Phys. JETP* **11,** 383.
AKHIEZER, I. A., POLOVIN, R. V. and TSINTSADZE, N. L. (1960) *Soviet Phys. JETP* **10,** 539.
ALEKSIN, V. F. and KHODUSOV, V. D. (1970) *Ukr. Fiz. Zh.* **15,** 1021.
ALFVÉN, H. (1949) *Phys. Rev.* **75,** 1732.
ALFVÉN, H. (1950) *Cosmic Electrodynamics,* Oxford University Press.
ALFVÉN, H. and FÄLTHAMMER, K. G. (1963) *Cosmic Electrodynamics,* Oxford University Press.
AL'TSHUL', L. M. and KARPMAN, V. I. (1965) *Soviet Phys. JETP* **20,** 1043.
AMER, S. (1958) *J. Electr. Contr.* **5,** 105.
ANDERSON, J. E. (1963) *Shock Waves in Magnetohydrodynamics,* MIT Press, Cambridge, Mass.
ANDRONOV, A. A. and TRAKHTENGERTS, V. YU. (1964) *Soviet Phys. JETP* **18,** 698.
ANDRONOV, A. A., VITT, A. A. and KHAĬKIN, S. É. (1959) *Theory of Oscillations,* Fizmatgiz, Moscow.
ANGELEĬKO. V. V. (1968) *Ukr. Phys. J.* **13,** 123.
ANGELEĬKO, V. V. and AKHIEZER, I. A. (1968) *Soviet Phys. JETP* **26,** 433.
ANGELEĬKO, V. V. and AKHIEZER, I. A. (1969) *Ukr. Phys. J.* **73,** 1451.
ANGELEĬKO, V. V. and KITSENKO, A. B. (1965) *Ukr. Fiz. Zh.* **10,** 16.
APPLETON, E. V. (1927) *URSI Proceedings,* Washington meeting.
APPLETON, E. V. and BARNETT, M. A. (1925) *Electrician* **94,** 398.
ARTSIMOVICH, L. A. (1963) *Controlled Thermonuclear Reactions,* Fizmatgiz, Moscow.
ASTRÖM, E. (1950) *Nature* **165,** 1019.
ASTRÖM, E. (1951) *Ark. Fys.* **2,** 442.
AUER, P. L. (1958) *Phys. Rev. Lett.* **1,** 411.
AXFORD, W. I. (1961) *Phil. Trans. Roy. Soc.* A **253,** 301.
BABENKO, K. I. and GEL'FAND, I. M. (1958) *Nauch. Dokl. Vyssh. Shk.-fiz.-mat. Nauki* No 1, 12.
BABYKIN, M. V., GAVRIN, P. P., ZAVOĬSKIĬ, E. K., RUDAKOV, L. I., SKORYUPIN, V. A. and SHOLIN, G. B. (1964) *Soviet Phys. JETP* **19,** 49.
BABYKIN, M. V., ZAVOĬSKIĬ, E. K., RUDAKOV, L. I. and SKORYUPIN, V. A. (1962) *Nucl. Fusion Suppl.* **3,** 1073.
BAKAĬ, A. S. (1970) *Nucl. Fusion* **10,** 53.
BAKAĬ, A. S. (1971) *Soviet Phys. JETP* **32,** 66.
BAKSHT, F. G. (1964) *Soviet Phys. JETP* **8,** 878.
BALESCU, R. (1963) *Statistical Mechanics of Charged Particles,* Interscience, New York.
BARANTSEV, R. G. (1962) *Soviet Phys. JETP* **15,** 615.
BARMIN, A. A. (1961) *Soviet Phys. Doklady* **6,** 374.
BARMIN, A. A. and GOGOSOV, V. V. (1961) *Soviet Phys. Doklady* **5,** 961.
BATCHELOR, G. K. (1950) *Proc. Roy. Soc.* A **201,** 405.
BAUM, F. A., KAPLAN, S. A. and STANYUKOVICH, K. P. (1958) *Introduction to Cosmic Gas Dynamics,* Fizmatgiz, Moscow.
BAUSSET, M. (1963) *Comptes Rendus Acad. Sci.* **257,** 372.
BAZER, J. (1958) *Astroph. J.* **128,** 686.
BAZER, J. and ERICSON, W. B. (1959) *Astroph. J.* **129,** 758.
BAZER, J. and FLEISCHMAN, O. (1959) *Phys. Fluids* **2,** 366.
BELEN'KIĬ, S. Z. (1945) *Dokl. Akad. Nauk SSSR* **48,** 172.
BELEN'KIĬ, S. Z. (1958) *Lebedev Inst. Proc.* **10,** 5.
BEREZIN, YU. A. (1961) *Soviet Phys. Doklady* **5,** 670.
BEREZIN, YU. A. and KARPMAN, V. I. (1964) *Soviet Phys. JETP* **19,** 1265.
BEREZIN, YU. A. and KARPMAN, V. I. (1967) *Soviet Phys. JETP* **24,** 1049.
BERNSTEIN, I. (1958) *Phys. Rev.* **109,** 10.
BERNSTEIN, I. and ENGELMAN, F. (1966) *Phys. Fluids* **9,** 937.
BERNSTEIN, J. (1968) *Elementary Particles and their Currents,* Freeman, San Francisco.
BERZ, F. (1956) *Proc. Phys. Soc.* B **69,** 939.
BHATNAGAR, P. L., GROSS, E. P. and KROOK, M. (1954) *Phys. Rev.* **94,** 511.
BLUDMAN, S. A., WATSON, K. M. and ROSENBLUTH, M. N. (1960a) *Phys. Fluids* **3,** 741.
BLUDMAN, S. A., WATSON, K. M. and ROSENBLUTH, M. N. (1960b) *Phys. Fluids* **3,** 747.
BOGDANKEVICH, L. S., RUKHADZE, A. A. and SILIN, V. P. (1962) *Radiofizika* **5,** 1093.
BOGOLYUBOV, N. N. (1946) *Zh. Eksp. Teor. Fiz.* **16,** 691.
BOGOLYUBOV, N. N. (1962) *Studies Stat. Mech.* **1,** 5.

BOHACHEVSKY, I. O. (1962) *Phys. Fluids* **5,** 1456.
BOHM, D. and GROSS, E. (1949a) *Phys. Rev.* **75,** 1851.
BOHM, D. and GROSS, E. (1949b) *Phys. Rev.* **75,** 1864.
BOHM, D. and PINES, D. (1951) *Phys. Rev.* **82,** 625.
BOHR, N. (1948) *Kgl. Danske Vid. Selsk., Mat-Fys. Medd.* **18,** No. 8.
BOOKER, H. G. (1935) *Proc. Roy. Soc.* A **150,** 267.
BORN, M. and GREEN, H. S. (1947) *Proc. Roy. Soc.* A **188,** 10.
BRAGINSKII, S. I. (1965) *Rev. Plasma Phys.* **1,** 205.
BRAGINTSEV, S. I. and KAZANTSEV, A. P. (1958) *Plasma Physics and Controlled Thermonuclear Reactions* **3,** 24.
BRIGGS, R. J. (1964) *Electron Stream Interaction with Plasmas*, MIT Press.
BUDKER, G. I. (1956) *Atomnaya Energiya* **1,** 9.
BUNEMAN, O. (1958) *Phys. Rev. Lett.* **1,** 104.
BUNEMAN, O. (1959) *Phys. Rev.* **115,** 503.
BUNEMAN, O. (1962) *J. Nucl. Energy* C **4,** 111.
BURT, P. and HARRIS, E. G. (1961) *Phys. Fluids* **4,** 1412.
BUSEMANN, A. (1942) *Luftfahrtforschung* **19,** 137.
BUTLER, D. C. (1965) *J. Fluid Mech.* **23,** 1.
CABANNES, H. (1957) *Comptes Rendus Acad. Sci.* **245,** 1379.
CABANNES, H. (1960a) *Comptes Rendus Acad. Sci.* **250,** 1968.
CABANNES, H. (1960b) *Rev. Mod. Phys.* **32,** 973.
CABANNES, H. (1963) *Comptes Rendus Acad. Sci.* **257,** 375.
CABANNES, H. and STAEL, C. (1961) *J. Fluid Mech.* **10,** 289.
CALLEN, H. B. and WELTON, T. A. (1951) *Phys. Rev.* **83,** 34.
CAVALIERE, A. (1962) *Nuovo Cim.* **23,** 440.
CHAPMAN, S. and COWLING, T. G. (1953) *Mathematical Theory of Non-uniform Gases*, Cambridge University Press.
CHERKASOVA, K. P. (1961) *Prikl. Mat. Teor. Fiz.* **6,** 169.
CHERKASOVA, K. P. (1965) *Izv. Akad. Nauk SSSR, Mekh.* **5,** 146.
CHU, C. K. (1964) *Phys. Fluids* **7,** 1349.
CHU, C. K. (1967) *Proc. Symp. Appl. Math.* **18,** 1.
CHU, C. K. and TAUSSIG, R. T. (1967) *Phys. Fluids* **10,** 249.
COHEN, I. M. and CLARKE, J. H. (1965) *Phys. Fluids* **8,** 1278.
COMISAR, G. G. (1962) *Phys. Fluids* **5,** 1590.
COURANT, R. (1962) *Partial Differential Equations*, Interscience, New York.
COURANT, R. and FRIEDRICHS, K. (1948) *Supersonic Flow and Shock Waves*, Academic Press, New York.
COWLEY, M. D. (1967) *J. Plasma Phys.* **1,** 37.
COWLING, T. G. (1957) *Magneto-hydrodynamics*, Interscience, New York.
DAVIES, L., LÜST, R. and SCHLÜTER, A. (1958) *Zs. Naturf.* **13a,** 916.
DAVYDOV, B. I. (1936) *Zh. Eksp. Teor. Fiz.* **6,** 463.
DAVYDOV, B. I. (1937) *Zh. Eksp. Teor. Fiz.* **7,** 1069.
DEMETRIADES, S. T. and ARGYROPOULOS, G. S. (1966) *Phys. Fluids* **9,** 2136.
DEMUTSKIĬ, V. P. (1962) *Soviet Tech. Phys.* **6,** 1014.
DEMUTSKIĬ, V. P. and POLOVIN, R. V. (1961) *Soviet Phys. Tech. Phys.* **6,** 302.
DERFLER, H. (1961) *J. Electr. Contr.* **11,** 189.
DEUTSCH, R. V. (1963) *Prikl. Mat. Teor. Fiz.* No. 1, 38.
DNESTROVSKIĬ, YU. N. (1963) *Nucl. Fusion* **3,** 259.
DNESTROVSKIĬ, YU. N. and KOSTOMAROV, D. P. (1959) *Khar'kov Conf. Proceedings.*
DNESTROVSKIĬ, YU. N. and KOSTOMAROV, D. P. (1961) *Soviet Phys. JETP* **13,** 986.
DNESTROVSKIĬ, YU. N. and KOSTOMAROV, D. P. (1962) *Soviet Phys. JETP* **14,** 1089.
DNESTROVSKII, YU. N., KOSTOMAROV, D. P. and PISTUNOVICH, V. I. (1963) *Nucl. Fusion* **3,** 30.
DÖRING, W. (1943) *Ann. Physik* **43,** 421.
DOUGHERTY, J. P. and FARLEY, D. T. (1960) *Proc. Roy. Soc.* A **259,** 79.
DOYLE, P. H. and NEUFELD, J. (1958) *Phys. Fluids* **2,** 39.
DRUMMOND, J. E. (1958) *Phys. Rev.* **110,** 293.
DRUMMOND, W. E. and PINES, D. (1962) *Nucl. Fusion Suppl.* **3,** 1049.
DRUMMOND, W. E. and ROSENBLUTH, M. N. (1962) *Phys. Fluids* **5,** 1507.
DRUYVESTEYN, M. J. (1930) *Physica* **10,** 61.

DRUYVESTEYN, M. J. (1934) *Physica* **14,** 1003.
DUNLAP, R., BREHM, R. L. and NICOLLS, J. A. (1958) *Jet Propulsion* **28,** 451.
ERICSON, W. B. and BAZER, J. (1960) *Phys. Fluids* **3,** 631.
ERPENBECK, J. J. (1961) *Phys. Fluids* **4,** 481.
ERPENBECK, J. J. (1964) *Phys. Fluids* **7,** 1424.
FADEEVA, V. N. and TERENT'EV, N. M. (1961) *Tables of Values of the Probability Integral for Complex Argument*, Pergamon, Oxford.
FAĬNBERG, YA. B. (1961) *Atomnaya Energiya* **11,** 313.
FAĬNBERG, YA. B., KURILKO, V. I. and SHAPIRO, V. D. (1961) *Soviet Phys. Tech. Phys.* **6,** 459.
FARLEY, D., DOUGHERTY, J. and BARRON, D. (1961) *Proc. Roy. Soc.* A **263,** 238.
FELDMAN, S. (1958) *Phys. Fluids* **1,** 546.
FERMI, E. (1940) *Phys. Rev.* **57,** 485.
FERRARO, V. C. A. (1956) *Proc. Roy. Soc.* D **233,** 310.
FLETCHER, E. A., DORSCH, R. G. and ALLEN, H. (1960) *ARS J.* **60,** 337.
FOCK, V. A. (1964) *Theory of Space, Time and Gravitation*, Pergamon Press, Oxford.
FRANCIS, G. (1960) *Ionization Phenomena in Gases*, Butterworth, London.
FRIED, B. D. (1959) *Phys. Fluids* **2,** 337.
FRIEDRICHS, K. (1955) *Bull. Am. Math. Soc.* **61,** 485.
GALEEV, A. A. and KARPMAN, V. I. (1963) *Soviet Phys. JETP* **17,** 403.
GALEEV, A. A., KARPMAN, V. I. and SAGDEEV, R. Z. (1965) *Nucl. Fusion* **5,** 20.
GALEEV, A. A. and ORAEVSKIĬ, V. N. (1963) *Soviet Phys. Doklady* **7,** 988.
GALITSKIĬ, V. M. and MIGDAL, A. B. (1958) *Plasma Physics and Controlled Thermonuelear Reactions* **1,** 161.
GAPONOV, A. V. (1961) *Soviet Phys. JETP* **12,** 232.
GARDNER, C., GOERTZEL, H., GRAD, H., MORAWETZ, C., ROSE, M. and RUBIN, H. (1958) Geneva Conference Paper No 374.
GARDNER, C. S. and KRUSKAL, M. D. (1964) *Phys. Fluids* **7,** 700.
GEFFEN, N. (1963) *Phys. Fluids* **6,** 566.
GEL'FAND, I. M. (1959) *Usp. Mat. Nauk* **14,** 87.
GERMAIN, P. (1960a) *Rech. Aéronaut.* No 74, 13.
GERMAIN, P. (1960b) *Rev. Mod. Phys.* **32,** 951.
GERSHMAN, B. N. (1953a) *Zh. Eksp. Teor. Fiz.* **24,** 659.
GERSHMAN, B. N. (1953b) *Zh. Eksp. Teor. Fiz.* **24,** 453.
GERSHMAN, B. N. (1955) *Andronov Festschrift*, p. 599.
GERSHMAN, B. N. (1958a) *Radiofizika* **1,** 3.
GERSHMAN, B. N. (1958b) *Radiofizika* **1,** 49.
GERSHMAN, B. N. (1960) *Soviet Phys. JETP* **11,** 657.
GERTSENSHTEĬN, M. E. (1952) *Zh. Eksp. Teor. Fiz.* **23,** 669.
GERTSENSHTEĬN, M. E. (1954) *Zh. Eksp. Teor. Fiz.* **27,** 180.
GINZBURG, V. L. (1940) *Zh. Eksp. Teor. Fiz.* **10,** 601.
GINZBURG, V. L. (1954) *Usp. Fiz. Nauk* **51,** 343.
GINZBURG, V. L. (1960) *Soviet Phys. Uspekhi* **2,** 874.
GINZBURG, V. L. (1970) *Propagation of Electromagnetic Waves in a Plasma*, Pergamon Press, Oxford.
GOGOSOV, V. V. (1961a) *Soviet Phys. Dokl.* **5,** 1160.
GOGOSOV, V. V. (1961b) *J. Appl. Math. Mech.* **25,** 678.
GOGOSOV, V. V. (1961c) *J. Appl. Math. Mech.* **25,** 148.
GOGOSOV, V. V. (1962a) *Soviet Phys. Dokl.* **6,** 971.
GOGOSOV, V. V. (1962b) *J. Appl. Math. Mech.* **56,** 88.
GOGOSOV, V. V. (1962c) *Soviet Phys. Dokl.* **7,** 10.
GOLANT, V. E. (1963) *Soviet Phys. Tech. Phys.* **8,** 1.
GOLDSWORTHY, F. A. (1958) *Rev. Mod. Phys.* **60,** 1062.
GOLITSYN, G. S. (1959) *Soviet Phys. JETP* **8,** 538.
GORDEEV, G. B. (1952) *Zh. Eksp. Teor. Fiz.* **23,** 660.
GORDEEV, G. B. (1954a) *Zh. Eksp. Teor. Fiz.* **27,** 19.
GORDEEV, G. B. (1954b) *Zh. Eksp. Teor. Fiz.* **27,** 24.
GOULD, R. W., O'NEIL, T. M. and MALMBERG, J. H. (1967) *Phys. Rev. Lett.* **19,** 2191.
GRAD, H. (1960) *Rev. Mod. Phys.* **32,** 830.
GRADSHTEĬN, I. S. and RYZHIK, I. M. (1966) *Tables of Integrals, Sums, Series, and Products*, Academic Press, New York.

GREBENSHCHIKOV, S. E., RAIZER, M. D., RUKHADZE, A. A. and FRANK, A. G. (1961) *Soviet Phys. Tech. Phys.* **6**, 381.
GREENBERG, O. W., SEN, H. K. and TREVE, Y. M. (1960) *Phys. Fluids* **3**, 379.
GREIFINGER, C. (1960) *Phys. Fluids* **3**, 662.
GREIFINGER, C. and COLE, J. D. (1961) *Phys. Fluids* **4**, 527.
GROSS, E. P. (1951) *Phys. Rev.* **82**, 232.
GROSS, R. A. (1959) *ARS J.* **29**, 63.
GROSS, R. A. (1965) *Rev. Mod. Phys.* **37**, 724.
GROSS, R. A., CHINITZ, W. and RIVLIN, T. J. (1960) *J. Aero. Space Sci.* **27**, 283.
GROSS, R. A. and OPPENHEIM, A. K. (1959) *ARS J.* **29**, 173.
GUREVICH, A. V., PARIISKAYA, L. B. and PITAEVSKIĬ, L. P. (1966) *Soviet Phys. JETP* **22**, 449.
GUSTAFSON, W. A. (1960) *Phys. Fluids* **3**, 732.
HAEFF, A. V. (1948) *Phys. Rev.* **74**, 1532.
HAEFF, A. V. (1949) *Proc. IRE* **37**, 4.
HARRIS, E. G. (1957) *Phys. Rev.* **108**, 1357.
HARRIS, E. G. (1959) *Phys. Rev. Lett.* **2**, 34.
HARRIS, E. G. (1961) *J. Nucl. Energy* C **2**, 138.
HAYES, W. D. (1958a) *Fundamentals of Gas Dynamics*, Princeton Univ. Press, Chap. 4.
HAYES, W. D. (1958b) *Fundamentals of Gas Dynamics*, Vol. 3, *High-Speed Aerodynamics and Jet Propulsion*, Princeton Univ. Press, p. 417.
HEALD, M. and WHARTON, S. (1965) *Plasma Diagnostics with Microwaves*, J. Wiley, New York.
HERLOFSON, N. (1950) *Nature* **165**, 1020.
HIRSCHFELDER, J. O. and CURTISS, C. F. (1958) *J. Chem. Phys.* **28**, 1130.
HOFFMANN, F. and TELLER, E. (1950) *Phys. Rev.* **80**, 692.
HOLWEGER, H. (1963) *Zs. Astroph.* **56**, 269.
HU, P. N. (1966) *Phys. Fluids* **9**, 89.
HUANG, K. (1963) *Statistical Mechanics*, J. Wiley, New York.
ICHIMARU, S. (1962) *Ann. Phys.* **20**, 78.
ICHIMARU, S. and NAKANO, T. (1968) *Phys. Rev.* **165**, 231.
ICHIMARU, S., PINES, D. and ROSTOKER, N. (1962) *Phys. Rev. Lett.* **8**, 231.
IMAI, I. (1960) *Rev. Mod. Phys.* **32**, 992.
IMSHENNIK, V. S. and MOROZOV, A. I. (1961) *Soviet Phys. Tech. Phys.* **6**, 464.
IORDANSKIĬ, S. V. (1959) *Soviet Phys. Dokl.* **3**, 736.
ISRAEL, W. (1960) *Proc. Roy. Soc.* A **559**, 129.
JACKSON, E. A. (1960) *Phys. Fluids* **3**, 786.
JEFFREY, A. and TANIUTI, T. (1964) *Non-Linear Wave Propagation with Applications to Physics and Magnetohydrodynamics*, Academic Press, New York.
JOHN, F. (1955) *Plane Waves and Spherical Means Applied to Partial Differential Equations*, Interscience, New York.
KADOMTSEV, B. B. (1957) *Soviet Phys. JETP* **5**, 771.
KADOMTSEV, B. B. (1958) *Plasma Physics and Controlled Thermonuclear Reactions*, Vol. 4, p. 364.
KADOMTSEV, B. B. (1965) *Plasma Turbulence*, Academic Press, New York.
KADOMTSEV, B. B. and PETVIASHVILI, V. I. (1963) *Soviet Phys. JETP* **16**, 1578.
KADOMTSEV, B. B. and POGUTSE, O. P. (1968) *Soviet Phys. JETP* **26**, 1146.
KALMAN, G. and RON, A. (1961) *Ann. Phys.* **16**, 118.
VAN KAMPEN, N. G. (1955) *Physica* **21**, 949.
KANTROVICH, A. R. and PETCHEK, G. E. (1958) *Magnetohydrodynamics*, Atomizdat, Moscow, p. 11.
KAPLAN, S. A. (1965) *Interstellar Gas Dynamics*, Pergamon Press, Oxford.
KAPLAN, S. A. and STANYUKOVICH, K. P. (1954) *Dokl. Akad. Nauk SSSR* **95**, 769.
KARPLYUK, K. S., ORAEVSKIĬ, V. N. and PAVLENKO, V. P. (1969) *Ukr. Phys. J.* **13**, 796.
KARPMAN, V. I. (1964a) *Soviet Phys. Dokl.* **8**, 919.
KARPMAN, V. I. (1964b) *Soviet Phys. Tech. Phys.* **8**, 715.
KARPMAN, V. I. and SAGDEEV, R. Z. (1964) *Soviet Phys. Tech. Phys.* **8**, 606.
KATO, Y. and TANIUTI, T. (1959) *Prog. Theor. Phys.* **21**, 606.
KAUTZLEBEN, H. (1958) *Hydromagnetische Theorie des Plasmas*, Akademie Verlag, Berlin.
KAZANTSEV, A. P. (1963) *Soviet Phys. JETP* **17**, 865.
KELLOGG, P. J. and LIEMOHN, H. (1960) *Phys. Fluids* **3**, 40.
KEMP, N. H., GERMAIN, P. and GRAD, H. (1960) *Rev. Mod. Phys.* **32**, 958.

KHALATNIKOV, I. M. (1954) *Zh. Eksp. Teor. Fiz.* **27**, 529.
KHALATNIKOV, I. M. (1957) *Soviet Phys. JETP* **5**, 901.
KIRIĬ, YU. A. and SILIN, V. P. (1969) *Soviet Phys. Tech. Phys.* **14**, 583.
KIRKWOOD, J. G. (1946) *J. Chem. Phys.* **14**, 180.
KIROCHKIN, YU. A. (1962) *Radiofizika* **5**, 1104.
KISELEV, M. I. and KOLOSNITSYN, N. I. (1960) *Soviet Phys. Dokl.* **5**, 246.
KITSENKO, A. B. (1963) *Soviet Phys. Dokl.* **7**, 632.
KITSENKO, A. B. and GAPONTSEV, B. A. (1965) *Interactions of Charged Particle Beams with a Plasma*, Kiev, p. 131.
KITSENKO, A. B. and STEPANOV, K. N. (1961a) *Soviet Phys. Tech. Phys.* **6**, 127.
KITSENKO, A. B. and STEPANOV, K. N. (1961b) *Ukr. Fiz. Zh.* **6**, 297.
KITSENKO, A. B. and STEPANOV, K. N. (1962) *Soviet Phys. Tech. Phys.* **7**, 215.
KITSENKO, A. B. and STEPANOV, K. N. (1963) *Plasma Physics and Controlled Thermonuclear Fusion*, Kiev, **2**, 144.
KITSENKO, A. B. and STEPANOV, K. N. (1964) *Nucl. Fusion* **4**, 272.
KLIMONTOVICH, YU. L. (1967) *Statistical Theory ot Non-Equilibrium Processes in a Plasma*, Pergamon Press, Oxford.
KOCHINA, N. N. (1959) *Soviet Phys. Dokl.* **4**, 521.
KOGAN, M. N. (1959a) *J. Appl. Math. Mech.* **23**, 92.
KOGAN, M. N. (1959b) *J. Appl. Math. Mech.* **23**, 784.
KOGAN, M. N. (1960a) *J. Appl. Math. Mech.* **24**, 129.
KOGAN, M. N. (1960b) *Izv. Akad. Nauk SSSR, Mekh.-Mashinostr.* **3**, 143.
KOGAN, M. N. (1960c) *J. Appl. Math. Mech.* **24**, 773.
KOGAN, M. N. (1962) *Magnetohydrodynamics and Plasma Dynamics*, Riga **2**, 55.
KOLOMENSKIĬ, A. A. (1956) *Soviet Phys. Dokl.* **1**, 133.
KOMAROVSKIĬ, L. V. (1961) *Soviet Phys. Dokl.* **5**, 1163.
KONDRATENKO, A. N. and STEPANOV, K. N. (1968) *Ukr. Fiz. Zh.* **13**, 1515.
KONTOROVICH, V. M. (1959) *Soviet Phys. JETP* **8**, 851.
KOROBEĬNIKOV, V. P. (1959) *Soviet Phys. Doklady* **3**, 739.
KOROBEĬNIKOV, V. P. (1960) *Prikl. Mat. Teor. Fiz.* **2**, 47.
KOROBEĬNIKOV, V. P. and RYAZANOV, E. V. (1959) *Dokl. Akad. Nauk SSSR* **124**, 51.
KOROBEĬNIKOV, V. P. and RYAZANOV, E. V. (1960) *J. Appl. Math. Mech.* **24**, 144.
KÖRPER, K. (1957) *Zs. Naturf.* **12a**, 815.
KORTEWEG, D. J. and DE VRIES, G. (1895) *Phil. Mag.* **39**, 422.
KOVNER, M . S. (1960a) *Radiofizika* **3**, 631.
KOVNER, M. S. (1960b) *Radiofizika* **3**, 746.
KOVNER, M. S. (1961a) *Radiofizika* **4**, 765.
KOVNER, M. S. (1961b) *Radiofizika* **4**, 1035.
KOVNER, M. S. (1961c) *Radiofizika* **4**, 444.
KOVRIZHNYKH, L. M. and RUKHADZE, A. A. (1960) *Soviet Phys. JETP* **11**, 615.
KOZLOV, B. N. (1960) *Atomnaya Energiya* **8**, 135.
KRASOVITSKIĬ, V. B. and STEPANOV, K. P. (1963) *Radiofizika* **6**, 1036.
KRASOVITSKIĬ, V. B. and STEPANOV, K. P. (1964) *Soviet Phys. Tech. Phys.* **9**, 786.
KROOK, M. (1959) *Ann. Phys.* **6**, 188.
KUBO, R. (1957) *J. Phys. Soc. Japan* **12**, 570.
KULIKOVSKIĬ, A. G. (1958a) *Soviet Phys. Dokl.* **2**, 269.
KULIKOVSKIĬ, A. G. (1958b) *Soviet Phys. Dokl.* **3**, 507.
KULIKOVSKIĬ, A. G. (1959) *Soviet Phys. Dokl.* **3**, 743.
KULIKOVSKIĬ, A. G. (1966) *J. Appl. Math. Mech.* **30**, 180.
KULIKOVSKIĬ, A. G. and LYUBIMOV, G. A. (1959) *Izv. Akad. Nauk SSSR, Mekh.-Mashinostr.* **4**, 130.
KULIKOVSKIĬ, A. G. and LYUBIMOV, G. A. (1960) *Soviet Phys. Dokl.* **4**, 1195.
KULIKOVSKIĬ, A. G. and LYUBIMOV, G. A. (1961) *J. Appl. Math. Mech.* **25**, 171.
KULIKOVSKIĬ, A. G. and LYUBIMOV, G. A. (1962) *Magnetohydrodynamics*, Fizmatgiz, Moscow.
KURILKO, V. I. and MIROSHNICHENKO, V. I. (1963) *Plasma Physics and Controlled Thermonuclear Fusion*, Kiev, **3**, 161.
LAMPERT, M. A. (1956) *J. Appl. Phys.* **27**, 5.
LANDAU, L. D. (1937) *Zh. Eksp. Teor. Fiz.* **7**, 203 (Collected Papers, Pergamon Press, Oxford, 1965, p. 163).
LANDAU, L. D. (1946) *J. Phys. USSR* **10**, 25. (Collected Papers, Pergamon Press, Oxford, 1965, p. 445).

LANDAU, L. D. and LIFSHITZ, E. M. (1957) *Soviet Phys. JETP* **5**, 512.
LANDAU, L. D. and LIFSHITZ, E. M. (1959) *Fluid Mechanics*, Pergamon Press, Oxford.
LANDAU, L. D. and LIFSHITZ, E. M. (1960) *Electrodynamics of Continuous Media*, Pergamon Press, Oxford.
LANDAU, L. D. and LIFSHITZ, E. M. (1969) *Statistical Physics*, Pergamon Press, Oxford.
LANDAU, L. D. and LIFSHITZ, E. M. (1971) *The Classical Theory of Fields*, Pergamon Press, Oxford.
LANGMUIR, I. (1926) *Proc. Nat. Acad. Sci.* **14**, 627.
LARISH, E. and SHEKHTMAN, I. (1959) *Soviet Phys. JETP* **35**, 203.
LARKIN, A. I. (1960) *Soviet Phys. JETP* **10**, 186.
LASSEN, H. (1927) *Elektr. Nachr. Tech.* **4**, 324.
LAX, P. (1957) *Comm. Pure Appl. Math.* **10**, 537.
LEHNERT, B. (1959) *Nuovo Cim. Suppl.* **13**, 59.
LEONTOVICH, M. A. (Ed.) (1965) *Reviews of Plasma Physics*, Vol. 1, Academic Press.
LEONTOVICH, M. A. (Ed.) (1966) *Reviews of Plasma Physics*, Vol. 2, Academic Press.
LEONTOVICH, M. A. (Ed.) (1967) *Reviews of Plasma Physics*, Vol. 3, Academic Press.
LEONTOVICH, M. A. (Ed.) (1968) *Reviews of Plasma Physics*, Vol. 4, Academic Press.
LEONTOVICH, M. A. (Ed.) (1970) *Reviews of Plasma Physics*, Vol. 5, Academic Press.
LEONTOVICH, M. A. and RYTOV, S. M. (1952) *Zh. Eksp. Teor. Fiz.* **23**, 246.
LESSEN, M. and DESHPANDE, N. (1967) *J. Plasma Phys.* **1**, 463.
LEVIN, M. L. and RYTOV, S. M. (1967) *Theory of Equilibrium Thermal Fluctuations in Electrodynamics* Nauka, Moscow.
LIGHTHILL, M. J. (1960) *Phil. Trans. Roy. Soc.* A **252**, 397.
LINDER, B., CURTISS, C. and HIRSCHFELDER, J. (1958) *J. Chem. Phys.* **28**, 1147.
LINDHARD, J. (1954) *D. Kgl. Danske Vid. Selsk. Mat.-Fys. Medd.* **28**, No 8.
LOMINADZE, D. G. and STEPANOV, K. N. (1964a) *Soviet Phys. Tech. Phys.* **8**, 976.
LOMINADZE, D. G. and STEPANOV, K. N. (1964b) *Nucl. Fusion* **4**, 281.
LOMINADZE, D. G. and STEPANOV, K. N. (1965) *Soviet Phys. Tech. Phys.* **9**, 1408.
LOVETSKIĬ, E. E. and RUKHADZE, A. A. (1966) *Lebedev Institute Proc.* **32**, 206.
LUDWIG, D. (1961) *Comm. Pure Appl. Math.* **14**, 113.
LUNDQUIST, S. (1949) *Phys. Rev.* **76**, 1805.
LUNDQUIST, S. (1952) *Arkiv Fys.* **5**, 297.
LUR'E, K. A. (1964) *Soviet Phys. Tech. Phys.* **8**, 664.
LÜST, R. (1955) *Zs. Naturf.* **10a**, 125.
LYNN, Y. M. (1962) *Phys. Fluids* **5**, 626.
LYUBARSKIĬ, G. YA. (1958) *Ukr. Fiz. Zh.* **3**, 567.
LYUBARSKIĬ, G. YA. (1961) *Soviet Phys. JETP* **13**, 740.
LYUBARSKIĬ, G. YA. (1962a) *Usp. Mat. Nauk* **17**, 183.
LYUBARSKIĬ, G. YA. (1962b) *J. Appl. Math. Mech.* **26**, 761.
LYUBARSKIĬ, G. YA. and POLOVIN, R. V. (1958) *Ukr. Fiz. Zh.* **3**, 567.
LYUBARSKIĬ, G. YA. and POLOVIN, R. V. (1959a) *Soviet Phys. JETP* **8**, 351.
LYUBARSKIĬ, G. YA. and POLOVIN, R. V. (1959b) *Soviet Phys. JETP* **9**, 902.
LYUBARSKIĬ, G. YA. and POLOVIN, R. V. (1959c) *Soviet Phys. JETP* **8**, 901.
LYUBARSKIĬ, G. YA. and POLOVIN, R. V. (1960) *Soviet Phys. Dokl.* **4**, 977.
LYUBARSKIĬ, G. YA. and POLOVIN, R. V. (1961) *Plasma Physics and Controlled Thermonuclear Fusion*, Kiev, p. 79.
LYUBIMOV, G. A. (1959a) *Izv. Akad. Nauk SSSR, Mekh.-Mashinostr.* **5**, 9.
LYUBIMOV, G. A. (1959b) *Soviet Phys. Dokl.* **4**, 526.
LYUBIMOV, G. A. (1959c) *Soviet Phys. Dokl.* **4**, 510.
MACDONALD, W. M., ROSENBLUTH, M. N. and CHUCK, W. (1957) *Phys. Rev.* **107**, 350.
MAKHAN'KOV, V. G. (1964) *Soviet Phys. Tech. Phys.* **8**, 673.
MAKHAN'KOV, V. G. and RUKHADZE, A. A. (1962) *Nucl. Fusion* **2**, 177.
MAKHAN'KOV, V. G. and SHEVCHENKO, V. I. (1965) *Plasma Physics and Controlled Thermonuclear Fusion*, Kiev, p. 190.
MALIK, F. B. and TREHAN, S. K. (1965) *Ann. Phys.* **32**, 1.
MALMFORS, K. G. (1950) *Arkiv Fys.* **1**, 569.
MALYSHEV, I. P. (1961) *Izv. Akad. Nauk SSSR, Mekh.-Mashinostr.* No 3, 182.
MARSHALL, W. (1955) *Proc. Roy. Soc.* A **233**, 367.
MCCUNE, J. E. (1965) *Phys. Rev. Lett.* **15**, 398.
MCCUNE, J. E. and RESLER, E. L. (1960) *J. Aero-Space Sci.* **27**, 493.

McLafferty, G. H. (1960) *ARS J.* **30,** 1019.
Mikhaĭlovskiĭ, A. B. (1965) *Nucl. Fusion* **5,** 122.
Mikhaĭlovskiĭ, A. B. (1968) *Soviet Phys. Tech. Phys.* **12,** 993.
Mikhaĭlovskiĭ, A. B. and Pashitskiĭ, E. A. (1965) *Soviet Phys. Dokl.* **10,** 209.
Mimura, I. (1963) *Raketnaya Tekhn. Kosmon.* **10,** 40.
Mitchner, M. (1959) *Phys. Fluids* **2,** 162.
Moiseev, S. S. (1967) *Proc. VIIth Int. Conf. Phenomena in Ionized Gases*, Belgrade, p. 645.
Moĭseev, S. S. and Sagdeev, R. Z. (1963) *J. Nucl. Energy* **C 5,** 43.
Montgomery, D. (1959) *Phys. Rev. Lett.* **2,** 36.
Morozov, A. I. (1958) *Plasma Physics and Controlled Thermonuclear Reactions*, Moscow, **4,** 331.
Morse, P. and Feshbach, H. (1953) *Methods of Mathematical Physics*, McGraw-Hill, New York.
Mott-Smith, M. H. (1951) *Phys. Rev.* **82,** 885.
Muckenfuss, C. (1960) *Phys. Fluids* **3,** 320.
Nakano, H. (1954) *Prog. Theor. Phys.* **15,** 77.
Nakano, H. (1957) *Prog. Theor. Phys.* **17,** 145.
Nedospasov, A. V. (1968) *Soviet Phys. Uspekhi* **11,** 174.
Neufeld, J. and Ritchie, H. (1955) *Phys. Rev.* **98,** 1632.
Neufeld, S. and Doyle, P. H. (1961) *Phys. Rev.* **121,** 654.
von Neumann, J. (1943) Progress Report No. 1140 on *The Theory of Shock Waves*, NDRC, Div. 8.
Nexsen, W. E., Cummins, W. F., Coengsen, F. H. and Sherman, A. E. (1960) *Phys. Rev.* **119,** 1457.
Nicholls, H. W. and Schelling, J. C. (1925a) *Nature* **115,** 334.
Nicholls, H. W. and Schelling, J. C. (1925b) *Bell System Tech. J.* **4,** 215.
Noerdlinger, P. D. (1960) *Phys. Rev.* **118,** 879.
Nyquist, H. (1928) *Phys. Rev.* **32,** 110.
Oleĭnik, O. A. (1957 *Usp. Mat. Nauk* **12,** 3.
O'Neil, T. M. and Gould, R. W. (1968) *Phys. Fluids* **11,** 134.
O'Neil, T. and Rostoker, N. (1965) *Phys. Fluids* **8,** 1109.
Oraevskiĭ, V. N. (1963) *Nucl. Fusion* **4,** 293.
Oraevskiĭ, V. N. and Sagdeev, R. Z. (1963) *Soviet Phys. Tech. Phys.* **7,** 955.
Ozawa, Y., Kaij, I. and Kito, M. (1962) *J. Nucl. Energy* **C 4,** 271.
Pakhomov, V. I. (1965) *High-frequency Plasma Properties*, p. 189.
Pakhomov, V. I., Aleksin, V. F. and Stepanov, K. N. (1962) *Soviet Phys. Tech. Phys.* **6,** 856.
Pargamanik, L. E. (1948) Khar'kov Thesis.
Peierls, R. E. (1955) *Quantum Theory of Solids*, Oxford University Press.
Pekarek, L. (1968) *Soviet Phys. Uspekhi* **11,** 188.
Penney, W. G. (1951) *Proc. Roy. Soc.* A **204,** 1.
Penrose, O. (1960) *Phys. Fluids* **3,** 258.
Petelin, V. I. (1961) *Radiofizika* **4,** 455.
Petviashvili, V. I. (1964) *Soviet Phys. Dokl.* **8,** 1218.
Pierce, J. E. (1947) *Proc. IRE* **35,** 111.
Pierce, J. E. (1948) *J. Appl. Phys.* **19,** 231.
Pierce, J. E. (1949) *J. Appl. Phys.* **20,** 1060.
Pines, D. and Bohm, D. (1952) *Phys. Rev.* **85,** 338.
Pistunovich, V. I. (1963) *Atomnaya Energiya* **14,** 72.
Polovin, R. V. (1957) *Soviet Phys. JETP* **4,** 290.
Polovin, R. V. (1959) *Soviet Phys. JETP* **9,** 675.
Polovin, R. V. (1960) *Soviet Phys. JETP* **11,** 1113.
Polovin, R. V. (1961a) *Ukr. Fiz. Zh.* **6,** 32.
Polovin, R. V. (1961b) *Soviet Phys. JETP* **12,** 326.
Polovin, R. V. (1961c) *Soviet Phys. Uspekhi* **3,** 677.
Polovin, R. V. (1961d) *Prikl. Mat. Teor. Fiz.* **6,** 3.
Polovin, R. V. (1961e) *Soviet Phys. JETP* **12,** 699.
Polovin, R. V. (1962) *Soviet Phys. Tech. Phys.* **6,** 889.
Polovin R. V. (1963a) *Plasma Physics and Controlled Thermonuclear Fusion*, Kiev, **3,** 169.
Polovin, R. V. (1963b) *Ukr. Fiz. Zh.* **8,** 709.
Polovin, R. V. (1963c) *Ukr. Fiz. Zh.* **8,** 1283.
Polovin, R. V. (1963d) *Soviet Phys. Tech. Phys.* **8,** 184.
Polovin, R. V. (1964) *Soviet Phys. Tech. Phys.* **9,** 205.

Polovin, R. V. (1965a) *Differential Equations*, Vol. 1, p. 499.
Polovin, R. V. (1965b) *Magnitnaya Gidrodin*. No. 2, 19.
Polovin, R. V. (1965c) *Ukr. Fiz. Zh.* **10,** 1045.
Polovin, R. V. (1965d) *Soviet Phys. Tech. Phys.* **9,** 1390.
Polovin, R. V. and Akhiezer, I. A. (1959) *Ukr. Fiz. Zh.* **4,** 677.
Polovin, R. V. and Cherkasova, K. P. (1962a) *Soviet Phys. Tech. Phys.* **7,** 475.
Polovin, R. V. and Cherkasova, K. P. (1962b) *Soviet Phys. JETP* **14,** 190.
Polovin, R. V. and Cherkasova, K. P. (1963) *Plasma Physics and Controlled Thermonuclear Fusion*, Kiev, **2,** 196.
Polovin, R. V. and Cherkasova, K. P. (1966a) *Magnitnaya Gidrodin*. No. 1, 3.
Polovin, R. V. and Cherkasova, K. P. (1966b) *Soviet Phys. Uspekhi* **9,** 278.
Polovin, R. V. and Cherkasova, K. P. (1967) *High-frequency Behaviour of a Plasma*, Naukova Dumka, Kiev, p. 84.
Polovin, R. V. and Demutskiĭ, V. P. (1960) *Ukr. Fiz. Zh.* **5,** 3.
Polovin, R. V. and Demutskiĭ, V. P. (1961) *Soviet Phys. JETP* **13,** 1229.
Polovin, R. V. and Demutskiĭ, V. P. (1963) *Plasma Physics and Controlled Thermonuclear Fusion*, Kiev, **2,** 190.
Polovin, R. V. and Lyubarskiĭ, G. Ya. (1958) *Ukr. Fiz. Zh.* **3,** 571.
Polovin, R. V. and Lyubarskiĭ, G. Ya. (1959) *Soviet Phys. JETP* **8,** 351.
Post, R. F. and Rosenbluth, M. N. (1965) *Phys. Fluids* **8,** 547.
Post, R. F. and Rosenbluth, M. N. (1966) *Phys. Fluids* **9,** 730.
Rappoport, V. O. (1960) *Radiofizika* **3,** 737.
Rayleigh, Lord (1906) *Phil. Mag.* **11,** 117.
Reed, S. G. (1952) *J. Chem. Phys.* **20,** 1823.
Repalov, N. S. and Khizhnyak, N. A. (1968) *High-frequency Plasma Properties*, Kiev, p. 90.
Reuter, G. E. H. and Sondheimer, E. H. (1948) *Proc. Roy. Soc.* **A 195,** 336.
Ribaud, G. (1959) *ARS J.* **29,** 876.
Rolland, P. (1965) *Phys. Rev.* **140,** B 776.
Romanov, Yu. A. and Filippov, G. F. (1961) *Soviet Phys. JETP* **13,** 87.
Romazashvili, R. R. and Rukhadze, A. A. (1962) *Soviet Phys. Tech. Phys.* **7,** 467.
Rosenbluth, M., Coppi, B. and Sudan, R. N. (1968) *Proc. Third Int. Conf. Plasma Phys. Controlled Thermonuclear Fusion*, CN 24/E 14.
Rosenbluth, M. and Rostoker, N. (1962) *Phys. Fluids* **5,** 776.
Rosenkilde, C. E. (1965) *Astroph. J.* **141,** 1105.
Rostoker, N. (1961) *Nucl. Fusion* **1,** 101.
Roth, J. R. (1967) *Phys. Fluids* **10,** 2712.
Roth, J. R. (1969) *Phys. Fluids* **12,** 260.
Rowlands, J., Sizonenko, V. L. and Stepanov, K. N. (1966) *Soviet Phys. JETP* **23,** 661.
Rozhdestvenskiĭ, B. L. (1960) *Usp. Mat. Nauk* **15,** 59.
Rukhadze, A. A. (1962) *Soviet Phys. Tech. Phys.* **7,** 353.
Ryazanov, E. V. (1959a) *Soviet Phys. Dokl.* **4,** 554.
Ryazanov, E. V. (1959b) *J. Appl. Math. Mech.* **23,** 260.
Rytov, S. M. (1953) *Theory of Electrical Fluctuations and Thermal Radiation*, Moscow.
Sachs, R. G. (1946) *Phys. Rev.* **69,** 514.
Sagdeev, R. Z. (1958a) *Plasma Physics and Controlled Thermonuclear Reactions*, Moscow, **1,** p. 384.
Sagdeev, R. Z. (1958b) *Plasma Physics and Controlled Thermonuclear Reactions*, Moscow, **4,** p. 384.
Sagdeev, R. Z. (1959) *Problems of Magnetohydrodynamics and Plasma Dynamics*, Riga, p. 63.
Sagdeev, R. Z. (1961) *Proc. Symposium Electromagnetism and Fluid Dynamics of Gaseous Plasmas*, New York, p. 443.
Sagdeev, R. Z. (1962) *Soviet Phys. Tech. Phys.* **6,** 867.
Sagdeev, R. Z. (1966) *Rev. Plasma Phys.* **4,** 23.
Sagdeev, R. Z. and Shafranov, V. D. (1958) *Plasma Physics and Controlled Thermonuclear Reactions*, Moscow, **4,** 430.
Sagdeev, R. Z. and Shafranov, V. D. (1961) *Soviet Phys. JETP* **12,** 130.
Salpeter, E. E. (1960a) *Phys. Rev.* **120,** 1528.
Salpeter, E. E. (1960b) *Geophys. Res.* **65,** 1851.
Salpeter, E. E. (1960c) *Geophys. Res.* **66,** 982.
Salpeter, E. E. (1961) *Phys. Rev.* **122,** 1663.

SALTANOV, N. V. and TKALICH, V. S. (1961) *Izv. Akad. Nauk SSSR, Mekh.-Mashinostr.* No. 6, 26.
SAMOKHIN, M. V. (1963a) *Soviet Phys. Tech. Phys.* **8,** 498.
SAMOKHIN, M. V. (1963b) *Soviet Phys. Tech. Phys.* **8,** 504.
SARASON, L. (1965) *J. Math. Phys.* **6,** 1508.
SEARS, W. R. (1960) *Rev. Mod. Phys.* **32,** 701.
SEDOV, L. I. (1967) *Mechanics of Continuous Media,* Moscow State University, Part II.
SEGRÉ, S. (1958) *Nuovo Cim.* **9,** 1054.
SELIGER, R. L. and WHITHAM, G. B. (1968) *Proc. Roy. Soc.* A **305,** 1.
SEN, H. K. (1952) *Phys. Rev.* **88,** 816.
SEN, H. K. (1956) *Phys. Rev.* **102,** 5.
SEVERNYĬ, A. B. (1961) *Soviet Astr.-AJ* **5,** 299.
SHAFRANOV, V. D. (1957) *Soviet Phys. JETP* **5,** 1183.
SHAFRANOV, V. D. (1958a) *Plasma Physics and Controlled Thermonuclear Reactions,* Moscow, **4,** 416.
SHAFRANOV, V. D. (1958b) *Plasma Physics and Controlled Thermonuclear Reactions,* Moscow, **4,** 426.
SHAFRANOV, V. D. (1958c) *Soviet Phys. JETP* **7,** 1019.
SHAFRANOV, V. D. (1967) *Rev. Plasma Phys.* **3,** 1.
SHAPIRO, A. H., HAWTHORNE, W. R. and EDELMAN, G. M. (1947) *J. Appl. Mech.* **14,** 317.
SHAPIRO, V. D. and SHEVCHENKO, V. I. (1962) *Soviet Phys. JETP* **15,** 1053.
SHAPIRO, V. D. and SHEVCHENKO, V. I. (1968) *Soviet Phys. JETP* **27,** 635.
SHARIKADZE, D. V. (1959) *J. Appl. Math. Mech.* **23,** 1356.
SHERCLIFF, J. A. (1960) *J. Fluid Mech.* **9,** 481.
SHEVCHENKO, V. I. (1963) *Plasma Physics and Controlled Thermonuclear Fusion,* Kiev, **2,** 156.
SILIN, V. P. (1952) Lebedev Thesis.
SILIN, V. P. (1955) *Lebedev Proceedings* **6,** 200.
SILIN, V. P. (1959a) *Soviet Phys. JETP* **8,** 870.
SILIN, V. P. (1959b) *Radiofizika* **2,** 198.
SILIN, V. P. (1964a) *Soviet Phys. JETP* **18,** 559.
SILIN, V. P. (1964b) *Prikl. Mat. Teor. Fiz.* No. 1, 32.
SILIN, V. P. (1967) Appendix to the Russian edition of Balescu (1963).
SILIN, V. P. and RUKHADZE, A. A. (1961) *Electromagnetic Properties of Plasmas and Plasma-like Media,* Atomizdat, Moscow.
SIMON, A. (1965) *Plasma Physics,* IAEA, Vienna, p. 163.
SINGHAUS, H. E. (1964) *Phys. Fluids* **7,** 1534.
SIROTINA, E. P. and SYROVATSKIĬ, S. I. (1961) *Soviet Phys. JETP* **12,** 521.
SITENKO, A. G. (1964) *Proc. Conf. New Techniques,* Moscow.
SITENKO, A. G. (1966) *Ukr. Fiz. Zh.* **11,** 1161.
SITENKO, A. G. (1967) *Electromagnetic Fluctuations in a Plasma,* Academic Press, New York.
SITENKO, A. G. and GURIN, A. A. (1966) *Soviet Phys. JETP* **22,** 1089.
SITENKO, A. G. and KAGANOV, M. I. (1955) *Dokl. Akad. Nauk SSSR* **100,** 681.
SITENKO, A. G. and KIROCHKIN, YU. A. (1960) *Soviet Phys. Tech. Phys.* **4,** 723.
SITENKO, A. G. and KIROCHKIN, YU. A. (1963) *Radiofizika* **6,** 469.
SITENKO, A. G. and KIROCHKIN, YU. A. (1964) *Soviet Phys. Tech. Phys.* **8,** 1008.
SITENKO, A. G. and KIROCHKIN, YU. A. (1966) *Soviet Phys. Uspekhi* **9,** 430.
SITENKO, A. G. and KOLOMENSKIĬ, A. A. (1956) *Soviet Phys. JETP* **3,** 410.
SITENKO, A. G., NGUEN VAN CHONG, and PAVLENKO, V. I. (1970a) *Nucl. Fusion* **10,** 259.
SITENKO, A. G., NGUEN VAN CHONG, and PAVLENKO, V. I. (1970b) *Sov. Phys. – JETP* **31,** 738.
SITENKO, A. G., NGUEN VAN CHONG and PAVLENKO, V. I. (1970c) *Ukr. Fiz. Zh.* **15,** 1372.
SITENKO, A. G. and RADZIEVSKIĬ, V. N. (1966) *Soviet Phys. Tech. Phys.* **10,** 903.
SITENKO, A. G. and STEPANOV, K. N. (1957) *Soviet Phys. JETP* **4,** 512.
SITENKO, A. G. and STEPANOV, K. N. (1958) *Proc. Phys.-Math. Faculty,* Khar'kov Univ. **7,** 5.
SITENKO, A. G. and TSZYAN' YU-TAĬ (1963) *Soviet Phys. Tech. Phys.* **7,** 978.
SIVUKHIN, D. V. (1966) *Magnitnaya Gidrodin* **1,** 35.
SIZONENKO, V. L. and STEPANOV, K. N. (1965) *Plasma Physics and Controlled Thermonuclear Fusion,* Kiev, **4,** 93.
SIZONENKO, V. L. and STEPANOV, K. N. (1966) *Soviet Phys. JETP* **22,** 832.
SIZONENKO, V. L. and STEPANOV, K. N. (1967a) *Nucl. Fusion* **7,** 131.
SIZONENKO, V. L. and STEPANOV, K. N. (1967b) *Ukr. Fiz. Zh.* **12,** 535.
SIZONENKO, V. L. and STEPANOV, K. N. (1968) *Ukr. Phys. J.* **13,** 628.

SMERD, S. F. (1955) *Nature* **175,** 279.
SOLOUKHIN, R. I. (1960) *Soviet Phys. Uspekhi* **2,** 546.
SPITZER, JR., L. (1956) *Physics of a Fully Ionized Gas*, Interscience, New York.
STANYUKOVICH, K. P. (1955a) *Dokl. Akad. Nauk SSSR* **103, 73.**
STANYUKOVICH, K. P. (1955b) *Izv. Akad. Nauk SSSR*, ser. fiz. **19,** 639.
STANYUKOVICH, K. P. (1959) *Soviet Phys. JETP* **8,** 358.
STEFANOVICH, A. E. (1962) *Soviet Phys. Tech. Phys.* **7,** 462.
STEPANOV, K. N. (1958a) *Soviet Phys. JETP* **7,** 892.
STEPANOV, K. N. (1958b) Khar'kov thesis.
STEPANOV, K. N. (1959a) *Soviet Phys. JETP* **8,** 808.
STEPANOV, K. N. (1959b) *Soviet Phys. JETP* **8,** 195.
STEPANOV, K. N. (1959c) *Khar'kov Preprint.*
STEPANOV, K. N. (1959d) *Ukr. Fiz. Zh.* **4,** 678.
STEPANOV, K. N. (1960) *Soviet Phys. JETP* **11,** 192.
STEPANOV, K. N. (1962a) *Plasma Physics and Controlled Thermonuclear Reactions* **1,** 45.
STEPANOV, K. N. (1962b) *Plasma Physics and Controlled Thermonuclear Fusion*, Kiev, p. 52.
STEPANOV, K. N. (1963a) *Plasma Physics and Controlled Thermonuclear Fusion*, Kiev, **2,** 164.
STEPANOV, K. N. (1963b) *Radiofizika* **6,** 403.
STEPANOV, K. N. (1963c) *Soviet Phys. Tech. Phys.* **8,** 177.
STEPANOV, K. N. (1965) *Soviet Phys. Tech. Phys.* **9,** 1653.
STEPANOV, K. N. and KITSENKO, A. B. (1961) *Soviet Phys. Tech. Phys.* **6,** 120.
STIX, T. (1957) *Phys. Rev.* **106,** 1146.
STIX, T. (1958) *Phys. Fluids* **1,** 308.
STIX, T. (1962) *Theory of Plasma Waves*, McGraw-Hill, New York.
STURROCK, P. A. (1957) *Proc. Roy. Soc.* A **242,** 277.
STURROCK, P. A. (1959) *Phys. Rev.* **112,** 1488.
STURROCK, P. A. (1960) *Phys. Rev.* **117,** 1426.
SUDAN, R. N. (1963) *Phys. Fluids* **6,** 57.
SYROVATSKIĬ, S. I. (1953) *Zh. Eksp. Teor. Fiz.* **24,** 622.
SYROVATSKIĬ, S. I. (1956) *Proc. Lebedev Inst. Phys.* **8,** 13.
SYROVATSKIĬ, S. I. (1957) *Usp. Fiz. Nauk* **62,** 247.
SYROVATSKIĬ, S. I. (1959) *Soviet Phys. JETP* **8,** 1024.
TAMADA, K. (1962) *Phys. Fluids* **5,** 871.
TAMM, I. E. and FRANK, I. M. (1937) *Dokl. Akad. Nauk SSSR* **14,** 107.
TANIUTI, T. (1958a) *Prog. Theor. Phys.* **19,** 69.
TANIUTI, T. (1958b) *Prog. Theor. Phys.* **19,** 749.
TANIUTI, T. (1962) *Prog. Theor. Phys.* **28,** 756.
TANIUTI, T., YAJIMA, N. and OUTI, A. (1966) *J. Phys. Soc. Japan* **21,** 757.
TAUB, A. H. (1948) *Phys. Rev.* **74,** 328.
TAUSSIG, R. T. (1967) *Phys. Fluids* **10,** 1145.
TAYLOR, G. I. and MACCOLL, J. W. (1933) *Proc. Roy. Soc.* **A 139,** 298.
THOMPSON, W. B. and HUBBARD, J. (1960) *Rev. Mod. Phys.* **32,** 714.
TIDMAN, D. A. (1958) *Phys. Rev.* **111,** 1439.
TIMOFEEV, A. V. (1961) *Soviet Phys. JETP* **12,** 281.
TIMOFEEV, A. V. and PISTUNOVICH, V. I. (1970) *Rev. Plasma Phys.* **5,** 401.
TITCHMARSH, E. C. (1937) *Theory of Fourier Integrals*, Oxford University Press.
TODD, L. (1964) *J. Fluid Mech.* **18,** 321.
TODD, L. (1965) *J. Fluid Mech.* **21,** 193.
TODD, L. (1966) *J. Fluid Mech.* **24,** 597.
TONKS, L. and LANGMUIR, I. (1929a) *Phys. Rev.* **33,** 195.
TONKS, L. and LANGMUIR, I. (1929b) *Phys. Rev.* **33,** 990.
TRUBNIKOV, B. A. (1958) *Plasma Physics and Controlled Thermonuclear Reactions* **3,** 104.
TRUBNIKOV, B. A. (1965) *Rev. Plasma Phys.* **1,** 105.
TSYTOVICH, V. N. (1962a) *Soviet Phys. Dokl.* **7,** 43.
TSYTOVICH, V. N. (1962b) *Soviet Phys. JETP* **15,** 320.
TSYTOVICH, V. N. (1963) *Soviet Phys. JETP* **17,** 643.
TSYTOVICH, V. N. (1970) *Non-linear Effects in a Plasma*, Plenum Press, New York.
TURCOTTE, D. L. and CHU, C. K. (1966) *Zs. Angew. Math. Phys.* **17,** 528.

TWISS, R. Q. (1951a) *Proc. Phys. Soc.* B **64**, 654.
TWISS, R. Q. (1951b) *Phys. Rev.* **84**, 448.
VEDENOV, A. A. (1958) *Soviet Phys. JETP* **6**, 1165.
VEDENOV, A. A. (1962) *Atomnaya Energiya* **13**, 5.
VEDENOV, A. A. (1965) *Theory of a Turbulent Plasma*, VINITI, Moscow.
VEDENOV, A. A. (1967) *Rev. Plasma Phys.* **3**, 229.
VEDENOV, A. A., VELIKHOV, E. P. and SAGDEEV, R. Z. (1961a) *Nucl. Fusion* **1**, 82.
VEDENOV, A. A., VELIKHOV, E. P. and SAGDEEV, R. Z. (1961b) *Soviet Phys. Uspekhi* **4**, 332.
VEDENOV, A. A., VELIKHOV, E. P. and SAGDEEV, R. Z. (1962) *Nucl. Fusion Suppl.* **2**, 465.
VLASOV, A. A. (1938) *Zh. Eksp. Teor. Fiz.* **8**, 291.
VLASOV, A. A. (1945) *Scientific Publ. Moscow State Univ.* **75**, 2.
VLASOV, A. A. (1950) *Many-particle Theory*, Gostekhizdat, Moscow.
WALKER, L. R. (1955) *J. Appl. Phys.* **25**, 131.
WALSH, J. M., SHREFFER, R. G. and WILLIG, F. J. (1953) *J. Appl. Phys.* **24**, 349.
WANG, H. S. C. (1963) *Phys. Fluids* **6**, 1115.
WANG, H. S. C. and LOJKO, M. S. (1963) *Phys. Fluids* **6**, 1458.
WATSON, G. N. (1958) *Theory of Bessel Functions*, Cambridge University Press.
WEIBEL, E. S. (1959) *Phys. Rev. Lett.* **2**, 83.
WEITZNER, H. (1961a) *Phys. Fluids* **4**, 1238.
WEITZNER, H. (1961b) *Phys. Fluids* **4**, 1245.
WHITHAM, G. B. (1959) *Comm. Pure Appl. Math.* **12**, 113.
WILHELMSSON, H. (1961) *Phys. Fluids* **4**, 335.
WOLTJER, L. (1959) *Proc. Nat. Acad. Sci.* **45**, 69.
WRIGHT, H., WIGINTON, C. L. and NEUFELD, J. (1965) *Phys. Fluids* **7**, 1375.1
YAKIMENKO, V. L. (1963) *Soviet Phys. JETP* **17**, 1032.
YANENKO, N. N. (1956) *Dokl. Akad. Nauk SSSR* **109**, 44.
YAVORSKAYA, I. M. (1958) *Soviet Phys. Dokl.* **2**, 273.
YAVORSKAYA, I. M. (1959) *Problems of Magnetohydrodynamics and Plasma Dynamics*, Riga, p. 175.
YVON, J. (1935) *La Théorie Statistique des Fluides*, Hermann, Paris.
ZABUSKY, K. and KRUSKAL, M. (1965) *Phys. Rev. Lett.* **15**, 240.
ZASLAVSKIĬ, G. M. (1970) *Statistical Irreversibility in Non-Linear Systems*, Nauka, Moscow.
ZAVOĬSKIĬ, E. K. (1963) *Atomnaya Energiya* **14**, 57.
ZAYED, K. E. and KITSENKO, A. B. (1968) *Plasma Phys.* **10**, 147.
ZEL'DOVICH, YA. B. (1940) *Zh. Eksp. Teor. Fiz.* **10**, 542.
ZEL'DOVICH, YA. B. (1957) *Soviet Phys. JETP* **5**, 919.
ZEL'DOVICH, YA. B. and KOMPANEETS, A. S. (1955) *Detonation Theory*, GITTL, Moscow.
ZEL'DOVICH, YA. B. and RAĬZER, YU. P. (1957) *Usp. Fiz. Nauk* **63**, 613.
ZEL'DOVICH, YA. B. and RAĬZER, YU. P. (1966) *Physics of Shock Waves and High-temperature Hydrodynamic Phenomena*, Academic Press, New York.
ZHELEZNYAKOV, V. V. (1959) *Radiofizika* **2**, 14.
ZHELEZNYAKOV, V. V. (1960a) *Radiofizika* **3**, 57.
ZHELEZNYAKOV, V. V. (1960b) *Radiofizika* **3**, 180.
ZHELEZNYAKOV, V. V. (1961a) *Radiofizika* **4**, 618.
ZHELEZNYAKOV, V. V. (1961b) *Radiofizika* **4**, 849.
ZHELEZNYAKOV, V. V. (1970) *Radio Emission of Sun and Planets*, Pergamon Press, Oxford.
ZHILIN, YU. L. (1960) *J. Appl. Math. Mech.* **24**, 794.
ZUMINO, B. (1957) *Phys. Rev.* **108**, 1116.

Glossary

A-wave:	Alfvén wave
$\alpha_\alpha = (kv_\alpha/\omega_{B\alpha})^2 = (kp_\alpha)^2$ (cf. p. 231 of Vol. 1)	
B:	magnetic induction
$B^{(e)}$:	external magnetic induction
$B^{(p)}$:	internal (self-consistent) magnetic field
$b=B/B$	
C_s:	s-particle correlation function
CS-wave:	cyclotron-sound wave
c_p:	specific heat at constant pressure
c_s:	sound velocity
c_v:	specific heat at constant volume
$D(k, p)$:	function defined by eqn. (4.2.1.11)
$\mathcal{D}$:	probability density
E:	electrical field
$E^{(e)}$:	external electrical field
$E^{(p)}$:	internal (self-consistent) electrical field
e:	elementary charge ($-e$ is the electron charge)
e:	polarization vector
e_α:	charge of particle of type α
F:	single-particle distribution function
$F^{(M)}$:	Maxwell distribution function
FE-wave:	fast extra-ordinary wave
FMS-wave:	fast magneto-sound wave
FS:	fast sound wave
f_s:	s-particle distribution function
G:	pair correlation function
g:	plasma parameter given by eqn. (1.2.1.3)
$\mathcal{H}$:	Hamiltonian
I-wave:	ionization wave
$\mathcal{I}$:	collision integral

j:	current density
$j^{(e)}$:	external current density
$j^{(p)}$:	current density in plasma
L:	Coulomb integral
$\mathcal{L}$:	differential operator given by eqn. (1.1.3.7)
l:	mean free path
l_a:	penetration depth
M:	Mach number
MS-wave:	magnetized sound wave
m:	mass of plasma particle
m_0:	mass of neutral particle
m_e:	electron mass
m_i:	ion mass
N_i:	total number of particles of kind i
n:	refractive index
n_α:	density of particles of kind α
$n'_{\alpha 0}$:	beam density
O-wave:	ordinary wave
p::	principal value symbol
$\mathcal{P}$	pressure
p^*:	effective pressure
Q, q:	energy flux density
q:	dissipated energy density
R_i:	Riemann invariant
R_v:	Reynolds number
R_σ:	magnetic Reynolds number, Lundquist number
r_0	electron radius
r_D:	Debye radius; usually electron Debye radius
r_{Di}:	ion Debye radius
S:	entropy
SE-wave:	slow extra-ordinary wave
SMS-wave:	slow magneto-sound wave
SS-wave:	slow sound wave
s:	entropy density
T:	temperature (in energy units)
T_0:	temperature of neutral particles
T_e:	electron temperature
T_i:	ion temperature
T^{ik}:	energy-momentum tensor

U:	group velocity
u:	hydrodynamical velocity
$u_{\alpha\beta}$:	velocity deformation tensor
V:	volume
V:	phase velocity
$\bar{v}$:	mean thermal velocity
v_+:	velocity of fast magneto-sound waves
v_-:	velocity of slow magneto-sound waves
v_A:	Alfvén velocity
v_e:	electron thermal velocity
v_{gr}:	group velocity
v_i:	ion thermal velocity
v_s:	ion sound velocity
v_{se}:	electron sound velocity
v_{ph}:	phase velocity
w:	enthalpy
$w(z)$:	function given by eqn. (5.2.2.2)
Ze:	ion charge
z_l:	quantity given by eqn. (5.2.2.3)
z_l':	quantity given by eqn. (6.2.1.2)
$\beta=\sqrt{(1-u^2/c^2)}$	
$\beta_e=v_e/c$	
γ:	adiabatic index; ratio of specific heats $(=c_p/c_v)$
$\gamma(k)$:	damping or growth rate
ε:	dielectric constant
ε:	internal energy per unit mass
ζ:	viscosity coefficient
η:	viscosity coefficient
θ:	temperature
$\theta[x]$:	step function given by eqn. (2.1.6.15′)
$\varkappa$:	thermal conductivity
$\varkappa$:	damping coefficient
$\boldsymbol{\varkappa}=h/k$	
Λ_{ik}:	matrix elements defined by eqn. (4.3.1.10′)
$\lambda=k_x v_\perp/\omega_{B\alpha}$	(see p. 228 of Vol. 1)

ν- ν_1- ν_2:	viscosity coefficients
ν_e:	Coulomb collision frequency
ν_m:	magnetic viscosity
$\Pi_{\alpha\beta}$:	momentum current density tensor
π:	electromagnetic energy flux density
π-$_{ij}^{(\alpha)}$:	polarizability tensor
$\pi_{\alpha\beta}$:	viscous stress tensor
ξ_e:	ratio of electron pressure to magnetic pressure (see eqn. 5.6.1.14))
ϱ, ϱ_e:	charge density
ϱ^e:	external charge density
ϱ_m:	mass density
ϱ^p:	charge density in plasma
ϱ_α:	Larmor radius of particle of type α
σ:	electrical conductivity
σ:	scattering cross-section
$\sigma_{\alpha\beta}$:	stress tensor
τ:	relaxation time
φ:	electrostatic potential
φ_k:	Fourier transform of φ
φ_{ij}:	two-particle electrostatic potential
ω_{Be}:	electron gyro-frequency
$\omega_{B\alpha}(=e_\alpha B/m_\alpha c)$:	gyro-frequency of particle of type α
ω_p:	plasma frequency, Langmuir frequency

Index*

*In order to increase the usefulness of the index we have included references to Volume 1: 1.299 refers to p.299 of volume 1, 67 to p.67 of Volume 2.